ROGER STEVENSON
JANUARY, 1992

Energy
Deskbook

by
Samuel Glasstone

VNR VAN NOSTRAND REINHOLD COMPANY
NEW YORK   CINCINNATI   TORONTO   LONDON   MELBOURNE

Preface Copyright © 1983 by Van Nostrand Reinhold Company Inc.

Library of Congress Catalog Card Number: 82-24864
ISBN: 0-442-22928-3

Manufactured in the United States of America

Published by Van Nostrand Reinhold Company Inc.
135 West 50th Street, New York, N.Y. 10020

Van Nostrand Reinhold Publishing
1410 Birchmount Road
Scarborough, Ontario MIP 2E7, Canada

Van Nostrand Reinhold
480 Latrobe Street
Melbourne, Victoria 3000, Australia

Van Nostrand Reinhold Company Limited
Molly Millars Lane
Wokingham, Berkshire, England

15 14 13 12 11 10 9 8 7 6 5 4 3 2 1

Library of Congress Cataloging in Publication Data

Glasstone, Samuel, 1897–
    Energy deskbook.

    "June 1982."
    "DOE/IR/05114-1 (DE82013966)."
    1. Power resources – Handbooks, manuals, etc.
2. Power (Mechanics) – Handbooks, manuals, etc.
I. Title.
TJ163.235.G53   1983   621.042'0321   82-24864
ISBN 0-442-22928-3

PREFACE

The long-term future of the United States, and indeed of the world, depends on a stable supply of energy at a reasonable price. In the past, energy sources were taken for granted with the tacit assumption that they would always be available. In recent years, however, this attitude has undergone a change.

The public may be lulled into complacency when, as a result of various circumstances, energy supplies are plentiful, but this condition can be only temporary. Reserves of the fossil fuels — coal, petroleum, and natural gas — which are now the major energy sources are limited and cannot be replaced. Consequently, even with the most careful conservation measures, the reserves will inevitably be depleted and energy will become less available and more expensive. Means must therefore be devised for the optimum use of fossil fuels and other energy sources must be developed.

In view of this situation, responsible citizens, especially scientists, engineers, instructors, students, and administrators, should be informed on various aspects of energy. Although there are many technical books and reports dealing in detail with individual energy sources, there appeared to be a need for a convenient, single-volume reference work covering all energy sources. The purpose of the *Energy Deskbook,* which combines the features of a glossary and an encyclopedia, is to satisfy this need.

Energy-related terms are defined and current and potential energy sources and their utilization are described. The material is presented at a low technical level, with emphasis on general principles, which are not difficult to understand, rather than on technology. To serve the interests of a broad range of readers, numerical quantities are given in both common and metric (International System) units. The table of contents, with some four hundred entries, is intended for use also as an index. In conclusion, the reader's attention is called to the indication of cross references in the text by words in boldface type.

Samuel Glasstone

ACKNOWLEDGMENT

This book was written under contract to the U.S. Department of Energy, and I am indebted to many people, most of whom are present or former members of the staff of the Department or its contractors, for help in preparing the manuscript. Thanks are due to Joseph G. Gratton who initiated the Energy Deskbook project, to Robert F. Pigeon for advice and assistance in administrative and other matters, and to those staff members of the Department of Energy (Headquarters and Technology Centers) and of the Oak Ridge National Laboratory who reviewed parts of the manuscript and made valuable comments. I also wish to thank Joseph G. Coyne, Manager, Technical Information Center, and his staff, in particular, Irene D. Keller, Director, Publishing Division, and her associates, who were responsible for the editing, composition, proofreading, artwork, design, and layout of the book. I am especially grateful to Elizabeth B. Howard, Oak Ridge National Laboratory, for the library research which provided much of the material used in this *Deskbook*. Finally, I have to thank my wife Kathleen for her understanding and encouragement.

Preparation of the *Energy Deskbook* was sponsored by the Technical Information Center, U. S. Department of Energy, under contract No. AC05-76IR05114.

Samuel Glasstone

CONTENTS

Energy Deskbook

A

ACID GASES

Specifically the gases hydrogen sulfide (H_2S), carbon dioxide (CO_2), and carbon oxysulfide (COS), commonly present in natural and synthetic fuel gases, which must be removed before the gas can be distributed. They are called acid gases because solutions in water are weakly acidic. For methods used for absorbing acid gases, see the summaries under **Desulfurization of Fuel Gases.**

ADIP (SHELL) PROCESS

A process for removing **acid gases**, especially hydrogen sulfide and to a lesser extent carbon dioxide and carbon oxysulfide, from fuel gases by chemical absorption. The absorber is an alkanolamine, namely, diisopropanolamine (DIPA), which is weakly alkaline. The general details are similar to those described for chemical absorption processes in the **desulfurization of fuel gases** (see Fig. 26).

AGGLOMERATING BURNER PROCESS

A Battelle Columbus Laboratories (with Union Carbide Corp.) process for **coal gasification** with steam and air; the product is an **intermediate-Btu fuel gas.** The process is carried out in two stages: (1) combustion of coal and **char** in air (in a combustor) to provide the heat required for (2) gasification of the coal by reaction with steam (in a gasifier). The off-gas (or flue gas) from the combustion stage is discharged and does not enter the gasifier;

hence the product gas is not diluted with inert nitrogen from the air. It is thus possible to make an intermediate-Btu fuel gas from coal, using air rather than oxygen gas for combustion.

The conditions in the combustor (or burner) are such that the coal ash forms free-flowing agglomerates (or pellets). The agglomerates are transferred continuously from the combustor to the gasifier, thereby supplying the heat required in the latter. After giving up part of its heat in the gasifier, the agglomerated ash is returned to the combustor for reheating (Fig. 1).

Char from the gasifier and crushed coal are fed with air to the bottom of the **fluidized-bed** combustor. [Caking coal may have to be pretreated by heating in air to 750°F (400°C) at atmospheric pressure.] Combustion of the coal and char produces a temperature of about 2000°F (1095°C); the pressure is roughly 8 atm (0.8 MPa). The hot ash agglomerates pass to the fluidized-bed gasifier into which crushed coal and superheated steam are injected. (Because of the high proportion of agglomerating ash, pretreatment of caking coal may not be necessary here.) The temperature in the gasifier is around 1800°F (980°C), and the pressure is the same as in the combustor. The carbon in the coal (and char) interacts with steam to produce hydrogen and carbon monoxide in a heat-absorbing reaction.

The composition of the product gas from the gasifier varies with the ratio of steam to carbon. After removal of sulfur (see **Desulfurization of Fuel Gases**), a typical gas might contain 50 volume percent of hydrogen, 30 percent of carbon monoxide,

1

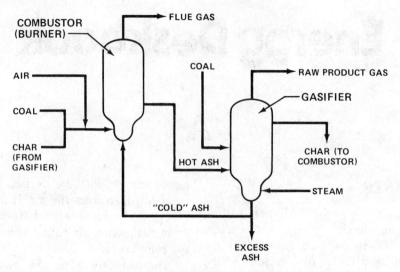

Fig. 1   Agglomerating burner process.

and a few percent of methane (dry basis) as fuel constituents; the remainder is mostly inert carbon dioxide. The **heating value** is around 300 Btu/cu ft (11 MJ/cu m).

ALCOHOL FUELS

A group of compounds with the general formula ROH, where R is a hydrocarbon radical. Alcohols may be regarded as being derived from certain (nonaromatic) **hydrocarbons** RH by replacement of a hydrogen (H) atom by a hydroxyl (OH) group. The alcohols of potential fuel interest are the two simplest members of the group: methanol or methyl alcohol (CH_3OH), related to **methane** (CH_4), and ethanol or ethyl alcohol (C_2H_5OH), related to ethane (C_2H_6). In addition, the higher alcohols propanol or propyl alcohol (C_3H_7OH), which exists in two structural forms, and butanol or butyl alcohol (C_4H_9OH), which has four forms, may be useful as fuel additives in small proportions. Methanol and ethanol (and other alcohols) are flammable liquids which produce substantial amounts of heat energy upon combustion in air. The heat may be used in a boiler to generate steam for conversion into mechanical energy in a turbine or it may be converted directly into mechanical energy in an **internal-combustion** (automobile) **engine** or in a **gas turbine.**

One of the earliest internal-combustion (Otto) engines is reported to have been operated with ethanol, and alcohols have been used for many years as a component of fuels for racing cars. The alcohols have long been considered and even utilized, especially in wartime, as automotive fuels, but with the ready availability of gasoline from petroleum, interest in alcohol fuels waned. It has now been revived, however, because of the rising cost of gasoline and the realization that petroleum oils will inevitably become scarcer and even more expensive in the future. Ethanol is important in this respect because it can be made from agricultural products which are renewable.

Methanol

Production. Methanol is now produced (for nonfuel use) almost exclusively by a chemical synthetic process first developed in Germany in the mid-1920s. Previously it had been made by the destructive distillation (or **pyrolysis**) of wood, that is, by heating wood in the absence of air and condensing the vapors; consequently, methanol was commonly called wood alcohol. The synthetic process involves the reaction of hydrogen gas (H_2) with carbon monoxide (CO) and/or carbon dioxide (CO_2) in the presence of a catalyst; thus,

$$CO + 2H_2 \rightarrow CH_3OH$$

$$CO_2 + 3H_2 \rightarrow CH_3OH + H_2O$$

The **synthesis gas** mixture of hydrogen and carbon oxides for the production of methanol is usually made from a **fossil** (primary) **fuel** (i.e., natural gas, petroleum product, or coal). Hence, methanol is a secondary fuel.

In the original (high-pressure) process, the synthesis gas at a pressure of about 300 atm (30 MPa) was passed over a catalyst, consisting mainly of zinc and chromium oxides ($ZnO-Cr_2O_3$) at a temperature in the vicinity of 660°F (350°C). This catalyst is not easily poisoned (i.e., inactivated) by small amounts of sulfur compounds and other impurities that may be present in the synthesis gas. On the other hand, methanol is not the only reaction product, and in order to improve the yield it was necessary to add carbon dioxide from an outside source to the synthesis gas.

The high-pressure synthesis has now been almost completely replaced by the low-pressure process introduced around 1967 by Imperial Chemical Industries, Ltd., in England. The gas pressure may be as low as 50 atm (5 MPa) and the reaction temperature is 480 to 570°F (250 to 300°C). The copper-based catalyst is selective for methanol formation, and the addition of carbon dioxide to the synthesis gas is optional. Because of this selectivity, the methanol product requires less purification than that from the high-pressure method. However, the copper catalyst is very sensitive to sulfur and other poisons.

Several modifications of the copper-based-catalyst process for methanol production have been described. In the medium-pressure method of Vulcan-Cincinnati, Inc., the product, trade-named Methyl-Fuel, consists of methanol with a few percent of ethanol, propanol, and butanol. The process efficiency is claimed to be better and the costs lower than for other synthetic processes. The presence of small amounts of the higher alcohols in the methanol may be advantageous, as will be seen later.

Production of Synthesis Gas. The synthesis gas for methanol production is usually made by the **steam reforming** of methane (CH_4) in **natural gas.** The natural gas, which must be freed from sulfur and other catalytic poisons, is mixed with steam and passed over a nickel catalyst in a reactor heated externally to about 1650°F (900°C); the reaction is

$$CH_4 + H_2O \rightarrow CO + 3H_2$$

The product contains more hydrogen ($3H_2$ to 1 CO) than is required for methanol production ($2H_2$ to 1 CO); hence, carbon dioxide may be added to utilize the extra hydrogen. Because of the purity of the steam reforming feed gas, the product is suitable for methanol synthesis with a copper catalyst.

Partial oxidation of methane provides a possible alternative for obtaining methanol synthesis gas from natural gas. A mixture of natural gas and a limited amount of oxygen, insufficient to cause complete oxidation to carbon dioxide and water, is heated to about 2460°F (1350°C). Ideally, the reaction should be

$$CH_4 + \frac{1}{2}O_2 \rightarrow CO + 2H_2$$

yielding a gas with the $2H_2$ to 1CO ratio required for methanol production. In practice, however, other reactions also take place at the high process temperature and the final gas composition is somewhat different.

Synthesis gas for methanol manufacture can also be made by steam reforming of paraffin hydrocarbons other than methane; these include **refinery gases** (i.e., ethane, propane, and butane), either alone or mixed, and light **naphthas** from petroleum distillation. However, methane has been the preferred source of synthesis gas because of the convenience and availability of natural gas with a very low sulfur content.

Instead of using natural gas or a petroleum product, methanol synthesis gas can be made from coal or even municipal wastes (see **Municipal Waste Fuels**). Essentially any process that yields a gas consisting mainly of carbon monoxide and hydrogen (see **Intermediate-Btu Fuel Gas**) can serve the purpose. If the ratio of H_2 to CO in the gas is too low for efficient methanol synthesis, additional hydrogen can be obtained by subjecting the mixture to the **water-gas shift reaction**.

Properties. Methanol is a colorless liquid of mild odor and low viscosity (i.e., it flows readily). Its **specific gravity** is approximately 0.8 at ordinary temperature, compared to about 0.7 for gasoline. Pure methanol boils at a single temperature, 148°F (64.6°C) at atmospheric pressure, whereas gasoline, being a mixture of hydrocarbons, boils over a temperature range of roughly 90 to 400°F (32 to 205°C).

As with gasoline, complete combustion of methanol in oxygen (from air), produces carbon dioxide and water (H_2O) by the reaction

$$CH_3OH + \frac{3}{2}O_2 \rightarrow CO_2 + 2H_2O$$

Theoretically, the reaction requires a mass of air 6.45 times the mass of the methanol; this is called the stoichiometric mass ratio. For gasoline, the corresponding ratio is 14 to 15. Thus, much less air is required for the combustion of a given mass of methanol than for the same amount of gasoline.

The heat of combustion (or **heating value**) depends on the state (liquid or vapor) of the methanol and water. For liquid methanol as fuel and water vapor as product, the heating value at 77°F (25°C) is 8600 Btu/lb (20 MJ/kg) or 57,000 Btu/gal

(16,000 MJ/cu m) compared to approximately 19,000 Btu/lb (44 MJ/kg) or 115,000 Btu/gal (32,000 MJ/cu m) for average gasoline. The heating value of methanol is thus roughly half that of gasoline on either an equal weight or equal volume basis.

The latent heat of vaporization of methanol (i.e., the heat that must be supplied to change the liquid into vapor at a specified temperature) is 503 Btu/lb (1.17 MJ/kg) at 77°F (25°C). The heat of vaporization of gasoline at the same temperature is roughly 150 Btu/lb (0.35 MJ/kg). The difference is important.

The **octane number** of pure methanol as the fuel in a **spark-ignition** (internal-combustion) **engine** is high. Different investigators have reported research octane numbers (RON) in the range of 106 to 115 and motor octane numbers (MON) from 88 to 92. The antiknock effect of methanol as an additive to gasoline is substantial, as shown by the data in the table. The MON is more representative of the antiknock value of the fuel in a modern automobile designed to minimize air pollution emissions.

	Unleaded gasoline	Gasoline with 10 percent methanol
RON	91	95.5
MON	82.5	84.5

Applications. Potential fuel uses of methanol are (1) as an additive, up to 15 percent by volume, to gasoline for spark-ignition engines, (2) in the essentially pure ("neat") state as a gasoline substitute in such engines, and (3) as the fuel for steam boilers and gas turbines.

1. Road tests made during the early 1970s with methanol–gasoline blends in several different automobiles led to conflicting conclusions, but the contradictions now appear to be resolved.

In pre-1973 cars, little attention was paid to emission control and carburetors were adjusted to operate on rich gasoline-air mixtures (i.e., the proportion of air was below the ideal mass stoichiometric ratio), so that not all the gasoline was utilized. Since unit mass of methanol requires less air for combustion than gasoline, the blended fuel gave a leaner mixture (i.e., with excess air) in the same carburetor. The result was an improved mileage (i.e., miles per gallon of fuel), due to increased gasoline burning, and smaller proportions of hydrocarbons, carbon monoxide, and **nitrogen oxides** (NO_x) in the exhaust. Later automobiles, however, have been designed to operate on relatively lean gasoline-air mixtures in order to reduce undesirable emissions. The additional leaning effect of methanol in a blend may then decrease the mileage efficiency.

There is possibly a decrease in the nitrogen oxide emissions because of the lower combustion temperatures of methanol-gasoline blends. (Nitrogen oxides are formed by the interaction of nitrogen and oxygen in air at high temperatures.) But catalytic converters are still required to meet federal emission standards. Furthermore, the compression ratios of modern, low-emission automobile engines are not high enough to benefit from the increased octane numbers of the methanol-gasoline blends.

Engines and carburetors could be designed to take advantage of the leaning and antiknock effects of methanol in blends, but the blended fuel would have to be widely accepted to make this economically possible. Furthermore, a number of difficulties associated with methanol-gasoline blends would have to be overcome.

One problem arising from the lean character of the air-fuel mixture is difficult starting and engine stalling and surging. Another problem is that, since the vapor pressure of the blend is higher than that of gasoline alone at the same temperature, there is a marked tendency for vapor lock (i.e., blockage of the fuel line by vapor) to occur. This effect could possibly be reduced

by removing the more volatile constituents (e.g., butane and pentane) from the gasoline in the blend. There would, however, be a decrease in the fuel mileage and anti-knock characteristics.

A basic difficulty in using methanol-gasoline blends, especially at low temperatures, is that quite small amounts of water cause the liquid to separate into two layers. The lower layer consists mainly of methanol together with the water and a small proportion of gasoline. If the two layers should be present in an automobile fuel tank, the feed to the carburetor would consist of the lower (i.e., methanol-rich) layer. But, as will be seen shortly, an engine designed to use a methanol-gasoline blend cannot function satisfactorily on methanol alone. The presence of quite small amounts of water in the blended fuel system would therefore cause operational failure.

If methanol-gasoline fuels were to be used, special precautions would have to be taken to preclude entry of water. This would be difficult since small quantities of water are commonly present in gasoline storage and transportation tanks. Moreover, methanol tends to absorb moisture from humid air. A possible solution to this problem may be to add a small proportion of higher alcohols (e.g., propanol and butanol), such as are present in the commercial product Methyl-Fuel mentioned earlier. The methanol-gasoline blend can then tolerate a larger proportion of water without separation into two layers.

2. The problem of layer separation in the presence of water does not arise with neat methanol. However, because the air-to-fuel stoichiometric ratio for methanol is much less than for gasoline, neat methanol is not suitable for standard automobile engines. Either carburetors with larger jets or direct fuel injection would be required for satisfactory operation. However, engines modified in these ways exhibited cold-starting difficulties due to the high latent heat of vaporization of methanol.

Several ways have been proposed to facilitate cold starting. One is to preheat the air entering the carburetor. A possible alternative is to add the volatile hydrocarbons butane and/or pentane to the methanol. The most reliable approach is probably to use a special starting system in conjunction with fuel injection; a quantity of quick-starting fuel (e.g., gasoline) is injected through a separate cold-start valve.

An additional drawback to methanol fuel is that the heating value is about half that of the same volume of gasoline. The higher octane number of methanol would permit the use of an engine with a higher compression ratio and greater efficiency. Nevertheless, for a given travel distance, methanol would require a larger automobile fuel tank. Furthermore, bulk transportation and storage systems would have to be increased correspondingly.

3. Neat methanol has been found to be a satisfactory substitute for fuel oil in an unmodified steam boiler. The methanol burned efficiently with a steady flame. The flame temperature is lower than with oil and consequently the combustion (flue) gases contain less nitrogen oxides. Because synthetic methanol is essentially free of sulfur compounds, so also are the flue gases. Thus, methanol or Methyl-Fuel, which is reported to be cheaper, could serve as the boiler fuel for electric power generation. It could also be used as a substitute for oil in home heating and for natural gas in a variety of agricultural drying processes. The main disadvantage of methanol is its low heating value compared to an equal volume of fuel oil.

Methanol has also been tested successfully as the fuel in a gas turbine. Carbon monoxide in the exhaust gas was higher than with No. 2 distillate oil (see **Fuel Oils**) but was still below the permissible maximum. Nitrogen oxides, on the other hand, were less with methanol. Apart from the need to pump a larger volume (or mass) of methanol, because of its lower heating value, the only significant problem arose in

connection with lubricating the fuel pump. When oil is pumped, the oil itself serves as the lubricant; but with methanol, additional (external) lubrication is required.

Another possible application of methanol is in **fuel cells** for the direct generation of electricity. The methanol can be used as a source of hydrogen by steam reforming or it might be supplied directly to the negative electrode of the cell. The liquid nature of methanol makes it more convenient than a gaseous fuel.

Methanol and Other Fuels. Instead of shipping **liquefied natural gas** (LNG) from overseas in special tankers at very low temperatures, it has been proposed to convert the methane first into methanol. The liquid methanol could then be transported in conventional oil tankers and stored and distributed like petroleum products. The methanol could be used as a fuel in the various ways already described. Shipping, storage, and distribution could then be achieved without the special precautions demanded by LNG.

Methanol can be converted into gasoline by the **Mobil M-Gasoline process.** Hence, methanol can be regarded as an intermediate stage in the production of gasoline from synthesis gas made from natural gas or coal.

Ethanol

Ethanol (or ethyl alcohol) is the alcohol present in alcoholic beverages. Consequently, the term "alcohol" is commonly used as a synonym for ethanol. It is also called grain alcohol because it is frequently made from grain, especially barley and corn. Like methanol, ethanol is a secondary fuel, since it is obtained either from plant products, which store solar energy, or from petroleum refinery gases.

Production. Ethanol is manufactured either (1) by biochemical processes or (2) by chemical synthesis. Beverage alcohol is invariably made by the biochemical method. Some industrial ethanol is obtained in the same way, but most has been produced chemically. Chemical synthesis requires ethylene, a petroleum product, and since the purpose of using ethanol as a fuel is to save petroleum, fuel ethanol should be made preferably from renewable sources by biochemical methods.

1. The basic reaction in the biochemical methods is called alcoholic fermentation. It is the decomposition in the absence of air of simple hexose sugars (i.e., sugars with the general formula $C_6H_{12}O_6$, containing six carbon atoms per molecule) in aqueous solution by the action of an enzyme (i.e., a natural catalyst) present in yeast; thus,

$$C_6H_{12}O_6 \rightarrow 2C_2H_5OH + 2CO_2$$

with ethanol (C_2H_5OH) and carbon dioxide gas (CO_2) as the products. An enzyme produced by the microorganism *Zymomonas mobilis* is more effective than that in yeast for alcoholic fermentation, but its commercial application has not yet been developed.

The raw materials for the biochemical production of ethanol are natural compounds of carbon, hydrogen, and oxygen, called carbohydrates (see **Biomass Fuels**). The carbohydrates fall into three general categories; in order of increasing complexity, they are sugars, starches, and cellulose. The carbohydrate is first converted into a mixture of simple hexose sugars by hydrolysis (i.e., breakdown by water) in the presence of a suitable catalyst. The resulting sugars are then fermented with yeast to form ethanol.

a. The sugars commonly used in alcohol production are monosaccharides (or single sugars) and disaccharides (or double sugars). Monosaccharides, such as glucose and fructose, found especially in sweet fruits, are directly fermentable to ethanol. The disaccharides ($C_{12}H_{22}O_{11}$) must first be converted into monosaccharides before fermentation. The most important disaccharide is ordinary sugar (e.g., cane or beet sugar) known technically as sucrose. It is changed into a mixture of glucose and dextrose (or invert sugar) by hydrolysis catalyzed by an enzyme found in most yeast varieties. A major source of sugars for industrial fermentation ethanol manufacture is blackstrap molasses, a by-product of cane sugar production. However, the cereal plant sweet sorghum is regarded as a promising future source of ethanol.

b. Starches present in grains (e.g., corn, barley, rye, etc.) and potatoes are an important source of beverage alcohol. They are degraded by enzyme action with water, first into disaccharides, mainly maltose, and then into the fermentable sugar glucose. The first stage is accomplished by a starch-splitting enzyme usually obtained from barley malt. Malt is made by allowing barley to soften and germinate in water; it is then heated sufficiently to stop the sprouting without destroying the enzyme. An alternative commercial source of the starch-splitting enzyme is a mold of the *Rhizopus* genus. In the second stage, the maltose is converted into glucose and then into ethanol by means of enzymes in yeast.

c. Cellulose occurs in all plant products, especially in wood and cotton. Various wood wastes are the major source of cellulose for conversion into fermentable sugars, chiefly glucose, but paper from municipal wastes could also be used. The most highly developed method for breaking down cellulose is by hydrolysis with a mineral acid catalyst. Modifications in the procedure, involving increased temperature and pressure and more concentrated acid, require more expensive equipment, but acceleration of the reaction and the higher yields of simple sugars are said to lead to lower overall costs.

Cellulose in water can also be converted into glucose by enzymes produced by a fungus belonging to the genus *Trichoderma*. The process is simple and inexpensive, but it is relatively slow; furthermore, the cellulose should be clean and finely milled. Lignin, the noncarbohydrate component of wood, interferes with the enzymatic degradation of cellulose; hence, methods for prior removal of lignin are being studied. Since most of the lignin is separated from the wood pulp in paper manufacture, the paper from municipal wastes might prove a useful source of cellulose for ethanol production by enzyme action.

In addition to roughly 50 percent of cellulose, wood contains, on the average, about 25 percent of hemicelluloses. Unlike cellulose, which is a combination of glucose (six-carbon) units, hemicellulose is composed mainly of (and breaks down into) five-carbon sugars, especially xylose. These sugars are not fermented by ordinary yeasts; hence, the hemicellulose is generally removed from wood fiber and discarded before the remaining cellulose is used for ethanol production. However, research has indicated that enzymes present in certain yeast varieties can convert xylose (as well as glucose) into ethanol.

Fermentation of simple sugars to ethanol in the absence of air can be conducted either in batches or continuously. In the common batch process, yeast is added to the sugar solution and is allowed to stand in the absence of air for about 2 days at a temperature of 85 to 95°F (30 to 35°C). The fermentation ceases when the liquid contains about 12 percent by volume of ethanol; distillation then yields a mixture of ethanol and water, together with small amounts of other substances (see below). In the continuous process the fermentable sugar solution is fed at a steady rate into a vessel containing yeast; the temperature is maintained at around 90°F (32°C). The ethanol formed (and some water) is removed continuously by distillation under reduced pressure.

The ethanol-water mixture distilled from the fermentation process is concentrated by successive fractional **distillation.** The final product then contains 95 volume percent of ethanol (190 proof); this is "pure" commercial alcohol. The remaining 5 percent of water cannot be removed by simple distillation, and special procedures are used to reduce the water content as may be required for blending the ethanol with gasoline.

The common method for removing the remaining water is to add benzene to the 95 percent product and distill the mixture. The first distillation fraction contains the benzene together with all the water and some alcohol; the subsequent distillate, known as "absolute" alcohol is 99+ percent ethanol (198+ proof). Although the benzene process is effective, it requires a substantial amount of energy. Consequently, less energy-intensive alternative procedures are being studied; these include various drying agents, extraction with solvents, and the use of membranes permeable to water or ethanol but not both.

Ethanol is the main product of sugar fermentation, but small amounts of higher alcohols and other substances are also formed. The distillation fraction in the high-temperature boiling range containing these impurities is called fusel oil; it consists largely of propanols, butanols, and especially pentanols (amyl alcohols). Fusel oil is not of direct interest as a fuel, but it might perhaps be a useful source of an additive to alcohol–gasoline blends. (About 5 percent of tertiary butyl alcohol has been added to a commercial variety of unleaded gasoline to increase the octane number and to prevent carburetor icing.)

2. The chemical synthetic process for the manufacture of ethanol is based on the hydration of (i.e., the addition of a water molecule to) the hydrocarbon gas ethylene (C_2H_4); thus,

$$C_2H_4 + H_2O \rightarrow C_2H_5OH$$

The ethylene is usually made by the **steam cracking** of petroleum refinery gases (see **Petroleum Products**); consequently, synthetic ethanol depends on the availability (and cost) of petroleum. Fermentation alcohol, on the other hand, is made from plant products.

In the older, indirect hydration method, the ethylene gas is reacted with concentrated sulfuric acid at 122 to 176°F (50 to 80°C) and a pressure of 10 to 14 atm (1 to 1.4 MPa). The products are ethyl hydrogen sulfate and diethyl sulfate which are decomposed by water to produce ethanol and dilute sulfuric acid. The ethanol is distilled off in the usual manner, and the sulfuric acid is concentrated for reuse.

Synthetic ethanol is now commonly made by the direct hydration process. A mixture of ethylene gas and steam at roughly 750°F (400°C) and 70 atm (7 MPa) is passed over a phosphoric acid catalyst supported on an inert (silica-based) solid. The product, consisting principally of ethanol and water vapor, is condensed and purified by fractional distillation.

Properties. Ethanol is a colorless liquid of low viscosity and a slight, characteristic odor. Its specific gravity is roughly 0.8 at ordinary temperatures. Pure (100 percent) ethanol boils at 173°F (78.4°C) under atmospheric pressure. The latent heat of vaporization at 77°F (25°C) is 395 Btu/lb (0.92 MJ/kg).

Complete combustion of ethanol in oxygen (or air) produces carbon dioxide and water. The heat of combustion of liquid ethanol, with water vapor as the product, is 11,500 Btu/lb or 76,000 Btu/gal (26.6 MJ/kg or 21,000 MJ/cu m); these are roughly two-thirds of the corresponding values for gasoline. The stoichiometric air-to-ethanol mass ratio for complete combustion is 9.0. In a conventional spark-ignition internal combustion engine, the octane numbers of neat ethanol and of blends with gasoline are similar to those of methanol. (Characteristic properties of methanol, and of gasoline for comparison, were given earlier.)

Applications. Neat ethanol, containing at least 90 percent of alcohol, can be (and has been) used as an automotive fuel. Because of the smaller stoichiometric ratio than for air-gasoline, some adjustment is required of the carburetor of a standard internal-combustion engine. The main advantage of ethanol as a fuel is that it can be produced from plant matter and is consequently a renewable energy source. In Brazil, where ethanol can be made relatively cheaply from sugar cane or manioc root, serious consideration is being given to the extensive use of 95 percent ethanol as a fuel.

Apart from the present high cost per unit of heat energy, the main disadvantages of neat alcohol fuel are the same as those of neat methanol: lower heating value than an equal volume of gasoline and difficult starting when cold [below about 45°F (7°C)]. The latter problem may be solved by adding a separate cold-starting system, as proposed for methanol.

Blends of gasoline with up to about 20 volume percent of ethanol can serve as fuel in many automobile engines without carburetor adjustment. Such blends, with fermentation alcohol, have been used in Brazil and elsewhere for several years to provide support for local agriculture and to conserve gasoline. In the United States consideration of ethanol-gasoline blends as fuel for internal-combustion engines dates back to the early years of the 20th century. The low cost of gasoline compared with that of ethanol made such blends uneconomic at the time, but the steadily increasing cost of crude oil may change the situation.

In order to find a use for spoiled and surplus grain, the state of Nebraska initiated in 1971 a program to develop and test ethanol-gasoline fuel blends. Since 1974, a mixture of 90 volume percent of unleaded gasoline and 10 percent ethanol, called *gasohol*, has become available as an automobile fuel in many parts of the United States. Among the factors that make gasohol economically competitive with unleaded gasoline are the following: exemption (at least temporarily) from the federal gasoline tax and reduction (or removal) of local taxes in some states; sale of the high-protein solid residue remaining after fermentation (distillers' dried grains) for cattle feed; and fermentation of moldy, sprouting, or low-grade (distressed) grain which has little monetary value.

Although ethanol is more expensive than methanol, the ethanol-gasoline blends have some advantages. The lower vapor pressure of ethanol at ordinary temperatures reduces the tendency for vapor lock, and the higher air-to-fuel stoichiometric ratio makes the ethanol-gasoline blends more compatible with unmodified automobile engines.

The problem of separation into two layers in the presence of water that arises with methanol-gasoline blends also occurs with ethanol blends. The latter can tolerate a somewhat larger proportion of water without separation at a given temperature; nevertheless, precautions must be taken to minimize the entry of moisture at all stages of blending and distribution of gasohol fuel. In particular, almost pure ethanol, at least 98.5 percent (197 proof), is used for blending with gasoline. The water tolerance of ethanol-gasoline blends could be improved by addition of a few percent of higher alcohols, possibly from the fusel oil by-product of the fermentation process. (Ethanol alone does not form two layers with water, and 90 percent alcohol can serve as automotive fuel, as stated earlier.)

Several programs have been conducted to compare the performance in automobiles of unleaded gasoline and gasohol in normal operation. Some drivers have claimed increased miles-per-gallon ratings, in spite of the lower heating value of the blend, and better antiknock characteristics. Others, however, have reported less favorable experiences. The differences are due, at least in part, to the differences in the air-to-fuel ratios and in compression ratios, as described earlier in connection with methanol-gasoline blends. Undoubtedly, automobile engines could be designed to take full advantage of gasohol, but they would be less efficient for gasoline alone.

ALKAZID (M AND DIK) PROCESSES

Processes for removing **acid gases** (hydrogen sulfide and carbon dioxide) from fuel gases by chemical reaction with a weakly alkaline solution. In the Alkazid M process, for the removal of both hydrogen sulfide and carbon dioxide, the absorber is an aqueous solution of the potassium salt of methylaminopropionic acid. In the Alkazid DIK process the absorber is a solution of the potassium salt of dimethylaminoacetic acid which removes hydrogen sulfide preferentially. The general principles of operation are similar to those described for chemical absorption processes in the **desulfurization of fuel gases** (see Fig. 26).

ALKYLATION

In general, the replacement of a hydrogen atom in a molecule by an alkyl radical, that is, a grouping of carbon and hydrogen atoms with the general formula C_nH_{2n+1}, such as methyl (CH_3), ethyl (C_2H_5), etc. In petroleum refining, alkylation is a process in which simple olefin **hydrocarbons,** in particular propylene (C_3H_6) and butylenes (C_4H_8), combine with the branched chain isoparaffin isobutane (i-C_4H_{10}) in the presence of a catalyst. The product, known as alkylate, is a mixture of isoparaffins mostly with seven to nine carbon atoms in the molecule. The mixture boils in the **gasoline** range and has a high **octane number** when used as a fuel for a **spark-ignition** (automobile) **engine.** The process is called alkylation because the net effect is the introduction of one or more simple alkyl radicals into isobutane.

The alkylation catalyst is either concentrated sulfuric acid or hydrofluoric acid (hydrogen fluoride). With sulfuric acid, the reaction temperature must be kept down to 35 to 45°F (2 to 8°C) and refrigeration is required. When hydrogen fluoride is the catalyst, the operating temperature is 75 to 115°F (25 to 45°C) and refrigeration is not necessary. However, hydrofluoric acid is a

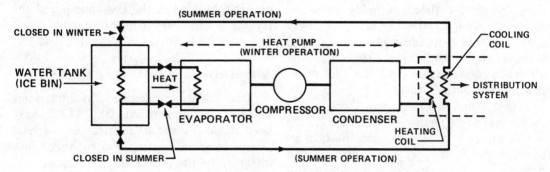

Fig. 2   Annual cycle energy system.

highly corrosive and toxic gas, and special handling is required.

ANILINE POINT

The temperature, usually expressed in °F, above which equal volumes of aniline and a liquid petroleum product (e.g., **gasoline, kerosine,** or **diesel** and **jet fuels**) are completely miscible, that is, they form a single layer (or phase). The aniline point is determined, in effect, by reversing the definition. A mixture of equal volumes of aniline and the petroleum product is heated until a completely miscible (one phase) solution is formed. The solution is then cooled slowly until separation into two phases is observed; this is taken as the aniline point.

The aniline point provides a qualitative indication of the relative proportions of paraffin and aromatic **hydrocarbons** in a petroleum product. Paraffins are miscible with aniline only at higher temperatures, whereas aromatic hydrocarbons become miscible at lower temperatures. Hence, petroleum products with high aniline points (e.g., 180°F) usually contain relatively small proportions of aromatic hydrocarbons, but those with lower aniline points (e.g., 110°F) are richer in these hydrocarbons.

ANNUAL CYCLE ENERGY SYSTEM (ACES)

A system for heating a building in winter and cooling it in summer by utilizing a **heat pump** in a manner that should result in a substantial saving (up to about 50 percent) in electrical energy requirements. The Annual Cycle Energy System (ACES) concept is being tested by the Oak Ridge National Laboratory in a house built for the purpose in Knoxville, Tennessee.

During the colder months, the heat pump provides space heating in the usual manner by way of the condenser section located in the air duct of the building's distribution system. However, instead of extracting heat from the outdoor air, as is usually done, the heat pump removes heat from a large, insulated tank of water. This is achieved by circulating an alcohol–water mixture through pipes in the water tank and a heat exchanger in the evaporator unit of the heat pump (Fig. 2). As the heat is withdrawn from the tank water, the temperature drops to 32°F (0°C) at which point freezing occurs. (Note that the alcohol–water mixture circulating through the pipes in the tank has a lower freezing point and does not freeze at 32°F.) Continued removal of heat results in more water freezing, but the temperature does not decrease further until all the water in the tank is frozen. Because the water is gradually converted into ice, the water tank is commonly referred to as an "ice bin."

In the summer, the heat pump is shut down and the alcohol–water mixture is now circulated through the ice bin and a cooling coil in the air duct of the distribution system. The cold alcohol–water mixture thus cools the air for air-conditioning purposes. Hence, as long as the ice lasts, the only energy requirement for space cooling is to operate the circulating pump for the alcohol–water mixture and the blower in the air-distribution system.

Apart from the heat pump, the major ACES cost item is the ice bin. Ideally, the bin should be of such a capacity that, in an average year, the ice produced in the winter should be sufficient to provide all the cooling required in the summer. For a well-insulated house of 1600 sq ft (170 sq m) floor area, a bin capacity of roughly 3500 cu ft (100 cu m) should be suitable for a latitude of 40°N and 3000 cu ft (85 cu m) for 35°N in the United States.

If the ice supply should be depleted before the end of the hot weather, the heat pump would be operated at night when the demand for electric power is at a minimum. Heat would be extracted from the ice bin and ice formed in the usual manner for cooling in the daytime. The heat from the condenser section would be discharged outdoors through a radiator rather than into the air-distribution system.

Energy savings in the ACES arise from the following circumstances:

1. The heat pump is operated (in the heating mode) during the winter only; except in the special case noted above, the heat pump is not used in the summer.

2. Heat transfer from the alcohol–water mixture to the working fluid in the evaporator section of the heat pump is more efficient than the usual transfer from air.

3. The temperature at which heat is taken up in the evaporator does not fall below 32°F (0°C), whereas the ambient air from which heat is extracted in a conventional heat pump is often lower.

As a consequence of the last two factors, the overall **coefficient of performance** (COP) of the heat pump in the ACES is significantly greater than that of a conventional system. Hence, a smaller quantity of electrical energy is required to produce the same amount of space heating. The limited operating period of the heat pump and the increased efficiency while operating result in substantial savings in energy.

Alternative Concepts

Two other proposals, having some features in common with the ACES, have been made for the economical use of heat pumps in space heating and cooling. They differ from the ACES in the respect that they are best applicable to a community or a large apartment complex rather than to a single residence.

A proposal originating at the Applied Physics Laboratory of the Johns Hopkins University is to use two large reservoirs, one containing water at about 60°F (15.6°C) and the other colder water at about 40°F (4.4°C). Each dwelling unit would have separate water supplies at the two temperatures. In the summer, the colder water would be used to provide space cooling; the discharged water, which has been warmed by taking up solar heat, would empty into the warmer reservoir. On the other hand, in the winter a water-to-air heat pump would withdraw heat from the warmer water for space heating; the residual, colder water would then be discharged to the cold reservoir for summer use. Should it be necessary, the latter reservoir could be cooled further by exposure to the air in winter.

An Argonne National Laboratory concept, called a modular heat pump system, is intended for a location, such as Chicago, where cold water can be obtained even in summer from a deep lake and warm water is available from an electric power plant condenser discharge (see **Waste Heat Utilization**). By varying the proportions of cold and warm water from these sources, water at a temperature between 55 and 65°F (13 and 18°C), warmer in winter and colder in summer, could be supplied to the individual dwelling units. Minimum operation of a water-to-air heat pump in each unit would then provide the relatively small temperature adjustments in winter and summer required to produce the desired space heating and cooling, respectively.

API (AMERICAN PETROLEUM INSTITUTE) GRAVITY

A quantity widely used in the petroleum industry to express the **specific gravity** (or **density**) of crude petroleum oils and petroleum products. The density in degrees API is given by the expression

$$\text{Degrees API} = \frac{141.5}{\text{sp gr } 60°F/60°F} - 131.5$$

where sp gr 60°F/60°F is the density of the oil at 60°F (15.6°C) divided by the density of water at the same temperature. There is no simple relationship between specific gravity (or density) and degrees API, but one increases as the other decreases. A specific gravity of 1.00, which is the normal value for water, corresponds to 10°API. A lower specific gravity is equivalent to a higher value on the API scale. Most crude petroleum oils have specific gravities in the range of 0.80 to 0.95, that is, 45 to 17°API.

ATGAS/PATGAS PROCESS

An Applied Technology Corporation process for **coal gasification** with steam and oxygen gas in a bath of molten iron. The product is an **intermediate-Btu fuel gas** (PATGAS process) which can be converted into a **high-Btu fuel gas** (ATGAS process). Any kind of coal, caking or noncaking, regardless of sulfur content, can be used without pretreatment.

In addition to providing a high temperature medium for gasifying the coal, the molten iron removes sulfur. By adding lime (calcium oxide) in the form of limestone (calcium carbonate), the buildup of sulfur in the molten iron is prevented by the formation of calcium sulfide. A molten slag, containing excess lime, calcium sulfide, and coal ash, floats to the surface of the molten iron and is removed continuously. It is treated with steam which regenerates the lime and releases the sulfur as hydrogen sulfide gas. The latter can be converted into elemental (solid) sulfur by the **Claus process.**

In the PATGAS process, crushed coal and limestone (or regenerated lime) are injected with steam below the surface of the molten iron in a gasifier vessel. The temperature is about 2500°F (1370°C), and the pressure is 3 to 4 atm (0.3 to 0.4 MPa). Oxygen gas is introduced through a separate nozzle. The coal is rapidly devolatilized (i.e., volatile matter is driven off), and the residual **char** dissolves in the molten iron where it reacts with steam and oxygen to produce hydrogen and carbon monoxide. The combustion of part of the carbon (char) in oxygen generates the heat required for the carbon–steam reaction and maintains the temperature of the molten iron.

The PATGAS product gas, with a heating value of about 320 Btu/cu ft (12 MJ/cu m), is essentially free from sulfur and can be used as fuel in a steam boiler or in a combustion **gas turbine.** In the ATGAS process, the gas is subjected to the **water-gas shift reaction** and **methanation** to produce a high-Btu **substitute natural gas.**

ATOM

The smallest (or ultimate) particle of a chemical element that can exist and still retain the characteristics of that element. The elements are the basic constituents of matter which cannot be subdivided by chemical means. Some 90 different elements are known to exist in nature, and several others, such as the **nuclear fuel** plutonium, have been made artificially. All substances consist of either single elements, mixtures of different elements, or chemical combinations (i.e., compounds) of two or more elements. No matter in what state it exists, an element is always made up of its characteristic atoms.

Atomic Nucleus

Atoms are so small that roughly 50 million atoms placed side by side would measure 1 centimeter (or about 120 million to an inch). In spite of its small size, the atom has an internal structure. Every atom consists of an extremely small central region or *nucleus*, which carries nearly all the mass of the atom, surrounded by a number of much lighter particles called *electrons*. The nucleus has a positive electric charge, whereas the electrons are negatively charged. In the normal atom, the positive charge on the nucleus is exactly balanced by the negative charges of the electrons. Hence, the atom as a whole has no net charge; that is, it is electrically neutral.

The atomic nucleus of an element is characterized by the numbers of protons and neutrons it contains. These are fundamental particles with the following properties:

The *proton* is a particle with a single positive charge; its mass is somewhat greater than unity (actually 1.007276 u) on the standard mass (or u) scale, described later. The protons are responsible for the positive charge of the nucleus, the magnitude of the charge being equal to the number of protons in the nucleus.

The *neutron*, as its name implies, is an electrically neutral particle (i.e., it has no electric charge). Its mass, which is also close to unity on the standard scale, is slightly greater than that of the proton, namely, 1.008665 u.

Atomic Number and Mass Number

Except for the nucleus of common (or light) hydrogen, which is a single proton, all atomic nuclei contain both protons and neutrons. The number of protons (or number of positive charges) in the atomic nucleus is called the *atomic number*; this is also equal to the number of electrons that normally surround the nucleus. Since protons and neutrons have masses close to unity, the total number of protons and neutrons is approximately equal to the mass of the nucleus (or of the atom, since the nucleus carries nearly all the mass). The number of protons plus neutrons in the nucleus is therefore known as the *mass number*.

The atomic number is the property that distinguishes one element from another. Thus, the simplest element, hydrogen, has an atomic number of 1; this means that nuclei of all hydrogen atoms contain one proton. At the other extreme is the element uranium with an atomic number of 92 (i.e., 92 protons in the nucleus); the uranium nucleus is the most complex of all elements found in significant amounts in nature. Elements with higher atomic numbers, such as neptunium (93), plutonium (94), and others, have been obtained by nuclear reactions starting with uranium.

Isotopes

Although all the nuclei of a given element contain the same number of protons, they may have different numbers of neutrons. Consequently, although they have the same atomic number (i.e., number of protons), they can have different mass numbers (i.e., protons plus neutrons). Such species, which have identical atomic numbers, and hence are the same element, but different mass numbers, are called *isotopes* of the given element.

The chemical properties of an element are determined almost entirely by the number and arrangement of the electrons surrounding the nucleus. All the isotopes of an element have the same number of atomic electrons, and their arrangement is usually affected to a slight extent only by the nuclear mass. Hence, all the isotopes of an element have essentially the same chemical properties. However, properties that are determined by the nuclear structure, espcially **radioactivity,** vary from one isotope to another of the same element.

All but 18 of the 90 naturally occurring elements exist in two or more isotopic forms. Most of the isotopes are stable (i.e., nonradioactive), but the isotopes of the heaviest elements, with high atomic and mass numbers, are generally radioactive. These radioactive (or unstable) isotopes exist in nature either because they break down (or decay) fairly slowly or because they are being continuously formed from parent (or predecessor) radioactive species.

Hydrogen Isotopes. The isotopes of hydrogen and uranium are of special interest for the release of **nuclear energy.** The element hydrogen exists in three isotopic forms, namely, hydrogen (or light hydrogen), symbol H; deuterium (or heavy hydrogen), symbol D; and tritium, symbol T. (Hydrogen is the only element whose isotopes are distinctive enough to show small chemical differences and to merit different names.) The compositions of the three hydrogen isotope nuclei are shown below.

	Hydro-gen (H)	Deute-rium (D)	Tritium (T)
Protons (atomic number)	1	1	1
Neutrons	0	1	2
Protons + neutrons (mass number)	1	2	3

Hydrogen and deuterium are found in all forms of water, but tritium is radioactive and exists in nature in traces only. Although tritium is formed continuously by the action of cosmic rays in the atmosphere, it decays fairly rapidly. Hence, the concentration always remains very small. However, tritium can be made in quantity by the action of neutrons on lithium nuclei. Tritium obtained in this manner and deuterium extracted from water are expected to be the fuel materials for the production of nuclear **fusion energy.**

Uranium Isotopes. As found in nature, uranium consists mainly of two isotopes, uranium-235 and uranium-238, where 235 and 238 are the respective mass numbers. The compositions of the nuclei are as follows:

	Uranium-235	Uranium-238
Protons (atomic number)	92	92
Neutrons	143	146
Protons + neutrons (mass number)	235	238

Both uranium isotopes are radioactive, but they decay fairly slowly so that substantial quantities still exist on earth. The lighter, less-abundant isotope, uranium-235, which constitutes only about 0.71 percent by weight of the uranium in nature, plays an important primary role in the release of nuclear **fission energy.** The heavier isotope, uranium-238, plays an indirect role as the source of plutonium-239 which can also be used to produce energy by fission (see **Nuclear Fuel**). Although the isotopes of uranium have essentially identical chemical properties, they have different nuclear properties (e.g., fission and radioactivity).

Atomic Mass Unit

The mass unit, u, which is used to express atomic (or isotopic) and nuclear masses, is defined in terms of the mass of the most abundant isotope of the element carbon. The mass of the neutral atom of this isotope, with a mass number of 12, is set at exactly 12.0000 u. Thus the *atomic mass unit*, 1 u, is one-twelfth of the mass of an atom of the carbon-12 isotope. In terms of more conventional mass units, $1\ u = 1.660 \times 10^{-27}$ kilogram (or 1.660×10^{-24} gram).

B

BAG FILTER

A bag-like device made from a woven or felted fabric through which gas is passed for the removal of suspended particulate matter. A system of bag filters (or baghouse) consists of thousands of such bags in several independent compartments; each compartment can be isolated periodically to discharge accumulated solid.

Bag filters can be used for removing the fly ash from the stack gases of coal-fired steam plants (see **Steam Generation**). The bag material is selected to suit the operating temperature, which may range from about 200 to 550°F (95 to 290°C). Cotton or a synthetic fabric may be used at the lower temperatures and asbestos or fiberglass at the higher temperatures.

Bag filters differ in the direction of gas flow and in the method for discharging the collected solid (e.g., fly ash). In *shaker-type* units, the bags are suspended from a framework that is free to oscillate; gas flows from the inside out. All the bags in an isolated compartment are shaken periodically to dislodge the accumulated solid which falls to the bottom of each bag. In *reverse-flow* bags, the normal flow is also from the inside out. However, to dislodge the collected solid, the flow direction is reversed by means of an auxiliary fan. Shaker-type and reverse-flow bags are often used in the same baghouse. In *reverse-pulse* units, the gas flow is from the outside in. Periodically, a short pulse of compressed air is directed from top to bottom in the interior of the bag. The bag expands rapidly and discharges the solid deposited on the outside.

The installation costs of a baghouse are high, and the pressure drop through the bags requires additional power in the gas exhaust fans. On the other hand, the bag filter has some advantages over the **electrostatic precipitator** in removing fly ash in coal-fired plants. Bag filters are very efficient and do not suffer from the problems of electrostatic precipitators when the sulfur content of the coal is low. The bag filter is consequently advantageous for low-sulfur, high-ash coals, especially in states that have exceptionally stringent particle emission standards.

BARREL

As applied to petroleum and its products, a barrel (abbreviated as bbl) is defined as 42 U.S. gallons or 5.6 cu ft (159 liters or 0.159 cubic meters). In the United States the average mass (or weight) of a barrel of crude oil is commonly taken to be 300 pounds (136 kilograms), corresponding to an average **specific gravity** of about 0.85 or an **API gravity** of 35 (see **Petroleum**). There are thus roughly 6.7 barrels of crude oil in 1 short ton (2000 pounds), 7.3 barrels in 1 metric ton (1000 kilograms), and 7.4 barrels in 1 long ton (2240 pounds).

BATTELLE MOLTEN-SALT DESULFURIZATION PROCESS

A high-temperature process of the Battelle Pacific Northwest Laboratories for the **desulfurization of fuel gases** by absorption of hydrogen sulfide. The main absorber is calcium carbonate (limestone) dissolved in a molten mixture of lithium, sodium, and potassium carbonates at a temperature above 1200°F (650°C). The calcium carbonate removes the hydrogen sulfide and is converted into calcium sulfide. The carbonate is regenerated by blowing with steam and carbon dioxide; a gas rich in hydrogen sulfide produced at the same time can be converted into elemental (solid) sulfur by the **Claus process.**

BCR TRI-GAS PROCESS

A Bituminous Coal Research (BCR), Inc. process for **coal gasification** with steam and air; the product is a **low-Btu fuel gas.** (Substitution of oxygen gas for air would lead to an **intermediate-Btu fuel gas.**) The coal is gasified in three **fluidized-bed** stages, with solids being transferred continuously from one stage to the next hotter stage. The process can be used for both caking and noncaking coals.

Pulverized coal is fed to the devolatilizer (stage 1) where partial **pyrolysis** occurs at a temperature of 1200°F (650°C); the coal is devolatilized (i.e., volatile matter is driven off) by hot gas from stage 3 (Fig. 3). The first stage also serves to pretreat caking coals and thus permits them to be fluidized. The devolatilized coal passes to the gasifier (stage 2) where it is fluidized by the off-gas from stage 1.

Interaction of the carbon (in the coal) with oxygen (in the air) and steam introduced into stage 2 results in the major gasification of the coal to form carbon monoxide and hydrogen, at roughly 2000°F (1090°C). The residual **char** is finally fluidized and burned with air and steam at about 2100°F (1150°C) in the combustor (stage 3), thereby producing the hot gas for stage 1. The pressure is up to about 17 atm (1.7 MPa) in each stage.

Ash is discharged from stage 3 and the raw product gas is obtained from stage 2. After removal of entrained solids and hydrogen sulfide (see **Desulfurization of Fuel Gases**), the gas contains about 31 percent by volume of carbon monoxide and 16 percent of hydrogen (dry basis) as fuel constituents, with a little over 50 percent of inert nitrogen from the air. The **heating value** of the gas is roughly 150 Btu/cu ft (5.6 MJ/cu m).

BEAVON PROCESS

A process for removing small amounts of elemental sulfur and sulfur compounds, including hydrogen sulfide, carbon oxysulfide, carbon disulfide, and organic sulfides, from **Claus process** tail gas. A hot gas containing hydrogen and carbon monoxide, obtained by partial combustion of a fuel gas with air, is mixed with the tail gas and

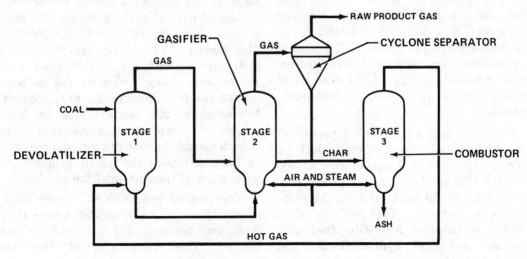

Fig. 3   BCR TRI-GAS process.

passed through a catalyst bed where the sulfur and sulfur compounds are converted, by the hydrogen and carbon monoxide, into hydrogen sulfide. The resulting gas is cooled by direct contact with a slightly alkaline solution and the hydrogen sulfide is finally removed as solid sulfur by the **Stretford process.**

BENFIELD PROCESS

A process for removing **acid gases** (hydrogen sulfide, carbon dioxide, and carbon oxysulfide) from fuel gases by chemical absorption in a weakly alkaline medium (see **Desulfurization of Fuel Gases**). The pressurized impure feed gas enters the bottom of a column where it meets a downflow of an aqueous solution of the absorber, potassium carbonate containing an amine and proprietary additives. The temperature in the absorber column is usually about 230°F (110°C), and the pressure is in the range of 7 to 135 atm (0.7 to 13.5 MPa). The general principles are similar to those described for chemical absorption processes in the **desulfurization of fuel gases** (see Fig. 26).

BERGIUS PROCESS

A process invented by F. Bergius around 1913 for the production of a **synthetic crude oil** (or syncrude) by the catalytic hydrogenation of coal with hydrogen gas at high temperature and pressure. The process was developed and used extensively in Germany, especially during World War II, to produce liquid fuels for internal combustion engines.

Pulverized coal mixed with a portion of the product oil was reacted with a hydrogen-rich gas at a temperature of 800 to 900°F (425 to 480°C) and a pressure of 200 to 700 atm (20 to 70 MPa) in the presence of a catalyst. The products after refining included a **high-Btu fuel gas, gasoline** (and light **hydrocarbons**), and various **fuel oils.**

More recent methods of **coal liquefaction** by direct hydrogenation are essentially modifications of the Bergius process. However, they operate at lower pressures with improved catalysts.

BI-GAS PROCESS

A Bituminous Coal Research, Inc. process for coal gasification with steam and oxygen; the immediate product is an **intermediate-Btu fuel gas** which can be converted into a **high-Btu fuel gas** (or **substitute natural gas**). The process can be used with caking or noncaking coals without pretreatment.

The gasifier consists of upper and lower stages (Fig. 4). A slurry of crushed coal in water enters at the bottom of the upper stage and is carried upward by **entrained flow** with steam injected simultaneously and hot gases from below. The temperature ranges from about 1900°F (1040°C) at the bottom to 1700°F (925°C) at the top of this stage. The pressure may be up to 100 atm (10 MPa). The coal is devolatilized (i.e., volatile matter is driven off), releasing **methane,** hydrogen, and other fuel gases. Carbon in the residual **char** reacts with steam to produce hydrogen and carbon monoxide.

Unconsumed char drops to the lower stage of the gasifier where it reacts with preheated oxygen gas and steam; the temperature may approach 3000°F (1650°C). The carbon (char)-steam reaction yields hydrogen and carbon monoxide which pass to the upper stage, whereas the carbon-oxygen reaction provides the heat required to maintain the temperatures in both stages. The temperature in the lower stage is high enough to melt the residual coal ash to form a liquid slag which is quenched with water at the bottom of the gasifier.

The product gas from the upper stage passes to a cyclone separator where char fines are removed and returned to the gasifier. The composition of the gas depends on the nature of the coal and the

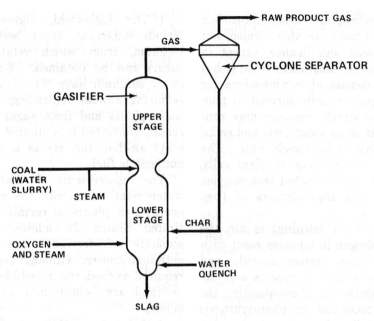

Fig. 4 B1-GAS process.

gasifier conditions. From a typical bituminous coal, the product gas after sulfur removal (see **Desulfurization of Fuel Gases**) might contain 32 volume percent of hydrogen, 29 percent of carbon monoxide, and 15 percent of methane (dry basis) as the fuel constituents; most of the remainder is inert carbon dioxide. The heating value is approximately 350 Btu/cu ft (13 MJ/cu m). By subjecting this gas to the **water-gas shift reaction** followed by **methanation,** the product is a high-Btu substitute natural gas.

See also **Foster–Wheeler Coal Gasification Process.**

BIOMASS FUELS

Fuels obtained from living matter, especially plants and plant products and residues. Coal, natural gas, and petroleum are fossil forms of biomass. The energy content of biomass fuels is derived from the process of photosynthesis occurring in green plants, both terrestrial and aquatic. These plants, which contain chlorophyll, are able to take up energy from solar radiation and synthesize compounds of carbon, hydrogen, and oxygen, such as carbohydrates, lignin, and others, from atmospheric carbon dioxide and water.

Of the many compounds produced as a result of photosynthesis, only those of possible interest as biomass fuels will be referred to here. The simplest common carbohydrate constituent of biomass is the sugar glucose, which is found to some extent in nearly all plants. Sucrose, commonly known as cane or beet sugar, is a combination of glucose and a related simple sugar fructose. There are several classes of more complex carbohydrates, the most important being starches and cellulose. The general term cellulose refers to the most complex carbohydrates; the name arises from the fact that cellulose is the major constituent of plant cell walls. It is thus found in all plant materials and particularly in wood and cotton fibers.

The photosynthetic process involves several successive stages, but the overall initial reaction is the formation of glucose, as represented by

$$6CO_2 + 6H_2O \xrightarrow{\text{Solar energy}} C_6H_{12}O_6 + 6O_2$$

Carbon dioxide / Water / Glucose / Oxygen gas

surface of open ocean waters are inadequate for large-scale, sustained kelp growth. It is planned, therefore, to pump water from a depth of 500 to 1000 ft (150 to 300 m) where the concentration of mineral nutrients is greater than it is near the surface. The cooling effect of the colder deep water could be advantageous because the surface waters in some locations may be too warm for optimum growth of the kelp. However, substantial amounts of energy are required for pumping the water.

Biomass Utilization Processes

Thermochemical Conversion. In a general sense, thermochemical conversion refers to the chemical reactions brought about by the use of heat alone or in the presence of air or other gases to release energy directly from biomass or to convert the biomass into clean gaseous and liquid fuels. The appropriate processes can be considered in the following three categories:

1. Direct combustion
2. Gasification
3. Liquefaction

1. The most direct way to use biomass as a fuel is by direct combustion (burning) in air. In the dry or moderately dry state (30 percent or less of moisture), any form of biomass will burn and generate heat. Bagasse, the fibrous material remaining after the extraction of the juice from sugarcane, is used as fuel in boilers to produce steam in sugar refineries. The main biomass materials suitable for direct combustion are wood and wood residues.

The fuel value of wood depends on its fiber (cellulose and lignin), resin, and moisture contents. (Lignin constitutes 20 to 30 percent of the dry weight of most woods; it is not a carbohydrate but is combustible.) Because of their high resin content, the softwoods (e.g., pine and fir) have higher **heating values** than the hardwoods (broadleaf trees). The average heating value of dry softwood is about 8400 Btu/lb (19.5 MJ/kg) and for hardwood it is about 8000 Btu/lb (18.6 MJ/kg). These values may be compared with 13,000 Btu/lb (30 MJ/kg) for average bituminous coal.

The sawmill, plywood, and paper-pulp industries already satisfy from 20 to 50 percent of their energy requirements by burning waste. The larger pieces of wood and bark are ground to a convenient size and burned, together with sawdust and wood shavings, in a boiler to produce steam. Conventional spreader stokers can be used (see **Steam Generation**). Most of the steam is utilized for process heat, with the remainder used for generating electricity if desired.

A few electric utilities burn wood and wood wastes in standby steam boilers for occasional use, and power plants in Oregon and Vermont use wood fuel on a regular basis. The utilization of wood and wood residues for large-scale electricity generation is presently limited by problems of collection, handling, and availability; furthermore, even in the dry state the heating value per ton is less than for bituminous coal, although it is about the same as for some subbituminous coals and lignite (see **Coal**). The most promising near-term potential for wood as a direct combustion energy source is in the forest-products industries. In particular, utilization in this way of logging residues left on the ground could make the industry self-supporting in energy.

2. The simplest biomass gasification process is **pyrolysis** (i.e., heating in the absence of air). Prior to the mid-1920s, when a synthetic process was invented, pyrolysis of wood at moderate temperatures was the sole source of methanol (wood alcohol). Modern pyrolytic processes for the conversion of biomass into fuels involve heating to higher temperatures, up to about 1650°F (900°C), leading to more extensive decomposition of the cellulose and lignin. Dry agricultural, wood, and paper residues, as well as municipal wastes from which metals and glass have been removed, have been converted into combustible gases, liquids (oils), and char by high-temperature pyrolysis.

The compositions and relative amounts of the various products depend on the nature of the biomass material and the pyrolysis conditions (e.g., the heating rate and the maximum temperature). In general, the products are similar to those from **coal pyrolysis.** The gas, which is much lower in sulfur than **coal gas,** consists mainly of carbon monoxide, methane, and hydrogen as the combustible components, with carbon dioxide as the chief inert diluent. The oils, which are condensed by passage of the pyrolysis vapors through water, contain substantial amounts of aromatic **hydrocarbons.** The residual **char** is mainly carbon and can be burned as a clean solid fuel with a low sulfur content.

More complete gasification of dry biomass results from heating to a temperature around 2000°F (1095°C) in the presence of a limited supply of oxygen (or air). Partial combustion of the carbon and hydrogen in the biomass generates enough heat to maintain the required reaction temperature. If oxygen is used, the product gas contains carbon monoxide and hydrogen as the main fuel components and carbon dioxide as an inert diluent; it has a heating value of about 500 Btu/cu ft (19.5 MJ/cu m). The gas can be used directly as an **intermediate-Btu fuel gas** or it can be converted into methane **(high-Btu fuel gas),** methanol, or a gasoline component. If limited combustion of the biomass occurs in air, instead of oxygen, the product is a **low-Btu fuel gas** best suited for local use.

In the hydrogasification process, the dry biomass is reacted with hydrogen gas at high temperature and pressure; the chief fuel products are the simpler gaseous hydrocarbons, mainly methane and smaller amounts of ethane and others. After removal of carbon dioxide by scrubbing with an alkaline medium, the resulting gas has a high heating value, somewhat higher than that of **natural gas** because of the greater ethane content.

3. Liquefaction of biomass (i.e., conversion mainly into liquid hydrocarbon fuels) is based on indirect hydrogenation. The complex cellulose and lignin molecules are broken down, oxygen is removed (as water and carbon dioxide), and hydrogen atoms are added. The product of these chemical reactions is a mixture of hydrocarbons which condenses to a liquid when cooled.

The liquefaction process involves heating the biomass with steam and carbon monoxide or hydrogen and carbon monoxide at a temperature in the range of 480 to 840°F (250 to 450°C) at a pressure of about 270 atm (27 MPa) in the presence of a catalyst. The biomass material does not need to have a low moisture content, as it generally does for gasification processes; in fact, water may be added to the feed.

Liquefaction of biomass by hydrogenation has been demonstrated successfully on a small scale with urban wastes, various agricultural residues, wood, cattle manure, and sewage sludge. A development plant is being operated for the U. S. Department of Energy at Albany, Oregon, using the steam-carbon monoxide process with wood as the raw material. It is designed to convert up to 3 short tons (2.7 metric tons) of wood residues into about 6.5 barrels of fuel oil daily.

Biochemical Conversion. Biochemical conversion is based on certain reactions occurring in the presence of enzymes (i.e., natural catalysts) supplied by living microorganisms. Two major biochemical processes are being investigated: (1) production of methane gas from any form of biomass with the possible exception of wood, and (2) formation of ethanol from wood (cellulose), starch, or sugar.

1. Methane is produced from biomass by anaerobic bacterial action (i.e., in the absence of air). The process is the same as occurs in nature in the formation of marsh gas by the decay of plants in swamps and stagnant waters. The conversion of cellulose into methane takes place in two stages, each catalyzed by a different enzyme supplied by a different microorganism.

The actual methane-forming bacteria in the final stage grow (and generate methane) at a slow rate; hence, the biomass material in the form of a water sludge is digested with the bacteria for several days in an airtight container. The digester temperature, maintained by self-generated heat, is commonly 85 to 110°F (30 to 45°C). The process may be expedited by operating at a somewhat higher temperature [about 120 to 150°F (50 to 65°C)]; the cost of the additional energy is offset by the greater throughput. The composition of the gas formed depends on the feed material and the digestion conditions, but as a general rule it contains about 50 volume percent of combustible methane; the remainder is carbon dioxide with small amounts of other impurities. The product gas can be used directly in a **gas turbine** or the carbon dioxide can be removed to yield a gas with a heating value similar to natural gas (i.e., about 1000 Btu/cu ft or 37 MJ/cu m).

Untreated wood is not easily converted into methane. The most useful biomass materials appear to be manure, algae, kelp, plant residues, and other organic substances with a high moisture content.

The production of methane from animal manure is being studied on a moderately large scale in the United States and in other countries. Small-scale conversion is already in extensive use. About 6 to 9 cu ft of gas can be produced per pound of manure (or 0.37 to 0.55 cu m/kg). An important economic aspect of the treatment of manure in this manner is that the digester residue may be used as a high-protein cattle feed or as a more concentrated fertilizer than the original manure.

2. The fermentation of sugars and starches to produce beverage alcohol (ethanol) is an enzyme process that has been practiced for many centuries. The production of ethanol from the carbohydrates in biomass is based on the same general principle. Agricultural or forest-product residues and waste paper are possible biomass materials for ethanol production. Details of the methods used for converting various carbohydrates into ethanol

and its use as a fuel are given in the section on **Alcohol Fuels.**

BOILING-WATER REACTOR (BWR)

A thermal **nuclear power reactor** in which ordinary (or light) water is the moderator and coolant, as well as the neutron reflector (see **Light-Water Reactor; Thermal Reactor**). The system pressure is high, but not as high as in a **pressurized-water reactor,** so that the water boils and steam is generated within the reactor core. In spite of doubts that existed at the time, the feasibility of the BWR concept was demonstrated in a series of experiments initiated at the Argonne National Laboratory in 1953. A full-scale commercial BWR started power generation in 1960. Roughly one-third of the nuclear power reactors operating or under construction in the United States in the early 1980s are BWRs.

Some of the design details of BWRs have changed in the course of time, but the general principles have remained the same in recent years. The following description applies, in particular, to BWRs with an electrical generating **capacity** of about 1000 megawatts built since the early 1970s (see **Watt**).

General Description

Core and Reactor Vessel. The core of a BWR contains more than 48,000 vertical fuel rods in an approximately cylindrical arrangement, 12 ft (3.65 m) high and 15.8 ft (4.8 m) across. The rods are combined in about 750 square 8 × 8 assemblies, with 62 fuel rods per assembly and two hollow rods near the center which fill with water to provide additional moderation (see Fig. 6). The fuel rods contain small cylindrical pellets of uranium dioxide with an average initial enrichment of about 2.6 percent in uranium-235 (see **Nuclear Fuel**). The pellets are packed in corrosion-resistant zirconium alloy (zircaloy) tubes to form 12-ft (3.75-m) long fuel rods roughly

0.49 in. (1.25 cm) in external diameter. The total mass of fuel in a BWR core is some 187 short tons (170,000 kg). Roughly one-fourth is removed after each year's operation and is replaced with fresh fuel.

The reactor core is surrounded by a steel "shroud" which extends well above the top of the core, and the whole is supported in a steel cylindrical reactor (or pressure) vessel (Fig. 5). A typical reactor

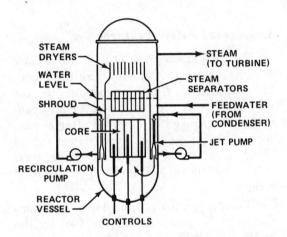

Fig. 5 Boiling-water reactor.

vessel has an internal height of some 72 ft (22 m) and a diameter of 21 ft (6.4 m); the wall thickness is 6 to 7 in. (15 to 18 cm). To permit controlled boiling within the reactor core, the pressure in the vessel is maintained at close to 71 atm (7.2 MPa).

Controls. The control elements in a BWR have a cruciform cross section; each element has four blades containing stainless steel tubes filled with the neutron poison (i.e., absorber) boron as boron carbide. The blades can move up and down in the spaces between the fuel assemblies, with one control element between four assemblies in most cases (Fig. 6). Some 180 (or so) such elements are distributed evenly throughout the core.

The controls of a BWR are inserted from the bottom of the core, rather than the top as in other reactors. This is convenient, because the space above the core is occupied by steam-water separators (see

below), and desirable, because the neutron absorber at the bottom of the core can compensate for the steam bubbles formed higher up in the core.

Coolant and Steam System

As the coolant water flows upward through the core, it removes the fission heat from the fuel rods and boils. The steam-water mixture leaving the top of the core contains about 14 percent by weight of steam; the remaining liquid water is recirculated as will be seen shortly. The wet steam enters a bank of water separators and then passes on to dryers in the upper part of the reactor vessel. In the separators and dryers most of the liquid water is removed and returned for recirculation. The relatively dry steam then proceeds to the turbines to generate electricity. The turbine condensate is returned to the reactor as feedwater (see below).

The water which has not been converted into steam in its passage through the core, together with the feedwater, is recirculated by means of two recirculation pumps. These pumps withdraw water from the annular region between the core shroud and the wall of the reactor vessel and force it through several jet pumps located in this region. The water jets draw in additional surrounding water and expel it into the volume at the bottom of the reactor vessel. The water is thus forced to circulate continuously through the core. The core shroud serves to separate the upward flow in the

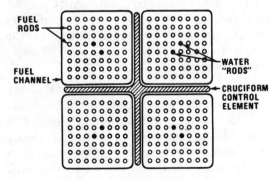

Fig. 6 BWR fuel assembly and control element (section).

core from the downward flow in the annular region (see Fig. 5).

Steam leaving the top of the reactor vessel at a temperature around 550°F (288°C) goes directly to the high-pressure stage of a two-stage turbine system which drives an electric generator (see **Steam Turbine**). Because of the relatively low steam temperature, the **thermal efficiency** for conversion into electricity is roughly 33 percent, compared with a steam temperature of 1000°F (540°C) and an efficiency of 40 percent for a modern, fossil fuel steam–electric plant (see **Electric Power Generation**).

The exhaust steam from the low-pressure stage of the turbine is converted to liquid in the condensers and this provides the reactor feedwater. Between the condenser and the reactor, the feedwater temperature is increased to 420°F (215°C) in a series of heaters using steam drawn from various intermediate sections of the turbines (see Fig. 160).

Control and Protection Systems

The neutron poison control elements are used mainly in reactor startup and shutdown. The control elements also serve to make the heat (or power) distribution as uniform as possible throughout the core. When the reactor is shut down, the control elements are fully inserted; for startup, the elements are gradually withdrawn and finally adjusted to maintain a desired power level. Small deviations from this level are corrected automatically by changes in the speed of the recirculation pumps; an increase (or decrease) in the water flow rate will increase (or decrease) the reactor power level.

The *reactor protection system* includes a large number and variety of instruments for monitoring the reactor and steam systems. If the instruments detect an abnormal condition, commonly referred to as a "transient," that cannot be corrected automatically by the control system, the reactor is immediately shut down. Such an emergency shutdown is called a reactor "scram" or "trip."

The control elements are actuated by hydrostatic pressure and are moved up (or down) by applying water pressure below (or above) a piston associated with each element. If the reactor has to be shut down rapidly in an emergency, the full hydrostatic pressure is applied automatically to force all the control elements upward into the core.

Engineered Safety Features

The engineered safety features are designed to prevent or minimize the escape of radioactive fission products to the environment as the result of a severe transient that persists or develops after a reactor trip. Among the more important of these features are the *emergency core-cooling system* (ECCS) and the *containment structure.*

When a reactor is shut down, heat continues to be generated in the fuel by the radioactive decay of the fission products (see **Radioactivity**). Consequently, unless there is an adequate supply of cooling water, the fuel elements may overheat and be damaged; the fission products would then be released. The purpose of the ECCS is to supply cooling water in the event of a partial or complete loss of normal coolant flow.

The ECCS of a BWR consists of three separate subsystems. If the coolant flow loss is small, the system pressure would drop to a moderate extent. An appropriate signal would then actuate the pumps of the *high-pressure core-spray system*, and water would be sprayed onto the fuel elements from above. In the event that the loss of coolant should cause a further pressure decrease, two other subsystems, namely, the *low-pressure core-spray* and *low-pressure injection systems*, should be able to remove decay heat generated in the fuel.

The containment structure is designed to prevent or delay the escape of radioactivity to the environment if there should be a release of fission products from the fuel rods. Boiling-water reactors have both a primary containment, which includes a *drywell* and a *wetwell*, and a secondary containment. The drywell contains the reactor and the recirculation pumps. In the older BWRs, the drywell is a steel cylindrical vessel, widening at the bottom like an electric light bulb. Below the drywell and connected with it by a number of wide pipes is the wetwell, a ring-shaped, steel vessel containing water.

In the more recent BWRs, the drywell is a large concrete cylinder with a domed top; it contains the reactor, pumps, and radiation shielding. The wetwell is a ring-shaped chamber in which water is retained between an interior concrete weir wall and a steel cylinder which constitutes the primary containment structure. Connection between the drywell and the wetwell is provided by a number of horizontal vents in the lower part of the drywell wall. The primary containment is enclosed in a concrete shield building, also called the reactor building, which represents the secondary containment structure.

If a break should occur in the reactor coolant system within the drywell, large volumes of steam would be released. The accompanying increase in pressure would depress the level of the water in the wetwell between the weir wall and the drywell wall. The vents in the latter would be uncovered, and the steam would be forced into the water and condensed. Not only does this lower the containment pressure and make the release of radioactivity less likely, but passage of the containment atmosphere through the water removes much of the radioactive material that may be present.

BOTTOMING CYCLE

A low-temperature **heat-engine** cycle in which part of the heat normally discharged in the condenser water from a steam turbine is converted into useful work (or electrical energy). As a general rule, the heat engine in a bottoming cycle would be a condensing (Rankine cycle) turbine similar in principle to a **steam turbine** but operating with a different working fluid at a much lower temperature and pressure (see **Vapor Turbine**).

The working fluid in a bottoming-cycle turbine would be a liquid of low boiling point (e.g., ammonia, propane, isobutane, or a Freon-type compound). It is vaporized in a **heat exchanger** utilizing heat from discharged steam-turbine condenser water at a temperature of roughly 105°F (40°C). The vapor at a moderate pressure (about 10 atm or 1 MPa) then operates the turbine. The discharged vapor is condensed and the liquid is returned to the heat exchanger for revaporization, and so on. A typical bottoming cycle would be similar to the system described in the section on **Ocean Thermal Energy Conversion.**

Because of the low temperature of the heat source in a bottoming cycle, the **thermal efficiency** (i.e., proportion of heat energy converted into useful mechanical work or electrical energy) is low. Hence, although the steam-turbine condenser discharge may contain 50 percent or more of the heat supplied by the fuel, the addition of a bottoming cycle would result in only a small increase in the total efficiency. Moreover, the small temperature differences across the heat exchanger and the bottoming-cycle condenser would require large and expensive equipment. In addition, a bottoming cycle needs an ample supply of very cold water to condense the working fluid. For these reasons, bottoming cycles are not receiving much attention at present.

BRAYTON (OR JOULE) CYCLE

A repeated succession of operations (or cycles) representing the idealized behavior of the working fluid in a **gas turbine** form of **heat engine.** The Brayton (or Joule) cycle is illustrated and described in Fig. 7. In this description each stage is assumed to have been completed before the next stage

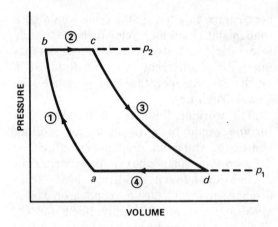

Fig. 7   Brayton cycle.

1. **Adiabatic compression of the working fluid (gas) along** ab**; the pressure increases from** p_1 **to** p_2 **accompanied by an increase in temperature.**

2. **Heat addition along** bc**, at the constant pressure** p_2**; the temperature and volume both increase.**

3. **Adiabatic expansion along** cd**; work is done by the expanding gas and its temperature and pressure decrease.**

4. **Heat removal (rejection) along** da**, at the constant pressure** p_1 **to restore the working gas to its initial condition.**

is initiated. However, in an actual engine there is a gradual, rather than a sharp transition from one stage to the next; hence, the sharp points in the figure would actually be rounded off.

In an open-cycle turbine (see Fig. 54) air is compressed adiabatically (see **Carnot Cycle**) in the compressor (stage 1) and partially heated. The air is heated further at constant pressure by burning gas or oil in the combustor (stage 2). The hot gas is expanded adiabatically in the turbine (stage 3) where mechanical work is done by the expanding gas. (In turbojet and similar aircraft engines, expansion occurs partly in the turbine and partly in the nozzle through which the gas is discharged.) The net (or available) work is the difference between the work done by the gas in stages 2 and 3 and the work done on it in stages 1 and 4. In an open-cycle gas turbine, stage 4 does not occur and the cycle is not closed. The turbine exhaust gas, at d in Fig. 7, is discharged and fresh air is

taken into the compressor from the atmosphere at a in each cycle.

In a closed-cycle gas turbine (see Fig. 56), the working fluid can be any convenient gas, but air is generally used. The air is compressed adiabatically and thereby heated (stage 1). It is heated further by passage at constant pressure through a **heat exchanger** (stage 2). The working fluid (air) is thus kept separate from the combustion gases. The hot, compressed air is then expanded adiabatically in the turbine where work is done and the temperature falls (stage 3). The turbine exit gas, at its initial pressure, is cooled (stage 4) before returning to the compressor at a, thus completing the cycle.

The **thermal efficiency** of the Brayton cycle (i.e., the fraction of the heat supplied in stage 2 that is converted into net mechanical work) is increased by having the temperature at c as high as possible (and at d as low as possible). An equivalent statement is that an increase in the compression ratio (i.e., volume at a divided by volume at b) increases the thermal efficiency.

BREEDER REACTOR

A nuclear reactor in which energy is released by fission while, at the same time, more fissile material is produced than is consumed (see **Fission Energy; Nuclear Power Reactor**). In addition to the fissile material, a breeder contains a fertile material which is converted into fissile material as a result of capturing neutrons in the reactor. In a breeder, the ratio of fissile nuclei produced from the fertile material to the fissile nuclei consumed in fission and nonfission reactions, called the *breeding ratio*, must be greater than one. If it is less than one, the reactor would be a converter.

When fissile nuclei undergo fission, two to three neutrons are released on the average per fission. The basic requirement for breeding is that, after allowing for neutrons lost by escape and by nonfission cap-

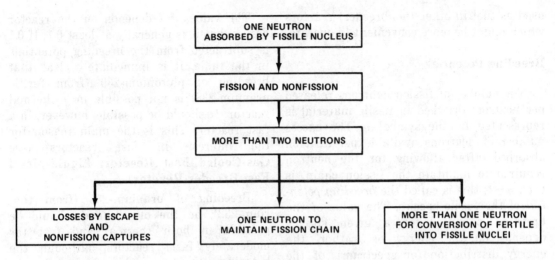

Fig. 8 Conditions for breeding.

ture in both fissile and nonfissile (other than fertile) species, more than two neutrons must be available for every neutron absorbed by fissile material. One of the available neutrons is required to maintain the fission chain, whereas the remainder (more than one) can be captured by the fertile species to generate new fissile material (see **Nuclear Fuel**). Then, on the average, for every fissile nucleus lost, more than one such nucleus can be produced (Fig. 8).

There are only two combinations of fertile and fissile species for which breeding is practical under the appropriate conditions. In these combinations, as given below, the fertile material occurs in substantial amounts in nature.

Fertile material	Fissile material
Uranium-238	Plutonium-239
Thorium-232	Uranium-233

The fissile materials plutonium-239 and uranium-233 do not exist naturally, except possibly in traces. Hence these substances must be produced artificially from the corresponding fertile species. In the earliest breeders, the fissile component is either uranium-235, obtained by isotope enrichment of normal uranium (see **Uranium Isotope Enrichment**), or plutonium-239 or uranium-233 made in other (possibly converter) reactors based on uranium-235.

However, the resources of uranium-235 are limited, so breeder reactors must ultimately utilize the same fissile species that they produce.

Resource Utilization

The prime objective of breeding is to make maximum use of the available natural resources of uranium and thorium. In nonbreeder (or converter) reactors less fissile material is produced than is consumed; hence, even if fissile material remaining in the fuel after reactor operation is recovered, the available supply of this material would eventually be exhausted. At that time, substantial quantities of fertile materials would still remain, but they could not be used directly as energy sources.

With breeders more fissile material would be produced than is consumed in operating the reactors. An excess of fissile material would then be available to operate breeder reactors for the conversion of more fertile into fissile species. In principle, therefore, it should be possible to utilize in this manner essentially all the economically accessible thorium and uranium, including the stockpile of depleted uranium tails from uranium-235 enrichment plants. When sufficient fissile material has been accumulated by breeding, part could be used as fuel in other (nonbreeder) reactors which might be more convenient to operate.

Breeding Potential

The number of fission neutrons released per neutron absorbed in fissile material is represented by the symbol η. The excess number of neutrons available per neutron absorbed, after allowing for the neutron required to maintain the fission chain, is then $\eta - 1$; this is called the *breeding potential* of the fissile species. The value of the breeding potential for a given species depends on the energy or, rather, the energy distribution (or spectrum) of the neutrons causing fission. (The actual, as distinct from the theoretical, breeding potential is affected to a small extent by the nature of the fertile species and the design of the reactor, but this may be neglected.) For present purposes, two broad neutron energy spectra may be considered: namely, those existing in **thermal reactors** and **fast reactors,** respectively. Approximate values for the breeding potentials $(\eta - 1)$ in these neutron spectra are given in the table.

Breeding Potential

Fissile species	Thermal spectrum	Fast spectrum
Plutonium-239	1.0	1.45
Uranium-233	1.25	1.3

The breeding potentials make no allowance for the neutrons lost by escape from the reactor and by parasitic (i.e., wasted) capture in coolant, fuel cladding, and structural and control materials. If the neutrons lost in these ways are l per neutron absorbed in fissile material, then $(\eta - 1) - l$ is the number of neutrons available for conversion of fertile into fissile nuclei. Hence, $(\eta - 1) - l$ is equivalent to the breeding ratio defined earlier. The condition for breeding is consequently that $(\eta - 1) - l$ should be greater than one.

The value of l depends on the reactor design, but it is generally at least 0.1. If 0.1 is subtracted from the breeding potentials in the table, it is immediately clear that breeding of plutonium-239 (from fertile uranium-238) is not possible in a thermal reactor. It should be possible, however, in a fast reactor. This is the main reason for the interest in fast reactors (see **Gas-Cooled Fast Reactor; Liquid-Metal Fast Breeder Reactor**).

Breeding of uranium-233 (from thorium-232), on the other hand, should be possible in both thermal and fast (or moderately fast) reactors, although the breeding ratio is expected to be less than in a plutonium-239 (fast) breeder. There is not much to be gained, as far as breeding uranium-233 is concerned, in using fast neutrons; consequently, only thermal reactors have been considered for this purpose, since they are easier to operate (see **Light-Water Breeder Reactor; Molten-Salt Breeder Reactor**).

Hitherto, however, there has been relatively little development of uranium-233 breeders. One reason is the difficulty arising from the gamma-ray activity that develops in the spent fuel and blanket (fertile) materials (see **Nuclear Fuel**). This has been partially overcome in the Molten-Salt Breeder Reactor concept, but it involves circulating a highly corrosive molten salt mixture at temperatures up to 1300°F (705°C).

Doubling Time

As a general rule, although not always, a breeder reactor consists of separate fuel and blanket components. The fuel rods in the core of a fast breeder might contain 15 to 20 percent of fissile species in a mixture with fertile material; the external blanket rods, which surround the core, contain fertile material only. In some designs, rods of fertile material, called the internal blanket, are included in the core.

During operation of the reactor, fissile material is generated in the blanket rods as well as in the fertile material in the fuel rods. From time to time, the fuel and blanket rods are removed from the reactor and transferred to a fuel reprocessing plant where the fissile and fertile materials are recovered separately for reuse, free from associated fission products (see **Nuclear Fuel Reprocessing**).

Since a breeder reactor produces more fissile material than it consumes, there will be an excess of this material after the requirements of the reactor for fresh fuel are met. Eventually, sufficient of this excess will be accumulated to supply the fuel for another similar breeder reactor. The *doubling time* of a breeder is defined as the operating time during which the reactor would produce excess fissile material equal to the amount required for a fresh core of a similar reactor plus the amount in the fuel cycle outside the reactor (i.e., in stockpile, reprocessing, and fabrication).

The doubling time is inversely related to the breeding ratio; the greater this ratio, the shorter the doubling time. If the purpose of a breeder is to accumulate fissile material at a maximum rate, then a short doubling time is desirable. With current plutonium-239 breeder designs, the doubling time is expected to be about 15 years, but it should be possible to reduce this eventually to nearer 10 years.

BRITISH GAS/LURGI SLAGGING GASIFIER PROCESS

A modification of the **Lurgi Dry Ash process** for **coal gasification** developed by the British Gas Corporation and sponsored by the Conoco Coal Development Company and other U. S. companies. A demonstration plant for the production of a **high-Btu fuel gas** (or **substitute natural gas**) is planned for the United States.

The main difference between the British Gas/Lurgi and the original Lurgi process is that in the former the temperatures in the gasifier are high enough to melt the residual coal ash and form a liquid slag. An advantage of the slagging process is that it can operate with either caking or noncaking coals without pretreatment. Furthermore, because of the high temperatures in the slagging gasifier, the carbon in the coal is utilized more efficiently and the steam requirement for the gasification process is decreased.

The British Gas/Lurgi gasifier is similar to that in the Lurgi Dry Ash process except that it does not have a rotating grate (see Fig. 81). Steam and oxygen gas are injected through nozzles near the bottom of the gasifier. Crushed and sized coal enters through a lock hopper at the top, and as it descends it is heated by and interacts with the rising hot gases. Molten slag is drawn off from the bottom and quenched.

The product gas leaving the top of the gasifier contains about 60 volume percent of carbon monoxide, 30 percent of hydrogen, and 8 percent of methane and other light hydrocarbons (dry basis) as fuel constituents. It has a **heating value** of about 380 Btu/cu ft (14 MJ/cu m) and could serve as an **intermediate-Btu fuel gas.** However, it is proposed to subject the gas to the **water-gas shift reaction** followed by **methanation** to yield a substitute natural gas.

BRITISH THERMAL UNIT

A unit quantity of energy, especially of heat (or thermal energy); it is commonly used to express the **heating value** of a fuel. The British thermal unit (Btu) is generally defined as the amount of energy as heat required to increase the temperature of 1 pound of water by 1°F at normal atmospheric pressure; this definition is adequate for most purposes. However, the heat required to raise the temperature of water by 1°F depends to some extent on the actual water temperature. Hence, when greater precision is desirable, the initial temperature is specified as 39.1°F (4.0°C) where water has its maximum density.

The International Table (or IT) Btu is defined as the equivalent of 251.996 IT calories, where 1 IT calorie is equivalent to 4.1868 **joules** (J). Hence, 1 Btu is equal to 1055 J = 1055 watt-sec = 0.293 watt-hour (see **Watt**).

BUREAU OF MINES CITRATE DESULFURIZATION PROCESS

A wet regenerative process of the U. S. Bureau of Mines for the **desulfurization of stack gases.** The absorber is a sodium citrate–citric acid solution which removes sulfur dioxide (SO_2) from stack (flue) gases

in a scrubber. The spent solution is treated with hydrogen sulfide (H_2S) gas, and reaction with the dissolved sulfur dioxide leads to the direct formation of elemental (solid) sulfur. At the same time the absorber solution is regenerated and recycled to the scrubber. The sulfur is removed by flotation with oil, followed by melting. The hydrogen sulfide required for producing the sulfur is regenerated by a mixture of steam and methane (**natural gas**) with part of the recovered sulfur. The remaining sulfur may have a commercial value (see **Desulfurization: Waste Products**).

C

CANDU (CANADIAN–DEUTERIUM–URANIUM) REACTOR

A thermal **nuclear power reactor** in which heavy water (99.8 percent deuterium oxide, D_2O) is the moderator and coolant as well as the neutron reflector (see **Heavy-Water Reactor; Thermal Reactor**). The CANDU reactor was developed (and is used extensively) in Canada, where a full-scale commercial reactor of this type first started operation in 1967. A few CANDU reactors are operating or under construction in some other countries, but there are none in the United States.

A basic design difference between the CANDU (heavy-water) reactor and **light-water reactors** (LWRs) is that in the latter the same water serves as both moderator and coolant, whereas in the CANDU reactor the moderator and coolant are kept separate. Consequently, unlike the pressure vessel of an LWR, the CANDU reactor vessel, which contains the relatively cool heavy-water moderator, does not have to withstand a high pressure. Only the heavy-water coolant circuit has to be pressurized to inhibit boiling in the reactor core.

General Description

Reactor Vessel and Core. The following description refers to a CANDU reactor with an electrical capacity of 600 megawatts (see **Watt**). The reactor vessel is a steel cylinder with a horizontal axis; the length is over 20 ft (6 m) and the diameter is nearly 26 ft (8 m). The vessel is penetrated by some 380 horizontal channels, called pressure tubes because they are designed to withstand a high internal pressure. The channels contain the fuel elements, and the pressurized coolant flows along the channels and around the fuel elements to remove heat generated by fission. Coolant flow is in opposite directions in adjacent channels. The high-temperature [about 590°F (310°C)] and high-pressure [100 atm (10 MPa)] coolant leaving the reactor core enters the steam generator, as described below. Roughly 5 percent of the fission heat is generated by fast neutrons escaping into the moderator, and this is removed by circulation through a separate **heat exchanger** (Fig. 9).

The fuel in the CANDU reactor is normal (i.e., unenriched) uranium dioxide as small, cylindrical pellets. The pellets are packed in a corrosion-resistant zirconium alloy (zircaloy) tube, nearly 19 in. (0.5 m) long and 0.5 in. (1.3 cm) diameter, to form a fuel rod (see **Nuclear Fuel**). The relatively short rods are combined in bundles of 37 rods, and 12 bundles are placed end-to-end in each pressure tube. The total mass of fuel in the core is 107 short tons (97,000 kg).

The CANDU reactor is unusual in that refueling (i.e., removal of spent fuel and replacement by fresh fuel) is conducted while the reactor is operating. A refueling machine inserts a fresh fuel bundle into one end of a horizontal pressure tube which is temporarily disconnected from the main coolant circuit. A spent fuel bundle is thus displaced at the other end and is removed. This procedure is carried out, like the

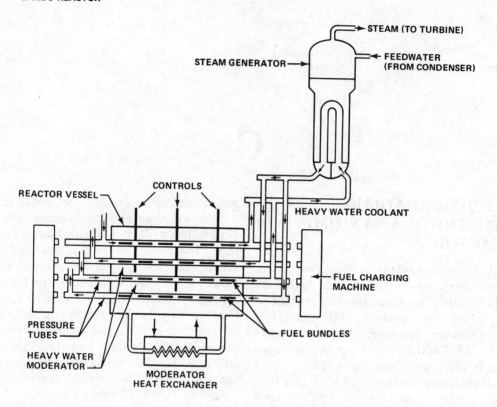

Fig. 9   CANDU reactor.

coolant flow, in opposite directions in adjacent channels.

Control and Protection Systems. The CANDU reactor has several types of vertical control elements. They include a number of strong neutron absorber (i.e., poison) rods of cadmium which are used mainly for reactor shutdown and startup. In addition there are other less strongly absorbing rods to control power variations during reactor operation and to produce an approximately uniform heat (power) distribution throughout the core. In an emergency situation, the shutdown rods would immediately drop into the core, followed if necessary by the injection of a gadolinium nitrate solution into the moderator. (Gadolinium is a very strong absorber of thermal neutrons.)

Steam System. The respective ends of the pressure tubes are all connected into inlet and outlet headers (or manifolds). The high-temperature coolant leaving the reactor passes from the outlet header to a steam generator of the conventional inverted-U type (see **Pressurized-Water Reactor**) and is then pumped back to the reactor by way of the inlet header. Steam is generated at a temperature of above 510°F (265°C). There are two coolant outlet (and two inlet) headers, one at each end of the reactor vessel, corresponding to the opposite directions of coolant flow through the core. Each inlet (and outlet) header is connected to a separate steam-generator and pump loop (see Fig. 9). A single pressurizer, of the type used in pressurized-water reactors, maintains an essentially constant coolant system pressure.

Safety Features. A break in a single pressure tube would result in some loss of coolant, but the particular tube could be disconnected and reactor operation would proceed with the other tubes. A more serious loss-of-coolant accident, with possible damage to the fuel and release of radioactive fission products (see **Boiling-Water Reactor**), would develop from a break in one of the coolant headers or in the pipes to or from the steam generators. An emergency core-cooling system would then supply additional coolant. The separate moderator system would also provide a substantial heat sink.

A concrete containment structure encloses the reactor vessel and the steam generator system. A water spray in the containment would condense the steam and reduce the pressure that would result from a large break in the coolant circuit.

CAPACITY

As applied to an electric generating plant or a turbine-generator, the rated load or **power** output; that is, the capacity is the maximum power for which the generating plant (or generator) is designed to operate continuously. The capacity is usually expressed in megawatts or kilowatts (see **Watt**). Large modern steam–electric plants (see **Electric Power Generation**), using either **fossil** or **nuclear fuel,** have electrical capacities in the vicinity of 1000 megawatts (i.e., 1 million kilowatts) per unit; two or three units may be located at a single power station.

CAPACITY FACTOR

The ratio of the electrical energy (e.g., in kilowatt-hours) delivered by a power plant over a period of time (e.g., 1 month or 1 year) to the amount that would have been generated had the plant operated continuously at its rated **capacity** for the whole period (i.e., rated capacity in kilowatts × number of hours in the time period). An equivalent definition is the ratio of the average **load** (e.g., in kilowatts) to the rated capacity of a power plant (also in kilowatts).

See also **Load Factor.**

CARBON ADSORPTION DESULFURIZATION PROCESSES

Dry processes for the desulfurization of stack gases. Several such processes have been developed, all involving adsorption (i.e., retention on the extensive surface of a highly porous material) of sulfur dioxide (SO_2) on specially treated ("activated") carbon. The sulfur dioxide removed from the stack (flue) gases by adsorption is oxidized by oxygen in air at a relatively low temperature [250 to 300°F (120 to 150°C)] to form sulfur trioxide (SO_3). Water (H_2O) vapor in the stack gas then converts this into sulfuric acid (H_2SO_4). A number of different methods, described below, have been proposed for regenerating the carbon adsorber.

In the Foster-Wheeler Corporation process, based on that of the German Bergbau-Forschung GmbH, the activated carbon is made from a **char** obtained from coal. The spent char, containing adsorbed sulfuric acid, is regenerated by heating to about 1500°F (815°C). The sulfuric acid is decomposed, and a gas rich in sulfur dioxide is driven off. This hot gas is passed through a reactor containing hot coal, where the sulfur dioxide is reduced by carbon in the coal to elemental sulfur. The latter can be recovered and sold (see **Desulfurization: Waste Products**).

Another method, developed in Germany and Japan, for removing the sulfuric acid from spent adsorber is to extract the acid with water. The resulting dilute (about 18 percent) acid has little market value, but it can be neutralized with lime or limestone to yield a high-quality gypsum.

In the Westvaco process, the spent adsorber is heated to drive off part of the sulfur dioxide which is then reduced by means of hydrogen gas to hydrogen sulfide. This is used to convert the sulfuric acid remaining on the adsorber to elemental sulfur.

CARBON DIOXIDE

A colorless, odorless noncombustible gas with the formula CO_2 present to the extent of about 335 parts per million by volume in the atmosphere (in the early 1980s). Carbon dioxide is formed by the combustion of carbon and carbon compounds, by respiration, which is a slow combustion, in animals (including man) and plants, and by the gradual oxidation of organic matter in the soil. Since fossil fuels (i.e., coal, petroleum, and natural gas) are carbon compounds, carbon dioxide is released to the atmosphere when these fuels or their products are burned to generate heat or are used to operate internal combustion engines. Carbon dioxide is also produced by burning plant products (e.g., wood) and alcohols (see **Alcohol Fuels**). On the other hand, carbon dioxide is removed from the atmosphere by green plants, which use the gas for photosynthesis in sunlight (see **Biomass Fuels**), and also by absorption in the ocean.

Prior to the industrial revolution, when fossil fuels were used to a limited extent, the amount of carbon dioxide released to the atmosphere was roughly the same as that removed in various ways. The carbon dioxide content thus remained approximately constant. However, in recent years the increased use of fossil fuels has been responsible for an increase in the amount of carbon dioxide in the atmosphere. Plausible predictions indicate that if fossil fuels continue to be the major energy sources, as they are now, the atmospheric content of carbon dioxide will be double the preindustrial value (about 275 parts per million) by the middle of the 21st century. Such a development is expected to have worldwide effects which would probably not be tolerable.

Carbon dioxide is not a noxious gas, but its presence in the atmosphere has important environmental consequences. Visible and other radiations from the sun that penetrate the atmosphere (see **Solar Energy**) are absorbed at the earth's surface. The surface warmed in this manner then emits infrared (i.e., longer wavelength) radiation, but this is absorbed by carbon dioxide (and water vapor) in the atmosphere. Hence, instead of being dissipated in space, the radiated energy serves to increase the temperature of the lower atmosphere and the earth's surface. The phenomenon is called the greenhouse effect because it is similar to the trapping of the sun's energy in a greenhouse.

An increase in the atmospheric carbon dioxide content, such as could occur from the increased use of fossil fuels, would probably increase the greenhouse effect, with associated increases in the earth's surface temperature especially in the polar regions. These temperature increases could have significant consequences, such as climatic changes, shifts in agricultural zones, partial melting of the polar icecaps, and flooding of coastal areas because of an increase in the ocean level. An increase in the atmospheric carbon dioxide content would take a long time to reverse, even if the use of fossil fuels ceased. Hence, carbon dioxide levels in the atmosphere and their potential effects are receiving careful study.

CARBON DIOXIDE ACCEPTOR PROCESS

A Conoco Coal Development Company process for **coal gasification** with steam and air; the product is an **intermediate-Btu fuel gas.** The process is designed for use with noncaking (i.e., subbituminous and lignite) coals. It is unusual in the respect that a large part of the heat required to gasify the coal is supplied by the chemical reaction between lime (calcium oxide), called the "acceptor," and carbon dioxide to form calcium carbonate; thus,

$$CaO + CO_2 = CaCO_3 + heat$$

Calcium   Carbon   Calcium
oxide   dioxide   carbonate

The calcium oxide acceptor is regenerated from the calcium carbonate (spent acceptor) by heating in a separate vessel; the heat is provided by the combustion of coal **char** in air. The calcium carbonate feed material is supplied as limestone. (Dolomite, a combination of calcium and magnesium carbonates, may be used instead of limestone; in that case, both calcium and magnesium oxides are acceptors.)

Pulverized coal is preheated and introduced near the bottom of the **fluidized-bed** gasifier, with steam as the fluidizing medium (Fig. 10). The coal is devolatilized (i.e., volatile matter is driven off) at 1500°F (815°C) and a pressure of 10 to 20 atm (1 to 2 MPa). The carbon in the residual char then reacts with steam to form hydrogen and carbon monoxide (see **BI-GAS Process**). The **water-gas shift reaction** also occurs to some extent, leading to the production of additional hydrogen (and carbon dioxide) from the carbon monoxide and steam. The combination of the carbon dioxide with the acceptor, as described above, supplies most of the heat for the gasifier.

The spent acceptor, together with unconsumed char, is transferred from the gasifier to the fluidized-bed regenerator; the fluidizing gas is air. The carbon in the char burns in air to produce a temperature of 1850°F (1010°C) at 10 to 20 atm (1 to 2 MPa) pressure. The heat released in this combustion reaction serves to regenerate the acceptor. The regenerated acceptor (calcium oxide) is then recycled to the gasifier where its high temperature contributes to the overall heat requirement. The off-gas (or flue gas), mainly carbon dioxide, leaving the regenerator passes to a cyclone separator to remove coal ash which contains (as calcium sulfide) some 80 percent of the sulfur originally present in the coal. The gas is discharged through a waste heat recovery system.

The product gas leaving the gasifier contains more than 50 volume percent of hydrogen, 16 to 18 percent of carbon monoxide, and 14 percent of methane (dry basis) as the fuel constituents; most of the remainder is inert carbon dioxide. The **heating value** of the gas is about 370 Btu/cu ft (14 MJ/cu m). Because of the unusually high ratio of hydrogen to carbon monoxide, roughly 3 to 1, the purified gas can be converted by **methanation** into a **high-Btu fuel gas** (or **substitute natural gas**) without the need for the **water-gas shift reaction** as an intermediate step.

CARBON MONOXIDE

A colorless, odorless but poisonous combustible gas with the formula CO. Carbon

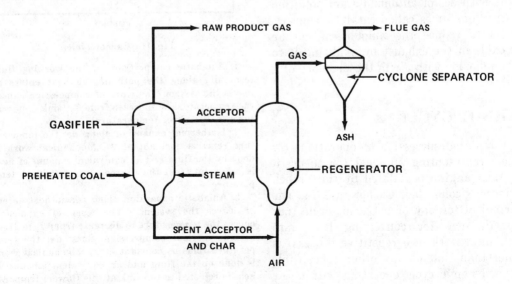

Fig. 10   Carbon dioxide acceptor process.

monoxide is produced in the incomplete combustion (in air or oxygen gas) of carbon and carbon compounds, such as **fossil fuels** (i.e., coal, petroleum, and natural gas) and their products (e.g., **liquefied petroleum gas, gasoline, diesel fuel, kerosine,** etc.). Combustion of carbon monoxide leads to the formation of **carbon dioxide** (CO_2), as also does the complete combustion of carbon and carbon compounds. Carbon monoxide (together with hydrogen) is commonly made by the action of steam on carbon, in **coal, coke,** or **char,** at high temperature; it is also formed by the action of carbon dioxide on carbon.

The **low-Btu** and **intermediate-Btu fuel gases** produced in **coal gasification** processes contain a substantial proportion of carbon monoxide as a fuel constituent. Its **heating value** is close to 320 Btu/cu ft (12 MJ/cu m). **Synthesis gas,** used in the manufacture of methanol (see **Alcohol Fuels**) and various liquid fuels (see **Fischer–Tropsch Process**) is a mixture of carbon monoxide and hydrogen gas.

In **internal-combustion engines,** combustion of the fuel is generally incomplete; consequently, the exhaust gas usually contains a few percent (or more) of carbon monoxide. This accounts for the poisonous nature of automobile engine exhaust. One of the purposes of automobile antipollution devices, sometimes called catalytic converters, is to reduce the amount of carbon monoxide in the exhaust by oxidizing it to carbon dioxide with air in the presence of a catalyst.

CARNOT CYCLE

A repeated succession of operations (or cycles) representing an ideal (or theoretical) **heat engine** postulated by the French engineer Nicolas Sadi Carnot in 1824. The **thermal efficiency** of a Carnot cycle (i.e., the efficiency for converting heat into mechanical work in a repetitive process) is higher than that of any other conceivable thermodynamic cycle operating within the same temperature range. Although the

Carnot cycle cannot be realized in practice, it is important because it indicates the maximum possible efficiency of practical heat engines. The Carnot cycle consists of two adiabatic stages alternating with two isothermal stages. In an *adiabatic* stage, no heat is taken up or rejected by the working fluid, but the temperature changes; it increases in an adiabatic compression but decreases in an adiabatic expansion. In an *isothermal* stage, the temperature remains constant; if expansion occurs, the temperature tends to fall and heat must be taken up from a "source" by the working fluid in order to maintain the constant temperature. Similarly, heat must be rejected to a "sink" during an isothermal compression. The four stages of the Carnot cycle are depicted and explained in Fig. 11, which

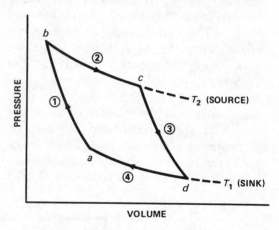

Fig. 11 Carnot cycle.

1. **Adiabatic compression of the working fluid** (e.g., air) along the path *ab*; no heat enters or leaves the system. The work of compression causes the temperature to increase from T_1 (sink temperature) to T_2 (source temperature).

2. **Isothermal expansion** along *bc*; the temperature remains constant at T_2. Mechanical work is done by the fluid and an equivalent amount of heat is taken up from the source at this (upper) temperature.

3. **Adiabatic expansion** along *cd*; no heat enters or leaves the system. The work of expansion causes the temperature to decrease from T_2 to T_1.

4. **Isothermal compression** along *da*; the temperature remains constant at T_1. Mechanical work is done on the fluid and an equivalent amount of heat is rejected to the sink at this (lower) temperature.

shows the pressure-volume relationships for a gas (e.g., air) as the working fluid.

After completion of stage 4, the working fluid is back at its initial state a and the Carnot cycle is complete. The net work done in the cycle is the sum of the work done by the working fluid in the expansion stages 2 and 3, minus the work done on the fluid in the compression stages 1 and 4.

The thermal efficiency of the Carnot cycle, defined as the fraction of the heat taken up in stage 2 that can be converted into net mechanical work in the cycle, is independent of the nature of the working fluid and is expressed by

$$\text{Thermal efficiency} = 1 - \frac{T_1}{T_2}$$

where the source (T_2) and sink (T_1) temperatures are expressed on the absolute (or thermodynamic) scale (i.e., $273.15 + t°C$ on the Kelvin scale or $459.67 + t°F$ on the Rankine scale). The efficiency of the Carnot cycle thus depends only on T_1/T_2, that is, the ratio of the temperature of the heat sink (T_1) to that of the heat source (T_2). The smaller the value of T_1/T_2 the more closely does the thermal efficiency approach unity (i.e., 100 percent). Hence, the efficiency of a Carnot cycle is increased by decreasing the lower (heat sink) temperature T_1 at which heat is rejected and increasing the upper (heat source) temperature T_2 at which heat is taken up.

CASINGHEAD GAS

The gas (and vapor) collected at the casinghead (i.e., the top of the casing) of an operating oil well. It is primarily a mixture of light paraffin **hydrocarbons,** including the gases **methane** (CH_4), ethane (C_2H_6), propane (C_3H_8), and butanes (C_4H_{10}), and the vapors of pentanes (C_5H_{12}), hexanes (C_6H_{14}), and heptanes (C_7H_{16}). Upon cooling at atmospheric pressure, the vapors condense to form casinghead (or **natural**) **gasoline**; this is similar to the lighter components of **straight-run gasoline.** If the

remaining gases are compressed, propane and butane are condensed to form **liquefied petroleum gas.** The residual methane and ethane mixture is a form of **natural gas.** The gas may be returned to the petroleum reservoir to improve recovery of the crude oil (see **Petroleum Production**).

CATACARB PROCESS

A process for the removal of **acid gases** from fuel gases by chemical absorption in a weakly alkaline solution of potassium carbonate, an amine, and proprietary additives. The process appears to be similar in principle to the well established **Benfield process.**

CATALYTIC RICH GAS (CRG) PROCESS

A process developed in the United Kingdom, utilizing catalytic **steam reforming** to convert a liquid petroleum product, **naphtha** in particular, to a **high-Btu fuel gas.** The Lurgi Gasynthan process in West Germany and the Methane Rich Gas (MRG) process in Japan are similar in principle but use different reforming catalysts.

The naphtha is first desulfurized by passing the vapor mixed with hydrogen gas over a heated nickel molybdate ($NiO-MoO_3$) or cobalt molybdate ($CoO-MoO_3$) catalyst; the hydrogen sulfide formed is removed in a conventional manner (see **Desulfurization of Fuel Gases**). The sulfur-free naphtha vapor is then decomposed by catalytic reforming with steam at a temperature of 840 to 1020°F (450 to 550°C) and a pressure up to 40 atm (4 MPa). The product consists of approximately 60 volume percent of methane, 20 percent of carbon dioxide, 19 percent of hydrogen, and 1 percent (or so) of carbon monoxide (dry basis). If this gas is subjected to **methanation** in two stages, first at a higher temperature and then at a lower temperature, part of the carbon dioxide and all the carbon monoxide are converted into **methane** and water vapor. Removal of the water and excess carbon dioxide leaves a high-Btu **substitute natural gas.**

CAT-OX DESULFURIZATION PROCESS

A dry process of the Monsanto Company for **desulfurization of stack gases.** The sulfur dioxide (SO_2) present in the stack (flue) gases is oxidized to sulfur trioxide (SO_3) by passage with air over a solid vanadium pentoxide catalyst at a temperature above 850°F (455°C). When the gases leaving the catalyst are cooled, the sulfur trioxide combines with water (H_2O) vapor in the stack gas to form a sulfuric acid (H_2SO_4) mist. This is removed by scrubbing with cool dilute sulfuric acid to yield a more concentrated (about 80 percent) solution which may have commercial value. The stack gases must be free from particulate matter (e.g., by passage through an **electrostatic precipitator**) to prevent rapid deterioration of the vanadium pentoxide catalyst.

CETANE NUMBER

A number indicating the ignition quality of a **diesel fuel.** A high cetane number represents a short ignition delay time, that is, a short time between injection of the fuel into a **diesel engine** cylinder and ignition (i.e., initiation of combustion) of the fuel. Fuels with high cetane numbers provide easier starting and smoother and quieter operation. Liquid normal (straight-chain) paraffins generally have higher cetane numbers than other **hydrocarbons** and are therefore desirable components of diesel fuels. (By contrast, liquid isoparaffins and aromatics have low cetane numbers.)

The cetane number scale is based on an assigned value of 100 to the paraffin hydrocarbon cetane or normal hexadecane ($C_{16}H_{34}$) and zero to the aromatic hydrocarbon alpha-methylnaphthalene ($C_{11}H_{10}$). The cetane number of a diesel fuel is the volume percent of cetane present in a blend with methylnaphthalene that would have the same ignition delay time as the given fuel. A diesel fuel with a cetane number of 40, for example, implies a delay time equal to that of a blend containing 40 volume percent of cetane.

The standard method for determining cetane number uses a single-cylinder, variable compression-ratio diesel engine. The compression ratio is adjusted until ignition occurs with a delay of 13° using the test fuel. From a series of diesel fuel blends having known cetane numbers, a fuel is found that gives a delay of 13° at the same compression ratio as the test fuel. The cetane number of the latter is thus identified.

Simpler methods are often used to estimate approximate cetane numbers. In one such procedure, which can be carried out with any diesel engine, the air intake is throttled until the engine misfires with the test fuel. The air intake pressure, measured with a vacuum gauge, can be related to the cetane number by comparison with fuel blends of known cetane numbers.

Instead of making engine tests, the ignition quality of a diesel fuel can be estimated from certain physical properties. The diesel index, for example, is defined by

$$\text{Diesel index} = \frac{\text{Aniline point (°F)} \times \text{API gravity}}{100}$$

(see **Aniline Point; API Gravity**). A high aniline point corresponds to a large proportion of paraffin hydrocarbons in the fuel; such a fuel will have a high diesel index and is expected to have a high cetane number because of its paraffin content. Another approximate indication of the cetane number is given by the aniline point divided by the mid-boiling point of the fuel (i.e., the temperature at which half the liquid distills when heated).

CHAR

A general name given to the residue from heating in the absence of air (see **Pyrolysis**) or incomplete combustion of an organic material, especially coal, petroleum, or wood. In the last case, the product is called charcoal. Char consists mainly (roughly 90 percent) of carbon, together with a small amount of volatile matter and the mineral elements initially present in the organic material. Because of its high carbon content, char can be used as a fuel; its **heating value** is about 14,000 Btu/lb (33 MJ/kg). Strictly speaking, **coke** is a form of char, but in connection with coal, the term char refers to the noncoherent product of low mechanical strength resulting from the low-temperature [840 to 1290°F; (450 to 700°C)] pyrolysis of coal, especially of noncaking coals, such as lignite, subbituminous coal, and anthracite. This material is often called semicoke.

CHEMICO (MAGNESIA) DE-SULFURIZATION PROCESS

A Chemico Corporation wet, regenerative process for the **desulfurization of stack gases**. The first stage is similar to the common **lime/limestone process**, except that the sulfur dioxide absorber is a slurry of magnesia (MgO) in water. The absorption of sulfur dioxide (SO_2) produces magnesium sulfite ($MgSO_3$) which is partially oxidized by air to sulfate ($MgSO_4$). Solid magnesium sulfite containing a small proportion of sulfate is separated and heated with **coke** (carbon); the magnesia absorber is regenerated, and sulfur dioxide is evolved. The latter may be liquefied for industrial use or it may be converted into elemental (solid) sulfur or sulfuric acid (see **Desulfurization: Waste Products**).

CHIYODA TWO-STAGE DESUL-FURIZATION PROCESS

A wet process developed in Japan for the **desulfurization of stack gases**.

Absorption of sulfur dioxide (SO_2) is achieved by scrubbing the stack (flue) gas with dilute (2 to 5 percent) sulfuric acid (H_2SO_4) containing ferric ion as a catalyst, followed by oxidation with air. The sulfur dioxide is thus converted into sulfuric acid. Part of the acid solution is treated with lime (CaO) or limestone ($CaCO_3$) to form calcium sulfate ($CaSO_4$), whereas the remainder is recycled to the scrubbers. A special feature of the process is that the calcium sulfate product is gypsum ($CaSO_4 \cdot 2H_2O$) in a form having commercial value.

CLAUS PROCESS

A process for oxidizing hydrogen sulfide gas into elemental (solid) sulfur; it is commonly used in the treatment of off-gases from refinery operations and from the desulfurization of fuel gases. The gas to be treated should have at least 10 volume percent of hydrogen sulfide. It is mixed with air containing the oxygen required for the reaction

$$\underset{\substack{\text{Hydrogen}\\ \text{sulfide}}}{2H_2S} + \underset{\substack{\text{Oxygen}\\ \text{(from air)}}}{O_2} \rightarrow \underset{\substack{\text{Water}\\ \text{(vapor)}}}{2H_2O} + \underset{\substack{\text{Sulfur}\\ \text{(vapor)}}}{2S}$$

in which the product is elemental sulfur (and water vapor).

The Claus process is conducted in two stages: a thermal (high-temperature) stage followed by a catalytic (low-temperature) stage. The mixture of air and gas to be processed is fed at a pressure just above atmospheric to a furnace at a temperature of 1100°F (595°C) or above. The exit gas is cooled to remove sulfur vapor by condensation to liquid which eventually solidifies. The residual gas is reheated to about 480°F (250°C) and passed through one or more catalytic reactors where nearly all the remaining hydrogen sulfide is oxidized. The sulfur formed is removed by condensation, as before. The catalyst is usually a type of alumina (Al_2O_3).

Small amounts of carbon oxysulfide (COS) and carbon disulfide (CS_2), as well as residual hydrogen sulfide and possibly some sulfur dioxide (SO_2), are commonly present in the Claus process off-gas. Since these substances are potential air pollutants, they are removed by further treatment in the **Beavon, Cleanair, SCOT,** or **Wellman–Lord SO_2 Recovery process**. The **IFP** and **Sulfreen processes** remove hydrogen sulfide, but apparently not other sulfur compounds, from the Claus off-gas.

CLEANAIR PROCESS

A process for removing small amounts of elemental sulfur and sulfur compounds, including carbon oxysulfide, carbon disulfide, and sulfur dioxide as well as hydrogen sulfide, from the **Claus process** off-gas. The operation is conducted in three stages: in the first stage carbon oxysulfide and disulfide are removed; in the next stage elemental sulfur is recovered, and all of the sulfur dioxide and part of the hydrogen sulfide are converted into elemental sulfur; in the final stage the remaining hydrogen sulfide is removed by the **Stretford process**.

CLEAN-COKE PROCESS

A U. S. Steel Corporation process for producing clean (low-sulfur) metallurgical **coke** from coal containing 2 to 3 percent of sulfur. Liquid and gaseous hydrocarbon fuels are obtained as by-products (see **Coal Gasification; Coal Liquefaction**).

The coal is divided into two roughly equal fractions. One fraction is carbonized (see **Coal Pyrolysis**) by heating at 1200 to 1400°F (650 to 760°C) at 8 to 10 atm (0.8 to 1 MPa) pressure in the absence of air. The coal is devolatilized (i.e., volatile matter is driven off) and partially desulfurized leaving a residue of **char**. The remainder of the coal is formed into a slurry with a **hydrocarbon** oil (see below) and reacted with hydrogen gas (hydrogenation) at 900°F (480°C) and 200 to 270 atm (20 to 27 MPa) pressure. The liquid condensed from the carbonization and hydrogenation products is distilled into several fractions, one of which is used in the slurrying operation (see **Distillation**). Another fraction is sent to the carbonizer where it forms a pitch coke that is mixed with the char obtained by direct carbonization of the coal. The mixture is formed into pellets using a third distillation fraction as the binder; a strong metallurgical coke is obtained by heating these pellets. A fourth fraction can be used as a fuel oil or as a source of chemical process intermediates (Fig. 12).

The gases obtained from the carbonization and hydrogenation stages contain most of the original sulfur content of the coal in the form of hydrogen sulfide. Treatment for sulfur removal (see **Desulfurization of Fuel Gases**) leaves a clean **intermediate-Btu fuel gas**, consisting mainly of hydrogen and methane; it is unusual in that it contains little or no carbon monoxide.

COAL

A natural, rock-like combustible material ranging in color from brown to deep black. The color variations are accompanied by variations in properties, such as carbon, hydrogen, and moisture (water) content and **heating value**. Coal can be burned directly as a fuel, or it can be converted into gaseous and liquid fuels (see **Coal Gasification; Coal Liquefaction**).

Coal originated from material which accumulated as a result of the decay of mostly woody plants over long periods of time. **Peat** deposits formed in this manner were buried by sand, silt, and mud and, in the course of millions of years, were subjected to high temperatures and pressures. The process called "carbonification" or "coalification" occurred whereby the soft peat, with its high moisture content, was slowly transformed into the harder, less moist material known as coal. About two-thirds of the world's coal is derived from plants which grew in the Carboniferous Period, from 280 to 350 million years ago. The remainder is from more recent plant material.

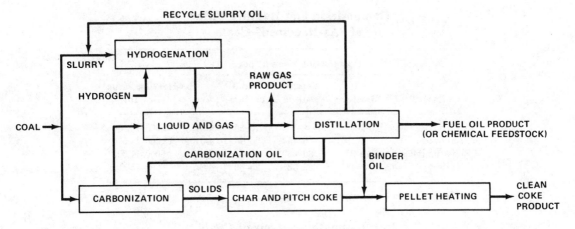

Fig. 12   Clean-coke process.

The "rank" of a coal indicates the degree of coalification. Lignite (including brown coal) has the lowest rank and this is followed by subbituminous coal, bituminous coal, and anthracite with successively increasing ranks. Each of these major ranks is further subdivided, but these sub-divisions are not important here. The rank assigned to a particular coal is determined largely by the fixed (or nonvolatile) carbon content and the heating value. Typical ("proximate") analyses, for the proportions of moisture, volatile matter [at a temperature of 1740°F (950°C)], fixed carbon, and mineral matter (ash) of "as-received" coal, as well as the heating values are given in the table.   (The data in this table differ from those published by the American Society for Testing and Materials (ASTM) because the latter are based on the analysis of coals free from moisture and mineral matter.)

As distinct from the proximate analysis, the "ultimate" analysis gives the proportions of five important elements in coal, namely, carbon, hydrogen, oxygen, nitrogen, and sulfur. Values for the four main ranks are quoted in the accompanying table for coal free from moisture and mineral matter (ash).

The heating value of coal is determined mainly by the combustion (or chemical reaction) of carbon with oxygen (in air). A subsidiary contribution is made by the hydrogen combined chemically with carbon in the coal. The proportion of carbon in coal generally increases with the rank, but the hydrogen content decreases, as seen in the ultimate analysis table. As a result of this combination of circumstances, the heating value per pound (or kilogram) of coal is a maximum for the low-volatile bituminous types.

A large proportion of the U. S. coal reserves is in the form of low-sulfur lignites in the Northern Great Plains states. However, use of this coal has been limited by its high moisture content and low heating value. The cost of transportation per heat unit thus restricts the radius within which economic use of lignite is possible. Other problems include freezing in railroad cars in cold weather and fouling of boiler tubes. Lignite is also very reactive with oxygen (from the air), making it susceptible to spontaneous combustion. A method is being developed for reducing the moisture content of lignite to about 15 percent and compressing it into hard pellets which are resistant to spontaneous combustion. In this form, lignite may find increasing use as a boiler fuel.

An important characteristic of certain coals is the tendency for small pieces to soften and adhere (or aggregate) upon heating. This is commonly referred to as

Composition and Heating Value of "As-Received" Coals

	Composition (weight percent)				
Rank	Moisture	Volatile matter	Fixed carbon	Ash	Heating value (Btu/lb)*
Anthracite	3–6	4–12	75–85	4–15	12,000–13,500
Bituminous	2–15	15–45	50–70	4–18	12,000–14,500
Subbituminous	10–25	30–45	31–55	3–12	8000–11,000
Lignite	35–45	22–32	25–30	4–15	6000–7500

*To convert to kJ/kg, multiply by 2.32.

Ultimate Analysis of Coals (Moisture and Ash Free)

	Weight percent				
Rank	Carbon	Hydrogen	Oxygen	Nitrogen	Sulfur
Anthracite	75–85	1.5–3.5	5.5–9	0.5–1.0	0.5–2.5
Bituminous	65–80	4.5–6.0	4.5–10	0.5–2.5	0.5–6.0
Subbituminous	57–69	5.5–6.5	15–30	0.8–1.5	0.2–2.0
Lignite	35–45	6.0–7.5	38–48	0.5–1.0	0.3–3.0

the caking property of coal. Most bituminous coals tend to cake, but anthracite, subbituminous coals, and lignite do not. Caking is one of the requirements of coal used for the production of coke. On the other hand, caking is a drawback in **fluidized-bed** processes for **coal gasification**. To prevent caking, bituminous coals are commonly pretreated by heating in air (or air and steam) at a temperature of about 750°F (400°C).

COAL CLEANING (OR BENEFICIATION)

The treatment of as-mined coal to separate extraneous mineral matter from the coal itself. The use of mining machinery has resulted in larger amounts of impurities in the coal, thus increasing the importance of cleaning (or beneficiation). Among the advantages are a substantial decrease in the amount of ash remaining after the coal is burned and, in particular, a decrease of about 30 to 35 percent in the sulfur content resulting from the removal of a large proportion of the mineral pyrite (see **Coal: Sulfur Content**). On the other hand, a significant amount of coal is lost with the separated mineral matter. Moreover, the common cleaning methods are wet processes, and water must be removed from the cleaned coal prior to shipment. This is done with vibrating screens, centrifuges, or hot combustion gases (thermal dryers).

The main methods used for cleaning coal, in order of importance in the United States, are jigs, dense media, concentrating tables, and froth flotation. All except the last depend on the difference in density between coal, **specific gravity** (sp gr) 1.2 to 1.4, and foreign material, sp gr from 1.8 (shale) to 5.0 (pyrite). Other cleaning procedures, such as electrostatic and magnetic separations, are being studied; these have the advantage of being dry methods so that the problem of moisture removal does not arise.

Jigs

In the wet jig method, the coal is fed onto a perforated submerged screen. Water is continuously forced up and down through the screen, either by periodically dropping and raising the screen (movable-screen jig) or by pulsing water through it (fixed-screen jig). The pulsating motion of the water causes the material on the screen, regardless of size, to stratify with the lighter component (coal) at the top and the heavier (mineral) at the bottom. The clean coal is carried off by the water flow, and the mineral matter is removed from the bottom of the screen.

The dry (or pneumatic) jig process is similar to the fixed-screen method except that air pulses rather than water pulses are used. The pneumatic jig is particularly suited to cleaning the smaller particles ("fines") which may constitute from 10 to 30 percent of the coal mined (see also the froth flotation process described below).

Dense Media

The dense-medium method, also called the sink–float method, depends on the fact that in a liquid medium of intermediate density (e.g., sp gr about 1.6) coal will float while the dense minerals will sink. The coal is thus removed from the top and the mineral waste from the bottom of a tank through which the liquid medium is circulated. The sink–float medium consists of a suspension of a finely ground dense mineral (e.g., magnetite or fine sand) in water; such a suspension behaves like a true dense liquid. Any magnetite remaining attached to the coal or waste is removed by magnetic separators. Sand is removed by washing.

In the dense-medium cyclone method of coal cleaning, separation of lighter from heavier material is aided by centrifugal force. A suspension of coal in the heavy medium is introduced tangentially into a **cyclone separator**. The more dense (i.e., mineral) material is forced toward the sides of the cone and collects at the bottom, whereas the coal, being less dense, remains suspended and leaves with the medium at the top of the separator.

Concentrating Table

A concentrating (or shaking) table consists of a large deck, 12 to 16 ft (3.7 to 4.9 m) long and 4 to 6 ft (1.2 to 1.8 m) wide, oscillating in a direction at right angles to the water flow. The surface is covered with riffles to form a series of grooves running across the table which is tilted slightly in both directions. The finely ground coal is suspended in water and fed onto the upper end of the table. As the suspension travels along the table, the heavier mineral particles are trapped by the riffles, whereas the lighter (coal) particles are washed over the end of the table. The motion and slope of the table cause the mineral matter to be carried along the grooves to the side of the table where it is removed.

Pneumatic concentrating tables utilize air instead of water as the separating medium. Air blown up through holes in the riffled table and the motion of the table together cause the material to be segregated into an upper, less dense layer (coal) and a lower, denser layer (mineral matter). The coal moves in a direction perpendicular to the riffles, whereas the heavier impurities travel along the grooves, just as when water is used.

Froth Flotation

The froth flotation process serves primarily for cleaning coal fines. The coal is suspended in water through which air is bubbled. The air bubbles tend to attach themselves to the coal particles rather than to the mineral matter. As a result, the coal, which would normally sink in water, rises to the surface as a froth; the froth is removed, and the clean coal is recovered. The mineral waste falls to the bottom and is discharged. A hydrocarbon oil added to the water facilitates adhesion of air bubbles to the coal particles, and a foaming agent, usually a complex alcohol, enhances foam formation.

See also **Desulfurization of Coal.**

COAL (OR TOWN) GAS

A fuel gas made from coal and formerly distributed by pipeline for use by industry and homes in the general vicinity of the gas production plant.   With the wide availability of natural gas, coal gas intended for distribution is no longer made in the United States. However, coke-oven gas, formed as a by-product in **coke** manufacture, is similar to coal gas; it is not distributed but is used as a fuel in or near the coke-oven facility.

Coal gas is made by heating coal to a temperature of about 1650°F (900°C) in the absence of air; this process is called **pyrolysis** (or carbonization). The gaseous (and vapor) products are scrubbed with water to remove ammonia, condensible vapors, and tar, leaving a combustible (coal or coke-oven) gas. The gas contains roughly 50 volume percent of hydrogen, 30 percent of methane, and 7 percent of carbon monoxide (dry basis) as fuel components, and a few percent each of inert nitrogen and carbon dioxide. The **heating value** is approximately 480 Btu/cu ft (18 MJ/cu m).

COAL GASIFICATION

A general term used to describe the production of fuel gas from coal. In the manufacture of **coal gas** by **pyrolysis**, roughly 30 percent of the carbon was converted into gas while most of the remainder was left as coke. In gasification processes of current interest, the aim is to convert a much larger proportion of the carbon into fuel gases. Other objectives of coal gasification are (1) to produce a fuel (e.g., for electric power generation) that is more environmentally acceptable than coal, and (2) to make a **substitute natural gas** for pipeline distribution.

When coal is burned as a fuel, the sulfur present appears as **sulfur oxides** in the stack (flue) gases. As a general rule, excess sulfur oxides must be removed from the stack gases before they are discharged to the atmosphere (see **Desulfurization of Stack Gases**). In coal gasification, however, the sulfur is usually converted into hydrogen sulfide, nearly all of which is removed before the gas is burned. The volume of fuel gas is considerably less than that of the stack gases to be treated for the same heat generation and, furthermore, desulfurization of fuel gases is more highly developed than the removal of sulfur from stack gases.

The major combustible (or fuel) products of coal gasification are hydrogen (H_2), carbon monoxide (CO), and methane (CH_4). The **heating value** of methane is close to 1000 Btu/cu ft (37 MJ/cu m), and the values for hydrogen and carbon monoxide are both almost 320 Btu/cu ft (12 MJ/cu m). The heating value of an actual coal gasification product depends on the proportions of these fuel constituents and of the inert gases nitrogen (from the air) and carbon dioxide.

Fuel gases are commonly classified as having high-Btu, intermediate-Btu, or low-Btu heating values. **High-Btu fuel gases** have heating values of 950 to 1050 Btu/cu ft (35 to 39 MJ/cu m); they consist largely of methane. **Intermediate-Btu fuel gases** have heating values of roughly 300 to 450 Btu/cu ft (11 to 17 MJ/cu m). On a dry basis they generally contain 65 to 70 percent of carbon monoxide and hydrogen in various proportions, and 5 to 15 percent of methane; the remainder is mostly inert carbon dioxide.

As a general rule, an intermediate-Btu fuel gas can be converted into a high-Btu gas (or substitute natural gas) by subjecting it to the **water-gas shift reaction** followed by **methanation**. Furthermore, an intermediate-Btu gas, consisting mainly of carbon monoxide and hydrogen, is a **synthesis gas**. It can be used for the production of liquid **hydrocarbon** fuels (see **Fischer–Tropsch Process**)   or   methanol (see **Alcohol Fuels**); the latter can be used

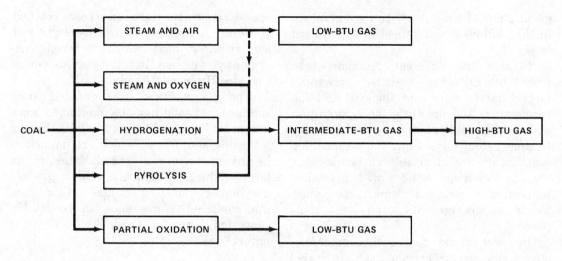

Fig. 13   Coal gasification processes.

directly as a fuel or converted into gasoline (see **Mobil M-Gasoline Process**).

Most of the **low-Btu fuel gases** are similar to intermediate-Btu gases diluted with a roughly equal volume of inert nitrogen gas; in some cases the fuel component is almost entirely carbon monoxide with little or no hydrogen. The heating values are in the range of 130 to 200 Btu/cu ft (4.8 to 7.4 MJ/cu m).

The basic schemes for coal gasification are outlined below and also in Fig. 13. Titles of various gasification processes in use or under development are given in the sections on **Intermediate-Btu Fuel Gases** and **Low-Btu Fuel Gases**. Underground (or in-situ) gasification of coal, that is, without mining, is described in the following section.

Steam Gasification Processes

The majority of coal gasification processes are based on the interaction of carbon (in coal) with steam according to the reaction

$$\underset{\text{Carbon}}{C} + \underset{\text{Steam}}{H_2O} = \underset{\substack{\text{Carbon}\\\text{monoxide}}}{CO} + \underset{\text{Hydrogen}}{H_2}$$

$$- 5000 \text{ Btu/lb C (11.6 MJ/kg)}$$

Heat is absorbed in this reaction, and it is usually provided, directly or indirectly, by partial combustion, i.e.,

$$C + \frac{1}{2}O_2 = CO + 4000 \text{ Btu/lb C}$$

$$(9.2 \text{ MJ/kg})$$

or complete combustion, i.e.,

$$C + O_2 = \underset{\substack{\text{Carbon}\\\text{dioxide}}}{CO_2} + 14,000 \text{ Btu/lb C}$$

$$(33 \text{ MJ/kg})$$

of carbon in oxygen (from air or oxygen gas). Some of the hydrogen formed in the reaction with steam may interact directly with the carbon to form methane; thus,

$$C + 2H_2 = \underset{\text{Methane}}{CH_4} + 3300 \text{ Btu/lb C}$$

$$(7.7 \text{ MJ/kg})$$

In the steam gasification of coal, which generally takes place at a temperature of 1200 to 1800°F (650 to 980°C) and a pressure up to 35 atm (3.5 MPa) or so, the proportions of steam and oxygen (or air) are adjusted so that there is a balance between the heat absorbing (endothermic) and heat-producing (exothermic) reactions given above. In this way, the temperature in the gasifier is maintained at the desired level.

Because the different reactions take place to different extents, depending largely on the nature of the coal and the temperature and pressure, the composition and heating value of the product gas vary to some extent. However, the combustible constituents are invariably carbon monoxide and hydrogen, together with a smaller proportion of methane. Some carbon dioxide is usually present as an inert component.

In most steam gasification processes, steam and air (or oxygen gas) are introduced simultaneously into the hot coal. If the gasification occurs in air, nitrogen from the air remains as an inert diluent. The product is then a low-Btu fuel gas. If oxygen gas is used instead of air, the resulting gas contains little or no nitrogen and is an intermediate-Btu fuel gas.

A few processes that use air are exceptional in the respect that the carbon–steam and carbon–air (oxygen) reactions are conducted in separate vessels, called the gasifier and combustor, respectively. Heat produced in the combustor is then transferred to the gasifier by a suitable solid carrier. The nitrogen from the air used in the combustor is discharged as flue gas and does not enter the gasifier. Hence, the product is an intermediate-Btu gas although combustion of the carbon has taken place in air.

Hydrogasification Processes

An alternative type of coal gasification involves the direct reaction of hydrogen gas with carbon to form methane. Hydrogasification is conducted by heating coal at a temperature of about 1500 to 1800°F (815 to 980°C) in the presence of a hydrogen-rich gas at a pressure of 70 to 100 atm (7 to 10 MPa). The reaction of carbon with hydrogen is accompanied by heat liberation (see above); hence no other heat source need be included in the gasifier once the desired

operating temperature has been reached. However, production of the hydrogen-rich gas requires heat which is most conveniently supplied by the combustion of coal (see **Hydrogen Production**).

The gas produced by hydrogasification contains, in addition to methane, some unreacted hydrogen together with carbon monoxide and dioxide which are impurities in the hydrogen-rich gas. Because of its large methane content, the product gas has a higher heating value than an intermediate-Btu gas and can be readily converted into a high-Btu substitute natural gas.

Other Gasification Processes

As a result of the high reaction temperatures, a certain amount of **coal pyrolysis** (i.e., decomposition by heat) occurs in all coal gasification processes. In a few cases, pyrolysis is the major reaction, as it is in the manufacture of coal (or coke-oven) gas. The product is relatively rich in hydrogen and methane and has an intermediate (or higher) heating value.

Another coal gasification procedure is based on the partial oxidation of the carbon in coal with air to form carbon monoxide; the product is a low-Btu gas. The fuel constituents are carbon monoxide (mainly) and hydrogen from coal pyrolysis; at least 50 volume percent of the gas consists of inert nitrogen and carbon dioxide. The nitrogen may be eliminated and the heating value of the gas increased by using oxygen instead of air in the process.

COAL GASIFICATION: UNDERGROUND (IN SITU)

Gasification of coal in place (i.e., without mining) by interaction with air to yield a **low-Btu fuel gas** or with a mixture of steam and oxygen to produce an **intermediate-Btu fuel** gas (see **Coal Gasification**). Underground (also called "in-situ") gasification could make available as an energy source large reserves of coal in the United States that are dangerous, difficult, or uneconomic to mine.

Underground gasification of coal was suggested in the United Kingdom in 1868 and in Russia in 1888. Some experiments were conducted in England between 1949 and 1959, but the most extensive development, starting in 1933 and reaching its peak in the late 1960s, occurred in the U.S.S.R. An underground coal gasification process was patented in the United States in 1909; tests were made in Alabama from 1948 through 1958 and in Kentucky some 10 years later, but the projects were abandoned. In recent years, however, interest in underground coal gasification has been revived under U. S. Government sponsorship.

In simple terms, underground gasification involves drilling two vertical wells roughly 60 ft (18 m) apart from the surface into a coal seam. One well, called the injection well, is for introducing air (or oxygen gas and steam) and the other, called the production well, is for withdrawal of the product gas. A permeable path through which gases can flow is then formed between the wells near the bottom of the coal bed. As explained below, the method of making the permeable path, which is an important aspect of the gasification process, depends on the local conditions.

The coal at the bottom of the injection well is ignited in a suitable manner (e.g., by burning gas), and large volumes of air (or oxygen–steam mixture) are introduced at a pressure of a few atmospheres. The coal undergoes combustion in the air (or oxygen) and generates the temperature at which several reactions occur between the carbon (in the coal), oxygen (from the air or oxygen gas), and steam (injected or from water in the coal).

The gases produced in these reactions and also by coal pyrolysis consist mainly of hydrogen, carbon monoxide, and methane as the fuel components, with nitrogen (if air is used) and carbon dioxide as inert constituents. The gas travels through the previously formed permeable path to the production well where it is withdrawn. As the coal is consumed, the combustion front moves in the same direction as the gas flow; this is called the forward gasification mode (Fig. 14A).

The gasification procedure just described is essentially the same in all in-situ coal gasification processes, but the method for forming the permeable path varies. In coal seams with some natural

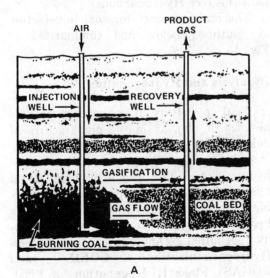

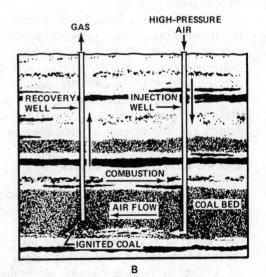

Fig. 14 Underground coal gasification. A: Forward gasification; B: Reverse combustion.

permeability, successful linkage of the wells has been achieved by reverse combustion (Fig. 14B). Air at high pressure is injected into one well, and the coal is ignited at the bottom of the other well. The combustion front then moves in the reverse direction to the airflow, producing a permeable path as it proceeds. In due course, the two wells are linked in this manner and gasification can be initiated by reversing the direction of air (or oxygen-steam) injection. At the same time, the pressure of the injected air (or oxygen-steam) is decreased and the flow rate is increased.

The combination of reversed combustion linkage with forward gasification has been tested successfully in both horizontal and sharply sloping subbituminous coal beds in Wyoming. In the latter case, recovery of the coal by commercial mining is not practical, and in-situ gasification seems the only approach to the utilization of coal in such sloping beds.

Deep, thick coal seams may not have sufficient natural permeability to permit linkage of wells by reverse combustion. Among alternative procedures under investigation are fracturing by explosives and linkage of vertical injection and production wells by directional drilling to make connection within the coal bed (see **Petroleum Production**).

Major problems have been encountered in implementing the general principles described above. They include combustion control, leakage through cracks and fissures, swelling of bituminous coals when heated leading to blockage of flow passages, surface subsidence as coal is consumed, and interference by groundwater.

COAL LIQUEFACTION

A general term used to describe conversion of coal into a mixture of liquid **hydrocarbons**. The mixture may be equivalent to a **synthetic crude oil** (syncrude) which can be used directly as a fuel or refined to yield a variety of products similar to those produced from natural petroleum (see **Petroleum Refining**). Some gaseous hydrocarbons, including methane, are also formed, and these can be used directly as fuel or converted into a **substitute natural gas** for pipeline distribution. Approximately 2.5 to 3 **barrels** of liquid fuel should be obtained per ton (0.44 to 0.53 cu m per 1000 kg) of coal.

The composition of coal is variable, but most coals contain somewhat less than one hydrogen atom per atom of carbon. In natural crude oil, however, there are almost two atoms of hydrogen per carbon atom. Consequently, all modern coal liquefaction methods involve addition of hydrogen in some form to the coal. The hydrogen (or hydrogen-rich) gas required is generated by the action of steam on coal or a coal product (see **Hydrogen Production**).

Oil made from coal differs from most petroleum crude oils in having a higher content of aromatic hydrocarbons, such as benzene and related compounds. These substances have industrial uses, and small proportions are beneficial in gasoline. However, for more general fuel use, the characteristics of coal syncrude can be changed by further reaction with hydrogen gas (e.g., by **hydrotreating**) in the presence of a catalyst, thereby converting the aromatic hydrocarbons into cycloparaffins and paraffins (see **Hydrocarbons**).

The basic schemes for coal liquefaction are outlined below and summarized in Fig. 15.

Pyrolysis and Hydrogenation

Coal **pyrolysis** (i.e., heating in the absence of air), as in the manufacture of **coke** (or **coal gas**), yields a light oil containing a large proportion of aromatic hydrocarbons. The production of liquid hydrocarbons can be improved by combining pyrolysis with hydrogenation using a hydrogen-rich gas. Examples of pyrolysis-hydrogenation techniques are the **Coalcon Hydrocarbonization**, **COED** (and **COGAS**), **Flash Hydrogenation** (or Flash Hydropyrolysis), and **Garrett Coal**

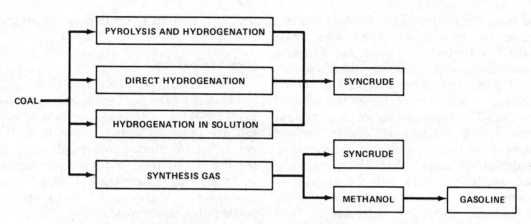

Fig. 15   Coal liquefaction processes.

Pyrolysis processes. Pyrolysis and hydrogenation may be conducted simultaneously or in separate stages.

Direct Hydrogenation

The crushed coal is formed into a slurry with some of the product oil and hydrogenated with a hydrogen-rich gas under pressure, usually in the presence of a catalyst. High hydrogen gas pressures and longer reaction times favor the production of lighter oils, whereas lower pressures and shorter times yield more of the heavier components. The **Bergius, COSTEAM, H-Coal, Synthoil,** and **Zinc Chloride Catalytic processes** are examples of direct hydrogenation of coal as a slurry. The Disposable Catalyst Hydrogenation process is similar except that a paste of pulverized coal and oil is used instead of a slurry.

Hydrogenation in Solution

Two different approaches are used for the hydrogenation of coal in solution. In one (**Consol Synthetic Fuel, Exxon Donor Solvent,** and **Pott–Broche processes**), the coal is partially dissolved in a hydrogen donor (i.e., hydrogen-rich) liquid solvent which adds hydrogen to the coal at a high temperature. The donor solvent is made by hydrogenating one of the liquid products of the process. In the other

approach (**Solvent Refined Coal process**), the coal is partly dissolved in a product oil in the presence of hydrogen-rich gas at high temperature and pressure.

Use of Synthesis Gas

The coal is converted into a mixture of carbon monoxide and hydrogen gases, called **synthesis gas,** by a suitable method of coal gasification that yields an **intermediate-Btu fuel gas.** By passage of the synthesis gas over an appropriate catalyst, the product can be either a mixture of liquid hydrocarbons (syncrude) or methanol (see **Alcohol Fuels**). The **Fischer–Tropsch process** is the prime example of coal liquefaction by way of synthesis gas. A further development is the conversion of methanol into gasoline by the **Mobil M-Gasoline process**.

COAL: MINERAL MATTER (AND COAL ASH)

Material consisting of the elements in coal other than carbon, hydrogen, oxygen, nitrogen, and sulfur. This material arises from the original plant cells and from nonorganic matter deposited by water during and after coal formation. Commercial coals also contain extraneous minerals introduced from adjacent rocks in the mining operation. Most of the coal mined in the United States contains from 3 to 15 percent of mineral matter. The major inorganic materials present are clay and shale (aluminosilicates), limestone and dolomite (carbonates), and pyrite (iron disulfide).

The ash remaining after coal is burned contains much but not all of the mineral matter. The compositions of the mineral matter and residual ash differ because some of the more volatile mineral elements, which are potentially hazardous, are emitted, wholly or in part, with the stack (flue) gases; these elements include arsenic, antimony, cadmium, mercury, and selenium. The coal ash remaining consists mainly (93 to 98 percent) of oxygen compounds of silicon, aluminum, iron, calcium, magnesium, titanium, sodium, potassium, and sulfur (as sulfate). The composition ranges of the residual ash from most U. S. coals, expressed as the percentages of the various oxides, are given in the table.

Composition of Ash from U. S. Coals

Oxide	Percent Range	Percent Average (approx.)
Silicon (SiO_2)	30–55	40
Aluminum (Al_2O_3)	15–30	23
Iron (Fe_2O_3)	10–30	20
Calcium (CaO)	2–20	6
Magnesium (MgO)	0.3–4	2
Titanium (TiO_2)	0.6–2	1
Sodium and Potassium (Na_2O,K_2O)	1–4	2
Sulfur (as sulfate)	0.5–6	3
Trace elements	2–4	3

On the average, about 3 percent of coal ash consists of many elements in small amounts which differ widely in different coals. These are commonly referred to as trace elements because the individual quantities are small, ranging from a few parts per billion to about one part per 1000 (0.1 percent). Most of the trace elements in coal are present in about the same relative proportions as in the earth's crust, but a few (e.g., beryllium, boron, germanium, and uranium) sometimes occur in larger proportions.

Uses of Coal Ash

The average ash content of U. S. coals is about 10 percent. Consequently, in the early 1980s, at least 45 million tons of coal ash are produced annually, mostly in the generation of electric power. Roughly 20 percent of this total is utilized for various purposes, as mentioned below. The remainder is stored under water in ash ponds at the power plants.

Because of its pozzolanic (cement-like) properties, coal ash can replace up to 30 percent of portland cement in concrete; the product is said to require less hand finishing than conventional cement mixes. Coal ash is also used as construction fill, road base (mixed with asphalt), soil conditioner, and in the manufacture of lightweight brick and other structural materials. Efforts are being made to find other uses for coal ash; among the possibilities are the recovery of iron and aluminum, which are present in substantial proportions.

COAL MINING

The removal of coal from its natural location in the earth, as free as possible from extraneous rock and soil. In the United States coal is obtained mainly by underground mining or by surface mining. Surface mining is generally economically feasible provided the overburden of rocks and soil covering the coal seam is less than about 200 ft (60 m) thick. When the overburden is thicker but the coal is at too shallow a depth to permit safe underground mining, coal can often be removed by auger mining. In 1980, roughly 35 percent of the coal mined in the United States was obtained by underground mining, more than 60 percent by surface mining, and about 3 percent by the auger method.

Underground Mining

Most underground mining in the United States is conducted by the "room-and-pillar" method. Parallel tunnels (or galleries) are cut into the coal seam with others at right angles. These tunnels are gradually widened by removal of the coal, leaving thick pillars required to support the overburden. Sometimes the coal pillars are removed, and the roof is allowed to cave in. On the average, only about 50 to 60 percent of the coal in a seam can be recovered by the room-and-pillar technique.

Mechanical equipment, now used almost exclusively in underground mining, is either of the cyclic (conventional) or continuous type. In cyclic mining, the following operations are carried out in sequence at the coal face: (1) slots are cut in the coal seam to allow for expansion during blasting, (2) holes are drilled, a safe explosive material is inserted and then fired, and (3) the coal broken up by the blast is removed in trucks or by conveyor belt. The operations are then repeated at an adjacent location. In recent years, this cyclic operation has been gradually replaced by continuous mining which now accounts for much of the coal produced from underground mines in the United States. A single continuous mining machine rips the coal from the exposed face, without blasting, and loads it for transportation.

Longwall mining is an alternative to the room-and-pillar method which has been used extensively in Europe. It is especially suited to mines with large roof stresses and where resources are limited and maximum recovery of the coal is essential. Since the mid-1960s, longwall mining has come into increasing use in the United States. The machinery is more expensive than that used in room-and-pillar mining, but the operating costs per ton of coal mined are lower.

In a typical procedure, two parallel tunnels are constructed about 300 to 600 ft (90 to 180 m) apart at right angles to the main tunnel. The longwall (continuous) mining machine moves back and forth along the exposed coal face between the parallel tunnels ripping away the coal which drops onto a conveyor belt. Self-advancing hydraulic jacks support the roof in the area being mined; in mined-out areas, the roof is allowed to collapse in a controlled manner or support may be provided by rock fill. The proportion of coal recovered is substantially larger than in room-and-pillar mining.

Longwall mining is of special interest for its potential use in thick coal seams. Normally, coal cannot be extracted completely from seams more than about 25 ft (7.5 m) thick. But mining of thicker seams may be possible by the longwall multilift technique in which two or more levels are mined by a longwall machine in successive vertical layers starting from the top.

Surface Mining

There are two broad categories of surface (or strip) mining of coal: area mining in fairly level country and contour (or collar) mining in hilly terrain. The details of surface mining vary with the local conditions, but in general they involve removal of the overburden by power shovels, draglines, or bucket wheels. The exposed coal seam is fragmented by blasting, and the broken coal is removed. Between 80 and 90 percent of the coal in place can be recovered by surface mining.

Surface mining is now regulated in accordance with the Surface Mining and Reclamation Act of 1977. One requirement of the Act is that, before mining is commenced in any location, the topsoil be scraped off and stored. When the area is mined out, the land must be regraded (as far as is practical) to the approximate original contour and the topsoil replaced. The land must then be seeded to establish a vegetative cover capable of self regeneration.

In *area mining*, the overburden is removed in a long straight strip (or cut) and is stacked to form a spoil bank parallel to the cut. After the coal has been extracted from the first cut, a second cut is made alongside it; the overburden is then placed in the space left by removal of the coal from the first cut. This procedure is continued until the coal is mined out; the area then consists of a succession of spoil banks and a final open cut (Fig. 16). The latter must then be filled, the area regraded, and the topsoil replaced.

horizontally into the coal seam for distances up to about 300 ft (90 m). Coal is removed as the auger head is withdrawn and is carried to trucks (or stockpile) by a conveyor belt. With a single auger head, coal recovery may be less than 50 percent

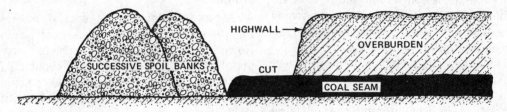

Fig. 16   Strip mining.

There are two basic procedures for *contour mining* in hilly areas: haulback mining and mountaintop leveling. In haulback mining, a cut is made at the lowest level at which the coal outcrop occurs. Subsequent cuts then follow the contour of the land at successively higher levels. All the overburden spoil, except that from the initial cut, is carried back to fill the space left by coal removal. The initial cut is used as a haul road for trucks.

Mountaintop leveling permits recovery of coal located near the tops of hills, ridges, and knolls. The first cut is made by removing overburden from the top, and the spoil is deposited in an adjacent valley. Extraction of the coal proceeds from the top downward, with the spoil from each level being dumped at the level above from which the coal has been removed. When the area is mined out, the mountaintop is not necessarily restored to its original contour. It may be shaped to a gently rolling terrain before the topsoil is replaced.

Auger Mining

Auger mining is frequently used as an adjunct to surface mining when the overburden becomes too thick for economic removal, but it is also utilized in other situations. An essential requirement is that the coal seam should be roughly horizontal and the face exposed. Powered, screw auger heads, up to 8 ft (2.4 m) in diameter, bore

because of the coal left between the holes. The use of closely spaced multiheaded augers, however, has made it possible to reduce the residual coal and thus increase the recovery.

COAL PYROLYSIS (OR CARBONIZATION)

The chemical decomposition of coal by heating in the absence of air; it is also called destructive distillation. The main volatile products are a gas, a condensible light oil, and tar; the solid residue is **coke** or **char**. The gas consists of hydrogen, methane (and other light **hydrocarbons**), and carbon monoxide as the combustible constituents, together with a few percent of inert gases (see **Coal Gas**). Gaseous ammonia is also produced from the nitrogen in the coal and is removed by scrubbing with water.

The relative amounts of the various products depend on the nature (or rank) of the coal (see **Coal**) and the pyrolysis temperature. Coals of low rank, with high contents of water and volatile matter, yield more gas and water vapor than high-rank coals, and the coke is of lower strength. High pyrolysis temperatures favor the production of high-strength cokes. As a general rule, more oil and tar are produced at the higher temperatures.

Low-temperature coal pyrolysis is not much used in the United States; most pyrolysis is conducted at high temperatures, in the range of about 1650 to 2100°F (900 to 1150°C), for the production of metallurgical coke. Typical weight percentages of the main products from high-temperature pyrolysis of coal with an average content (30 percent) of volatile matter are given below.

Coke	75 percent
Gas	15
Tar	3
Light oils	1

One short ton (2000 lb) of coal produces about 10,000 cu ft of gas (310 cu m gas per 1000 kg coal) with a **heating value** of roughly 500 Btu/cu ft (18.6 MJ/cu m).

COAL: SULFUR CONTENT

The amount (or proportion) of sulfur present in various combined forms in as-mined coal. This quantity is important because of the undesirable effects of sulfur in coal at various stages of production and use.

When coal is burned, the sulfur is converted into **sulfur oxides**, sulfuric acid, and sulfates, which are harmful to man, animals, and plants and can cause corrosion of metals and other structural materials. Sulfur compounds in mine spoil (waste) banks inhibit the growth of vegetation, and sulfuric acid resulting from the oxidation of sulfur compounds by air contributes to surface water pollution. Furthermore, sulfur in metallurgical coke, made from coal, lowers the quality of iron and steel products.

Sulfur is found in coal in both inorganic (mineral) and organic (plant origin) forms. The inorganic sulfur is mainly present as the mineral pyrite with a small proportion of marcasite; these are different crystalline forms of iron disulfide (FeS_2) and are generally referred to simply as "pyrite." Small quantities of another iron sulfide, called pyrrhotite, which is magnetic, is also often

associated with coal. The organic sulfur, on the other hand, does not occur as a separate material; it is bound to the carbon atoms in the coal structure and is more difficult to remove than pyrite sulfur. The relative amounts of inorganic and organic sulfur vary with the source and nature (rank) of the coal (see **Coal**). The inorganic sulfur may constitute about 40 to 80 percent of the total sulfur; a rough general rule is that, on the average, coal contains equal proportions of inorganic and organic sulfur. One of the objectives of **coal cleaning** (beneficiation) is to decrease the sulfur content by removal of mineral matter (see also **Desulfurization of Coal**).

In addition to pyrite and organically bound sulfur, coal, especially if it has been subjected to weathering by exposure to air and moisture, contains small amounts of sulfur as sulfates. The most important are iron (ferrous) sulfate ($FeSO_4 \cdot 7H_2O$), formed by weathering of pyrite, and gypsum ($CaSO_4 \cdot 2H_2O$), produced by chemical interaction with limestone (calcium carbonate).

The sulfur content of most coals in the United States ranges from 0.2 to 6 percent. The approximate distribution of identified resources according to sulfur content and rank is given in the table. The large proportion of low-sulfur lignite is partially discounted by the low heating value of this fuel. Thus, more lignite must be consumed than other fuels to generate the same amount of heat.

Approximate Distribution of Sulfur Content in U. S. Coals

Rank	Sulfur content range (percent)			
	0–0.7	0.8–1.0	1.1–3.0	3.1+
	Percentage of total in each range			
Anthracite	96.5	0.6	2.9	
Bituminous	14.3	15.2	26.2	44.3
Subbituminous	66.0	33.6	0.4	
Lignite	77.0	13.7	9.3	
U. S. average	46	19	15	20

Apart from anthracite, of which the resources are relatively small, subbituminous coals and lignite are the ranks with the lowest sulfur content. Such coals are found mainly in the western (Rocky Mountain and Northern Great Plains) states.

COAL TRANSPORTATION

The mode of conveyance of coal from the mine to the point of use. With a minor exception (see below), coal is transported in the United States by rail, water, road, or by a combination of these modes. Since 1955, about 70 to 75 percent of the coal leaving the mines has been carried by railroad cars. The coal is either loaded directly at the mine or it is carried to the railroad by motor truck. The cost of rail shipment has been reduced by the use of "unit trains"; loads, averaging 10,000 short tons (9 million kg) per shipment, originate at one place and are unloaded at a single destination at a distance of 150 to 450 miles (240 to 720 km). Shipment by water is the most economical means of transportation; hence, much coal is carried by rail to the Great Lakes and other inland (including tidal) waterways for transfer to barges or coastal vessels. Roughly 30 percent of the coal mined in the United States is eventually shipped by water.

Transportation of a slurry of roughly equal volumes of finely ground coal and water (or other liquid medium) is another possibility. A method for pumping coal with water was patented in 1891 and a system was operated in England in 1914. In the United States a 10-inch (0.25-m) diameter pipeline, 108 miles (172 km) long, operated in Ohio from 1956 to 1963. It was closed down because transportation by railroad was cheaper at the time.

The most successful coal-slurry transportation system is the Black Mesa Pipeline which carries coal a distance of 273 miles (440 km) from a mine near Kayenta, in northeast Arizona, to a 1500-megawatt electric power station in southeast Nevada. It has been in use since 1970. The pipe,

with a diameter of 12 in. (0.3 m) or 18 in. (0.45 m) in different parts, has a daily capacity of over 15,000 short tons (13.6 million kg) of coal, equivalent to about 150 railroad cars. The system has four pumping stations—one at the mine and three booster stations at distances of 60 to 80 miles (100 to 130 km) along the length of the line. In full operation, the daily consumption of water is about 2.7 million gallons (10,000 cu m).

Coal pipelines present no environmental problems after construction. The Black Mesa Pipeline, for example, is buried 3 ft (0.9 m) underground. For long distances and large throughputs of coal, the cost of pipeline transportation, including capital costs, is estimated to be less than by rail. Operational reliability depends mainly on the pumping stations; station bypassing is possible in the event of a pump failure. An alternative power source should be available in case the primary source should fail.

The major potential future use of pipelines in the United States may be to transport low-sulfur coal from mines in the western states to midwestern markets. However, the areas in which the mines are located are relatively arid, and the availability of water would be a major problem. A proposed long-term solution is to replace water by methanol made from coal near the mine. The methanol would then be used as an additional fuel where the pipeline delivers the coal (see **Alcohol Fuels**).

COALCON HYDROCARBON-IZATION PROCESS

A Coalcon Company (Union Carbide Corporation and Chemico) process for **coal liquefaction** by simultaneous **pyrolysis** (or carbonization) and hydrogenation. **Methane** and other light **hydrocarbon** fuel gases are formed at the same time. The process can be used with a variety of coals, but caking coals may have to be pretreated by heating in air at 750°F (400°C).

Crushed coal, pretreated if necessary, is fed to a **fluidized-bed** reactor where it undergoes pyrolysis at a temperature of roughly 1000°F (540°C) in the presence of hydrogen gas at a pressure of about 40 atm (4 MPa). The gases and vapors leaving the reaction vessel are cooled to yield separate liquid and gaseous products. The **char** remaining is reacted with steam and oxygen gas in a separate vessel to obtain the hydrogen required for the hydrocarbonization process (see **Hydrogen Production**).

The liquid (oil) product, like all coal liquefaction products, contains a large proportion of aromatic hydrocarbons. It may be treated as a **synthetic crude oil** (syncrude) and refined to produce light and heavy **fuel oils**. The gaseous product of the process is a **high-Btu fuel gas** which can serve as a **substitute natural gas**.

COED (CHAR–OIL–ENERGY DEVELOPMENT) PROCESS

An FMC Corporation process for **coal liquefaction** using **pyrolysis** and hydrogenation; in addition to oil, fuel gases are formed and a **char** remains. The oil is a mixture of **hydrocarbons** which can be treated as a **synthetic crude oil** (syncrude), and the gaseous product is an **intermediate-Btu fuel gas**. If the sulfur content is sufficiently small, the char can be used as a solid boiler fuel. In any event, the char can be gasified with steam and air (see **COGAS Process**).

The crushed and dried coal is fed to the first stage of a series of **fluidized-bed** reactors. The temperature in each stage is just below that at which particles would start to agglomerate and defluidize the bed. Hence, any type of coal, caking or noncaking, can be used in the process. The number of stages and operating temperatures would vary with the caking properties of the coal. For a highly caking bituminous coal, four stages would be used with typical (approximate) temperatures of 650°F (340°C), 850°F (455°C), 1000°F (540°C), and 1600°F (870°C), respectively. The pressure throughout is slightly above atmospheric.

The first stage is heated by gases from an external burner, whereas in each of the next two stages heat is provided by hot gases from the following stage (Fig. 17).

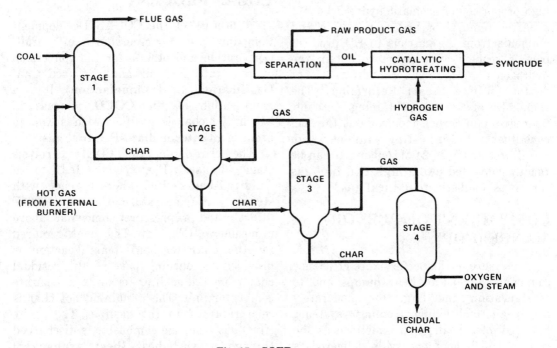

Fig. 17   COED process.

Stages 2 and 3, which are shown separately for clarity, may be combined, one above the other, in a single vessel. In each stage a fraction of the volatile matter is released by pyrolysis, and the remaining char proceeds to the next higher temperature stage. Thus, the gases (and vapors) released and the residual char move in opposite directions. The hot gas for stage 3, which contains a higher proportion of hydrogen, is generated in the last stage by reaction of part of the carbon in the char with steam and oxygen (see **Hydrogen Production**).

The off-gas from the first stage is not a fuel gas; it is discharged after passage through a waste-heat (**heat exchanger**) steam boiler to utilize its heat content. The gases and vapors from the second stage are cooled to form liquid (oil) and gaseous products. The oil, separated from gas and suspended solid particles, is treated with hydrogen gas at a temperature of 750°F (400°C) and a pressure of about 200 atm (20 MPa) in a **fixed-bed** catalytic reactor. In this **hydrotreating** process, sulfur, oxygen, and nitrogen impurities are removed and the composition of the oil is changed; the product is a syncrude with a moderately high proportion of aromatic hydrocarbons.

After removal of hydrogen sulfide and ammonia from the gaseous COED product, part is reacted with steam to obtain the hydrogen gas required for hydrotreatment of the oil (see **Steam Reforming**). The remainder of the gas, containing methane, hydrogen, and carbon monoxide as the fuel constituents, has a **heating value** of 550 to 650 Btu/cu ft (20 to 24 MJ/cu m). It can be readily converted into a **high-Btu fuel gas** for use as a **substitute natural gas**.

COEFFICIENT OF PERFOR-MANCE (COP)

A measure of the performance efficiency commonly applied to **heat pumps** and to **refrigeration** and large (or moderately large) air-conditioning (cooling) systems. The COP of a heat pump is defined as the rate at which heat energy is delivered to (in the heating mode) or removed from (in the cooling mode) the surroundings divided by the rate of energy input; that is,

$$COP = \frac{\text{Rate of heat delivery or removal}}{\text{Rate of energy input}}$$

where the energy input rate is the electric **power** required to operate the compressor and the blower (or blowers) in the air circulation system. Both the rate of heat delivery (or removal) and the energy input rate are commonly given in **watts**, so that the COP is a pure ratio (i.e., it has no units). For refrigeration and air-conditioning systems, the COP is defined in the same manner, except that it refers only to heat removal.

For a properly designed system, the COP should be greater than unity; that is, the machine should deliver (or remove) more heat energy than the electrical energy consumed in its operation. Commercial heat pumps and refrigeration systems generally have a COP in the range of 2 to 3.

See also **Energy Efficiency Ratio**.

COGAS PROCESS

A process of the COGAS Development Company, a partnership of FMC Corporation and others, for the conversion of coal into fuel oil and gas (see **Coal Gasification; Coal Liquefaction**). It is a modification of the **COED process** in which the char is gasified with steam to produce an **intermediate-Btu fuel gas**.

The char from the COED pyrolysis stage 3 (see Fig. 17) is fed to a **fluidized-bed** gasifier where it reacts with steam at a temperature of about 1600°F (870°C) and a pressure somewhat above atmospheric (Fig. 18). The heat required for the **char** (carbon)–steam reaction is provided by burning part of the residual char from the gasifier in air in a separate combustor unit. The remaining hot char is then returned to the gasifier. The waste (flue) gas from the combustor is discharged through a waste-heat (**heat exchanger**)

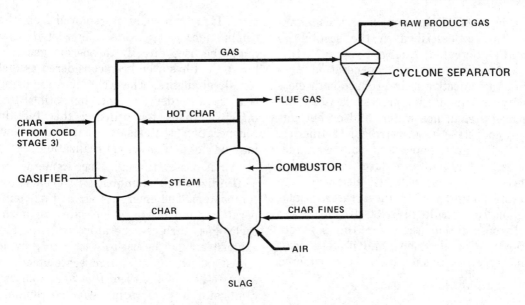

Fig. 18   COGAS process.

steam boiler, or it may serve as the fluidizing and heating gas for the first COED stage.

Because of the high temperature in the combustor [over 3000°F (1650°C)], the mineral ash remaining from combustion of the char forms a molten slag which flows to a quench tank. By conducting the gasification and combustion reactions in separate vessels, inert nitrogen from the air does not dilute the fuel gas produced.

Some of the gas from the gasifier is returned for fluidizing and heating the char in COED stage 3. The remainder is mixed with the **pyrolysis** gas from stage 2, and part is used to produce the hydrogen required for treating the COED oil (from stage 2). The rest is the gaseous product of the COGAS process. After sulfur removal (see **Desulfurization of Fuel Gases**), the gas typically contains about 58 volume percent of hydrogen, 31 percent of carbon monoxide, and 4 percent of methane (dry basis) as the fuel constituents. The **heating value** is approximately 350 Btu/cu ft (13 MJ/cu m). Application of the **water-gas shift reaction** followed by **methanation** yields a **high-Btu fuel gas** (or **substitute natural gas**).

COGENERATION

A procedure for generating electric power and useful heat in a single installation; the useful heat may be in the form of steam, hot water, or hot air. In a **heat engine** (i.e., a device for converting heat into mechanical work), part of the heat taken up from a "source" (e.g., a burning fuel) at a higher temperature must be discharged to a "sink" at a lower temperature. In a cogeneration system, the mechanical work is converted into electrical energy in an **electric generator**, and the discharged heat, which would otherwise be dispersed to the environment, is utilized in an industrial process or in other ways. The net result is an overall increase in the efficiency of fuel utilization.

Three general types of cogeneration systems may be distinguished.

1. The major purpose of the system is to generate electricity (i.e., by a utility in a central power station) with heated water as a by-product. This type of cogeneration is referred to in the section on **Waste Heat Utilization**.

2. The system is designed to supply both the electrical and heat requirements

of a large building complex or a community. This is described in the section on **Total (Integrated) Energy System**.

3. In most existing cogeneration systems, the objective is to produce both electricity and industrial process heat in the form of steam, hot water, and/or hot air. Many industries (e.g., petroleum refineries, chemical plants, paper and pulp manufacturers, and others) use large amounts of process heat as well as electricity to operate pumps, etc. In these cases, cogeneration can result in a 30-percent lower fuel consumption than for the separate generation of electricity and process heat. This type of cogeneration is considered here.

Gas Turbines and Diesel Engines

If natural gas or a suitable petroleum product is available as fuel, it may be used either in a **gas turbine** or a **diesel engine** to drive an electric generator. These **prime movers** have a fairly high **thermal efficiency** (i.e., proportion of heat supplied that is converted into useful work or electrical energy) because they take advantage of the high temperatures of the combustion gases that are the heat source. Instead of being discharged to the atmosphere, the hot exhaust gases, at temperatures above 600°F (315°C), pass to a waste-heat (**heat exchanger**) boiler. Water under pressure, flowing through finned tubes surrounded by the hot gases, is converted into steam at the desired temperature (and pressure).

Process steam is usually in the temperature range of 300 to 430°F (150 to 220°C). If the temperature of the steam from the waste-heat boiler is higher than needed, its temperature may be decreased, and useful energy recovered, by passage through a steam-turbine generator (see **Combined-Cycle Generation**). The exhaust steam from the turbine is used as process steam or to produce hot (near boiling) water.

To conserve natural gas and petroleum products, it would be desirable to substitute coal for these fuels in cogeneration sys-

tems. If fly ash could be removed from the combustion gases, finely powdered coal might be used directly to operate gas turbines. Coal has even been considered as fuel for diesel engines. The techniques for using coal require development, but other less direct methods for utilizing this fuel in cogeneration systems, as outlined below, may be closer to practical realization.

Air is passed through pipes immersed in a **fluidized-bed combustion** furnace (or combustor). The emerging air at high temperature (and pressure) operates an open-cycle gas turbine-generator system. The exhaust air can be used directly to provide industrial heat or it can generate steam in a waste-heat boiler. The flue gas from the combustor is used to preheat the combustion air.

Another method of cogeneration is to burn the coal in air in a combustor (fluidized bed or conventional) which serves as the heat source for a closed-cycle gas-turbine generator. The heat source and the working gas (e.g., air, carbon dioxide, or helium) are then independent. The hot flue gases from the combustor generate process steam in a waste-heat boiler.

A third possibility is to convert the coal into a **low-Btu fuel gas** and to use this as fuel for a gas turbine (open or closed cycle) or diesel engine. The exhaust combustion gases from the turbine or engine are then utilized in a waste-heat boiler to produce steam or hot water.

Steam Turbines

In using a **steam turbine** for cogeneration, the exhaust steam is not condensed, as it is in conventional steam-turbine generators (see **Electric Power Generation**), but is withdrawn at the desired pressure for process use. The turbines are thus of the noncondensing (or back-pressure) type in which the steam leaves the turbine at a significant pressure rather than at the very low pressure in a condensing turbine. Steam is discharged from a back-pressure turbine at a higher temperature than from a condensing turbine; consequently, the former has a lower thermal efficiency for electric power generation. However, this is not important because the discharged steam is required for an industrial process. Furthermore, elimination of condenser and cooling water requirements reduces the capital and operating costs of power generation.

COG (COAL–OIL–GAS) REFINERY

A concept for combining the **BI-GAS, H-Coal,** and **Solvent Refined Coal-I (SRC-I) processes** to produce clean (low-sulfur, low-ash) solid and liquid fuels and a **high-Btu fuel gas** from coal. Both caking and noncaking coals can be used without pretreatment.

About 25 percent of the coal, together with undissolved solid residue from the SRC-I process, is gasified by the BI-GAS process (Fig. 19). The ratio of hydrogen to carbon monoxide in the product gas is then increased by the **water-gas shift reaction**. After removal of **acid gases** (hydrogen sulfide and carbon dioxide), part of the product provides the hydrogen-rich gas for the SRC-I and H-Coal processes (see below). The remainder of the gas is subjected to **methanation** to yield a high-Btu gas (or **substitute natural gas**).

The other 75 percent of the coal is "solubilized" by the SRC-I process using the hydrogen-rich gas referred to above; the solid and liquid products are then separated. The residual solid (undissolved coal and ash) is mixed with coal for the BI-GAS process, as mentioned earlier. After recovery of solvent oil, part of the liquid is solidified to produce the solvent refined coal. The remaining heavy oil is broken down and hydrogenated by the H-Coal method to yield a light refinery liquid from which **gasoline** and light **fuel oils** may be obtained by distillation. The off-gases from the solvent recovery and hydrogenation operations may be converted into high-Btu fuel gases.

COKE

The porous, dark gray, coherent solid remaining after coal has been heated (carbonized or pyrolyzed) in the essential absence of air (see **Coal Pyrolysis**) at a temperature in the vicinity of 1800°F (about 1000°C). It consists mainly of carbon (87 to 92 percent), together with ash (5 to 10 percent), originally present as mineral matter in the coal, and about 1 or 2 percent of residual volatile matter. Roughly, one-fifth of the coal consumed in the United States has been utilized in the production of coke for metallurgical industries; this represents the major use of coal after electric power generation. About 90 percent of the coke is consumed in blast furnaces for producing pig iron; it serves partly as a

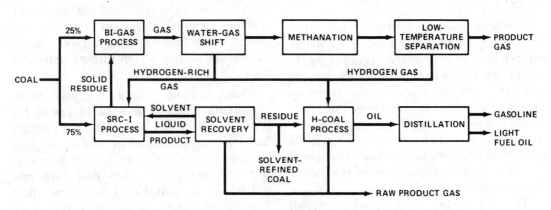

Fig. 19   COG refinery.

fuel (by combustion of the carbon in air) and partly to reduce the iron oxide in the ore to metallic iron.

The presence of sulfur in coke has an adverse effect on the properties of the iron and steel products; furthermore, some of the sulfur is converted into **sulfur oxides** which can cause atmospheric pollution. Hence, much of the low-sulfur coal mined in the eastern United States is used for coke production. However, the **Clean-Coke process** may make possible the manufacture of metallurgical coke from coals containing 2 to 3 percent of sulfur.

Bituminous coals generally tend to soften and agglomerate (or cake) when heated (see **Coal**) and so are used in the production of coherent coke. The carbonization of noncaking coals, such as lignite, subbituminous coal, and anthracite, leads to a noncoherent residue, commonly referred to as **char**. Coke is generally made from a blend of two or more different varieties of coal. The high-volatile bituminous coals have good caking properties, but the coke produced lacks the strength required to support the mass of iron ore, limestone, and coke in a blast furnace. The inclusion of roughly one-third of a low-volatile bituminous coal (or anthracite) with the high-volatile coal, increases the strength of the resulting coke.

See also **Petroleum Coke**.

COMBINED-CYCLE GENERATION

The production of electricity using two or more **heat engines** in tandem as **prime movers** to operate **electric generators**. In a heat engine (i.e., a device for converting heat into work) part of the heat taken up from a "source" (e.g., a burning fuel) at a higher temperature is discharged to a heat "sink" at a lower temperature. In a combined-cycle system, the heat discharged from one heat engine serves as the source for the next engine.

The net result is a greater overall operating temperature range (i.e., between the initial heat source and the final sink) than is possible with a single heat engine. The **thermal efficiency** of the combined system (i.e., proportion of heat converted into useful work or electrical energy) is thus greater than for the two heat engines operating independently. Less fuel is then required to generate a given amount of electrical energy.

As a general rule, combined-cycle systems have two heat-engine stages, the second stage being a conventional, condensing **steam turbine**. The highest inlet steam temperature for a turbine is limited by the properties of materials to about 1000°F (540°C). However, the flame temperatures in a boiler burning **fossil fuel** may be more than 3000°F (1650°C). Consequently, in a combined-cycle system, the steam turbine is preceded by a **topping cycle** heat engine which can utilize heat at higher temperatures. The working fluid leaves the topping cycle at a sufficiently high temperature to generate steam for the steam turbine.

Because the technology is well developed, the **gas turbine** is most commonly used as the topping cycle engine. The turbine exhaust gases, at a temperature of 1110°F (600°C) or more, pass to a waste-heat (**heat exchanger**) boiler where steam is produced from water under pressure flowing through finned tubes surrounded by the hot gases. If it is desirable, the steam temperature may be increased in a subsidiary heater before it enters the steam turbine. In addition to using fuel more efficiently, a combined gas and steam turbine generating system requires less condenser cooling water for a given electrical output than a steam-turbine generator alone.

Gas turbines generally use liquid **petroleum** products or **natural gas** as the fuel, but coal would be a preferred primary fuel for combined-cycle generation because of its availability. One possibility is to convert the coal into a **low-Btu fuel gas**, which is a suitable gas-turbine fuel. Another approach is to burn coal with compressed air in a pressurized,

fluidized-bed combustion furnace. After removing particulate matter, the hot combustion gases are mixed with compressed air that has been heated by passage through pipes immersed in the fluidized bed. The mixture of hot pressurized gases then drives the gas turbines.

Other topping cycles are being studied as alternatives to the gas turbine in combined-cycle systems. The most important are **magnetohydrodynamic (MHD) conversion** and the **potassium-vapor cycle**. In both cases, combustion of coal would provide the heat for conversion into electrical energy.

At least 50 percent of the heat energy supplied to a steam turbine is discharged at a fairly low temperature, averaging about 105°F (40°C), in the condenser cooling water (see **Electric Power Generation**). A possible way of converting part of this heat into useful work (e.g., electrical energy) is by means of a **bottoming cycle** in which the heat discharged by the steam turbine is used in a **vapor turbine** with a working fluid of low boiling point. However, the economics of this form of cogeneration is uncertain.

COMBUSTION ENGINEERING PROCESS

A Combustion Engineering, Inc. process for **coal gasification** with air only; the product is a **low-Btu fuel gas**. Superheated steam and, if desired, process steam at a lower temperature are by-products. Any type of coal, caking or noncaking, may be used.

The process is conducted in a tall cylindrical gasifier with combustor and reductor regions, where coal gasification occurs, at the bottom (Fig. 20). Pulverized coal (with some **char** fines separated from the product gas) is fed to the combustor region where it burns in air at atmospheric pressure to form carbon monoxide and dioxide. The temperature in the combustion zone is 3000°F (1650°C) or more. Additional pulverized coal is introduced into the reductor region just above the combustor. The coal, entrained in the hot combustion gases, is first devolatilized (i.e., volatile matter is driven off); the char then reduces the carbon dioxide to the fuel gas carbon monoxide. The coal ash melts at the high temperature of the combustor and is

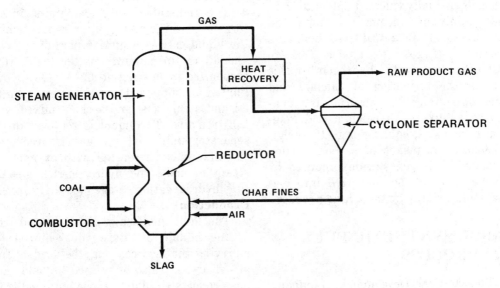

Fig. 20   Combustion Engineering process.

drawn off as slag from the bottom of the gasifier.

The gas leaving the reductor region at a temperature of 1700 to 1800°F (930 to 980°C) enters the long upper section of the gasifier which is lined with water tubes (see **Steam Generation**). Steam is generated and is subsequently superheated by the hot gas leaving the gasifier. After further heat recovery, the gas passes through a dryer, a **cyclone separator** (to remove char fines), and a scrubber. The char from the separator is returned to the combustor region of the gasifier.

The gas leaving the scrubber is treated for sulfur removal (see **Desulfurization of Fuel Gases**) and then constitutes the product of the process. Typically, it might contain 25 volume percent of carbon monoxide and 10 percent of hydrogen (from the coal) as the fuel components; nearly all the remainder is inert nitrogen (from the air). The **heating value** of the gas is about 120 Btu/cu ft (4.4 MJ/cu m).

CONOCO DESULFURIZATION PROCESS

A Conoco high-temperature process for the **desulfurization of fuel gases** by chemical absorption of hydrogen sulfide. The absorber is partially calcined dolomite, consisting mainly of calcium carbonate and magnesium oxide in a **fluidized bed** at a temperature of 1500 to 1700°F (815 to 925°C). Reaction with hydrogen sulfide results in the formation of calcium and magnesium sulfides. The absorber is regenerated as a mixture of calcium and magnesium carbonates, equivalent to the original dolomite, by means of carbon dioxide and steam. Hydrogen sulfide produced at the same time can be converted into elemental (solid) sulfur by the **Claus process**.

CONSOL SYNTHETIC FUEL (CSF) PROCESS

A Conoco Coal Development Company (formerly Consolidated Coal Company) pro-

cess for **coal liquefaction**. The process is conducted in two stages: (1) the soluble part of the coal is extracted with a hydrogen-donor (i.e., hydrogen-rich) solvent, and (2) the extract is hydrogenated with a hydrogen-rich gas. The solid residue is carbonized by heating, and the resulting **char** is used to produce the hydrogen-rich gas. The products of the process are light, middle, and heavy **fuel oils** and a **high-Btu fuel gas** (or **substitute natural gas**). Any type of coal, caking or noncaking, can be used.

A preheated slurry of the crushed coal in the solvent oil, which is made in the process, is fed to the extractor at about 750°F (400°C) temperature and 10 atm (1 MPa) pressure (Fig. 21). Part of the coal forms a solution in the oil, and this is separated from the undissolved solid. The liquid passes to a **distillation** unit from which light and middle distillate fractions are collected. The residual heavy oil is hydrogenated at a temperature near 850°F (455°C) and a pressure up to 200 atm (20 MPa) by means of a hydrogen-rich gas in the presence of a catalyst. The hydrogenated oil is distilled to yield light, medium, and heavy fuel oil products. The middle distillate is used as the medium for slurrying the feed coal.

The undissolved solid, remaining in the form of a thick slurry after separation of the liquid extract, contains part of the coal and all of the mineral matter (ash). It is carbonized by heating at 925°F (495°C) in a limited air supply; the products are a liquid oil and a fuel gas, and a char (mixed with ash) remains. The liquid is recycled to the separator unit, and the gas is recovered; the char may be reacted with oxygen and steam to make the hydrogen-rich gas for the hydrogenation stage (see **Hydrogen Production**).

The light and heavy distillate and part of the middle distillate not required for slurrying the feed coal may be used as fuel oils. They may also be refined to yield various products similar to those obtained from **petroleum refining**. The off-gases from

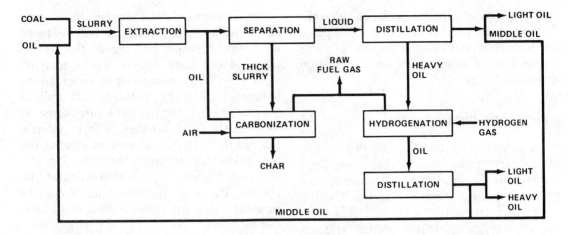

Fig. 21   CONSOL synthetic fuel process.

the distillation, hydrogenation, and carbonization reactions consist mainly of **methane** and other light paraffin **hydrocarbons**. After removal of hydrogen sulfide (see **Desulfurization of Fuel Gases**), the product is a clean, high-Btu fuel gas.

COOLING LAKE (OR POND)

A natural or man-made body of water used to transfer heat from warmed water to the atmosphere; the water is thus cooled and can be reused in a heat-removal process. The main use for cooling lakes is to cool the warmed water leaving the **steam-turbine** condenser of an **electric power generation** plant. The warmed water is discharged to the lake at one end, usually near the surface, and is withdrawn at the other end of a flow path. Cooling lakes often have internal dikes (or guides) to direct the water flow and extend the path length. A continuous system of parallel canals can serve the same purpose as a cooling lake. Several days are usually required for the water to circulate from one end to the other of a lake or canal system.

The warm water loses heat to the atmosphere mainly by vaporization (see **Cooling Tower**). The rate of cooling thus depends, among other factors, on the relative humidity of the air (i.e., the amount of water vapor in the air relative to the amount the air would hold if it were saturated at the existing temperature). In a region of high humidity, when the air approaches saturation with water vapor, a large lake would be required to compensate for the slow cooling rate. Conversely, where the air is usually dry, a smaller lake would be adequate. For a large (1000 electrical megawatts) steam-electric plant (see **Watt**), the lake area might range from 1000 to 3000 acres (400 to 1200 hectares) according to the type of steam plant (e.g., fossil fuel or nuclear) and the atmospheric conditions. (One hectare is 10,000 square meters.)

The cooling efficiency of a cooling lake (or canal system) can be enhanced by breaking up the water into small droplets by pumping it through nozzles to form sprays. The increase in the water surface exposed to the air results in a more rapid loss of heat by evaporation as well as in other ways. Effective cooling can thus be achieved with a lake of smaller area than when sprays are not used.

Due to the continuous evaporation of the lake (or canal) water, chemical additives and minerals normally present in the water tend to become more concentrated. To prevent deposition of scale on the condenser surfaces, part of the water must be discharged as *blowdown* and replaced with fresh water. As a general rule, an adjacent flowing-water body is required to disperse the blowdown and to supply makeup water to the lake to compensate for evaporation and other losses.

COOLING TOWER

A structure for transferring heat to the atmosphere from water that has been warmed in a heat-removal process; the water is thus cooled for reuse. Cooling towers are frequently used in closed-cycle (or circulating) condenser cooling systems for **steam turbines** in **electric power generation** plants. In these systems, the warmed water leaving the condenser passes through a cooling tower where heat is removed; the cooled water is then recirculated to the condenser. (See also **Cooling Lake.**)

There are three general types of cooling towers: wet, dry, and wet/dry. In *wet towers*, also called *evaporative towers*, the air and the water to be cooled are in contact. Cooling is caused mainly by evaporation of the water and partly by direct heat transfer. (When a liquid turns into vapor it absorbs the latent heat of vaporization from the surroundings and cooling occurs.) In *dry* (or *nonevaporative*) *towers*, the water or steam to be cooled and the air are not in direct contact and cooling results entirely by the transfer of heat across a separating surface. A *wet/dry tower* has a wet section and a dry section which can be combined in various ways.

Wet Towers

In wet towers, the warmed water flows over a packing (or fill) consisting of long, narrow slats of wood, plastic, or asbestos-cement supported in a rectangular or circular structure. The water is broken up into droplets and film offering a large exposed surface to air passing through the fill. Wet towers differ from one another primarily in the way in which the airflow (or draft) is produced.

In *mechanical-draft towers*, fans may either push (forced draft) or pull (induced draft) air through and across the wetted fill. Induced-draft towers are commonly used when large quantities of water are to be cooled. A single induced-draft unit is generally a rectangular structure, some 40 to 60 ft (12 to 18 m) high; a large electric fan, up to 28 ft (8.5 m) in diameter, at the top draws air through the fill (Fig. 22). The cooled water collects in a basin at the bottom. Up to 20 individual units may be combined into one tower with a single collecting basin.

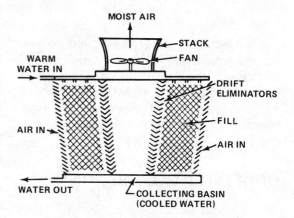

Fig. 22   Mechanical-draft cooling tower.

In circular mechanical-draft towers, as many as 20 fans are arranged within an interior circle. These fans draw air by induced draft through a surrounding outer ring containing the fill. Towers of the circular type are said to require less space than a combination of rectangular towers with the same cooling capacity.

In *natural-draft cooling towers* the air is drawn through the fill by the natural draft produced by a tall, reinforced-concrete structure which acts like a chimney. Natural-draft towers have an approximately hyperbolic profile and so are sometimes known as hyperbolic towers (Fig. 23). To generate sufficient draft, the towers are 300 to 540 ft (90 to 165 m) high; the circular base, containing the fill, may be 250 to 300 ft (75 to 90 m) in diameter. The open

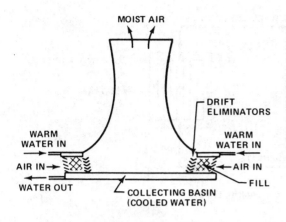

Fig. 23   Natural-draft cooling tower.

top of the tower has about two-thirds the diameter of the base. Natural-draft cooling towers are more expensive to build than forced-draft towers, but since they have no fans, the operating and maintenance costs are lower and they operate quietly.

Hybrid towers, also called fan-assisted, natural-draft towers, combine the favorable features of mechanical and natural-draft towers. A hybrid tower is similar in shape to a natural-draft tower but is about half as high and two-thirds the diameter. Fans arranged vertically around the base containing the fill or horizontal fans at the tower top supplement the natural draft by forced draft or induced draft, respectively.

Water is lost from a wet cooling tower, mainly by evaporation and to a small extent by *drift* (i.e., water droplets entrained by the air leaving the tower). These losses are replaced continuously by fresh water. Some replacement is also required to compensate for *blowdown*; this is water that is discharged to prevent the accumulation of chemical additives and minerals that might deposit as scale on the condenser surfaces. For a wet tower used to cool steam-turbine condenser water, the total makeup is about 2.5 to 3 percent of the water passing through the tower in a given time. For a large (1000 electrical megawatts) steam-electric plant (see **Watt**), the replacement rate is roughly 11,500 gal (45,000 liters) of water per minute for a

fossil-fuel plant and some 50 percent more for a nuclear power plant. About one-fifth of the total is accounted for by blowdown and most of the remainder by evaporation.

Dry Towers

The simplest type of dry cooling tower is a water-to-air **heat exchanger**, similar in principle to an automobile radiator. The water to be cooled is pumped through finned pipes or other channels with large surfaces, and air is drawn past them by mechanical or natural draft. Heat is transferred from the warm water through the pipe walls to the moving air and is dissipated. Heat-exchanger dry towers have been used to cool industrial process water, but they do not appear practical for cooling condenser water from steam-electric plants.

The *air-cooled condenser* dry towers have been preferred for use with steam turbines. They operate by using ambient air as the coolant to condense the exhaust steam, either directly or indirectly. In the *direct* condensing cycle, the steam leaving the turbine is passed through a system of finned condenser pipes (or other channels) over which air is drawn. The air removes heat from the steam so that it is condensed, and the condensate is returned as feedwater to the steam generator.

The most efficient dry tower is based on the Heller system with an *indirect* condensing cycle. The steam leaving the turbine is condensed by bringing it into contact with jets or sprays of cool water from previously condensed steam. Part of the resulting warmed water is returned as feedwater to the steam generator while the remainder is cooled by passage through a tower containing finned pipes over which air is drawn. The cooled water leaving the tower is recirculated to the condenser sprays (Fig. 24).

The great advantage of dry condensing systems is that they do not require any cooling water. The location of an electric power plant is thus independent of the availability of makeup water and a means for disposing of blowdown. On the other hand, a dry tower is more expensive than

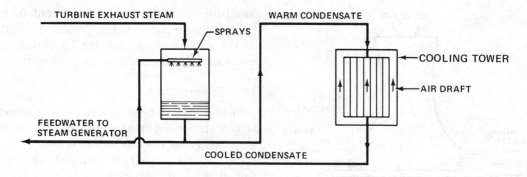

Fig. 24   Heller dry cooling tower.

an equivalent wet tower; furthermore, the temperature of the condensate is higher with a dry tower under the same atmospheric conditions. The efficiency of a turbine for converting heat into mechanical work (or electricity) is thus less with a dry cooling tower than a wet tower (see **Heat Engine**).

Wet/Dry Towers

Various combinations of wet and dry systems have been proposed for cooling the water leaving a steam-turbine condenser. A typical design consists of a lower evaporative wet tower and an upper heat-exchanger dry tower. A fan at the top draws air in simultaneously through both sections. The relative airflows through the wet and dry sections can be adjusted as required.

The warmed condenser discharge water is first air-cooled by passage through finned pipes in the upper (dry) section, and it then flows down over the fill in the lower (wet) section for further cooling. In cold weather, most of the air is drawn through the upper section to take advantage of dry cooling; there is then a decrease in evaporation loss and also in the need for blowdown. However, when the air is warm, wet cooling by evaporation is more effective.

COPPER OXIDE DESULFUR-IZATION PROCESSES

Dry processes for the **desulfurization of stack gases** by oxidizing the sulfur diox-
ide (SO_2) and absorbing the resulting sulfur trioxide (SO_3). Methods developed by the Royal Dutch/Shell Group (licensed to Universal Oil Products), Exxon Corporation (with Babcock and Wilcox), and the U. S. Bureau of Mines are apparently similar in using copper oxide (CuO) on an alumina base as the chemical absorber. The stack (flue) gases are passed with air over the copper oxide at a temperature of 650 to 750°F (345 to 400°C); the sulfur dioxide in the gas combines with oxygen from the air and copper oxide to form solid copper sulfate ($CuSO_4$). Treatment of the spent absorber with hydrogen gas at the existing temperature regenerates the copper oxide absorber. The gas produced at the same time is rich in sulfur dioxide which may be recovered for industrial use or converted into elemental (solid) sulfur or sulfuric acid (see **Desulfurization: Waste Products**).

COSTEAM PROCESS

A Pittsburgh Energy Technology Center (U. S. Department of Energy) process for **coal liquefaction** by direct hydrogenation. It is designed to utilize reactive, low-rank coals with a high moisture content, such as lignite (see **Coal**). The coal is reacted with a mixture of hydrogen and carbon monoxide (**synthesis gas**) or carbon monoxide alone. In the latter case, hydrogen is generated by the **water-gas shift reaction** between the carbon monoxide and steam from the large amount of water in the coal. The products are low-sulfur liquid and gaseous fuels. A special feature of the COSTEAM process is that, unlike other direct hydrogenation processes, it does not require a catalyst.

The crushed coal is formed into a slurry with part of the product oil and pumped into a stirred reactor with synthesis gas or carbon monoxide. The gas is made in a separate generator from either fresh or unreacted residual coal (see below). Hydrogenation of the coal and other reactions occur at a temperature of about 800°F (430°C) and a pressure of 270 atm (27 MPa).

All the products from the reactor pass to a receiver where the gas is separated from the oil and residual solids consisting of unreacted coal and ash. The gas contains most of the sulfur in the coal as hydrogen sulfide which can be removed (see **Desulfurization of Fuel Gases**); the clean gas can serve as a fuel. After separating the solids, the product oil, low in sulfur and ash, provides a clean boiler fuel as well as the coal slurrying medium.

CRACKING

Processes for breaking down heavier (i.e., high molecular weight, low volatility, viscous) **hydrocarbons** obtained in **petroleum refining** or from **synthetic crude oil** into lighter (i.e., lower molecular weight, more volatile, less viscous) hydrocarbons. Cracking makes it possible to increase the yield of **gasoline, kerosine,** and **diesel fuel** from crude oils. The three main types of cracking processes are (1) thermal cracking using heat only, (2) catalytic cracking with a catalyst, and (3) hydrocracking with hydrogen gas and a catalyst.

Thermal Cracking

Cracking by heat alone at about 1000°F (540°C) was used extensively at one time in refinery operations for the treatment of petroleum products distilling above the gasoline range. This operation, conducted under a high pressure, has now been largely replaced by catalytic cracking, but two modifications of thermal cracking, namely, visbreaking and coking, are still used to some extent to crack **heavy crude oils** and the **residual oil** remaining after the distillation of petroleum.

Visbreaking is a relatively mild form of thermal cracking; the operating temperature is about 900°F (480°C) and the pressure is 5 to 25 atm (0.5 to 2.5 MPa). The name "visbreaking" arises from the breakdown of the complex hydrocarbons in the viscous residual fuel oil into a mixture of simpler hydrocarbons with a lower viscosity.

Coking, on the other hand, is an extreme form of thermal cracking. Several coking techniques have been developed; in general, the temperatures are higher and the pressures lower than for visbreaking. The main products from the coking of heavy and residual oils are a gasoline fraction, various distillates, which can be subjected to further treatment (e.g., catalytic cracking), and **petroleum coke**.

Catalytic Cracking

The temperatures in catalytic cracking are similar to those in conventional thermal cracking [about 1000°F (540°C)] but the pressures are lower (slightly above atmospheric). A mixture of silica (mainly) and alumina was the common catalyst at one time, but in recent years natural and artificial zeolites (aluminosilicates of sodium or calcium) have come into general use. Most catalytic cracking is conducted in **fluidized-bed** systems with a fine powder catalyst. A wide variety of feed materials boiling above the gasoline range can be treated by catalytic cracking.

The gasoline distilled from the cracked product has substantial proportions of isoparaffin and aromatic hydrocarbons; as a result it has a higher **octane number** than gasoline from thermal cracking. Catalytic cracking also produces the olefin hydrocarbons propylene (C_3H_6) and butylenes (C_4H_8) which are used in the **alkylation** process in petroleum refining.

The catalyst used in catalytic cracking gradually becomes coated with a layer of carbon which impairs its activity. However, the catalyst can be regenerated by burning off the carbon in air continuously in a separate vessel. The heat produced in this way is utilized to maintain the high temperature required for the cracking reactions.

Hydrocracking

In hydrocracking, the feedstock is passed with hydrogen gas through a catalyst bed at a somewhat lower temperature than in catalytic cracking [650 to 800°F (340 to 425°C)] but at a much higher pressure 100 to 150 atm (10 to 15 MPa). The hydrocracking catalyst consists of a cracking catalyst (silica-alumina or a zeolite) that provides support for a hydrogenation catalyst (nickel, tungsten, palladium, or platinum). Unlike the catalyst for cracking alone, the hydrocracking catalyst retains its activity for long periods and requires regeneration at extended intervals only.

Hydrocracking has been applied to a wide variety of feedstocks, from light distillates (**naphthas**) to heavy oils. The products include hydrocarbon gases, especially isobutane which is required for the alkylation process, through gasoline (relatively high in isoparaffins and aromatics), to middle distillate fuel oils. The gasoline has good antiknock properties, but the octane number is not as high as from catalytic cracking.

The hydrocracking catalyst is readily poisoned (i.e., rendered inactive) by nitrogen compounds in the feedstock. In petroleum refineries that treat crude oils high in nitrogen, hydrocracking is carried out in two stages; the temperatures and pressures are much the same as given above in both stages, but the catalysts are different. The main purpose of the first stage is to convert the nitrogen in the feedstock to ammonia (NH_3) gas which is readily absorbed in an acidic solution. Sulfur is also removed by conversion into hydrogen sulfide (H_2S). The catalyst in the first stage is a combination of a regular cracking catalyst and a metal sulfide. The product from this stage is then subjected to the hydrocracking treatment described above.

See also **Steam Cracking**.

CYCLONE SEPARATOR

A device used mainly for removing particulate matter suspended in a gas. It can also be used to separate heavier from lighter particles suspended in a liquid.

In the cyclone separator the dust-laden gas is injected tangentially into the top of a cylindrical chamber with a conical lower section. A vortex is formed in which the gas swirls around the interior of the chamber. Because of their inertia, the suspended particles are unable to follow the motion of the gas; consequently, they are thrown by centrifugal force against the walls of the chamber from which they fall into the conical collector at the bottom. The cleaned gas leaves through a central exit at the top of the chamber. Cyclone separators are more effective when small; hence, many small-diameter units are commonly combined in a single system.

Even at best, cyclone separators are not very efficient for small particles, but they are inexpensive. Consequently, they are often used to supplement another method for removing fly ash from the stack gases of a coal-burning steam plant (see **Steam Generation**). A practical combination is a cyclone separator followed by an **electrostatic precipitator**.

D

DENSITY

The mass (or weight) of a unit volume of any substance at a specified temperature. The numerical value of the density at a given temperature depends on the units in which the mass and volume are expressed. In the common ("English") system, the mass is in pounds (or ounces) and the unit volume is usually 1 cubic foot (or 1 cubic inch). The density of pure water at 60°F is then 62.4 lb/cu ft. In the U. S. metric system the mass is in grams and the unit volume is taken as 1 milliliter or 1 ml (i.e., one-thousandth part of 1 liter) for liquids and solids and 1 liter for gases. The density of water is then slightly less than 1.00 gram/ml at 15.6°C (60°F). In the International (SI) System of units, density is expressed in kg/cu m; the density of water in these units is about 1000 kg/cu m.

See also **Specific Gravity**.

DESULFURIZATION

Removal of sulfur from a fuel (coal, oil, or gas), either before use or from the combustion products (flue or stack gases) after use, in order to limit the amount of **sulfur oxides** discharged to the environment. To meet the U. S. Environmental Protection Agency's air quality standards, high-sulfur coals (or their combustion products) must be treated in an appropriate manner; the possible treatments fall into the following categories.

1. Remove the sulfur from the fuel before use.

2. Remove the sulfur, mainly as sulfur dioxide, from the stack (or flue) gases before discharge to the atmosphere.

3. For coal, in particular, treat with hydrogen to convert the sulfur into hydrogen sulfide which is then removed; the form of the coal is changed in the process, but its fuel properties are retained or improved.

1. Cleaning of coal removes part of the inorganic (or mineral) sulfur (see **Coal Cleaning**), but some remains. Processes are under development for removing most of the inorganic sulfur from coal, and others may also reduce the organic sulfur content (see **Desulfurization of Coal**). Even removal of the inorganic sulfur alone would be useful for certain eastern U. S. coals in which most of the sulfur is in this form. In any case, partial desulfurization of coal before burning would be advantageous in easing the stack gas desulfurization problem (see below).

The sulfur in liquid petroleum fuels (or in crude oil) is generally removed by reaction with hydrogen gas in the presence of a catalyst at a moderately high temperature and pressure (see **Desulfurization of Liquid Fuels**). The sulfur is converted into hydrogen sulfide which can be absorbed in various established ways (see **Desulfurization of Fuel Gases**). In both natural and synthetic fuel gases the sulfur is present as hydrogen sulfide and is removed directly (see **Natural Gas**).

2. The removal of sulfur dioxide from stack gases is the most practical current approach to the burning of high-sulfur

71

fuels. A few methods are in actual use and others are under development (see **Desulfurization of Stack Gases**). Partial desulfurization of coal (by removal of the inorganic sulfur) followed by treatment of the stack gases seems to be economically promising.

3. A number of processes are being studied (or used) for converting coal into clean (i.e., low-sulfur) solid, liquid, or gaseous fuels (see **Coal Gasification; Coal Liquefaction**). In each case, part of the sulfur remains in the ash (as sulfate) and is discarded, and much of the rest is converted into hydrogen sulfide; this can be absorbed as mentioned earlier. The final product is then a clean fuel. An advantage of this approach is that the volume of gas to be treated is much less than in stack gas desulfurization.

DESULFURIZATION OF COAL

Decrease in the sulfur content of coal to a level below that attainable by conventional **coal cleaning**. Cleaning procedures can remove a substantial proportion of the inorganic (mineral) sulfur, but the remainder, both inorganic and organic, is sufficient to prevent many eastern and midwestern coals from meeting existing air quality standards for sulfur dioxide emissions in stack (flue) gases. In some cases, removal of the inorganic sulfur alone from the sulfur remaining in the coal after cleaning would be adequate to meet the standards, and methods under development, as described below, should be capable of doing this. Removal of organic sulfur from coal, which is more difficult, is also being studied (see **Coal: Sulfur Content**).

Processes for treating coal to remove much of the sulfur remaining after conventional mechanical or physical cleaning are of three general types: (1) chemical treatment, (2) froth flotation, and (3) magnetic separation. These processes leave the coal essentially unchanged in form. Another approach, not considered here, is to convert the high-sulfur coal into a low-sulfur fuel of a different form (see **Desulfurization**).

Chemical Treatment

In the Meyers (TRW Systems and Energy) process the crushed coal is leached with an aqueous solution of ferric sulfate at a temperature in the range of 195 to 265°F (90 to 130°C). It is claimed that 90 to 95 percent of the pyritic (inorganic) sulfur can be extracted without changing the physical form of the coal. The organically bound sulfur is unaffected. The pyrite (FeS_2) is converted into elemental sulfur, which must be separated, and the ferric sulfate [$Fe_2(SO_4)_3$] is reduced to ferrous sulfate ($FeSO_4$). Treatment of the latter with air or oxygen gas results in the regeneration of the ferric sulfate.

Another chemical process, proposed by the Kennecott Copper Corporation, involves leaching a water slurry of powdered coal with oxygen gas at temperatures in the range of 210 to 265°F (100 to 130°C) at pressures up to about 20 atm (2.0 MPa). Under these conditions, the pyrite is oxidized to ferrous and ferric sulfates which pass into solution. The residual coal is filtered and washed. The leach solution and water washings are treated with lime or limestone; the precipitate of calcium sulfate and hydrated iron oxides formed is separated from the water and sent to a disposal area. A somewhat similar process of the Pittsburgh Energy Technology Center (U. S. Department of Energy) is based on the treatment of a coal–water slurry with air (rather than oxygen gas) under pressure at about 390°F (200°C).

The most practical way to remove organically bound sulfur from coal appears to be by oxidation of the sulfur compounds to a gaseous or soluble product. The processes just described, in which oxygen gas or air at a moderately high temperature and pressure are used primarily to remove inorganic sulfur, also remove some of the organic sulfur by oxidation. Other more effective oxidizing agents are being sought to serve this purpose; they include chlorine gas and nitrogen dioxide.

In the Battelle (Columbus Laboratories) hydrothermal process, the pulverized coal is slurried with an aqueous alkaline solution, either sodium or calcium hydroxide. The slurry is then heated to 435 to 660°F (225 to 350°C) at a pressure of 24 to 170 atm (2.4 to 17 MPa) to extract the sulfur as sodium or calcium sulfide. In tests with several eastern coals, from 90 to 99 percent of the pyritic sulfur was removed; in some cases there was also a substantial decrease in the organic sulfur. The alkaline leach solution can be regenerated with steam and the sulfur removed as hydrogen sulfide (see **Desulfurization of Fuel Gases**).

Froth Flotation

The Pittsburgh Energy Technology Center has reported on a two-stage froth flotation process for removing pyrite. The first stage is similar to that used in ordinary coal cleaning, but the coal floated by the froth still contains some pyrite. This is removed in the second stage in which the roles of the coal and pyrite are reversed; the pyrite particles are floated, whereas the coal sinks to the bottom. To do this, a "depressant," consisting of a hydrophilic (i.e., water attracting) colloid, is added to the water thereby reducing the tendency of air bubbles to become attached to the coal. At the same time, the addition of potassium amyl xanthate, a pyrite "collector," facilitates attachment of air bubbles to the pyrite in an acidic solution. Air is then bubbled through the coal-pyrite suspension in the presence of a frothing agent. The pyrite is floated and removed; the coal, essentially free from pyrite, falls to the bottom of the tank.

Magnetic Separation

Pyrite is weakly magnetic, but it is commonly associated with the more strongly magnetic mineral pyrrhotite, also an iron sulfide. Coal itself is nonmagnetic. Hence, it should be possible to remove pyrrhotite and associated pyrite by passing the finely crushed coal through a strong magnetic field. In one proposed method, a high-gradient magnetic field is used to separate iron sulfides from a slurry of pulverized coal in water.

DESULFURIZATION OF FUEL GASES

Removal of sulfur compounds from fuel gases. These gases may contain hydrogen sulfide and organic sulfides (mercaptans) which have unpleasant odors and are corrosive. Moreover, the **sulfur oxides** formed during the combustion of the fuel gas are atmospheric pollutants. Many different procedures have been proposed for removing sulfides from common fuel gases, including **natural gas, liquefied petroleum gas,** etc.; these procedures are also applicable to **coal gasification** products. The methods fall into two general categories: physical absorption and chemical absorption. In the physical methods, hydrogen sulfide dissolves in a liquid absorber and there is no chemical interaction. In the chemical methods, of which three types may be distinguished, a chemical reaction occurs between the hydrogen sulfide and the absorber. The three categories are: weakly alkaline absorbers, oxidation methods, and high-temperature methods.

Physical Absorption Methods

The physical methods depend on the fact that the solubility of a gas in a liquid is greater at higher than at lower pressure. The hydrogen sulfide is extracted from the fuel gas by an organic liquid solvent under a pressure of up to 70 atm (7 MPa). When the pressure is reduced, most of the dissolved hydrogen sulfide is released in gaseous form; the hydrogen sulfide remaining in solution can be removed by passing nitrogen gas through the solution or by heating it. The residual solvent, essentially free from sulfur, is recycled to the absorber (Fig. 25).

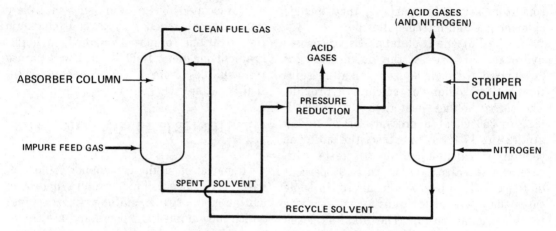

Fig. 25   Desulfurization of fuel gas by physical absorption.

The physical absorption methods listed in the table are described in separate sections under the indicated titles. In the **Rectisol process** advantage is taken of the greater solubility of hydrogen sulfide at lower temperatures. In all the procedures the sulfur is recovered as hydrogen sulfide gas; this may be converted into elemental (solid) sulfur by the **Claus process**.

Physical Absorption Methods for Desulfurization of Fuel Gases

Process	Absorber solvent
Fluor Solvent	Propylene carbonate
Purisol	N-methyl pyrrolidone
Rectisol	Methanol [below 0°F (-18°C)]
Selexol	Polyethylene glycol dimethyl ether
Sulfinol	Sulfolane (tetrahydrothiophene dioxide) mixed with an alkanolamine and water

Chemical Absorption Methods

Weakly Alkaline Absorbers. The most common chemical absorption methods depend on the use of an aqueous solution of a weakly alkaline material which combines with (and removes) the acidic hydrogen sulfide from the fuel gas. The absorbers generally operate at temperatures in the range of 100 to 200°F (38 to 93°C) and at any con-

venient pressure. The solution containing the sulfur is pumped from the absorber column to a regenerator (or stripper) column where the hydrogen sulfide is removed by heating in a steam-heated "reboiler." The stripped solution is then returned to the absorber column, by way of a heat exchanger if necessary (Fig. 26).

The chemical absorption processes given in the table are described in separate sections. In all cases the sulfur is recovered as hydrogen sulfide gas.

Desulfurization of Fuel Gases with Weak Alkaline Solutions

Process	Absorber in aqueous solution
Adip (Shell)	Alkanolamine [diisopropanolamine (DIPA)]
Alkazid M	Potassium methylamine propionate
Alkazid DIK	Potassium dimethylamine acetate
Benfield Catacarb	Potassium carbonate and "additives"
Fluor Econamine	Alkanolamine [diglycolamine (DGA)]
MDEA	Alkanolamine [methyldiethanolamine (MDEA)]
SNPA-DEA and DEA	Alkanolamine [diethanolamine (DEA)]
Sulfiban	Alkanolamine [monoethanolamine (MEA)]
Vacuum Carbonate	Sodium carbonate

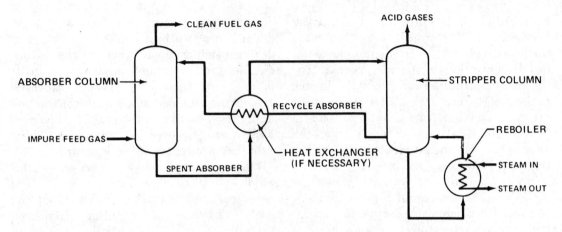

Fig. 26 Desulfurization of fuel gas by chemical absorption.

Oxidation Methods. In these methods the hydrogen sulfide is absorbed from the fuel gas in a weakly alkaline (sodium carbonate) aqueous solution and is then oxidized to form elemental (solid) sulfur; the latter is removed by filtration or froth flotation. The oxidizing material is regenerated in a separate vessel by oxidation of the reduced form with air. Three aqueous oxidation processes are outlined in the table and are described more fully in separate sections.

Oxidation Methods for Desulfurization of Fuel Gases

Process	Oxidizer
Giammarco Vetrocoke	Sodium arsenate
Stretford	Sodium vanadate
Takahax	Sodium 1,4-naphthoquinone -2-sulfonate

High-Temperature Methods. Before a fuel gas can be desulfurized by any of the methods referred to above, it must be cooled to about 100 to 200°F (38 to 93°C). Coal gasification processes are often designed for on-site use of the resulting gas as a fuel; a high-temperature hydrogen sulfide removal process might then be advantageous. Three such processes, which are described under the indicated titles, are listed in the table.

High-Temperature Methods for Desulfurization of Fuel Gases

Process	Absorber
Battelle Molten Salt	Calcium carbonate (in a carbonate melt)
Conoco	Calcium carbonate and magnesium oxide
Iron Oxide	Iron oxide (oxidizer)

DESULFURIZATION OF LIQUID FUELS

Removal of sulfur, usually present as various organic (carbon) compounds, from liquid fuels (e.g., raw **gasoline, kerosine,** and **distillates**) obtained in **petroleum refining**. During combustion of the fuel, the sulfur is converted into noxious **sulfur oxides**. Desulfurization may then be required to permit discharge of the combustion products to the atmosphere without violating air quality standards. Many sulfur compounds are also objectionable because they have an unpleasant odor and are corrosive to metals.

The most common procedure for removing sulfur from liquid fuels is by hydrodesulfurization, sometimes called "hydrofining." In fact, almost any hydrogenation treatment has the same result (see **Hydrotreating**). The liquid is reacted with hydrogen gas in the presence of a catalyst; hydrogen sulfide formed from the sulfur compound can be converted into elemental (solid) sulfur by the **Claus process**. The reaction temperature is generally in the range of 550 to 850°F (290 to 450°C) and the gas pressure is from about 8 to 100 atm (0.8 to 10 MPa). As a rough rule, the heavier the fuel being treated, the higher the temperature and pressure. The common catalyst is initially cobalt molybdate ($CoO-MoO_3$) supported on a predominantly alumina base; however, during the operation, the actual catalyst is apparently a mixture of cobalt and molybdenum sulfides. In most cases, catalytic hydrodesulfurization removes 80 to 95 percent of the sulfur, but it is possible to remove up to 99 percent if necessary.

In principle, crude oil could be desulfurized before refining, but vanadium, nickel, and other elements present in the oil would soon render the catalyst ineffective. A more practical approach is to treat the separate distillation products. The lighter distillates are generally low in sulfur, as well as in the impurity elements vanadium, nickel, etc., and are easy to desulfurize. Only the much smaller quantities of heavy oils and distillation residues are difficult to treat.

See also **Sweetening Processes**.

DESULFURIZATION OF STACK (FLUE) GASES

Reduction in the amount of **sulfur oxides**, mainly sulfur dioxide (SO_2) with a few percent of sulfur trioxide (SO_3), in the combustion (or flue) gases from a coal- or oil-burning plant prior to discharge to the atmosphere through a stack. Many processes have been proposed for achieving this objective. As a general rule, these processes can remove from 80 to 90 percent of the sulfur oxides from a combustion gas containing 0.2 to 0.3 percent of these oxides. This is usually sufficient to permit emission standards to be met with coal containing up to 3 percent sulfur.

Stack gas treatment processes may be divided into two broad categories: wet and dry, depending upon whether the sulfur dioxide (and trioxide) absorber is in a liquid or dry solid form. The wet processes are further divided into "nonregenerative" and "regenerative" (also called "regenerable" or "recovery") types. All dry processes are regenerative in nature.

In nonregenerative processes, the absorber (lime or limestone) is not recovered and must be replaced continuously. Absorption of the sulfur oxides converts it into a sludge consisting mainly of solid calcium sulfite, calcium sulfate, and water. It has no commercial value and must be disposed of without adverse environmental effects. Since they are simple, at least in principle, the commercial development of nonregenerative processes is ahead of that of other sulfur oxide removal techniques.

In regenerative processes, wet or dry, the absorber is recovered for reuse. In the regeneration stage the absorbed sulfur is sometimes converted into sulfur dioxide which has commercial value. Alternatively, the sulfur dioxide can be converted into elemental (solid) sulfur or sulfuric acid, both of which are useful products (see **Desulfurization: Waste Products**).

Wet Processes

In wet processes, the sulfur oxides are removed from the stack gases by scrubbing with an aqueous solution or slurry. In order to avoid vaporization of the water and associated problems, the gas must be cooled before it enters the scrubber. Several different types of scrubbers have been designed for achieving intimate contact between the gas and the scrubbing (absorber) liquid. These include spray towers, in which the gas flows through sprays of liquid, and towers containing small solid spheres (or other shapes) over which the liquid flows while exposed to the stack gases. Both countercurrent systems, with gas and liquid flowing in opposite directions, and cocurrent systems, with both flows in the same direction, have been proposed. The preferred scrubber design depends on the nature of the absorber and the details of the absorption process.

Small droplets of liquid suspended in the gas are removed by a demisting section in the upper part of the scrubber. Finally, after one or more scrubbing stages, the cleaned gas is reheated to restore its buoyancy before discharge through a stack.

Although liquid–gas scrubbing is simple in principle, several problems arise in practice. In the first place, deposition of scale, especially with a slurry absorber, decreases the efficiency of the scrubber. There is also a tendency for the demister to be blocked by solid particles. In addition, the flowing solution or slurry causes corrosion and erosion of the equipment.

Some wet processes for desulfurization of stack gases, which are described separately under the titles given in the left column, are listed in the accompanying table.

Wet Processes for Desulfurization of Stack Gases

Process	Absorber
Nonregenerative	
Lime/Limestone	Lime or limestone slurry
Regenerative	
Bureau of Mines Citrate	Citrate solution
Chemico	Magnesia slurry
Chiyoda Two-Stage	Dilute sulfuric acid
Double (or Dual) Alkali	Alkali or sulfite solution
Molten Salt	Molten carbonates
Stone and Webster/Ionics	Sodium hydroxide solution
TVA Ammonia	Ammonia solution
Wellman-Lord SO_2 Recovery	Sulfite solution

Dry Processes

The equipment for dry desulfurization scrubbing of stack gases is generally simpler than for wet scrubbing. However, reaction of sulfur oxides with a dry absorber is much slower than with a solution or even a slurry. To overcome this drawback, dry scrubbers must be large in order to expose a large surface area of solid absorber to the stack gases. The processes included in the table are described separately under the titles in the left column.

Dry Processes for Desulfurization of Stack Gases

Process	Absorber
Carbon Adsorption	Activated carbon
Cat-Ox	Oxidation
Copper Oxide	Oxidation
Limestone Injection	Limestone

See also **Fluidized-Bed Combustion.**

DESULFURIZATION: WASTE PRODUCTS

The sulfur-containing products obtained in the treatment of coal, fuel gases (natural and synthetic), liquid fuels, and stack (flue) gases to reduce the sulfur content (see **Desulfurization**). Depending on the process used, the waste may be calcium sulfite–sulfate sludge, for which there is at present little or no use, liquid sulfur dioxide, solid elemental sulfur, or sulfuric acid, for which there are potential markets. The quantity of waste produced from a specified amount of primary fuel (e.g., coal or crude oil) depends on the sulfur content of the fuel and the emission standards which must be met, as established by the U. S. Environmental Protection Agency.

A large proportion of eastern and midwestern bituminous coal contains from 1 to 3 percent of sulfur; for illustrative purposes, an average of 2 percent will be assumed. Suppose the coal is burned in a steam-electric power station with a rating of 1000 megawatts (MW), where 1 MW is equal to 1000 kilowatts (see **Watt**), operating at 70 percent **capacity**; the stack gases are treated for sulfur removal to the extent required to meet the accepted emission standards. The approximate quantities of four alternative waste products generated annually would then be as given in the table. (Fuel with lower or higher sulfur content than the 2 percent assumed here would produce smaller or larger amounts, respectively.)

Annual Desulfurization Products from 1000-MW Electric Power Plant Using 2 Percent Sulfur Coal (70 Percent Capacity)

Product (alternatives)	Short tons (thousands)	Kilograms (millions)
Calcium sulfite-sulfate sludge		
Lime absorber	400	360
Limestone absorber	500	450
Sulfur dioxide (100%)	70	63
Sulfur (elemental)	35	32
Sulfuric acid (95%)	110	100

High-sulfur coal can be converted into a low-sulfur gaseous, liquid, or solid fuel (see **Coal Gasification; Coal Liquefaction**) before it is burned. The sulfur is then removed as hydrogen sulfide which is generally converted into elemental sulfur by the **Claus process.** The amount of sulfur produced would be the same as that given in the table, assuming the same quality of the stack gas emission.

The simplest way to dispose of calcium sulfite-sulfate sludge is in ponds. An impervious clay lining may be required to minimize pollution of surface and ground waters by undesirable substances (e.g., magnesium salts). For the 1000-MW power plant described above, the annual sludge production would occupy roughly 200 or 250 acre-feet, respectively, depending on whether the absorber is lime or limestone. It appears that with a substantial number of such coal-burning plants (or their equivalent in smaller plants) in operation, pond disposal is not a long-term solution.

An alternative is to partially dewater the sludge and mix it with fly ash residue from coal to form a landfill material. Another possibility is to produce a pozzolanic (cement-like) material by adding a small proportion of lime to a mixture of calcium sulfite–sulfate sludge and fly ash. The product is said to have many potential uses in the construction of reservoirs, roads, dams, etc. The presence of magnesium in the sludge would not be deleterious; hence, less pure and cheaper lime or limestone could be used as the sulfur dioxide absorber.

Although sulfur dioxide, sulfur, and sulfuric acid have many industrial uses, the amounts recovered from the desulfurization of stack and fuel gases and liquid fuels would greatly exceed the demand. This is already the case in western Canada where hydrogen sulfide is removed from "sour" **natural gas** and converted into elemental sulfur.

Liquid sulfur dioxide and sulfuric acid are both difficult to store or transport. Solid sulfur, however, could be stockpiled in the hope that new uses can be found for this substance. Some possibilities being investigated are a road-paving material by mixing sulfur with asphalt and sand; a concrete substitute made by adding sand or aggregate to molten sulfur and allowing the mixture to cool; and coating water-soluble fertilizer material (e.g., urea) with sulfur to provide slow release in the ground.

The problem of by-product disposal would be greatly alleviated by efficient mechanical cleaning of coal which removes a substantial proportion of the inorganic (pyrite) sulfur (see **Coal Cleaning**). A possible decrease of 40 to 50 percent in the sulfur content of the coal would correspondingly decrease the amount of by-product. However, the pyrite and other mineral matter removed from the coal would still need to be disposed of, perhaps in mined-out areas.

DIESEL CYCLE

A repeated succession of operations (or cycle) representing the idealized behavior of the working fluid in the **diesel engine** form of **heat engine**. The Diesel cycle is illustrated and described in Fig. 27. In the description each stage is assumed to have been completed before the next stage is initiated. However, in an actual engine there is a gradual rather than a sharp transition from one stage to the next; hence, the sharp points in the figure would actually be rounded off.

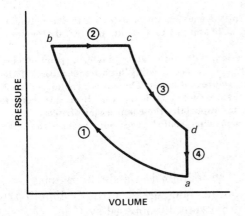

Fig. 27 Diesel cycle.

1. **Adiabatic compression of the working fluid (gas) along** *ab*; **the temperature and pressure are increased.**

2. **Heat addition along** *bc* **at constant pressure; the gas temperature and volume are increased.**

3. **Adiabatic expansion along** *cd*; **work is done by the expanding gas and the temperature and pressure decrease.**

4. **Heat removal (rejection) along** *da* **at constant volume; the pressure and temperature decrease, and the gas is restored to its initial condition.**

In a diesel engine (see Fig. 28), air is drawn into a cylinder where it is compressed adiabatically (see **Carnot Cycle**) by the inward motion of a piston and thereby heated (stage 1). Just prior to maximum compression, fuel is injected and it burns rapidly in the very hot compressed air; heat is thus added to the working fluid at essentially constant pressure (stage 2). The hot combustion gases expand adiabati-

cally and in doing so push back the piston and mechanical work is done (stage 3). The cycle is not completed in a diesel engine (i.e., stage 4 is absent) because the gases are exhausted to the surroundings at *d*. Fresh air is drawn in from the atmosphere to start the next cycle at *a*.

The net work done in a Diesel cycle is the difference between the work done by the working fluid in stages 2 and 3 and the work done on the fluid in stage 1. The **thermal efficiency** (i.e., the fraction of the heat supplied in stage 2 that is converted into net mechanical work) is increased by increasing the temperature at *c* and by decreasing that at *d*. An equivalent statement is that an increase in the compression ratio (volume at *a* divided by volume at *b*) and a decrease in the cutoff ratio (volume at *c* divided by volume at *b*) increase the thermal efficiency. The minimum value of the cutoff ratio is unity.

DIESEL ENGINE

An **internal-combustion engine** in which the fuel is ignited by injecting it into air that has been heated to a high temperature by rapid compression; hence, diesel engines are also called compression-ignition engines. The concept of ignition compression was patented by Rudolf Diesel in 1892 and first demonstrated in an engine some five years later. The compression-ignition engine is a **heat engine** (i.e., one that converts heat partially into mechanical work) operating on an approximation to the idealized **Diesel cycle** in which combustion of the fuel, that is, the heat addition stage, occurs at essentially constant volume.

The diesel engine can use a wide variety of fuels, ranging from **natural** (or other fuel) **gas** to fairly heavy petroleum **distillate** oils which are cheaper than **gasoline**. High-speed diesel engines use lighter fuels than do those operating at lower speeds (see **Diesel Fuel**). The heavier fuels require longer times to be injected and to vaporize prior to combustion and hence are more suited to low-

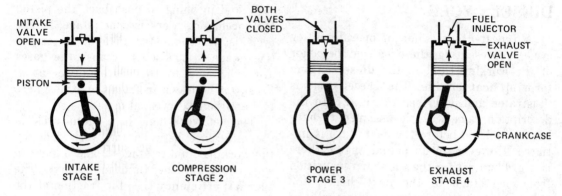

Fig. 28   Four-stroke diesel engine.

1. *Intake.* The piston moving downward (i.e., out of the cylinder) draws air into the cylinder by way of the open intake valve. The exhaust valve is closed.

2. *Compression.* The intake valve is closed, and the piston moving upward (i.e., into the cylinder) compresses the air. The pressure is increased to about 35 to 40 atm (3.5 to 4 MPa), and the air temperature rises to 840 to 930°F (450 to 500°C).

3. *Power.* Just before the point of maximum compression, with both valves closed, a spray of very small droplets of fuel is injected into the top of the cylinder. At the existing high temperature of the air, the fuel burns rapidly and produces extremely hot compressed gases. The gases expand and push back the piston; this is the power stroke in which mechanical work is done. Not all of this work is available, however, since part is utilized in the other strokes, especially the compression stroke.

4. *Exhaust.* The piston moving upward pushes the somewhat cooled gases out through the open exhaust valve.

speed engines. Dual-fuel diesel engines can burn fuels ranging from oil alone to a mixture of gas and only 5% fuel oil; a certain amount of oil is desirable to provide stable ignition and smooth combustion.

Because they have to withstand high internal pressures, diesel engines are generally strong and heavy. Moreover, they can be built in large sizes for high-power operation. Consequently, diesel engines have been used extensively for such heavy-duty purposes as railroad locomotives, ship propulsion, road-building and farm machinery, electric generators, freight trucks, and buses. However, since diesel engines can make efficient use of fuels that are cheaper than gasoline, they are being utilized increasingly in automobiles.

Like the conventional gasoline engine, the diesel engine is a reciprocating engine; a piston moves to and fro (or in and out) in a cylinder within which combustion takes place. The reciprocating motion is converted into rotary motion by connecting the piston to a crankshaft. Several cylinders may be combined in a single engine, just as in automobiles or aircraft. In-line, V-type and horizontally opposed systems (i.e., with equal numbers of horizontal cylinders on each side of the crankshaft) are the most common. Because they can be built in large sizes to develop considerable power (more than 200 horsepower per cylinder), diesel engines often have only one cylinder, thereby simplifying construction. A **flywheel** on the crankshaft of a diesel engine smooths out irregularities in the rotation arising from intermittent power strokes.

Four-Stroke and Two-Stroke Engines

Diesel engines, like **spark-ignition** (or gasoline) **engines**, can operate on four-stroke or two-stroke cycles. (A stroke is an in or an out motion of the piston.) In the four-stroke cycle there are two in and two out motions (i.e., two rotations of the crankshaft) per cycle. However, only one of these four strokes is a power stroke; hence, there is only one power stroke for every two rotations of the crankshaft. In the two-stroke engine, on the other hand, there are one in and one out motion (i.e., one rotation of the crankshaft) per cycle. Consequently, there is one power stroke in each rotation of the crankshaft.

In starting a large diesel engine, the crankshaft is rotated by means of compressed air; for smaller engines a battery-operated electric motor (self-starter) is used. An electrically heated igniter serves to initiate fuel combustion and power generation when starting with a cold engine.

Four-Stroke Cycle. The cylinder of a four-stroke diesel engine has intake and exhaust ports (openings) that are opened and closed by valves to admit air and remove burned gases, as required. The valves are operated at the appropriate times by cams on a camshaft that is geared to the engine crankshaft. The principles of a four-stroke diesel engine are shown and described in Fig. 28.

The power production per cycle in a four-stroke system of a specific size can be increased by supercharging, that is, by raising the pressure of the air introduced into the cylinder in the intake stroke (see **Supercharger**). The mass of air is then greater than at ordinary atmospheric pressure. If the amount of fuel injected is increased in proportion, there is a corresponding increase in the power generated per cycle. The pressure of the intake air is generally not more than 1 atm (0.1 MPa) above atmospheric, but this may be sufficient to produce more than 50 percent increase in power.

The increased air pressure for supercharging can be obtained from a blower (or pump) driven directly from the engine or crankshaft or from a turbine-driven compressor (or turbocharger) that uses the hot diesel exhaust as the working gas.

Two-Stroke Cycle. The two-stroke diesel engine can be designed without valves and with only two ports in the cylinder wall; the ports are opened and closed when they are uncovered and covered, respectively, by the moving piston. The two-stroke diesel is thus even simpler than the two-stroke, spark-ignition engine which requires three ports. However, some increase in the intake air pressure is desirable in a two-stroke diesel engine for effective removal (or scavenging) of the burned gases during the exhaust stroke. The stages in the operation of the engine are given in Fig. 29.

The advantage of a two-stroke cycle in providing a power stroke for each revolution of the engine crankshaft, rather than one power stroke in two revolutions in a four-stroke cycle, is outweighed in a spark-ignition (gasoline) engine by the associated power losses. In two-stroke diesel engines, however, especially those operating at low and medium speeds, these losses are greatly decreased. There is no loss of fuel through the exhaust port because the fuel is not added until both ports are closed. Consequently, because of its design simplicity and increased power for a given engine speed, the two-stroke diesel engine is quite common, whereas the corresponding spark-ignition engine has found only limited use.

A minor disadvantage of the valveless diesel engine is that, since both intake and exhaust ports are open simultaneously, appreciable supercharging is not practical. Some supercharging can be achieved, however, in the uniflow design in which air is admitted through an intake port at the bottom of the cylinder and the burned gases are expelled through a timed valve at the top. The main objective of the uniflow system is to improve scavenging of the burned gases, but it also makes possible a degree of supercharging.

Fuel Injection and Speed Control

In the great majority of spark-ignition engines, the air intake and carburetor system provide an air-fuel mixture of essentially constant composition; the engine speed is then varied by adjusting the throt-

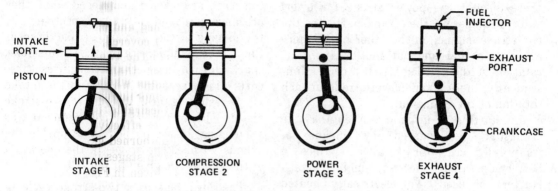

Fig. 29   Two-stroke diesel engine.

1. *Intake.* The piston is near the bottom of the down stroke (after stage 4), and both intake and exhaust ports are open. Air under a moderate pressure, roughly 0.3 atm (0.03 MPa) above atmospheric, enters through the intake port, sweeps out the burned gases through the exhaust port, and fills the cylinder.

2. *Compression.* The piston moves upward, closing both ports and compressing and heating the air to about the same pressure and temperature as in stage 2 of a four-stroke cycle (Fig. 28).

3. *Power.* Just before the top of the compression stroke, a fine spray of fuel is injected; the fuel burns in the hot air, and the resulting compressed high-temperature gases expand. The piston is forced downward in the cylinder, and mechanical work is done; not all of this work is available since part is used in the up (compression) stroke.

4. *Exhaust.* Toward the end of the power stroke, the intake and exhaust ports are uncovered by the piston moving downward. The exhaust gases are then expelled by the incoming, slightly pressurized air.

tle valve that controls the quantity of air (and fuel) drawn into the cylinder. In diesel engines, however, the air supply is not throttled and is constant in amount; the speed is varied by changing the quantity of fuel injected. This is possible because the temperature attained in the compressed air is so high that even very lean mixtures (i.e., mixtures containing much more air than is required theoretically for complete combustion of the fuel) can be burned satisfactorily.

A metered quantity of fuel is injected into the diesel cylinder as fine droplets obtained by forcing the liquid under high pressure through a spray nozzle with very small orifices. To achieve the maximum combustion efficiency, injection is timed to occur just before the top of the compression stroke (i.e., prior to maximum compression). Since the fuel is ignited almost immediately upon injection into the hot compressed air, the injection before maximum compression in a diesel engine is equivalent to ignition (or spark) advance in a spark-ignition engine.

Ignition Delay and Cetane Number

In practice, a short time, called the *ignition delay*, elapses between the start of fuel injection and ignition in a diesel engine. The ignition delay is usually not more than a few thousandths of a second (i.e., a few milliseconds), but a relatively long delay time may be accompanied by difficult starting from cold and rough and noisy operation. The property of a diesel fuel that affects ignition delay is expressed by the **cetane number**; an increase in the cetane number decreases the ignition delay, facilitates cold starting, and makes the engine run more smoothly.

The cetane number requirement of a diesel fuel depends on the characteristics of the engine. A short ignition delay is important in a high-speed engine; the fuel should then have a high cetane number. Large engines operating at low and medium speeds can use fuels with lower cetane numbers. Diesel fuels with the highest cetane numbers contain a large proportion of paraffin hydrocarbons; by contrast, these hydrocarbons have low **octane numbers** and cause knocking (i.e., explosive combustion) in spark-ignition engines. Thus noisy operation of a diesel engine is quite different from that in a gasoline engine.

Compression Ratio and Thermal Efficiency

At a given compression ratio (i.e., the ratio of the air volumes at the beginning and end of the compression stroke), the **thermal efficiency** (i.e., the proportion of the heat supplied by combustion of the fuel that is converted into useful work) of a Diesel cycle engine is expected to be less than for an **Otto cycle** (spark-ignition) engine. In both engine types, the efficiency (and power output) can be increased by increasing the compression ratio, but in the spark-ignition engine this ratio is limited by knocking.

In a diesel engine, where the fuel burns in the compressed air at high temperature, the combustion mechanism is different from that in a spark-ignition engine. Conventional knocking does not occur, and higher compression ratios are feasible. They are commonly in the range of from 15 to 20 or more, compared to a maximum of about 12 with high-octane gasoline in a spark-ignition engine. The increase in compression ratio results in some increase in the efficiency of the diesel engine.

Use in Automobiles

Diesel-powered automobiles, which have been made in Europe for many years, are attracting interest in the United States. The potential advantages are a high thermal efficiency, with at least 25 percent improvement in fuel economy, and the use of fuel that is cheaper than gasoline. Furthermore, because diesel engines operate with lean air–fuel mixtures (i.e.,

with an excess of air), combustion is generally more complete than in a gasoline engine. The hydrocarbons and carbon monoxide exhaust emissions, which accompany incomplete fuel combustion, might be low enough from a well-maintained diesel engine to meet existing standards without the use of catalytic converters.

On the other hand, the high combustion temperatures in a diesel engine result in higher levels of **nitrogen oxides** in the exhaust gases than in a gasoline engine. In addition, the exhaust contains undesirable aldehydes and particulate matter (soot), and it has an unpleasant odor.

Diesel engines have to withstand higher internal pressures than do gasoline engines; they are consequently stronger and heavier for a given power output. In the four-stroke diesel engine, as used in automobiles, the size (and weight) for a specified power may be decreased by turbocharging, as described earlier.

Other drawbacks of the diesel engine are difficult starting when cold, and often rough and noisy operation. However, further research and development should alleviate most of the problems associated with the use of diesel engines in automobiles, so that their potential advantages can be realized.

DIESEL FUEL

A mixture of liquid **hydrocarbons** used as fuel in **diesel** (compression ignition) **engines**. Diesel fuels are either various **distillates** obtained in **petroleum refining** operations or blends of such distillates with **residual oil**. The boiling range [390 to 680°F (200 to 360°C)] and **specific gravity** (0.82 to 0.92; 40 to 20°API) are higher than for gasoline; diesel fuels are also more viscous.

An important criterion of diesel fuel is the ignition quality as indicated by the **cetane number**. A high cetane number is desirable for easy starting and smooth operation. The cetane numbers of diesel fuels are usually in the range of 30 to 60.

Cetane improver additives (e.g., alkyl nitrates) are used to some extent.

The American Society for Testing and Materials (ASTM) distinguishes three main grades of diesel fuel as follows:

Grade No. 1-D. A relatively volatile (light) distillate, essentially identical with **kerosine**, for engine service requiring frequent speed and load changes (e.g., truck and tractor engines). This fuel has a minimum cetane number of 40 and is particularly desirable for cold weather operation.

Grade No. 2-D. A medium distillate oil of lower volatility and higher density and viscosity than No. 1-D for engines in industrial and heavy-mobile service (e.g., railroad engines) operating with high loads and fairly uniform speeds. The minimum cetane number is 40.

Grade No. 4-D. A blend of heavier distillates (i.e., higher boiling range, density, and viscosity than Grade No. 2-D) and residual oils. It is used in large engines, (e.g., for marine propulsion and electric power generation) operating at low and medium speeds with sustained loads. The minimum cetane number is 30.

Diesel engines are also being used to an increasing extent in private automobiles and buses. These engines require fuel with a high cetane number, possibly 60 or more.

DIRECT CONVERSION

The conversion of heat or other form of energy into electrical energy without intermediate conversion into mechanical work (i.e., without any moving components). The conventional **electric generator** is thus eliminated. The main approaches for converting heat directly into electricity are by **magnetohydrodynamic conversion, thermoelectric conversion,** and **thermionic conversion**. Solar energy can be converted into electrical energy by means of solar cells (see **Solar Energy: Photovoltaic Conversion**), and electricity can be generated directly from chemical energy in a battery (see **Fuel Cell**).

DISPOSABLE CATALYST HYDROGENATION

A process under study by the Pittsburgh Energy Technology Center (U. S. Department of Energy) and other organizations for **coal liquefaction** by direct hydrogenation using an inexpensive catalyst which does not have to be recovered and regenerated. The mineral matter (ash) in the coal is the basic catalyst which is activated by means of an iron compound. In addition to saving on the cost of the catalyst, it is claimed that less hydrogen is required than for other coal liquefaction processes.

Powdered coal is mixed with the catalyst and formed into a paste with oil produced in the process. The compressed and preheated paste is introduced into the reactor vessel with a hydrogen-rich gas and some recycled product gas under pressure. The gas and slurry from the hydrogenation stage are separated, and part of the gas is recycled to the reactor vessel. The remaining gas, consisting mainly of light **hydrocarbons**, is a fuel gas. After separating solids from the slurry, the oil is distilled to obtain a light liquid fuel and pasting oil. The residual heavy oil and the solids are processed to recover light and heavy **distillates** and pasting oil.

DISTILLATE

In general, the liquid obtained by condensation of the vapors formed in a **distillation** operation. In **petroleum refining**, the term distillate is usually applied to liquids boiling at temperatures higher than about 350°F (180°C), which is above most of the **gasoline** range. Petroleum distillates are commonly classified as light, medium (middle or intermediate), and heavy. The volatility decreases (i.e., the boiling range increases) and the density and viscosity (i.e., flow resistance) increase in this order. Light and medium distillates are used as **diesel fuel** and **jet fuel**; all three types can serve as **fuel (heating) oils**.

DISTILLATE FUEL OIL

A general term used in the petroleum refining industry for a range of distillates which include Nos. 1-D and 2-D **diesel fuels**, Nos. 1 and 2 heating oils, and No. 4 fuel oil [see **Fuel (Heating) Oil**].

DISTILLATION

A process in which a liquid is heated to produce a vapor; the vapor is then cooled and condensed to a liquid that is collected in a separate vessel. The condensed liquid is called the distillate. A simple form of distillation is used to separate a volatile liquid (i.e., one which boils and vaporizes at a moderate temperature) from a nonvolatile dissolved solid. An example is the production of potable (distilled) water from brackish or saline water.

The major industrial application of distillation is to separate the components of a mixture of liquids with different boiling points (or boiling ranges), such as crude petroleum. When the liquid mixture is heated, the more volatile (lower boiling point) components vaporize more readily and the first distillate contains a larger proportion of these components than did the original mixture. With continued heating, the temperature increases steadily and the less volatile components are distilled. A partial separation of the components of the mixture can thus be achieved by collecting successive distillate fractions separately.

Fractional Distillation

The procedure just described is a crude form of fractional distillation (also called fractionation or rectification). A more effective separation process in common use is fractional distillation with reflux. The vapor formed by heating the liquid mixture passes up a vertical fractionation column (or tower), down which flows part of the liquid that has condensed in the upper part of the column. The downflowing liquid is called the reflux. The vapor and liquid, moving in opposite directions in the column, are brought into intimate contact either by a suitable packing material (small ceramic rings or saddles), by a series of perforated horizontal plates, or in other ways. The repeated contact between vapor and liquid is equivalent to successive small-scale distillations; this improves the separation of the liquids with different volatilities.

Batch Distillation

In small industrial operations and in the laboratory, fractional distillation is conducted as a batch process. A quantity (or batch) of the liquid mixture to be separated is placed in the distilling vessel (or still) and heated. The vapor emerging from the top of the fractionation column is condensed and part is returned as reflux; the remainder of the liquid condensate (or distillate) is collected as product. The initial product contains the more volatile (lower boiling point) constituents of the mixture, but as distillation proceeds, successive fractions are collected in which components of decreasing volatility (increasing boiling point) predominate in turn. If greater purification is desirable, the separate fractions may be subjected to further distillation.

Continuous Distillation

Almost all large-scale industrial distillation, such as in **petroleum refining**, is carried out in a continuous manner. The mixture to be separated (e.g., crude oil) is heated in a furnace (or boiler), and the resulting vapor, often accompanied by some hot liquid, is fed continuously into the fractionation column about one-fourth of the way up from the bottom. Instead of collecting successive fractions in turn from the vapor at the top of the column, as in batch distillation, the separate fractions are removed continuously as liquids from different levels of the column (Fig. 30).

The temperature in the fractionation column is lowest at the top where the cool reflux liquid enters and it increases steadily to where the hot feed is intro-

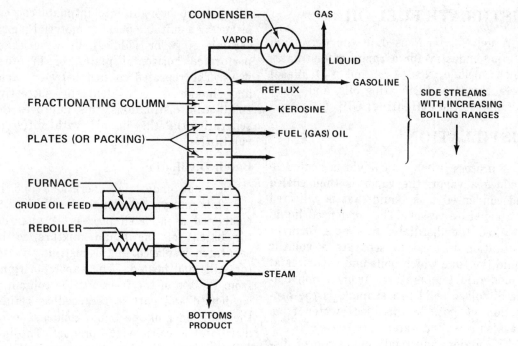

Fig. 30   Distillation column.

duced. The more volatile (lower boiling point) components of the feed condense where the temperature is lowest (i.e., near the top of the column) and the less volatile (higher boiling point) condense at successively lower levels. The various liquid fractions, boiling in different ranges, are drawn off as "side streams" at different levels.

The residue (i.e., the bottoms product or "bottoms") containing the least volatile (highest boiling point) components is also withdrawn continuously from the bottom of the still. If the original mixture contained gases which do not condense, as is often the case with petroleum, they are removed at the top of the column.

In the distillation of petroleum, the fractionation is improved by passing each side-stream condensate, other than the top one, through a "side stripper." This is a small fractionation column in which the more volatile components are vaporized with steam (see Steam Distillation, below). The purer liquid product is drawn off from the bottom of the stripper while the vapor is returned to the main column just above

the level of the particular side stream (Fig. 31).

Two additional features are common in petroleum distillation. One is to maintain a high temperature at the bottom of the column by continuously circulating the residual liquid through an external heater, called a "reboiler," and returning it at a level below that of the main feed entry. The second is to inject steam at the bottom of the still to enhance vaporization by steam distillation.

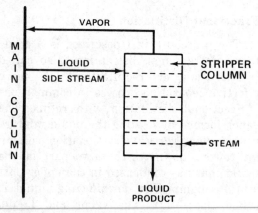

Fig. 31   Side-stripper in fractional distillation.

Steam Distillation

By passing steam through a liquid mixture that is being distilled, high boiling point components can be volatilized at temperatures well below their normal boiling points. Thus, relatively nonvolatile components may be distilled from a mixture at temperatures below those at which chemical decomposition would occur. Steam distillation is frequently used in the fractionation of petroleum, as mentioned above.

Vacuum Distillation

The boiling point of a liquid is reduced by lowering the pressure exerted on its surface. Hence, liquids of low volatility, which require high temperatures for vaporization at normal atmospheric pressure, can be readily vaporized at a lower temperature if the pressure is reduced. The process of distillation under reduced pressure is called vacuum distillation. It is, in a sense, an alternative to steam distillation and may be preferred when the liquid to be distilled has a very low volatility at normal atmospheric pressure. Vacuum distillation is commonly used to recover lubricating oils from petroleum.

Destructive Distillation

Destructive distillation is quite different from the distillation processes described above; it is actually a form of **pyrolysis** (i.e., decomposition by heat) in which the vapor products are condensed and collected. In the other types of distillation, there is little or no decomposition (or destruction) of the material being distilled.

DOUBLE- (OR DUAL-) ALKALI DESULFURIZATION PROCESSES

Several two-stage wet regenerative processes for the **desulfurization of stack gases**. The absorber of **sulfur oxides**, mainly sulfur dioxide (SO_2), is an aqueous solution of either an alkali (sodium hydroxide or ammonia) or a sulfite (sodium, ammonium, or magnesium sulfite). The addition of lime (CaO) or limestone ($CaCO_3$) to the spent absorber solution results in the regeneration of the absorber and the precipitation of solid calcium sulfite ($CaSO_3$) and sulfate ($CaSO_4$), which are disposed of as sludge (see **Desulfurization: Waste Products**).

If the absorber is a solution of sodium hydroxide (NaOH) or ammonia (NH_3), the interaction with sulfur dioxide in the stack (flue) gas leads to the formation of soluble sodium sulfite (Na_2SO_3) or ammonium sulfite [$(NH_4)_2SO_3$], respectively. Regeneration of the alkali absorber then requires the addition of lime rather than limestone. When the absorber is a sulfite solution, namely, sodium sulfite, ammonium sulfite, or magnesium sulfite ($MgSO_3$), the products are sodium bisulfite ($NaHSO_3$), ammonium bisulfite (NH_4HSO_3), or magnesium bisulfite [$Mg(HSO_3)_2$], respectively. Addition of limestone or lime (preferably) then causes precipitation of solid calcium sulfite (and some sulfate) and regeneration of the original sulfite absorber in solution.

The Thiosorbic process of the Dravo Lime Company, which uses magnesium sulfite solution as absorber and hydrated lime (i.e., lime and water) containing a small amount of magnesium hydroxide as regenerator, has been found to be effective for stack gas desulfurization.

The major advantage of the double-alkali systems is that absorption of sulfur dioxide by the solution is highly efficient and permits the use of simple scrubbers. Furthermore, there is no scaling in the scrubber as there is with lime or limestone slurries. On the other hand, the double-alkali process involves two stages: absorption, followed by the addition of lime (or limestone) for regeneration. There is also some loss of absorber due to formation of sodium (or ammonium) sulfate by oxidation of the sulfite in the absorber solution by the air. Accumulation of the sulfate is somewhat of a drawback, since discharge of the spent solution to the environment may not be permissible without treatment.

E

EBULLATED BED

A bed of small solid particles with streams of gas and liquid flowing upward through the bed; the name arises from the appearance which is like that of a boiling liquid. The flow velocities of gas and liquid are such as to maintain the particles in a state of random motion in the liquid medium. The level of the liquid in the container is above that of the suspended particles; hence, the latter are not carried away by the liquid leaving the container. A feature of the ebullated bed is that the solid particles occupy a larger volume than they do in a fixed (or stationary) bed. Consequently, the ebullated bed is sometimes called an *expanded bed*. An expansion of at least 10 percent is usually necessary to provide the benefit of enhanced contact between gas, liquid, and solid.

Ebullated beds are used in the petroleum refining industry to facilitate catalytic reactions between hydrogen gas and a liquid (e.g., in hydrocracking and **desulfurization of liquid fuels**). The solid particles then consist of the required catalyst. Ebullated beds have also been proposed for use in some **coal liquefaction** processes.

See also **Entrained Bed; Fixed Bed; Fluidized Bed.**

ELECTRIC GENERATOR

Any device for converting another form of energy into electrical energy. Special forms of electric generators, with no moving components, can convert heat or light (i.e., solar) energy into electricity (see **Direct Conversion**). However, in the great majority of electric generators, mechanical energy, in the form of rotational motion, is converted into electrical energy. The rotation is usually produced in a **turbine** driven by steam, water power, hot combustion gases, or wind. The combination of turbine and generator is often referred to as a *turbogenerator*. In some cases a **diesel engine** produces the rotational motion (see **Electric Power Generation**).

The conversion of mechanical energy of rotation into electricity is based on electromagnetic *induction*. An electric voltage (or electromotive force) is induced in a conducting loop (or coil) when there is a change in the number of magnetic field lines (or magnetic flux) passing through the loop. The change in flux may be brought about either by rotating the loop relative to a stationary magnetic field or by rotating the field relative to a stationary loop. If the loop is closed by connecting the ends through an external load, the induced voltage will cause an electric current to flow through the loop and load (e.g., a lamp, motor, etc.). Thus, rotational energy is converted into electrical energy.

Alternating-Current Generation

An *alternating current* is an electric current which changes continuously in such a way as to reverse its direction at regular intervals. (By contrast, a *direct current* always flows in the same direction.) A single sequence, in which the current changes from one direction to the other and returns to its initial state, is called a *cycle* (see Fig. 33). An alternating current consists of a succession of cycles, and the number of cycles (or reversals) per second is known as the *frequency*. Because there are advantages in its generation and transmission, electric utilities invariably generate alternating current. The common frequency in the United States is 60 cycles per second, usually written as 60 hertz (or 60 Hz).

The principle of the alternating current generator (or *alternator*) may be explained by considering a conducting (e.g., metal wire) loop (or coil) rotating in a stationary magnetic field. The axis of rotation is perpendicular to the direction of the field lines (also perpendicular to the plane of the paper). Suppose the loop is in the vertical position with the field lines passing through it in the horizontal direction, as represented by A in the section in Fig. 32.

As the loop is rotated clockwise, the *magnetic flux* (or field lines) through it decreases and becomes zero in position B, this decrease is accompanied by an increase in the loop voltage. In the subsequent rotation from B to C, the flux through the loop increases and the voltage drops to zero. Then, from C to D the flux decreases again and the voltage increases. But since the loop at C is reversed from its position in A, as indicated by the letters a and a', the direction of the induced voltage is reversed. Finally, from D to E, the flux increases and the voltage decreases until the original condition is restored (at E).

The variation in the loop voltage or of an electric current passing through a load across the loop during a single rotation is then as shown in Fig. 33; the letters A, B, C, D, E in this figure correspond to the loop positions in Fig. 32. The curve represents a single cycle of alternating current in which the current (or voltage) changes in direction and returns to its initial state. A flow of alternating current consists of a succession of such cycles, as already mentioned.

In the foregoing description the loop is assumed to rotate while the magnetic field is stationary. An exactly equivalent situation would arise if the loop were stationary and the field rotated. This may be envisioned from Fig. 32 by rotating the page so that the loop is always vertical. The magnetic flux changes through the loop, and the voltages induced in the loop are the same as when the loop is rotated.

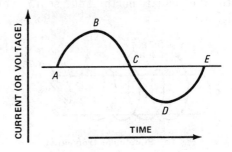

Fig. 33   **Alternating-current cycle.**

Generator Design. An electric generator consists essentially of an armature and a field structure. The *armature* carries the wire loop, coil, or other winding in which the voltage is induced, whereas the *field structure* (or *field winding*) produces the magnetic field. In small generators, the armature is usually the rotating component (or *rotor*); it is surrounded by the stationary field structure (or *stator*). In large generators, such as are used in commercial

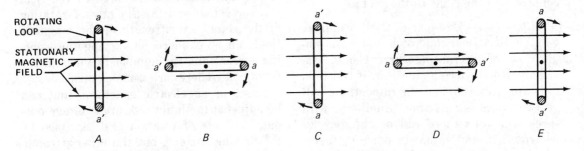

Fig. 32   **Principle of electric generator. (Loop, shown in section, is perpendicular to the plane of the page.)**

electric power plants, the situation is reversed. Because of the high voltages generated in the armature, it is more convenient to have the coils stationary; the field structure is then the rotor.

In some electric generators, such as magnetos used to a limited extent in **spark-ignition engines,** the field structure is a permanent magnet. As a general rule, however, the field structure is an *electromagnet*; this consists basically of a coil of several turns of insulated wire through which is passed a direct current. An iron core in the center of the coil serves to concentrate the lines of force. Such an arrangement is equivalent to a magnet with the usual north and south poles (Fig. 34).

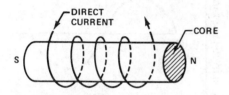

Fig. 34   Simple electromagnet.

The direct (or exciting) current for the electromagnet is provided by an *exciter* which is a small direct-current generator (see later) usually mounted on the same shaft as the main generator. In generators with rotating magnetic fields, as commonly used in power plants, the field is generated by a rotating field coil. The exciting current is then supplied by bringing out the ends of the coil to two slip rings which rotate with the coil. Electrical connection to the exciter is made through two stationary conducting "brushes" that remain in contact with the rings as they rotate.

Alternation Frequency. Each complete rotation of a field coil rotor induces a single cycle of alternating current in a stationary armature coil. However, if two field coils are spaced around the rotor, there will be two cycles per rotation. Similarly, with three field coils there will be three cycles per rotation, and so on. In general, therefore, the number of alternating current

cycles per rotation of the field coils is equal to the number of such coils.

Since each coil produces two magnetic (north and south) poles, the number of cycles per rotation is $N/2$, where N is the number of poles. If the field coils rotate R times per second, the number of cycles per second (i.e., the alternating current frequency) is $NR/2$ Hz. Consequently, a desired output frequency can be obtained by adjusting either the number of magnetic poles or the rate of rotation (or both). A generator of this type, in which the frequency is directly related to the speed of rotation of the rotor, is called a *synchronous generator.*

Steam turbines, such as are used in the majority of power plants, operate most efficiently at high speeds. The rotation rates are generally either 1800 or 3600 rpm (i.e., R is 30 or 60 rotations per second). The number of magnetic poles N required to generate 60-Hz current must then be either four or two, respectively. Thus two-pole and four-pole generators are in common use. In nuclear power plants, the steam turbines usually run at 1800 rpm and the generators are of the four-pole type. **Hydraulic turbines** in **hydroelectric power** plants, on the other hand, operate at slow speeds. A typical speed might be 120 rpm (i.e., 2 rotations/sec); 60 poles would then be required to produce 60-Hz alternating current.

Voltage. The maximum value of the alternating voltage generated in an armature coil depends on (1) the number of loops (or turns) in the coil, (2) the strength of the magnetic field produced by the field coil (or coils), and (3) the rate of rotation of the rotor. The strength of the magnetic field can be changed by varying the exciting current and the number of turns in the field coil. There are thus several design features of an electric generator that can be adjusted to obtain a desired voltage output. The rate of rotation is determined by the driving turbine, but the other parameters are variable. In commercial power plants, the generator output voltages are usually in the range of 11,000 to 30,000 volts.

Three-Phase Systems. In a generator with a single armature coil, which may have many turns, the alternating current output consists of a succession of cycles like that in Fig. 33; this is called a *one-phase* system. As a general rule, however, electric power is generated and transmitted by utilities as *three-phase* alternating current (see **Electric Power Generation**). The stationary armature then has three separate coils arranged 120° apart, as shown in section by *aa'*, *bb'*, and *cc'* in Fig. 35. For simplicity,

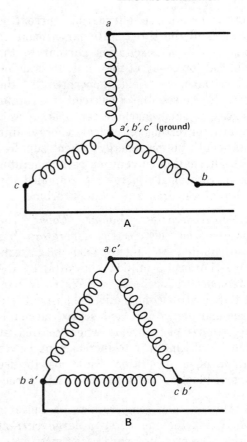

Fig. 36   Three-phase system. A: Y-connection; B: △-connection.

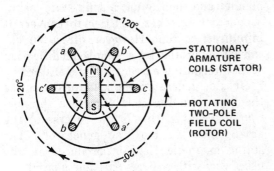

Fig. 35   Three-phase alternator.

a single (two-pole) rotor field coil is assumed. The voltage (or current) generated is then equivalent to three one-phase systems a third of a cycle apart. The result is a three-phase alternating current.

Each armature coil has two output terminals, but the number of conductors (or load wires) required to transmit three-phase current is reduced to three or four. One way is to connect together the terminals to *a'*, *b'*, and *c'* to form a single (or common) output terminal; this constitutes a Y-connected arrangement, as depicted schematically in Fig. 36A. Transmission load wires are then connected to the terminals from *a*, *b*, and *c* and to the common terminal from *a'*, *b'*, and *c'*. If the generator output is balanced so that the three phases are identical, the voltage at the common terminal, called the neutral terminal, is always zero. By connecting this to

ground, with a corresponding ground at the use point, the three-phase current can be transmitted over three load wires.

In the alternative three-phase △ connection, *a* is joined to *c'*, *b* to *a'*, and *c* to *b'* (Fig. 36B). There is no neutral point and no ground; only three load wires, connected to *ac'*, *ba'*, and *cb'*, are required to transmit the three-phase current. However, the Y connection is often preferred because the line voltage (i.e., the voltage between any two output connections) is $\sqrt{3}$ (i.e., 1.73) times the individual phase voltage. In the △-connected arrangement these two voltages are equal.

Generator/Motor Unit. In **pumped storage** systems the same machine functions either as an electric generator or an **electric motor** as may be required (see also **Tidal Power).** When serving as a motor, three-phase alternating current is supplied to the three stationary armature coils and direct excitation current to the field coil (or coils) on the rotor. Interaction of the magnetic fields generated by the currents causes the field coil to rotate. Hence, the same machine can operate as a motor, converting electrical energy into rotational motion energy when supplied with alternating current, or as a generator when rotational energy is supplied by means of a hydraulic (or other) turbine.

Superconducting Magnet Coils. The power output of electric generators has been limited to about 1500 megawatts (1500 MW or 1.5 million kilowatts) by the maximum practical size (see **Watt**). In particular, a strong magnetic field requires a large-diameter rotor, and severe mechanical stresses can develop when rotated at high speed. One way to increase the power output of a generator would be to use superconducting coils for the field structure.

At very low temperatures, a suitable superconductor can carry a large current with essentially no resistance (see **Superconductivity**). It is thus possible to produce a strong magnetic field in a rotor of small dimensions. The iron core which is used in conventional electromagnets is not required with superconducting coils. Apart from the construction of the rotor, all other features of the generator would be unchanged. The use of superconducting magnets should substantially increase the power output of a generator of moderate size.

Experimental generators with superconducting magnets, having power outputs up to 20 MW, have been built for testing. Steady increases in power are expected to lead in due course to 3000-MW machines for electric utilities. Some special lightweight generators with superconducting field coils have been constructed for use on aircraft.

The most suitable superconducting materials available in the early 1980s are a niobium–titanium alloy and a niobium–tin compound. These substances must be cooled with liquid helium to maintain the superconducting state. The field coils are designed to permit cooling by a continuous flow of liquid helium which enters and leaves by way of the rotor axis.

Direct-Current Generation

Direct current is required for electroplating and other commercial electrochemical processes and also for battery charging. Furthermore, direct-current (dc) motors are preferred for operation with variable loads, such as electric locomotives. The general tendency at present is to generate alternating current in the most efficient manner and then to use a solid-state (semiconductor) or other rectifier to convert it into direct current. However, direct current can be generated by a simple modification of an alternating-current (ac) generator.

In a dc generator, the armature must be the rotor and the field structure the surrounding stator; this is the reverse of the arrangement in most ac generators. The stator magnetic field is produced in the usual way with a direct exciting current.

For simplicity, consider a single armature coil. The open ends of the coil are connected to two halves of a split ring, which rotates with the coil. Two stationary brushes make contact with the half-rings and provide the output terminals. The arrangement, shown in Fig. 37, is called a *commutator*. During each rotation of the armature coil, the current changes direction, but at the same time the positions of the half-rings of the commutator are reversed. Hence, the current in the external load is always in the same direction. To avoid excessive sparking between the rotating half-rings and the brushes, the output is usually limited to about 1500 volts.

The current (and voltage) produced in this manner has a marked ripple as indicated by A in Fig. 38. In most practical dc generators, the armature consists of several coils, with the ends brought out to segments of a multiple commutator. There are also more than two poles to the stationary

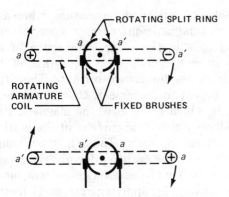

Fig. 37 Commutator principle.

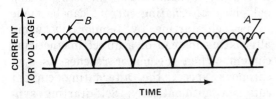

Fig. 38 Smoothing out ripples in rectified current.

field. The output voltage is then almost constant with a slight ripple that has little or no effect on the use of the direct current (*B* in Fig. 38).

Several different means have been used to provide the direct exciting current required for the field coils. Since the generator produces direct current, the output current can serve this purpose. In series-wound dc generators, the armature and field coils are connected in series, whereas in shunt generators the coils are in parallel. Compound generators make use of different combinations of series and parallel connections. When the external load is highly variable, the separately excited dc generator is preferred. The exciting current, which is independent of the load, is then provided by a separate dc generator as it is in an ac system.

In the United States, direct current is generated for local industrial use, rather than for distribution from a central power station. It is then simpler to utilize a **gas** (combustion) **turbine** or an **internal-combustion** (e.g., **diesel**) **engine**, rather than a steam turbine, to provide the rotational motion for the generator. In some circumstances, where precise power control is required, an ac motor is used to drive a dc generator; this is, in effect, a system for converting alternating into direct current. However, solid-state rectifiers (or converters) are now being used to an increasing extent.

ELECTRIC MOTOR

A machine for converting electrical energy, in the form of an electric current, into mechanical energy of rotation. The principle of an electric motor is that when an electric current is passed through a conductor located within a magnetic field, a force acts on the conductor so as to move it out of the field (or into a region where the field is weaker). In effect, the current in the conductor, called the *working current*, produces a magnetic field which interacts with the existing magnetic field in such a manner as to cause one of the components to move.

In an electric motor, the arrangement of the working-current conductor and the magnetic field are such that the interaction of the fields results in a rotational motion. Thus, the electrical energy carried by the working current is converted into rotational energy.

In a general (and sometimes in an actual) sense, an electric motor is the reverse of an **electric generator**. In a generator, an electric current is produced by rotation of a conductor in a stationary magnetic field (or in a stationary conductor by a rotating field). On the other hand, in a motor, rotational motion occurs when an electric current is passed through a conductor in a magnetic field. By appropriate design, the same machine, as used in **pumped storage** facilities, **tidal power** plants, and in some **electric vehicles**, can serve as either a motor or a generator; that is to say, it can convert electrical energy into mechanical energy of rotation or reversed to convert mechanical (rotational) energy into electrical energy.

Structurally, an electric motor, like a generator, consists of a fixed component, called the *stator*, within which rotates the moving component, called the *rotor* (see Fig. 35). Also, like a generator, a motor has an *armature*, which carries the working current, and an electromagnetic *field structure*, in which a magnetizing current produces the required magnetic field. In most alternating-current (ac) motors, the armature is the stator and the field structure is the rotor. However, in direct-current (dc) motors, the situation is reversed; the magnetic field is the stator and the armature is the rotor.

Several different types of electric motors have been built. A few of the more important types are described here, according to whether they operate with alternating or direct current.

Alternating-Current Motors

The most common ac motor is the *induction motor*. Three-phase (or any multi-phase) alternating current (i.e., the working current) is supplied to iron-cored coils (or windings) within the stator. As a result, a rotating magnetic field is set up, which induces a magnetizing current in the rotor coils (or windings). Interaction of the magnetic field produced in this manner with the rotating field causes rotational motion to occur.

Polyphase ac induction motors are self-starting and, especially in the larger sizes, have many industrial applications. Induction motors of lower power, with single-phase alternating current, are used in domestic appliances. However, they require a special armature coil or other means to make them self-starting.

Another important type of ac motor is the *synchronous motor*; as in the synchronous generator, the speed of rotation is directly related to the frequency of the working alternating current and the number of magnetic field poles. Thus, the speed of the motor remains constant, regardless of the load, provided the alternating-current frequency is constant.

In a synchronous motor, the working alternating current is supplied to iron-cored coils within the armature, whereas a direct (magnetizing) current energizes the field coils in the rotor. Interaction of the armature magnetic field with that of the rotor causes the latter to rotate. The situation is just the reverse of that in a generator, in which rotation of the magnetic field produces an electric current in the stationary armature coils. Thus, a synchronous machine can be designed for use either as a motor, by supplying electric current, or as a generator, by applying rotational motion to produce current.

Large synchronous motors using polyphase alternating current can be made to be self-starting by temporarily using alternating instead of a direct magnetizing current. When the motor reaches its synchronous speed, the alternating current cuts out automatically. Self-starting synchronous motors have also been designed for single-phase alternating current.

Direct-Current Motors

Because electric utilities supply alternating current, dc motors are not as common as ac machines. However, they are useful for heavy-duty operation, especially where variable speed is required. Thus, dc motors are used in diesel-electric and other electric locomotives. In these cases, the alternating current generated locally or supplied by a utility must be rectified to produce direct current. In some cases the direct current is supplied by a **storage battery**. Like a synchronous ac motor, a dc motor can act as either motor or generator, as required.

In the simple ac generator in Fig. 32, rotation of a coil in a steady magnetic field produces an alternating current in the coil. When a direct-current output is required, it is obtained by means of a commutator which reverses the current at regular intervals (see **Electric Generator**). Similarly, in a dc motor, the working direct current is supplied to the armature (rotor) coils through a commutator, so that the current undergoes periodic reversal. The action of a steady magnetic field on the alternating current in the armature coils then causes the armature to rotate continuously in the same direction.

Except for some small dc motors which use permanent magnets, the steady magnetic field is obtained by passing direct current through stationary iron-cored field coils. Direct-current motors are distinguished by the manner in which current is supplied to the magnetic field (stator) coils and the armature (rotor) coils.

In the *series motor*, the direct current flows in series through the field coils to the commutator connections of the armature coils. Hence, both field and armature currents are the same and will increase (or decrease) simultaneously. This permits the motor to operate smoothly at different speeds.

In the *shunt motor*, the field and armature currents are in parallel; that is to say, the direct current supply is divided so that part flows through the field coils and another, independent part flows through the armature coils by way of the commutator. In the operation of a shunt motor, the field current remains essentially constant, whereas the armature current varies with the load. As a result, the motor runs at almost constant speed at all loads.

Finally, the *compound motor* is a combination of the other two types. Part of the current flows through both the field and armature coils, as in a series motor, and another part flows through separate field coils, as in a shunt motor. The compound motor is used for high loads where some speed variation is required.

ELECTRIC POWER DISTRIBUTION

In the electric utility industry, *distribution* refers to the transportation of moderate and small blocks of power from a main (or bulk-power) substation to several secondary substations from which power is supplied to individual industrial, commercial, and residential users. The transportation of large blocks of power from a central generating station to the distribution system is called *transmission*, as distinct from distribution (see **Electric Power Transmission**).

The high-voltage transmission line from the power station terminates at the bulk-power substation fairly close to a load (or consumption) area. Here the voltage is reduced by step-down **transformers** to various levels at which the electric power is conveyed by feeder lines to several distribution substations. If the power is to serve a large industry, the voltage at the substation may be in the range of 23,000 to 138,000 volts. For smaller industries and commercial and residential users, voltages are lower (e.g., 4160 volts). Secondary feeder lines transmit the power from the distribution substations to secondary substations where the voltage is reduced further for distribution to individual users.

In addition to transformers for reducing the voltage, substations include voltage and **power-factor** regulators to assure steady operation of electrical equipment. Circuit breakers at the substation disconnect major feeder lines from the power system in the event of an overload that could cause damage.

Power from the main (bulk-power) substation to various subsidiary substations is generally supplied as three-phase alternating current over three phase conductors (see **Electric Power Generation**). Industrial motors, for example, almost invariably require three-phase power. Where this is not needed, however, as for lighting and residential purposes, the three phases in a Y-connected system are separated into three single-phase circuits, each using one phase conductor and a ground. In a △-connected three-phase system, each pair of the three conductors provides a single-phase circuit.

Homes are generally supplied with 120-volt current for lights and most appliances and also with 240-volt current for electric ranges, heat pumps, clothes dryers, and hot-water (or other) heaters. This dual supply is accomplished by a three-wire (single-phase) system. The 120-volt, single-phase current provides the input to the primary of a transformer which steps up the output to 240 volts. The secondary of the transformer has a central (or neutral) tap which is grounded (Fig. 39). Three wires,

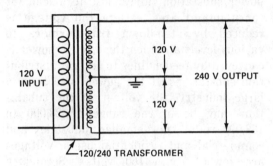

Fig. 39   120/240 volt transformer.

one from each terminal of the secondary and one from the central tap, provide the conductors to a residence. To obtain a 120-volt supply, connections are made to either end of the transformer secondary and to the central tap; for 240 volts, the connections are to both ends of the secondary.

Bare conductors, similar to those used for transmission lines, may also be used for distribution systems where it is safe to do so. Most distribution lines, however, consist of insulated cables which may be carried aboveground or installed in underground conduits, according to circumstances. For the higher distribution voltages, power cables are made of twisted copper conductors surrounded by rubber, synthetic, or other insulation; the whole is enclosed in a protective jacket or sheath. Simpler cables, with fewer conductors and thin insulation, are adequate for lower voltages.

ELECTRIC POWER GENERATION

The large-scale production of electricity at a central power station for transmission to a load (or demand) area and subsequent distribution to users (see **Electric Power Distribution; Electric Power Transmission**). A power station includes three major elements: (1) a primary energy

source, (2) a means for converting the primary energy into rotational energy, and (3) a generator for converting the rotational energy into electrical energy. The last item is discussed in the article on the **Electric Generator**; the other two are considered here.

Solar energy in its various forms (including **wind energy**) and **geothermal energy** can be converted into electricity, but these primary sources are still in the development stage. The primary energy sources in existing power plants are the **fossil fuels, nuclear fuels**, and **hydroelectric power** (i.e., the energy of flowing water). The last of these can be converted directly into rotational energy, but for the other primary sources the energy is first released as heat which is then converted (in part) into rotational energy.

Steam-Electric Plants

More than 75 percent of the electric power used in the United States is produced in **steam-electric plants**. In these plants, fossil or nuclear fuels generate heat which is used to produce steam at high temperature and pressure. The steam then serves to operate a **steam turbine**, thereby converting heat energy of the steam into rotational energy to drive an electric generator.

In *fossil-fuel plants*, combustion (or burning) of a fossil fuel in air in a furnace produces heat that generates steam in a boiler. The boiler is usually designed to deliver superheated steam, that is, steam at a temperature above the boiling point of water at the system pressure (see **Steam Generation**). The higher the initial steam temperature, the better the efficiency for converting the heat supplied into useful work (see **Thermal Efficiency**), but the strength of materials limits the practical temperature in a steam turbine to about 1000°F (540°C). By increasing the steam density (by an increase in pressure), the mass of steam supplied to a turbine of a given size is also increased. Hence, with a

higher steam pressure, a smaller turbine can provide a desired power output. The maximum initial turbine steam pressure is around 235 atm (24 MPa).

In *nuclear power plants* heat is obtained from **nuclear fuel** in a **nuclear power reactor** (or, in brief, a reactor) by the fission process (see **Fission Energy**). In some cases, steam is produced by boiling water directly in the reactor vessel by the heat generated (see **Boiling-Water Reactor**), whereas in others the fission heat is removed from the nuclear fuel by water maintained at a sufficiently high pressure to prevent boiling (see **Pressurized-Water Reactor**). Heat is then transferred from the hot, pressurized water to water at a lower pressure in a **heat exchanger**; at this lower pressure the water boils and produces steam.

Regardless of the manner in which the steam is generated, its temperature in nearly all existing nuclear power plants is about 555°F (290°C) and the pressure is close to 75 atm (7.5 MPa); both are substantially lower than in modern fossil-fuel plants. Consequently, the turbines in nuclear plants differ somewhat in design and operating speeds from those in conventional power plants. Reactors have been designed that will produce steam at higher temperatures and pressures but they have not yet been fully developed.

Steam Turbine and Condenser. Steam turbines for generating electricity are systems with either three (high, intermediate, and low pressure) or two (high and low pressure) sections. Steam reheating occurs between stages preceded by moisture separation when necessary (see **Steam Turbine** for further details). Like all **heat engines**, a steam turbine takes up heat from the steam at a high temperature, converts part of the heat into mechanical work, and discharges the remainder at a lower temperature.

In addition to a high initial steam temperature, a low discharge temperature is required to achieve efficient conversion of heat into work in a turbine (or other heat engine). The low temperature, usually within a few degrees of 105°F (40°C), is attained by passing the exhaust steam from the turbine through a condenser where it is cooled and converted into liquid water. The condensation of the exhaust steam also lowers the back pressure and thereby increases the steam (and heat) flow rate through the turbine. (In a **cogeneration** system the turbine exhaust steam would not be condensed but would be used for industrial purposes.)

A turbine condenser consists of a large number of tubes through which flows cooling water from an outside source. Turbine exhaust steam passing around the outside of the tubes is condensed, and the water is collected in a condensate tank. The condensate is heated in a series of feedwater heaters by steam drawn from various stages of the turbines. The heated feedwater is then pumped back to the steam generator (i.e., the boiler in a fossil-fuel plant or the equivalent in a nuclear plant).

Thermal Efficiency. The heat removed by the cooling water from the condensing steam represents the low-temperature heat rejection (or discharge) which is a feature of all heat-engine cycles. The rate of heat removal in the condenser depends primarily on the electrical output of the plant and on the proportion of the heat supplied by the fuel that is converted into electrical energy (i.e., the thermal efficiency).

Modern steam-electric plants, using fossil fuels and high-temperature (superheated) steam, have an average *gross* thermal efficiency, based on the total electric power generated, of about 42 percent compared with 33 percent for nuclear plants with their lower steam temperatures. However the *net* thermal efficiencies, after allowing for the power required to operate auxiliary equipment, are not so different. In a coal-burning plant, for example, substantial amounts of power are used to move the coal and ash; the net thermal efficiency is then 37 to 38 percent. The net efficiency of a nuclear plant is close to 32 percent. The efficiencies of older and smaller fossil-fuel plants are much the same as for nuclear plants.

Cooling Requirements. In a fossil-fuel power plant about 10 percent of the heat supplied by the burning fuel is lost in the hot combustion gases discharged through the stack. If the gross thermal efficiency is taken to be 42 percent, this leaves 48 percent of the heat to be removed by the cooling water in the condenser. In a nuclear plant, on the other hand, almost 67 percent of the fission heat generated must be removed in this way. For the same electrical output, the amount of heat rejected to the condenser in a nuclear power facility is about 50 percent greater than for a modern fossil-fuel plant.

The rate of cooling water flow through the condenser is determined by the heat rejection rate and the acceptable increase in the water temperature. The temperature increase varies to some extent with the local conditions, but it is usually in the range of roughly 14 to 40°F (8 to 22°C), with an average around 22°F (12°C). For this average value, a modern fossil-fuel plant, with a net electrical **capacity** of 1000 megawatts (MW) where 1 MW = 1000 kilowatts = 1 million watts (see **Watt**), would require a condenser cooling water flow rate of some 450,000 gal (1.7 million liters) per minute. For a nuclear power plant of the same electrical capacity, the flow rate would be about 1.5 times as great.

If a large body of water, such as a major river, a large lake, or the ocean, is accessible, the simplest way to provide the enormous volumes of cooling water required by a large steam-electric power plant is by a once-through (or open-cycle) system. The water is drawn continuously from the water body, pumped through the condenser, and discharged to the same water body.

Since January 1, 1970, the U. S. Environmental Protection Agency regulations have precluded the use of once-through cooling for all power plants of more than 500-MW electrical capacity except for situations where large volumes

of water are available and the warm condenser discharge would have no significant environmental (ecological) effects. Otherwise, completely or partially closed-cycle systems, requiring the use of **cooling towers** or **cooling lakes**, are mandatory. The warm water leaving the condenser is passed through the tower or lake where the water is cooled; it is then returned to the condenser for further heat removal. In a completely closed system, the same water, except for minor losses which must be replaced, is used over and over again.

Improved Heat Utilization. Even in the most efficient steam-electric plants, more heat is removed by the condenser cooling water and eventually dissipated to the surroundings than is converted into electrical energy. The proportion of the heat energy supplied that is lost in this way could be reduced by increasing the steam temperature and hence the thermal efficiency. However, the temperature is limited by the adverse effect on materials. Consequently, other means are being studied for more efficient utilization of the heat generated. Possible uses of the warm condenser discharge are for space heating of buildings and in various aspects of food production (see **Total Energy System; Waste Heat Utilization**).

In fossil-fuel plants, advantage could be taken of the very high combustion temperatures to generate power in a **topping cycle** (see **Combined-Cycle Generation**). The heat discharged from this cycle would be used to produce steam for a conventional steam-electric plant. Furthermore, part of the heat in the warmed condenser water from any steam turbine might serve to produce additional electric power in a **bottoming cycle.**

Hydroelectric Power

Hydroelectric plants, which produce 13 to 14 percent of the electric power generated in the United States, are described in the section on **Hydroelectric Power.** In these plants, the energy of flowing water drives **hydraulic turbines** connected to electrical generators. Hydraulic turbines operate at much lower speeds than do steam turbines; hence, the generators used to produce standard 60-Hz alternating current are of different design (see **Electric Generator**).

Gas Turbines and Diesel Engines

Almost 10 percent of the electric power generated in the United States is obtained from **gas turbines** (sometimes called combustion turbines) and **diesel engines.** Rotational motion to drive a generator is obtained directly from a gas turbine, whereas a diesel engine produces a reciprocating motion which is converted into rotational motion by means of a crankshaft, in the usual manner.

In the gas turbine, the hot pressurized gases, produced by the combustion of a fuel gas or light fuel oil in compressed air, drive a turbine. As a general rule, the turbine exhaust gas is discharged directly to the atmosphere; this reduces the thermal efficiency, but condenser and cooling water are eliminated. Gas turbine generators have been built with electrical outputs up to 100 MW.

The major use of gas turbines in commercial power plants is for special situations in which the demand is above normal, as will be seen later. The rapid startup (and shutdown) capability of a gas turbine is then advantageous. The capital cost per unit of electric power generated is relatively low but the fuel is expensive. However, if available nearby, **low-Btu fuel gas,** obtained from coal, would reduce the costs. For moderately small central power installations, gas turbines are more convenient, and may even be more economical to operate, than equivalent steam-electric plants.

In a simple open-cycle gas turbine, the exhaust gases are discharged at a high temperature and the heat loss is substantial. The overall efficiency can be increased by using a regenerator in the exhaust for heating the combustion air, but this adds to the complexity and cost of the system. Another way to utilize the heat content of the exhaust gases is in a combined-cycle generation system consisting of gas-turbine and steam-turbine generators. The steam for the latter is produced from the hot gas-turbine exhaust in a waste-heat (**heat exchanger**) boiler.

For electric power generation on a moderate scale, the **diesel engine** has the same advantages and disadvantages as the gas turbine. The capital costs of a diesel-generator set are low, but fuel costs are high; condenser and cooling water are not required, but direct discharge of the hot exhaust gas results in some loss of efficiency. This could be improved by utilizing the heat to produce steam in a waste-heat boiler. Electric generators operated by diesel engines are available with power outputs up to 10 MW.

Since diesel engines can be started quickly, they are often used to generate electricity in emergency situations (e.g., in hospitals, nuclear power plants, etc.) when the normal power supply fails. Diesel generators also provide electric power for locomotives and ship propulsion.

Three-Phase Alternating Current

Essentially all commercial power plants generate three-phase alternating current rather than direct current (see **Electric Generator**). The voltage of alternating current can be readily increased (for transmission) and decreased (for subsequent distribution) by means of **transformers;** this is not possible with direct current. Generation of alternating current is also more convenient in some respects. For example, a direct-current (dc) generator requires a commutator, but an alternating-current (ac) generator does not. Furthermore, in an ac generator, contact brushes are used only for the low-voltage, exciting current for the magnetic field, whereas in a dc generator brushes carry the much higher output voltage. Sparking is thus much more severe in the latter case. Transmission of electric power as direct current may have some advantages; the power is then generated as alternating current and converted into direct current prior to transmission (see **Electric Power Transmission**).

One advantage of three-phase (over one-phase) alternating current is that more power can be produced in a generator of a given size. Another is that large electric motors of the induction type, which are commonly used in industry, are self-starting and operate smoothly when designed for three-phase alternating current. Where single-phase current is adequate (e.g., for small motors and for lighting and heating), three independent circuits can be supplied from the three conductors (and a ground) of a three-phase system (see **Electric Power Distribution**). If single-phase alternating current were generated, six conductors would be required for the three circuits.

Load Variations

A problem in commercial power generation is that the **load** (or demand) varies in the course of the day as well as from day to day during the whole year. A fairly typical weekly load curve for an electric utility is shown in Fig. 40; as indicated, the load may be divided into three categories: base, intermediate, and peak. The *base load* is the demand that occurs at all times; it represents roughly 70 percent of the total electrical energy (in kilowatt-hours) generated. Over and above this is a variable *intermediate load* and, finally, for about 8 to 10 hours on workdays, there is a *peak load*.

In normal utility operation, the base load is supplied by the most modern (fossil-fuel, nuclear, or hydropower) generating units. These units are most efficient when operating continuously at maximum power. Part of the additional demand may be satisfied, as required, by power purchased from utilities in other time zones or where the climate is different. In general, however, the intermediate load is provided by the less efficient older and smaller fossil-fuel plants. When the demand cannot be satisfied by the base-load and intermediate-load units or by purchased power, *peak-shaving* (or *peaking*) *units* are used; these are commonly gas-turbine generators which can be started and shut down quickly.

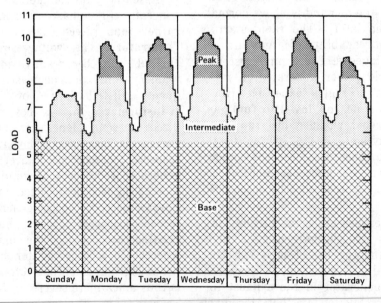

Fig. 40 Power plant load variation.

Although this operational scheme has proved reasonably satisfactory in the past, it may not be so in the future. Eventually the "older" units, now relied upon for the intermediate load, will be the larger plants of the present. These plants, especially nuclear plants, are unsuited to cyclic (i.e., periodic) operation. The preferred solution to this problem would be by what is called *load leveling*. The most efficient generators are operated continuously under a steady load; when the power generated exceeds the demand (e.g., at night), the excess would be stored for use at other times (e.g., in the day) when the situation is reversed. **Pumped** (hydroelectric) **storage** is already being used to some extent, and other means for storing electrical energy are being studied (see **Energy Storage**).

ELECTRIC POWER TRANSMISSION

In the electric utility industry, *transmission* refers to the transportation of large blocks of power over considerable distances (1) from a central generating station (see **Electric Power Generation**) to main substations close to major load (or consumption) centers or (2) from one central station (or power system) to another for load sharing. *Distribution* is the subsequent transportation of smaller blocks of power to individual users (see **Electric Power Distribution**).

Alternating-Current Transmission

Transmission Voltage. At the power station, the voltage of the three-phase alternating current output from the generator (see **Electric Generator**) is increased with a step-up **transformer** to the transmission voltage, mostly in the range of 138,000 to 765,000 volts. This is the effective (or root-mean-square) value of the alternating voltage of each phase. The high-voltage alternating current is usually transmitted by overhead conducting lines supported on steel towers. (Underground transmission is

described later.) The lines are made of a good electrical conductor, either copper or aluminum. However, as a result of electrical resistance, part of the power transmitted is inevitably dissipated as heat.

For a given electric power transmission, the proportion lost by resistance heating is decreased by increasing the line voltage. Consequently, a conductor can carry more power at a higher voltage without suffering mechanical deterioration by overheating. This is the main reason for the steady increase in transmission voltage over the years. An approximate rule is that the electric power that can be carried by a line is proportional to the square of the voltage. Hence, doubling the voltage would result in a roughly fourfold increase in power carrying capacity. The maximum power that can be carried by existing transmission lines is about 3000 megawatts (MW), where 1 MW = 1000 kilowatts = 1 million watts (see **Watt**).

Another advantage of high-voltage transmission is that a smaller land area is required for the lines for a given power carrying capacity. For example, a 345,000-volt line requires a path some 100 ft (30 m) wide, whereas for a 765,000-volt line, with about five times the power capacity, the path is only twice as wide.

Since 1969, when the first 765,000-volt transmission lines were introduced, research has been done on still higher voltages. A number of such ultra-high-voltage lines have been developed and one of 1,200,000 volts is being tested by the Bonneville Power Administration. However, increasing public objections to overhead power lines on environmental, esthetic, and other grounds make the future of ultra-high-voltage lines uncertain.

Conductor Design. Copper has been employed extensively for long-distance transmission lines because it has the highest electrical conductivity of the common metals. In recent years, however, aluminum has come into general use. It has a lower conductivity than copper and requires a thicker conductor to carry the same current (for a given voltage), but this is more than offset by the lighter weight and lower cost of the aluminum.

Transmission lines are made of many (up to 40 or more) separate strands twisted in a helical (or spiral) form. A central core, consisting of steel or strong aluminum alloy strands, adds mechanical strength. Among other advantages, a stranded line provides better alternating-current (ac) conductivity than a single conductor of the same total diameter. This is due to the skin effect, that is, the preference of alternating current to flow near the surface of a conductor. The stranded line has a larger surface area than an equivalent single line and hence is a better ac conductor.

Overhead transmission lines are generally bare (i.e., without an insulating sheath). They are supported by insulators at such a distance aboveground as not to constitute a hazard. Three well-separated lines, called phase conductors, carry three-phase current from the central power station to transformers at the main (bulk-power) substation.

Corona Loss. A high-voltage transmission line can lose significant amounts of power, especially in damp weather, in an electrical discharge called a *corona discharge.* One way to decrease corona discharge loss, although at increased cost, is by increasing the overall diameters of the phase conductors. Such expanded conductors can be made without a significant weight increase by introducing a paper-twine filler between the core and the outer conducting strands. The diameter can also be increased by arranging some wider strands near the core in such a way as to leave substantial air spaces in the interior of the line.

Another way to decrease corona loss is to use a multiple conductor, consisting of two or more subconductors, instead of a single phase conductor. Metal spacers keep the subconductors apart but make electrical contact so that they are effectively one line. Such multiple conductors are beneficial in other respects, including better cooling and lower inductance (see below) than single conductors.

Inductance and Capacitance. In a simple ac circuit which includes a purely resistive load only, the voltage and current are in phase; that is to say, the voltage and current cycles, as shown in Fig. 41, coincide

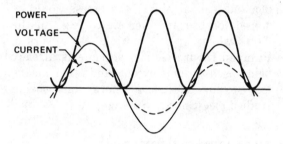

Fig. 41   **Alternating current and voltage in phase.**

in time. The **power** at any instant is equal to the product of the current and voltage. When these two quantities are in phase, they are simultaneously positive, zero, or negative. Consequently, the power is always positive or zero. (Note that a negative current and a negative voltage lead to positive power.) The average power carried by such a circuit is equal to $V_e I_e$, where V_e is the effective (root-mean-square) voltage and I_e is the effective current. For an ideal (or sinusoidal) alternating current, effective voltage and current are related to the maximum values V_m and I_m in a cycle by $V_e = V_m/\sqrt{2}$ and $I_e = I_m/\sqrt{2}$ (i.e., $V_e I_e = V_m I_m/2$).

In practice, ac circuits generally include inductance and capacitance in addition to resistance; as a result, the voltage and current are out of phase (i.e., the cycles do not coincide in time). *Inductance* arises from variations in the magnetic field produced by the alternating current; the varying field then induces a current which always opposes the main current. The effect is to delay the current relative to the voltage; the voltage is then said to lead the current while the current lags the voltage. *Capacitance* in an ac circuit is associated with variations in the electric field; it has the effect of introducing a voltage which delays the main voltage. The current then leads the voltage.

The current and voltage variations for a case of leading voltage are represented in Fig. 42. (If the voltage and current labels

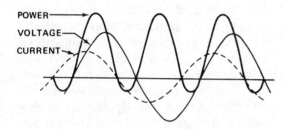

Fig. 42 Alternating current and voltage out of phase, with voltage leading.

were interchanged, the figure would represent a leading current.) Since the current and voltage are now out of phase, the product of these quantities (i.e., the power) is sometimes positive and sometimes negative in each cycle, as shown by the thicker curve in Fig. 42. (The combination of a positive voltage and a negative current, or vice versa, leads to a negative power.) Hence, the power flows in the positive (or desired) direction during parts of each cycle, but in other parts it flows in the negative (or opposite) direction, that is, back to the generator.

The net (or actual) power carried, equal to the difference between the positive and negative powers, is less than when the voltage and current are in phase. The ratio of the actual power to the maximum (in-phase) power, referred to as the apparent power, is called the **power factor**; thus,

$$\frac{\text{Actual power}}{\text{Apparent power } (V_e I_e)} = \text{Power factor}$$

(For a sinusoidal alternating current, the power factor is equal to the cosine of the phase angle difference between the current and voltage.) If the voltage and current are

in phase, the power factor is unity and the actual and apparent powers are equal (i.e., $V_e I_e$); however, when the voltage and current are out of phase, either as a result of inductance or capacitance, the power factor is less than unity, and the actual power carried by the ac circuit is less than the apparent (or maximum) power.

Different units are generally used to distinguish the actual power capability of an ac transmission line and the apparent power. The actual power is stated in watts, kilowatts, or megawatts. Since V_e is usually in volts and the current I_e in amperes, the apparent power $V_e I_e$ is expressed in volt-amperes (VA), kilovolt-amperes (kVA or 10^3 VA), or megavolt-amperes (MVA or 10^6 VA). The power factor is then expressed as the actual power in watts divided by the apparent power in volt-amperes.

As a rule, an ac circuit has both inductance and capacitance, as well as resistance. Since inductance results in a lagging current (leading voltage) and capacitance in a leading current (lagging voltage), they act in opposite senses. Thus, it is possible, although rarely experienced, for inductance and capacitance to balance each other so that voltage and current are in phase. In practice, however, one or the other usually predominates. In long-distance overhead ac transmission lines, inductance is the more important and the power carrying capability is consequently increased by introducing capacitance. This is done by adding devices called capacitors (or electric condensers) to the circuit.

Direct-Current Transmission

High-voltage overhead transmission of direct current, rather than alternating current, has certain advantages. (1) Two conductors can carry as much (or more) dc power than three similar phase conductors carrying ac power; the cost of installing the direct current (dc) system is thus less. (2) Direct current is carried throughout the conductor and not only near the surface (i.e., there is no skin effect); hence, conductor construction is simpler for direct current. (3) There is no inductance or capacitance in a dc line, and the power does not depend on a power factor. On the other hand, there are problems in dc generation, and the voltage cannot be increased or decreased by means of transformers (see **Electric Power Generation**).

The advantages of dc transmission can be combined with those of ac generation and voltage variability. Alternating current is generated and its voltage is increased by a transformer in the usual way. The high-voltage alternating current is then converted into high-voltage direct current (i.e., rectified) by means of a semiconductor (or other) device. The direct current is transmitted to a load center where it is reconverted to alternating current (i.e., inverted) for distribution at lower voltages to consumers.

Direct-current transmission is most economical over long distances when the costs of rectification and inversion are more than balanced by the lower cost of the transmission line. The longest dc line in the United States, from Oregon to Southern California, has a length of about 850 miles (1350 km). It consists of two conductors, one being 400,000 volts positive and the other 400,000 volts negative with respect to the ground. The maximum power carrying capacity is about 1450 MW at 800,000 volts.

Underground (Alternating-Current) Transmission

Underground transmission of ac power is generally much more expensive than overhead transmission, and hence it is used only when necessary, such as in or near populated areas. In a large city, the right-of-way for an overhead line, even if available, would be so expensive that underground transmission may be more economical. In the early 1980s, underground high-voltage transmission lines were only about 1 percent the length of overhead lines. But because of increasing objections to the environmental, visual, and other impacts of overhead lines and their supporting towers, underground transmission lines are receiving increasing attention.

Unlike overhead lines, underground conductors are insulated; they are then called *cables*. The insulation must be sufficient to prevent *breakdown*, that is, direct discharge to the ground. The most common type of high-voltage underground transmission cable consists of a central conductor made of many twisted strands of copper; this is wrapped in an insulating material of special paper impregnated with mineral oil. For three-phase ac transmission three such cables are inserted in a steel pipe and insulated from one another by oil under pressure. The pipe is usually laid in a deep trench and covered with soil. The maximum power capacity of such an underground line is about 650 MW at 550,000 volts.

In an overhead transmission line the inductance effect exceeds that of capacitance, but in an underground cable the reverse is true, largely because of the presence of the insulation. As a result of the increasing capacitance, there is a critical length beyond which the load factor is zero and the line will not transmit ac power. The critical length decreases with increasing line voltage and is generally 25 to 40 miles (40 to 65 km). Inductance to compensate for the capacitance must therefore be included at intervals along the line to bring the power factor closer to unity and maintain the power capacity of the underground cable.

Line Heating. Just as in an overhead transmission, there is a loss of power in underground lines as a result of resistance heating. In addition, however, the insulator introduces what is called *dielectric loss*; this loss appears as heat, and it increases rapidly with increasing ac line voltage. The dielectric loss may perhaps be decreased by using a synthetic material instead of oil-impregnated paper as insulator, but it cannot be eliminated.

Not only is a larger proportion of the power carried in an underground conductor converted into heat, but the rate of heat dissipation is less than in an overhead line. In the latter, heat is removed by the surrounding air, but in an underground line the electric insulating material is also a good thermal insulator. Transfer of heat from the cable to the ground is thus inhibited, and the temperature of the insulator rises. Deterioration of the insulator as a result of excessive heating places a limit on the power carrying capacity of an underground line. In a few cases, the power capacity has been increased by circulating the insulating oil through an external cooler.

Cable Improvement. The gas-insulated cable, which has a higher power rating, is a promising alternative to the conventional underground cable with oil-paper insulation. The line consists of two concentric (or coaxial) aluminum pipes which serve as the conductors; the space between the pipes is filled with sulfur hexafluoride (SF_6) gas under a pressure of a few atmospheres. The gas is a good electrical insulator, but the dielectric loss is small. Heating is thus less than in more conventional cables. In addition, the large surface area of the outer pipe permits good heat transfer to the surrounding ground. However, a very wide trench would be required for a three-phase line, thus increasing installation costs.

Another approach to improving underground transmission is to take advantage of the decrease in electrical resistance that results from lowering the temperature. The ac resistance of an aluminum or copper cable, for example, can be reduced by a factor of ten by cooling with liquid nitrogen to a temperature of about 77K (i.e., −321°F, −196°C). The decrease in resistance heating and the increased rate of heat removal by the circulating liquid nitrogen lead to a substantial increase in the power-carrying capacity of the line. This benefit is largely offset, however, by the cost of the nitrogen liquefaction system. (The possible use of still lower temperatures is described later in connection with superconducting transmission lines.)

Underground (Direct-Current) Transmission

Underground transmission of dc power has advantages over ac transmission in addition to those already mentioned for overhead lines. There is no dielectric loss in a dc cable and there is less resistance heating than in an ac conductor because there is no skin effect. A smaller proportion of the power is thus dissipated as heat. In addition, less electrical insulation is required to prevent breakdown to the ground in a dc cable than in an ac cable. The thermal insulation is thus decreased, and the heat generated can be more readily transferred to the ground. Hence, more power can be carried by an underground dc line than by a similar ac line.

Because of the costs of converting alternating to direct current (and back again), it is doubtful that underground transmission of direct current would be economic except for long distances. For transmissions of this type, superconducting lines, as described in the next section, might prove preferable.

Superconducting Transmission

Since the early 1960s, consideration has been given to the development of long-distance underground transmission lines using superconductors (see **Superconductivity**). At very low temperatures, these materials can carry large currents (alternating or direct) without resistance, and consequently with no resistance heating. A superconducting line should thus be capable of carrying the same (or more) power at a lower voltage and with thinner cable than a conventional underground line. The possibility of being able to decrease the line voltage is one of the main benefits of superconducting transmission.

Although resistance heating can be virtually eliminated and dielectric loss is greatly reduced by the lower line voltage, there is another source of power loss (and heating) in an ac superconducting cable. This loss, which increases rapidly with the current strength, arises from certain magnetic effects accompanying an alternating current in a superconductor. It has been found, however, that the magnetic loss is affected by the physical condition of the superconducting material and successful methods are being developed for reducing it.

So far, the most suitable superconductor for transmission lines appears to be the niobium-tin compound Nb_3Sn. It requires liquid helium as refrigerant to achieve the very low temperature (below about 10K, i.e., $-441°F$, $-263°C$) necessary to maintain the superconducting state with large currents flowing. In outline, a superconducting, underground ac cable might consist of two coaxial tubes around each of which is wound a spiral (or spirals) of thin Nb_3Sn tape or ribbon embedded in or coated with copper. The tubes are separated by a suitable insulating material. Liquid helium flows through the interior tube and around the outer tube in opposite directions.

Because some heat is inevitably generated by the alternating current and heat is gained from the surrounding soil, the temperature of the helium will rise. Suitable refrigeration (cryogenic) equipment would thus have to be located at intervals along the transmission line to maintain the necessary low temperatures. This equipment would add to the capital and operating costs of the line, but it should be offset by the improvement in power transmission efficiency.

For long-distance transmission, where the costs of ac rectification and dc inversion are a small proportion of the total costs, underground transmission of direct current by superconducting cables might have some special advantages. The magnetic heating which occurs with alternating current is absent from dc transmission; one consequence is that a simpler cable design might be satisfactory. Furthermore, operation at somewhat higher temperatures is possible, thereby decreasing refrigeration costs.

A major drawback to superconducting transmission of either ac or dc power is the need for liquid helium as refrigerant. Helium gas is relatively rare, expensive, and difficult to liquefy. Efforts are consequently being made to find a material that would be superconducting when cooled with liquid (or a slurry of solid and liquid) hydrogen. Hydrogen is a cheap material and is easier to liquefy than helium, but it cannot maintain as low a temperature. An interesting possibility is the use of a superconducting line to transport both electric power and liquid (or slurry) hydrogen which would be used as a fuel (see **Hydrogen Fuel**).

ELECTRIC VEHICLE

In the broad sense, a vehicle propelled by an electric motor with power supplied from a rechargeable **storage battery, fuel cell,** or any other source of electrical energy. In the common type of electric vehicle, to which most of the following description refers, electric current for the drive motor is provided by a storage battery. Direct current required to charge the battery is obtained by conversion of alternating current from an electric utility supply. The safe charging time varies with the design of the battery, but it is usually 5 to 8 hours.

Electric automobiles were in common use in the late 19th and early 20th centuries, but they suffered from the drawback of limited speed and driving range and long recharge time. Consequently, when the more versatile gasoline-powered engine was developed, the number of electric road vehicles declined. However, other battery-operated electric vehicles, such as mine locomotives, forklift trucks, golf carts, and wheelchairs, have continued in general use.

Interest in electric cars has been revived in recent years, largely because of the need to conserve petroleum products and reduce air pollution. Japan has been especially active in developing battery-powered passenger cars, delivery trucks, and even city buses. Electric vehicles of various types have been demonstrated in other countries. In the United States, a few thousand electrically-propelled passenger cars and some buses are in operation. The U.S. Postal Service has used electric vans for mail delivery, and the National Park Service has been testing a variety of electric vehicles.

The Electric and Hybrid Vehicle Research, Development, and Demonstration Act, passed by the U.S. Congress in 1976 and later amended, was intended to "promote the substitution of electric and hybrid vehicles for many gasoline- and diesel-powered vehicles . . . where such substitution would be beneficial." A hybrid vehicle is one in which more than one form of energy is supplied to the vehicle for propulsion. A hybrid electric vehicle, for example, might use batteries for city driving and a gasoline engine, which is most efficient at steady speeds, for highway travel.

The Act requires that the U.S. Department of Energy contract for the purchase or lease of specific numbers of electric or hybrid vehicles. The minimum specified performance standards for the earlier models are: speeds up to 50 miles (80 km) per hour, a driving range of 31 miles (50 km) on a single battery charge (124 miles or 200 km for hybrids), and acceleration to 31 miles (50 km) per hour within 15 seconds. Experience with these vehicles will provide the basic information for the development of more advanced designs.

One of the objectives of the Electric and Hybrid Vehicle Act is to conserve the limited petroleum resources and reduce the dependence on foreign supplies. This is possible because electricity can be generated from coal, nuclear fuel, or hydropower. The batteries would preferably be charged at night when the load (or demand) on the electric utility is normally at its lowest (see **Electric Power Generation**).

The use of electric vehicles would help to control atmospheric pollution, especially in cities where the road vehicle density is high. In spite of emission controls, **internal-combustion** (gasoline and diesel) **engines** emit pollutants when they operate, regardless of whether the vehicle is moving or not. The operation of electric vehicles, on the other hand, is essentially pollution free. Pollutants may be produced at the central-station plant when the electricity is generated, but these can be more readily controlled than the pollutants emitted from large numbers of vehicles.

Because there is a practical limit to the weight, and hence the electrical energy storage capacity, of batteries that can be carried, the driving range of an electric vehicle is necessarily restricted. Furthermore, batteries are most efficient in energy recovery when they operate at a low steady power, that is, at relatively low driving speeds. Thus, electric (battery) propulsion would appear to be most suitable for applications where driving involves short distances at moderate speeds.

Statistics show that in an urban area, such as Chicago, about 75 percent of the daily trips made by automobiles are of 31 miles (50 km) or less. Hence, an electric vehicle with the specifications given earlier should be capable of satisfying most urban travel requirements. Since automobiles have their lowest fuel efficiency (miles per gallon) in stop-and-go city driving, the use of electric vehicles could result in substantial savings in gasoline.

Regenerative braking is a feature of electrically propelled vehicles that is being developed. In a conventional vehicle, energy is wasted as heat caused by friction when the brakes are applied. In an electric vehicle, however, it should be possible to conserve most of the energy normally lost when a vehicle is braked. One method is to reverse the drive motor so that it acts as a generator producing electricity that can be returned to the battery (see **Electric Motor**). Another idea is to store the energy in a **flywheel**; this energy is then used to provide the additional power needed to accelerate the vehicle when it is restarted.

Batteries for Electric Vehicles

The weight and lifetime of the storage battery are essential aspects of electric vehicle design. The only battery sufficiently developed and available at a reasonable price in the early 1980s is the familiar lead–acid battery. Batteries of this type are used extensively in the electrical systems for starting, lighting, and ignition in automobiles and other vehicles. Common automobile batteries are, however, too heavy (for their energy capacity) and would have too short a lifetime for use in vehicle propulsion. (The lifetime of a battery is commonly stated in terms of the number of charge–discharge cycles it can sustain before failing.) Consequently, efforts are being made to improve the lead–acid battery (see **Storage Battery**).

A basic consideration for electric-vehicle batteries is the *specific energy*, sometimes called the *energy density*; it is the energy stored (or, more exactly, the energy delivered upon discharge) per unit mass of battery. The higher the specific energy, the smaller the mass (or weight) of the batteries required to deliver a certain amount of energy. This means that, for a vehicle of given weight, including the battery, the travel range is increased in proportion to the specific energy of the battery. Hence, a high specific energy is a desirable feature of batteries for electric propulsion.

The specific energy is usually expressed in watt-hours (W-hr) of energy delivered per kilogram (kg) or pound (lb) of battery mass (i.e., as W-hr/kg or W-hr/lb). Actually, the specific energy depends to some extent on the operating power (in **watts**), that is, on the rate at which the battery is discharged. Because of changes that occur in the battery materials, the specific energy decreases as the power (or discharge rate) is increased. The specific energy is generally stated for a discharge time of 5 or 6 hours, representing normal operating conditions.

The best lead–acid batteries available commercially in the United States in the early 1980s have a specific energy of roughly 40 W-hr/kg (18 W-hr/lb). Lead–acid batteries, with a specific energy of 60 W-hr/kg (27 W-hr/lb) have been reported from Japan, but they apparently have a relatively short charge–discharge cycle life when subjected to the severe operating conditions of an electric vehicle. However, it is expected that such a lead–acid battery with a satisfactory lifetime will be developed.

The range and load-carrying capability of an electric vehicle is dependent on the energy storage capacity (i.e., the total energy output) of the battery, expressed in kilowatt-hours. For a vehicle weighing 1000 kg (2200 lb), including motor, battery, and payload, about the smallest reasonable battery capacity is 15 kW-hr (i.e., 15,000 W-hr). If a lead–acid battery with a specific energy of 40 W-hr/kg is used, the mass of the battery would be 15,000/40 = 375 kg (827 lb). Hence, the battery would contribute almost 38 percent of the total vehicle weight. However, a specific energy of 60 W-hr/kg would decrease the proportion to 25 percent. This emphasizes the need for developing electric vehicle batteries with higher specific energies.

The speed of an electric vehicle is determined by the rate at which energy is drawn from the battery; this is the battery power, usually stated in kilowatts. (For an explanation of **power** in watts and energy in watt-hours, see **Watt**.) In the case considered above, the battery would probably be designed to operate at a sustained power of about 10 kW for steady driving. However, in order to provide additional power for acceleration and hill climbing, a temporary peak power of roughly 20 kW should be attainable without causing damage to the battery. The battery power is sometimes given in terms of the *specific power* (or *power density*) in watts per kg (or per lb). For a battery weighing 375 kg (see above), a sustained power of 10 kW corresponds to a specific power of 10,000/375 = 27 W/kg (or 12 W/lb).

Apart from demonstration and test projects, electric vehicles will use lead-acid batteries almost exclusively into the middle 1980s. Research in progress is aimed at both improving the characteristics of these batteries and developing new types of storage batteries with higher specific energies and longer charge-discharge cycle lives. Such batteries must be economically competitive with lead-acid batteries. Some of the possible alternatives, which are described in the section on the **Storage Battery**, are :

Nickel-iron (alkali)
Nickel-zinc (alkali)
Zinc-air (alkali)
Iron-air (alkali)
Zinc-chlorine (zinc chloride)
Lithium alloy-metal sulfide
  (high temperature)
Sodium-sulfur (high temperature)

The high-temperature batteries may be the best candidates for vehicle propulsion in the long term (after 1985). Demonstration vehicles with sodium-sulfur batteries have been operated in the United Kingdom and Japan. A road vehicle with a lithium-iron sulfide battery is to be tested in the United States in the early 1980s. Possible battery specifications for vehicles to be manufactured after 1985 are the following:

Energy storage, kW-hr	60
Specific energy, W-hr/lb (kg)	59 (130)
Sustained power, kW	24
Peak power, kW	60
Range per charge, miles (km)	125 (200)
Life, cycles	1000
Lifetime, years	4-5

Fuel Cell Batteries. Certain fuel cells are potential alternatives to storage batteries for electric vehicles. Such vehicles may have longer travel ranges than those with the most advanced storage batteries. The fuel cells of special interest are the aluminum-air, methanol-air, and the hydrogen-oxygen cell (see **Fuel Cell**).

The aluminum-air cell is of special interest for electric vehicle propulsion because of the high specific energy that is possible. An aluminum-air battery and associated equipment may weigh roughly the same as the gasoline engine and fuel in a medium-size automobile. A five-passenger electric vehicle is expected to have a travel range of at least 1000 miles (1600 km) before replacement of the aluminum electrodes is necessary. However, the aluminum hydroxide produced when the cell operates must be removed every 250 to 375 miles (400 to 600 km), and water must be added at similar intervals.

The range of a vehicle with a methanol-air battery depends on the volume of the fuel tank, that is, on the quantity of methanol that can be carried. The reaction of methanol with oxygen (in air) yields about half as much heat energy as gasoline (see **Alcohol Fuels**), but the efficiency of a fuel cell is greater than that of a gasoline engine. The travel range of the fuel-cell vehicle on a single tank of methanol might then be similar to that of an equivalent gasoline vehicle.

If fairly pure hydrogen gas were available at a reasonable price, the hydrogen-oxygen (air) fuel cell might be used for electric vehicle propulsion. The major problem would be storage of the fuel on the vehicle (see **Hydrogen Fuel**). However, the high efficiency of the fuel cell would compensate for the weight of the hydrogen storage system.

Hybrid Electric Vehicles

A simple type of hybrid electric vehicle would have a small gasoline engine in addition to a storage battery. Instead of drawing on the battery, the additional power required for acceleration (e.g., in starting, driving uphill, or passing) would be supplied by the gasoline engine. By avoiding high rates of discharge in this manner, the amount of stored energy recoverable from the battery (and the travel range of the vehicle) would be increased. The battery would also have a longer life.

An alternative possibility, mentioned earlier, is to use a more powerful gasoline engine with a storage battery. The battery would be used for city driving and the gasoline engine for sustained highway travel. A battery of moderate storage capacity would be adequate because it can be kept charged by a generator driven by the gasoline engine.

A different type of hybrid electric vehicle combines a battery drive with a flywheel drive. This is described under **Flywheel.**

ELECTROMAGNETIC RADIATION

Energy propagated as waves resulting from oscillating electric and magnetic fields. Electromagnetic radiations are characterized by their wave length or by their frequencies (i.e., number of wave lengths passing a fixed point per second). Familiar electromagnetic radiations, in order of increasing wavelength (or decreasing frequency) are gamma rays (see **Radioactivity**), X rays, ultraviolet, visible light, infrared, microwaves (including radar), and radio waves. All electromagnetic radiations, regardless of their wave length, travel in a vacuum with the same speed (i.e., the speed of light), close to 186,000 miles (3.00×10^8 m) per second.

According to modern theory, electromagnetic energy is carried by massless "particles" known as *photons*. The amount of energy carried by a photon of a particular radiation, called a *quantum*, is proportional to the frequency of the radiation. Hence, a photon of gamma rays (high frequency) carries a much larger quantum of energy than a photon of radio waves (low frequency).

Electromagnetic radiations travel through a vacuum without absorption (i.e., loss of energy), but they are absorbed when they pass through matter. The extent of absorption depends on the wavelength of the radiation and the nature of the matter.

When electromagnetic radiation is absorbed, its energy is eventually converted into heat energy.

For example, solar radiation is electromagnetic radiation covering a range of wavelengths (see **Solar Energy**). The radiation is not absorbed in the vacuum of space between the sun and the top of the earth's atmosphere. Partial absorption occurs in the atmosphere, and the remainder is absorbed by the earth. As a result of the increase in temperature, the earth's surface emits some of the energy as infrared electromagnetic radiation.

ELECTROSTATIC PRECIPITATOR

An arrangement utilizing the mutual attraction of opposite electric charges to remove dust or ash particles (or liquid droplets) suspended in a gas. Electrostatic precipitators are commonly used to separate fly ash from the flue gases of a coal-burning boiler plant before discharge through a stack (see **Steam Generation**).

In essence, an electrostatic precipitator consists of a number of vertical, parallel metal plates, generally 20 to 36 ft (6 to 11 m) high, about 1 ft (0.3 m) apart, arranged in a rectangular unit roughly 12 to 24 ft (3.7 to 7.3 m) on each side (Fig. 43). A large coal-burning steam plant requires several such units. Thin wires, 0.1 to 0.15 in. (2.5 to 3.8 mm) in diameter, are suspended vertically about 1 ft (0.3 m) apart between the plates. The wires are electrically insulated from the plates and are weighted at their lower ends to keep them in position. The gas from which the fly ash (or other particulate matter) is to be removed flows in a horizontal direction between the plates.

A direct-current high voltage, between 40,000 and 65,000 volts in most cases, is connected between the wires and the plates; the wires are attached to the negative side and the plates, which are grounded, to the positive side. An electrical glow discharge,

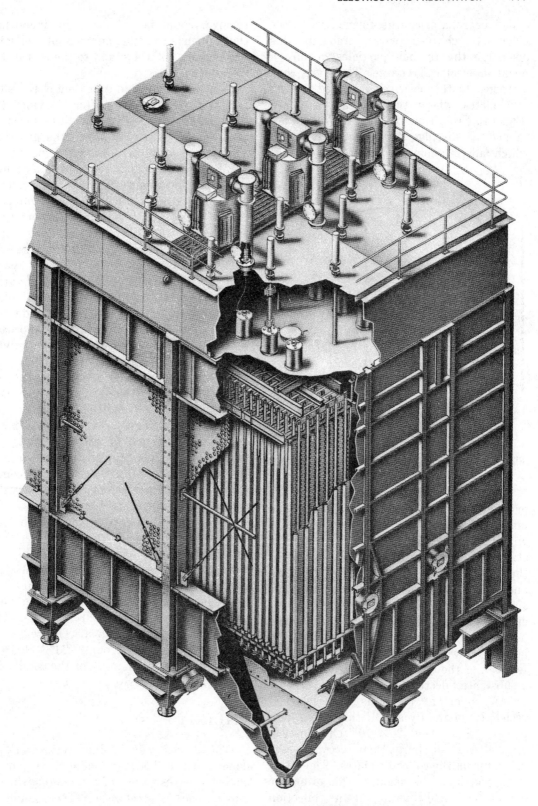

Fig. 43 Electrostatic precipitator (courtesy of Research Cottrell).

called a corona discharge, forms around the negatively charged wires, and in this discharge the fly ash particles acquire a negative electrical charge. They are then attracted to the positively charged (collector) plates where they remain attached (Fig. 44). From time to time, the plates are rapped to dislodge the accumulated ash which falls into hoppers below.

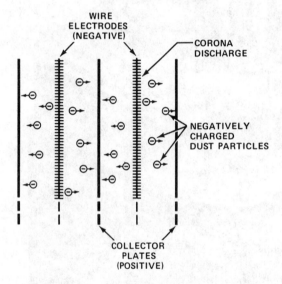

Fig. 44   Principle of the electrostatic precipitator.

Electrostatic precipitators are simple in principle, but their operation is sensitive to many variables; these include particle size and loading (i.e., mass per unit volume of stack gas), temperature, sulfur content of the coal, chemical composition of the ash, and others. When properly designed for specific conditions, the fly ash removal efficiency is more than 99 percent, thus permitting particulate emission standards to be met. However, a change in the characteristics of the coal being burned can cause a substantial decrease in efficiency.

An important factor is the electrical resistivity of the fly ash. If this is too high (greater than about 2×10^{10} ohm meter), an electrical discharge may occur in the ash layer on the collector plates. The effect is to neutralize the negative charge on the ash particles and thus impair the collection efficiency. The overall efficiency of fly ash

removal could be increased by increasing the dimensions and number of collector plates, but this would add to the cost of the installation.

Normally, the flue gases enter the electrostatic precipitator after leaving the combustion air heater and just prior to discharge through the stack; the gas (and ash) temperature is then in the vicinity of 300°F (150°C). This is referred to as "cold-side" precipitation. The fly ash removal efficiency is high, provided the sulfur content of the coal being burned is more than about 1.5 percent. However, with low-sulfur coal, the electrical resistivity of the ash particles is too high for satisfactory operation of the precipitator. The effect of variations in ash resistivity could be controlled by overdesigning the system, as indicated above, but a better economic approach appears to be to decrease the resistivity when required.

There are two general ways for decreasing the electrical resistivity of fly ash. One is by the use of a "conditioning agent." The most common agent is sulfur trioxide, added to the extent of about 20 parts per million of flue gas before it enters the precipitator. The sulfur trioxide becomes attached to the surface of the ash particles and decreases the resistivity.

The second general method is to take advantage of the decrease in electrical resistance of the ash at higher temperatures by use of "hot-side" precipitation. The precipitator is then located ahead of the economizer, where the average gas temperature is roughly 700°F (370°C). The resistivity of the fly ash also decreases at lower temperatures, around 220°F (105°C), but there is then a danger of the condensation of highly corrosive sulfuric acid.

ENERGY

The capacity for doing work; work is done when a body (or object) is moved against the resistance of a force (or when a moving body is accelerated). The amount of energy that can be utilized to do work depends on the conditions (see **Heat Engine**); consequently, a better definition of energy is the property of a system that decreases when work is done by an amount equivalent to the work done.

Energy can be manifested in various forms, such as mechanical, thermal, chemical, electrical, and nuclear. One form of energy can often be converted into another form (e.g., mechanical into electrical and vice versa, and thermal into mechanical and vice versa, etc.). It has been found that, without exception, when energy conversion occurs, the amount produced is exactly equivalent to the amount transformed; in other words, energy can be changed in character, but it cannot be created or destroyed. This is the basis of the law of conservation of energy, a fundamental law of nature that implies the impossibility of realizing so-called "perpetual motion."

An apparent exception to the law of conservation of energy arises with **nuclear energy,** where the production of energy is associated with a loss of mass (or quantity of matter). In the release of nuclear energy it appears that mass is converted into energy; it is also possible to convert energy into mass on the nuclear scale. However, there is an exact equivalence between mass and energy, so that mass may be thought of as a manifestation of energy.

Because of the equivalence of different forms of energy, they can all be expressed in the same units. In practice, different units are sometimes used for convenience; thus, electrical energy is expressed in kilowatt-hours (see **Watt**), heat energy in **British thermal units,** and mechanical energy in foot-pounds, but there are precise relationships between the different units which permit conversion from one energy unit to another. In the modern metric (SI) system of units, amounts of all forms of energy are stated in **joules** (J) or multiples of the joule, such as kilojoule (kJ) = 1000 J and megajoule (MJ) = 1,000,000 J.

Essentially all forms of energy, other than energy in transit, can be regarded as being either potential or kinetic. *Potential energy* arises from the position, state, or configuration of a body (or system). Water at a higher level has potential energy that can be converted into work by allowing it to flow to a lower level, as in **hydroelectric power** generation. A wound spring also has potential energy that is convertible into work when the spring is allowed to unwind. The chemical energy possessed by **fossil fuels** and their products (and oxygen), which can be converted into heat or other forms of energy, is a form of potential energy arising from the configuration of the **atoms** in the fuel (and oxygen).

Kinetic energy is energy of motion. A body of mass m moving with a velocity v has a kinetic energy of $\frac{1}{2} mv^2$ (in appropriate units). The kinetic energy of a rotating body is given by a similar expression. When water falls from a higher level to a lower level, the potential energy at the higher level is converted into kinetic energy of the flowing water at the lower level. This can be converted, in turn, into rotational kinetic energy in a **hydraulic turbine.**

The term *internal energy* has a specific meaning in thermodynamics, but the energy associated with the motions of molecules and atoms and the forces between them is sometimes called internal energy. However, this energy arises from kinetic and potential energies on the atomic scale.

Except for nuclear energy, almost all the energy used on earth originated in the sun. Some forms of **solar energy** are utilized directly, whereas others represent indirect usage; fossil fuels are derived from ancient plant and animal organisms which depended on solar energy for their development, and **biomass fuels** are based on plants of recent growth. Nuclear energy, on the other hand, is derived from materials (e.g., uranium and thorium) that have a purely terrestrial origin. **Geothermal energy** is also terrestrial because the temperatures in the earth result almost entirely from the decay of radioactive species (see **Radioactivity**). **Tidal power** (or energy) arises mainly from gravitational attraction by the moon.

ENERGY EFFICIENCY RATIO (EER)

A measure of performance efficiency applicable in particular to home air conditioners and to **heat pumps** operating in the cooling mode. It is defined as the rate at which heat is removed from the surroundings, expressed in **British thermal units** per hour, divided by the rate of energy input (or **power**) in **watts** required to operate the machine and associated air blower (or blowers); that is,

$$\text{EER} = \frac{\text{Rate of heat removed in Btu/hr}}{\text{Rate of energy input (power) in watts}}$$

The units of EER are Btu per hour per watt or Btu per watt-hour (Btu/watt-hr).

The relationship between the EER and the **coefficient of performance,** as used for heat pumps and large refrigeration systems, can be obtained by noting that 1 watt is equivalent to 3.41 Btu/hr. It then follows that

$$\text{EER (Btu/watt-hr)} = 3.41 \text{ COP}$$

The COP is a pure ratio and has no units.

The EER of home air conditioners and heat pumps is generally in the range of 7 to 8.5 Btu/watt-hr. The corresponding values of the COP would be 2 to 2.5.

ENERGY STORAGE

Means for storing energy in a readily recoverable form when the supply exceeds the demand for use at other times. Storage of primary fuels (e.g., coal, oil, and gas) is a form of energy storage, but the term generally applies to actual energy and to secondary fuels (e.g., hydrogen) rather than to primary fuels.

The effective utilization of intermittent and variable energy sources, such as solar energy and wind energy, often requires energy storage (see **Solar Energy: Direct Thermal Applications; Wind Energy Conversion**). If the intermittent energy is converted into electricity, as it is with solar photovoltaic cells (see **Solar Energy: Photovoltaic Conversion**) and in most cases of wind energy utilization, electrical energy in excess of the demand might be fed directly into a utility grid. If this is not possible, some form of energy storage would be required. Furthermore, where solar energy is used to produce steam in generating electric power (see **Solar Energy: Thermal Electric Conversion**) thermal (heat) energy storage is necessary for continuous operation. Storage of thermal energy is also desirable when solar energy is used for space and water heating.

In some circumstances, electricity may be generated, either on land or at sea, at a location that is too distant from a load (or consumption) center for conventional transmission lines to be used (see, for example, **Ocean Thermal Energy Conversion**). Means must then be found for both storing the energy and transporting it economically to a load center.

Electrically propelled vehicles, which are expected to come into increasing use, require some form of energy storage (see **Electric Vehicle**). Since the vehicle must carry its energy supply, equivalent to the gasoline in a conventional automobile, the storage system should be readily transportable.

Electric utilities generally use less efficient (intermediate and peaking) units, in addition to the more efficient base load equipment, to meet the additional demand for power during the daytime. With the availability of energy storage facilities, however, the less efficient units can be eliminated. The most efficient plants are operated continuously at the optimum (or rated) power level; excess electrical energy generated at night and during weekends is stored for use when the demand exceeds the base load. This procedure, called *load leveling,* can reduce the overall cost of generating electric power (see **Electric Power Generation**).

Methods for energy storage may be classified according to the form in which the energy is stored; the following categories appear to be the most important:

1. Mechanical Energy Storage
   Pumped hydroelectric storage
   Compressed air
   Flywheel
2. Chemical Energy Storage
   Storage batteries
   Hydrogen
   Reversible chemical reactions
3. Electromagnetic Energy Storage
4. Thermal Energy Storage
   Sensible heat
   Latent heat

Mechanical Energy Storage

Pumped Hydroelectric Storage. Electric power in excess of the immediate demand is used to pump water from a supply (e.g., lake, river, or reservoir) at a lower level to a reservoir at a higher level. When the power demand exceeds the supply, the water is allowed to flow back down through a **hydraulic turbine** which drives an electric generator. The overall efficiency of **pumped storage,** that is, the percentage of the electrical energy used to pump the water that is recovered as electrical energy, is about 70 percent.

Pumped hydroelectric storage is the most economical means presently available to electric utilities. It could also be used for storing electrical energy produced from solar or wind energy. There are relatively few suitable sites where there is a water supply at a lower level and a reservoir can be constructed at a higher level, but the use of natural or excavated underground caverns as lower reservoirs, now being developed, should greatly increase the number of possible sites.

Compressed-Air Storage. In a **gas turbine,** roughly 60 percent of the power output is consumed in compressing air for combustion of the gas. In the compressed-air storage system, electrical energy in excess of the demand is used to compress air which is stored in a reservoir for later use in a gas turbine to generate electricity. Compressed-air storage could serve for electric utility load leveling or for storing electrical energy generated from solar or wind energy. The overall recovery efficiency is estimated to be 65 to 75 percent.

In a conventional gas turbine, the compressor and turbine are connected (see Fig. 54). In a compressed-air energy storage system, however, the turbine and compressor are uncoupled so that they can operate separately. Furthermore, the electric generator, normally connected to the turbine, must also be capable of functioning as a motor (see **Electric Motor**) when electricity is supplied (Fig. 45).

Electric power in excess of the immediate demand is supplied to the motor/generator which drives the compressor; the compressed air, at about 70 atm (7 MPa), is stored in a suitable reservoir (see below). The air is heated during compression and may have to be cooled prior to storage to prevent damage to the reservoir walls. When additional power is needed to meet the demand, the compressed air is released and heated using gas or oil fuel. The hot compressed air is then expanded in a gas turbine connected to the motor/generator unit which now acts as a generator.

Compressed-air storage reservoirs would probably be too large and too expensive for aboveground construction; hence, underground reservoirs, preferably existing ones, are being considered. Among the possibilities are natural caverns, deep aquifers, depleted gas or oil reservoirs, mined-out rock or salt caverns, and abandoned mines.

A patent on compressed-air energy storage was issued in the United States in 1948 but there was little immediate interest. In view of the increasing costs of fuels and the need for conservation, the concept is attracting attention in several

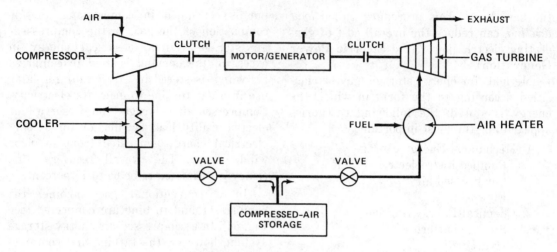

Fig. 45   Compressed-air energy storage.

countries, including the United States, and a commercial installation is in operation near Bremen, West Germany.

Flywheel Storage. A **flywheel** driven by an electric motor during offpeak hours stores mechanical (rotational) energy as its speed is increased. The rotation of the flywheel can be used to operate a generator to produce electricity when required. The same machine serves as both a motor, when electricity is supplied, and as a generator, when the armature is rotated by the flywheel. Flywheels could conceivably be used for electric utility peaking units, for storage of solar and wind energy, and for vehicle propulsion. The energy recovery efficiency is estimated to be up to 90 percent. Practical flywheel systems for such uses are being developed.

Chemical Energy Storage

Storage Batteries. When a **storage battery** is charged, by connecting it to a source of direct electric current, chemical changes occur in the battery materials. As a result, electrical energy is converted into stored chemical energy. When the battery is discharged, by connecting a load (e.g., a motor) between the terminals, the chemical reactions are reversed, and the stored chemical energy is reconverted into electrical energy. The energy recovery effi-

ciency of a storage battery varies with the type of battery and the rate of discharge, but 75 percent should be attainable. However, the efficiencies are often lower.

Batteries are modular in nature; that is to say, they are built up of individual units of moderate size. This means that the energy storage capacity can be varied over a wide range simply by varying the number of units that are connected together. Consequently, batteries are adaptable to any type of storage in which the energy input is in electrical form. Potential applications are utility **peak shaving** (and possibly load leveling), vehicle propulsion, and storage of electrical energy generated from wind energy or solar cells. The capability of rapid operation reversal, from charge to discharge, makes batteries especially convenient for electric utility applications. Moreover, they permit dispersed distribution by locating storage facilities near load centers.

By far the most common type of storage battery is the lead–acid battery used in the SLI (starting, lighting, and ignition) systems of essentially all automobiles and other road vehicles. These batteries are usually maintained in a fully charged condition by a generator which produces direct current when the engine is running. For utility applications or vehicle propulsion, the operating conditions are quite different.

In such uses, the battery is subjected to a deep (i.e., almost complete) discharge and then recharged, roughly once a day. When subjected to such repeated cycling, the lead-acid batteries used for automobile SLI systems have a short lifetime.

Studies in progress will undoubtedly result in the production of lead-acid batteries more suited to utility and vehicle applications. However, there are some basic limitations to lead-acid batteries (e.g., their heavy mass per unit of stored energy) that cannot be overcome. Consequently, several other kinds of storage batteries, using entirely different materials and some operating at high temperatures, are under development in the hope of finding a type (or types) that has a low (or moderate) cost, a long lifetime, and is lighter than the lead-acid battery. Further details are given in the section on **Storage Batteries.**

The battery requirements for vehicle propulsion are described under **Electric Vehicles;** the discussion here is concerned mainly with stationary (utility) applications. The main considerations are low (or moderate) cost and a long lifetime; the mass is less important than for vehicle propulsion. For stationary storage, a battery should be capable of at least 3000 deep discharges over a lifetime of 10 to 15 years. The discharge time for peak power supply would be 8 to 10 hours and the charge time roughly 10 hours. No existing storage battery can meet these requirements, but there is hope that one or more of the new types will eventually do so.

In order to understand and find solutions for the potential problems associated with storage batteries, a Battery Energy Storage Test (BEST) Facility is being operated by the Public Service Gas and Electric Company in New Jersey under the sponsorship of the U. S. Department of Energy and the Electric Power Research Institute. Plans are being made to supplement the BEST facility by the Storage Battery for Electric Energy Demonstration (SBEED) project to be completed in 1984. The SBEED facility, with a storage capac-

ity of 30,000 kilowatt-hours, will be connected to an electric utility system to provide operational experience. Because they are the only reliable storage batteries available at a reasonable cost, lead-acid batteries will be used, at least initially, in the BEST and SBEED facilities. Subsequently, more advanced battery types will be tested.

Central-station power plants usually generate and transmit alternating current (ac), but direct current (dc) is required for battery charging. Furthermore, upon discharge, the batteries produce dc which must be converted into ac for feeding into transmission lines. Hence, the battery test facilities will involve the testing of equipment for converting ac into dc and dc back to ac in an efficient manner.

Hydrogen Storage. Energy can be both stored and transported as hydrogen, which serves as a secondary fuel (see **Hydrogen Fuel**). The input energy, usually electrical but possibly thermal, serves to decompose water (H_2O) by electrochemical (or chemical) reaction into its constituent elements hydrogen and oxygen. These substances can then be recombined to release the stored energy as required. Instead of using the oxygen produced from water in this energy recovery process, oxygen from the air is commonly employed. The pure oxygen from water can then be sold for industrial applications (e.g., in iron and steel fabrication).

Hydrogen can be transported either as compressed hydrogen gas, as liquid hydrogen (at low temperature), or in the form of a solid compound with certain metals or alloys. Consequently, hydrogen may be useful as a means of storing and transporting energy generated in remote locations far from load centers.

The most convenient means for producing hydrogen and oxygen from water is by electrolysis, that is, by passing a direct electric current through water containing an acid or alkali to make it an electrical conductor. The input energy is then in the form of electrical energy. It may also be possible to decompose water by heat (i.e., with thermal energy input) as a result of a series of chemical reactions. Several such processes are under investigation (see **Hydrogen Production**).

The chemical energy in hydrogen (and oxygen) can be converted into thermal, mechanical, or electrical energy. One possibility is to burn hydrogen in air, in a manner similar to natural gas, to produce heat (thermal energy) for use in the home or in industry. Hydrogen can also serve as the fuel, in place of gasoline, in automobile, truck, and even aircraft engines.

Electrical energy can be obtained from hydrogen in several ways. For example, steam from a water boiler heated by burning hydrogen could be used to drive a conventional **steam turbine** with attached electric generator. Alternatively, hydrogen can provide the fuel for a gas (combustion) turbine which in turn drives a generator. The maximum overall efficiency (possibly 55 to 60 percent) for recovery of the input energy, however, would be obtained by means of a **fuel cell**; in such a cell, electrical energy is generated directly from hydrogen and oxygen. Fuel cells could conceivably be used in homes and apartment houses, by industry, for **peak shaving** by utilities, and in electric vehicles.

Reversible Chemical Reactions. In a general sense, a reversible chemical reaction is one that proceeds simultaneously in both directions. Of interest here are reversible reactions that occur predominately in one (forward) direction with the absorption of heat at a higher temperature, and predominately in the opposite (reverse) direction with the emission of heat at a lower temperature. The products of the forward reaction store thermal energy (heat) as chemical energy which can be recovered as thermal energy when the conditions are changed to permit the reverse reaction to occur. This kind of energy storage may be useful for storing high-temperature heat obtained by concentrating solar energy.

To be suitable for heat storage, the reaction system should involve materials that are inexpensive and not too difficult to handle. In addition, the forward and reverse reactions, as required, must be able to proceed at reasonable temperatures. A catalyst (or catalysts) may be needed to speed up the desired reaction, especially at the lower temperature.

The following are some reversible reaction systems that have been proposed for energy storage:

$$CH_4 + H_2O \rightleftharpoons CO + 3H_2$$

Methane, Water (vapor), Carbon monoxide, Hydrogen

$$CH_4 + CO_2 \rightleftharpoons 2CO + 2H_2$$

Methane, Carbon dioxide, Carbon monoxide, Hydrogen

$$NH_4HSO_4 \rightleftharpoons NH_3 + SO_3 + H_2O$$

Ammonium hydrogen sulfate, Ammonia, Sulfur trioxide, Water

In each case, the forward reaction (i.e., from left to right) predominates at the higher temperature accompanied by the absorption of heat; the reverse reaction (i.e., right to left) occurs preferentially at a lower temperature with the evolution of heat.

Strictly speaking, the decomposition of water into hydrogen and oxygen by the addition of energy and the recombination of hydrogen and oxygen to release energy, as described earlier, is an example of the use of a reversible process for energy storage. However, it differs in some respects from the processes considered here.

Electromagnetic Energy Storage

Electromagnetic energy storage requires the use of superconducting materials; these materials (metals and alloys) suddenly lose essentially all resistance to the flow of electricity when cooled below a certain very low temperature (see **Superconductivity**). If maintained below this temperature a superconducting metal (or alloy) can carry strong electric currents with little or no loss. Useful superconducting materials available commercially are a niobium-titanium (Nb-Ti) alloy at temperatures below −442°F (−263°C) and a compound of niobium and tin (Nb₃Sn) below −427°F (−255°C).

An electromagnetic field, such as is produced by an electric current flowing through a wire (or wire coil), can store energy. Under ordinary conditions, losses result from the resistance of the wire, and energy must be supplied continuously to maintain the current. But if the wire coil were made of a superconducting material and kept at the required low temperature, resistance losses would be very small and, once initiated, an electric current would remain almost constant. Electrical energy supplied as direct current to the wire coil would then be stored in the electromagnetic field. By attaching the coil to a load, the stored energy could be recovered as electrical energy (direct current). This is the basis of the Superconducting Magnetic Energy Storage (SMES) system under study for possible eventual use in electric utility load leveling and/or peaking applications.

A number of problems are associated with superconductivity storage. They include the operation and maintenance of a cryogenic (i.e., refrigeration) plant for producing the liquid helium required for the very low superconductivity temperatures. Furthermore, special structures would be needed to withstand the strong magnetic field of an SMES unit. It appears at present that superconductivity storage would, at best, be economic only in large installations. On the whole SMES is rated as being a promising but long-term prospect for energy storage.

Thermal Energy Storage

The need for storing thermal (or heat) energy arises in several situations, for example, in connection with the applications of solar and wind energy. The storage methods in common use are described in the sections dealing specifically with these applications. The discussion here will deal, therefore, with the general principles of heat storage as either sensible heat or latent heat.

Sensible Heat Storage. Sensible heat refers to thermal energy that results in an increase of temperature when added to a material (or decrease of temperature when taken from it). For example, water at temperatures above its freezing point and below its boiling point can store energy as sensible heat. So also can other liquids, air, and solids provided they do not change their form (e.g., by freezing, melting, or boiling). Water or ceramic bricks are used for heat storage on a large scale in industry, and either water or a combination of air and rock provide storage for heat derived from solar energy. Thermal energy storage as sensible heat in high-temperature steam can provide utilities with a means for generating electric power to satisfy peak demands.

Storage of cold water for use in space cooling (i.e., air conditioning) is a form of sensible heat storage. For example, refrigeration equipment operated by electric power could be used to cool water at night when the power demand is low; the cold water would then provide cooling in the daytime. Another possibility is to cool a reservoir of water in the winter by means of **heat pumps**; heat withdrawn from the water is used for space heating. The cold water in the reservoir would then provide cooling in the summer (see **Annual Cycle Energy System**).

If the temperature is too high for the use of water without pressurization, special hydrocarbon oils with high boiling points, either alone or mixed with rock, can serve to store sensible heat. Molten salt mixtures can be used for heat storage at very high temperatures. One such mixture is Du Pont's HITEC, containing sodium nitrite and sodium and potassium nitrates; it melts at 288°F (142°C) and can be used to store sensible heat up to about 1000°F

(540°C). Mixtures of sodium and potassium nitrates have been proposed for somewhat higher temperatures. For temperatures up to about 1500°F (815°C), molten fluorides (e.g., lithium fluoride and mixtures of sodium and magnesium fluorides and of sodium and zinc fluorides) are possibilities.

Latent Heat Storage. Latent heat is thermal energy that is stored in (and can be removed from) a substance or mixture when it undergoes a change of phase (e.g., of physical form) while the temperature remains unchanged. The heat that can be stored per unit mass (or volume) in this manner is usually several times greater than for sensible heat storage.

The phase change from solid to liquid, taking place at the melting point of the solid, is accompanied by the absorption of latent heat without a change in temperature. The heat is recovered when the process is reversed (i.e., the liquid is converted to solid) at the same temperature. Hence, solid and liquid phases of the same material when present together can store thermal energy as latent heat. Saturated steam (i.e., steam and moisture) and freezing water (i.e., water and ice) can be used for latent heat storage.

Eutectic mixtures of stable salts have been proposed for energy storage. (A eutectic is a mixture of two or more solid substances with a lower melting point than the substances alone or any other mixture.) A eutectic mixture melts and solidifies at a definite temperature, with the absorption and release, respectively, of latent heat, just like a pure substance. The material HITEC and the fluoride mixtures referred to above are actually eutectic mixtures; however, the compositions are selected because of the minimum melting point, rather than the capability of storing latent heat. At temperatures above the melting point, the liquids store sensible heat.

Another type of phase change, involving a change in chemical composition, that has been used for latent heat storage, is the transition between solid sodium sulfate decahydrate ($Na_2SO_4 \cdot 10H_2O$), on the one hand, and solid anhydrous sodium sulfate (Na_2SO_4) and an aqueous solution, on the other hand. At a temperature of 90°F (32°C), heat is stored when the $Na_2SO_4 \cdot 10H_2O$ phase changes to the Na_2SO_4 phase (plus solution) and can be recovered when the process is reversed (see **Solar Energy: Direct Thermal Applications**). Latent heat storage occurs only at 90°F where both solid phases can coexist. Sensible heat storage is possible, although to a smaller extent, either at lower temperatures (in $Na_2SO_4 \cdot 10H_2O$ alone) or at higher temperatures (in Na_2SO_4) with or without solution.

A somewhat similar phase transition between sodium thiosulfate pentahydrate ($Na_2S_2O_3 \cdot 5H_2O$) and the anhydrous salt ($Na_2S_2O_3$) plus solution occurs at 118°F (48°C). It has been considered for latent heat storage of solar energy for space heating.

ENTRAINED BED

A bed of solid particles suspended in a fluid (liquid or gas) flowing at such a rate that some of the solid is carried over (i.e., entrained) by the fluid. The flow rate is greater than in **fixed** or **fluidized beds.** Entrained beds are used in several **coal gasification** processes (e.g., **BI-GAS, Koppers–Totzek,** and **Texaco Processes**); the particles carried by the gas flow are usually **char** remaining from decomposition of the coal.

EXPANSION TURBINE (OR TURBOEXPANDER)

A **heat engine,** operating on an approximate **Brayton cycle,** in which the energy of a compressed gas is converted into mechanical work in a **turbine.** The expansion turbine is a type of **gas turbine** in which the working gas is at a much lower temperature than in a conventional gas turbine.

The gas expands in the turbine and does work; as a result the temperature and pressure fall and the gas is discharged at a lower temperature and pressure. Expansion turbines are utilized in the petroleum industry to recover energy from gases generated at high pressures. The decrease in temperature of the gas in passing through the turbine is also used to some extent for **refrigeration.**

EXXON COAL GASIFICATION PROCESS

An Exxon Corporation process for **coal gasification** with steam in the presence of a catalyst (potassium compound). An unusual feature of the process is that the heat required for the carbon (coal)-steam reaction is supplied by preheating the steam and recirculated gas in an external furnace which is independent of the gasifier vessel. Another special aspect is that the process leads directly to a **high-Btu fuel gas** (or **substitute natural gas**), consisting largely of **methane,** without the need for the **water-gas shift** and **methanation reactions.** Any type of coal can be used.

Coal and the potassium hydroxide catalyst are fed to the gasifier together with steam and recycled gas; the temperature is 1200 to 1300°F (650 to 705°C) and the pressure is about 34 atm (3.4 MPa). The catalyzed reaction of the coal and steam produces a gas consisting of hydrogen, carbon monoxide, and methane with a small proportion of inert carbon dioxide. The gas is passed through a waste-heat (**heat exchanger**) boiler, where process steam is generated, and then to a cryogenic (i.e., low-temperature) separator. Here the carbon dioxide is removed as a solid, and the methane is separated from the other gases as a liquid. At ordinary temperature this liquid vaporizes and becomes the gaseous product of the process (Fig. 46).

The hydrogen and carbon monoxide gases which remain after the cryogenic treatment are mixed with steam and preheated in a furnace to a temperature about 150°F (83°C) above that in the gasifier. The mixture is then injected into the gasifier to provide the steam and heat required for the process. By recirculating the hydrogen and carbon monoxide in this way, the catalytic formation of methane from coal and steam is favored. The **char** and ash residue removed from the gasifier contains the catalyst (as potassium carbonate) which is recovered by extraction with water.

EXXON DONOR SOLVENT (EDS) PROCESS

An Exxon Corporation process for **coal liquefaction** with a hydrogen-rich (donor) liquid solvent; a catalyst is not required.

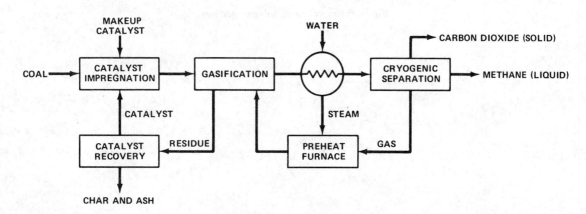

Fig. 46   Exxon coal gasification process.

The process can be used with almost any type of coal, and the liquid hydrocarbon products, mainly light and middle **distillate fuel oils,** can be varied according to circumstances.

The coal is ground and formed into a slurry with the hydrogen-donor solvent, which is one of the products of the process, and heated in an externally fired heater. The heated slurry is fed with preheated hydrogen gas to a reactor in which the temperature is about 800°F (425°C) and the pressure 120 atm (12 MPa). The coal is hydrogenated, and the resulting gaseous and liquid products are removed and separated by **distillation** (Fig. 47).

A portion of the gas, consisting mainly of simple paraffin **hydrocarbons,** is treated by **steam reforming** or partial oxidation to produce the hydrogen required for the process (see **Hydrogen Produc-** tion); the remaining gas can be used as a fuel gas. Part of the middle distillate is subjected to catalytic hydrogenation (i.e., with hydrogen gas in the presence of a catalyst) and is recycled as the donor solvent to the coal slurrying operation. The remainder and the light distillate are the major products of the EDS process. They may be treated and refined, if necessary, as in **petroleum refining,** for use as engine or boiler fuels.

The distillation residue (or "bottoms"), composed largely of unreacted coal and heavy oils, may be used for hydrogen production, or it can be reacted with steam and air in the **Flexicoking process.** In the latter case, the products are a **low-Btu fuel gas,** which is used to provide process heat, and a liquid that can be treated with the main liquid product of the EDS process.

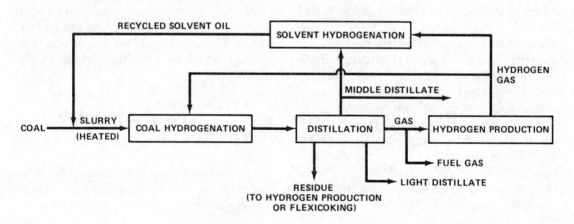

Fig. 47   Exxon donor solvent process.

F

FAST REACTOR

A nuclear fission reactor in which most of the fissions are caused by the absorption of fast neutrons (see **Nuclear Power Reactor**). The main purpose of fast reactors is to breed fissile plutonium-239 from fertile uranium-238 (i.e., produce more plutonium-239 than is consumed) while generating useful power. This type of breeding is not possible in a **thermal reactor** in which the fissions are caused mainly by slow neutrons (see **Breeder Reactor**). Thus, fast reactors are usually designed to be breeders.

Neutrons released in the fission process have high energies (i.e., they are fast neutrons). In order to prevent appreciable slowing down and maintain a high neutron energy distribution (i.e., a fast spectrum), the presence of elements of low mass number (see **Atom**) must be avoided as far as possible. These elements act as moderators to slow down the fast neutrons. Thus, in contrast to a thermal reactor, a fast reactor does not include a moderator.

For this reason, and also because fuel of higher enrichment (15 to 20 percent) in the fissile species is required (compared to 2.5 to 3 percent in most thermal reactors) to maintain the fission chain with fast neutrons, the core of a fast reactor is generally smaller than a thermal reactor of the same power output. The compact nature of the core makes heat removal more difficult. Water cannot be used as coolant because of its moderating properties, and the preferred coolant is liquid (i.e., molten) sodium (see **Liquid-Metal Fast Breeder Reactor**). Although the latter is more difficult to handle, it has some advantages over water as a heat-removal agent. Helium gas has been proposed as a possible alternative to sodium as coolant for fast reactors (see **Gas-Cooled Fast Reactor**).

In the earliest fast reactors, metal (elemental) fuels, rather than compounds, were used because they would give higher breeding ratios. However, the relatively low melting points of metallic uranium and plutonium and the dimensional changes that occur upon exposure to neutrons at high temperatures are drawbacks. Consequently, ceramic dioxide (UO_2 and PuO_2) fuels, with very high melting points, have come into general use (see **Nuclear Fuel**). The oxygen in the dioxides causes a limited slowing down of the neutrons, with the result that there is some sacrifice in the breeding ratio.

Higher breeding ratios may be possible with monocarbide (UC and PuC) fuels which are under development. Since they contain only one low-mass atom per molecule, compared with two for the dioxides, there is less slowing down of the fission neutrons. The larger proportions of fuel (uranium and plutonium) atoms are also advantageous.

FISCHER ASSAY

A laboratory method for determining the amount of oil that can be recovered from **oil shale** by heating in the absence of air; the method has also been applied to **coal**. In the Fischer assay of oil shale, a 100-gram sample of the shale is heated at a specified rate to a temperature of 932°F (500°C) in an aluminum retort. The temperature is maintained for 20 min and the vapors produced are cooled and condensed at 32°F (0°C). The volume of oil in the distillate is then measured. The assay is expressed as U.S. gallons of oil per (short) ton of shale (1 gal/ton = 4.2 liters/metric ton, where 1 metric ton = 1000 kg). Because the conditions used for producing oil from shale are generally different from those in the Fischer assay, the oil yield may be greater or less than indicated by the assay.

FISCHER–TROPSCH PROCESS

A process invented by F. Fischer and H. Tropsch in Germany in 1923 for **coal liquefaction,** based on the catalytic conversion of **synthesis gas** (i.e., a mixture of hydrogen and carbon monoxide) into mainly liquid and some gaseous **hydrocarbons.** Unlike the products of more direct coal liquefaction processes which include aromatics, the hydrocarbons made from synthesis gas are mainly paraffins and olefins and are more easily refined into **gasoline** and **diesel fuel.** In addition to hydrocarbons, some oxygenated compounds, such as methanol (see **Alcohol Fuels**), are produced from the synthesis gas.

The Fischer-Tropsch process was used in Germany in World War II to make gasoline and other fuels, but to a lesser extent than the **Bergius process.** However, since 1955, liquid fuels have been produced to an increasing extent at the SASOL plant in the Republic of South Africa by means of the Fischer-Tropsch process. A form of this process is being considered for coal liquefaction in the United States.

In South Africa, synthesis gas is made from coal with steam and oxygen by the **Lurgi Dry Ash process.** But since synthesis gas is essentially an **intermediate-Btu fuel gas,** several other processes could be used. The gas is treated for the removal of carbon dioxide, hydrogen sulfide, and other impurities (e.g., by the **Rectisol process**), and the clean synthesis gas passes to a catalytic reactor. Many different substances, including fuels, petrochemicals, and fertilizers, are made at the SASOL plant, but emphasis here will be on the fuel products.

Two catalytic procedures are used at SASOL (Fig. 48). In the Arge process, the synthesis gas, consisting of hydrogen, carbon monoxide, and a few percent of methane, is passed through a **fixed-bed** catalyst at a temperature of about 450°F (230°C) and 25 atm (2.5 MPa) pressure. The catalyst is reported to be based on iron with a small quantity of a "promoter." The product is cooled so that gas and liquid are separated.   Part of the gas is used as described below, while the liquid is a **synthetic crude oil** (or syncrude) which can be refined in the usual manner (see **Petroleum Refining**). The major products are diesel fuel and furnace oils, with a relatively small proportion of gasoline.

In the alternative Synthol (Kellogg) process, the purified synthesis gas, together

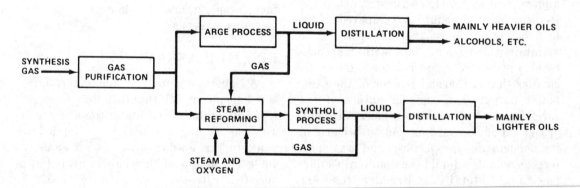

Fig. 48   Fischer–Tropsch process.

with residual gas from the Arge process, is first reacted with steam (**steam reforming**) and oxygen. This removes the methane and increases the ratio of hydrogen to carbon monoxide. The resulting gas is then injected into a **fluidized bed** of catalyst, said to be also based on iron with a promoter, at 625°F (330°C) temperature and 22 atm (2.2 MPa) pressure. The product oil is a syncrude which yields a large proportion of gasoline upon refining and distillation. The residual gas is recycled to the fluidized-bed catalyst by way of the steam reformer.

Methanol and other alcohols are produced, especially in the Synthol process. The methanol could be used as a fuel or it could be converted into gasoline by the **Mobil M-Gasoline process.**

FISSION ENERGY

A form of **nuclear energy** released in the fission (or splitting) of heavy nuclei (i.e., nuclei with high mass and atomic numbers) into two nuclei of intermediate mass (see **Atom**). The particular type of fission of interest for the generation of useful energy is that caused by absorption of a free neutron by an appropriate heavy nucleus. The fission process is accompanied not only by the release of energy but also by the emission of neutrons, called *fission neutrons*; two to three neutrons are emitted on the average per nucleus undergoing fission. The emitted neutrons can cause fissions in additional nuclei, thus liberating more neutrons which can cause further fissions, and so on. However, for a self-sustaining fission chain to be possible with the continuous release of energy, certain requirements must be met.

$$\text{Neutron} + \begin{matrix}\text{Heavy}\\\text{nucleus}\end{matrix} \rightarrow \begin{matrix}\text{Fission}\\\text{fragments}\end{matrix} + \begin{matrix}\text{Neutrons}\\\text{(2 to 3)}\end{matrix} + \text{Energy}$$

One requirement is that the heavy nuclei must be such that they can be fissioned by neutrons of any energy; these substances are referred to as *fissile species.*

The fission neutrons initially have a high kinetic (or motion) energy, but this decreases rapidly as a result of collisions with various nuclei. For a fission chain to be sustained, the neutrons of lower energy must still be capable of causing fission; this is possible only with a fissile species.

Substances with nuclei that can be fissioned by high-energy neutrons but not by neutrons of low or intermediate energy are said to be *fissionable* (but not fissile). Fissionable species (e.g., uranium-238) make some contribution to the energy released in a **nuclear power reactor,** but they cannot sustain a fission chain. The reason is that most of the fission neutrons have their energies rapidly reduced below the threshold value at which they can cause fission in the fissionable species.

All fissile substances are radioactive, and to be useful as practical energy sources they should have reasonably long half-lives so that they do not decay too quickly (see **Radioactivity**). Furthermore, they should exist in nature in moderate quantities, at least, or be capable of fairly easy production from available materials. Only three fissile species satisfy those conditions: uranium-233, uranium-235, and plutonium-239. (The latter is generally associated with several percent of fissile plutonium-241 which has a half-life of 14 years; this should be understood when reference is made to plutonium-239.) Uranium-235 constitutes 0.71 weight percent of all natural forms of uranium, but the other fissile species are made artificially.

The sources of plutonium-239 and uranium-233 are uranium-238 and thorium-232, respectively; substantial amounts of these substances, called *fertile species*, exist on earth. Both uranium-238 and thorium-232 are also fissionable species (see above), but this is less important than the fact that they can be converted into fissile species. By the capture of neutrons followed by two stages of beta-particle emission, uranium-238 is changed into plutonium-239 whereas thorium-232 forms uranium-233 (see **Nuclear Fuel**).

Critical Mass

Another requirement for a self-sustaining fission chain is concerned with the mass (or quantity) of fissile material. Although two to three neutrons are produced for every nucleus undergoing fission, not all of these fission neutrons are available to cause further fissions. Some of the fission neutrons are lost in various nonfission reactions, both with the fissile species and with other substances that may be present, whereas other neutrons escape. In order to sustain a fission chain, with the continuous release of energy, at least one neutron must be available to cause fission for every neutron previously absorbed in fission. (Note that a proportion of the neutrons interacting with fissile nuclei are captured without causing fission.)

In a small quantity of fissile material, the neutrons may be lost so rapidly by escape from the surface, compared with the rate at which they are produced by fission in the interior, that a self-sustaining chain will not be possible. If fission chains are initiated by introducing neutrons into this small quantity of fissile material, the chains will soon die away. The escape of fission neutrons from the surface, relative to their production by fission, can be decreased by increasing the size (or mass) of the fissile material. The minimum quantity of such material necessary for a self-sustaining fission chain is called the *critical mass*.

A system consisting of a critical mass of fissile material is said to be *critical*; fission chains initiated in such a system are just self-sustaining. If the mass is less than the critical value, the system is *subcritical* and fission chain reactions cannot be sustained. On the other hand, in a *supercritical* system the mass exceeds the critical amount. In such a system the number of fission chains increases with time.

The critical mass depends on several conditions, including the nature and proportions of the fissile species and of other substances that are present in the system. However, the critical mass has a definite value for specified conditions. The variation of the critical mass of uranium-235 with the conditions is illustrated by the following examples: it is about 2.2 lb (1 kg) when in the form of a salt dissolved in water, perhaps 100 lb (45 kg) or so as a bare sphere of metal, and 460 lb (210 kg) when present in about 33 tons (30,000 kg) of natural uranium arranged in a matrix of graphite. Natural uranium alone cannot become critical, no matter how large the mass, because it does not contain enough of the fissile isotope uranium-235.

If neutron absorbers are present, they remove some of the neutrons which would otherwise have been available for maintaining the fission chains. As a result, the mass of fissile material required for criticality is increased. A previously supercritical mass can thus be made critical, and a critical mass can be made subcritical by introduction of a neutron absorbing material. Removal of the absorber will reverse the situation. This is the basis of the method commonly used to control the fission rate (and power output) of a nuclear power reactor.

Fission Energy

The energy released in fission can be determined by means of the Einstein equation for the equivalence of mass and energy (see **Nuclear Energy**). The difference between the masses of the fissile nucleus and a neutron, on the one hand, and that of the fission end products, including beta particles (see below) and fission neutrons, on the other hand, is directly related to the total energy release. The individual masses are known, and from these the typical decrease in mass accompanying fission is found to be about 0.21 atomic mass unit (see **Atom**); this means that roughly 200 million electron volts (MeV) of energy are released per nucleus undergoing fission. (The electron volt is defined in the article on **Nuclear Energy**.) The equivalent in more familiar energy units is given later.

About 90 percent of the fission energy is released instantaneously; this is mainly kinetic (motion) energy of the two lighter nuclei (or *fission fragments*) formed by the fission of the heavy nucleus and, to a small extent, kinetic energy of the fission neutrons and energy of prompt gamma rays. These gamma rays are emitted by high-energy (excited) states of some of the fission fragment nuclei (see **Radioactivity**). The remaining roughly 10 percent of fission energy is present as internal energy of the fission fragment nuclei. It is released as energy of the beta particles (and associated neutrinos) and gamma rays emitted over a period of time by the radioactive fission fragments and their decay products.

The distribution of the energy released in the fission of a uranium-235 nucleus is given in the table. The numbers are approximately the same for all three fissile species referred to earlier. The neutrinos

Approximate Distribution of Energy per Fission

Instantaneous	*MeV*
Kinetic energy of fission fragments	168
Kinetic energy of fission neutrons	5
Prompt gamma rays	7
Delayed	
Beta particles from fission products	7
Neutrinos from beta decay	10
Gamma rays from fission products	6
	~200

are neutral, massless particles which carry about two-thirds of the energy associated with beta-particle decay; they travel very great distances through matter before losing their energy.

In a nuclear power reactor, designed for the practical utilization of fission energy, nearly all the energy, with the exception of the neutrino energy, is deposited as heat within the system. The neutrinos escape from the reactor carrying off about 5 percent of the fission energy, but this loss is largely compensated by energy released within the reactor by nonfission nuclear

reactions with neutrons. In view of the variations that occur in practice, a good approximation is to take 200 MeV per fission to be the amount of energy deposited in a reactor as heat.

Equivalents of the available heat energy of 200 MeV per fission are given below per kilogram and per pound of fissile material, assuming that it all undergoes fission. The combustion of 1 lb of bituminous coal produces about 13,000 Btu of heat (see **British Thermal Unit**); hence, 1 lb of fissile material can, in principle, generate as much heat as the combustion of 2.6 million lb of coal. In other words, complete fission of a certain mass of fissile material will produce as much heat energy as 2.6 million times that mass of bituminous coal. Fissile materials thus represent highly concentrated sources of energy.

Fission Energy Equivalents

Per pound	Per kilogram
10 million kW-hr	22 million kW-hr
34 billion Btu	81×10^{12} joules

The foregoing discussion is based on the assumption that all the fissile material consumed is fissioned. However, in a reactor, some of the fissile nuclei capture neutrons and are lost in nonfission reactions. Consequently, the heat energy generated per unit mass of fissile material consumed is less than given above. In the common types of commercial power reactors, in which uranium-235 is the main fissile material, about 86 percent of the nuclei consumed actually undergo fission. Hence, the heat energy produced is approximately 8.6 million kilowatt-hour (kW-hr) per lb (19 million kW-hr per kg) of uranium-235 consumed (see **Watt**).

Fission Neutrons

The number of neutrons released in fission of a given species depends to some extent on the energy of the neutrons causing the fission. For neutrons of low and moderate energy, such as are dominant in most power reactors, fission is usually accompanied by the emission of two or three neutrons. There are actually many different modes of fission (see below) and the average number of neutrons released per fission has a fractional value between 2 and 3. These average numbers for the three common fissile species under the conditions of a **thermal reactor** are given in the table. (For plutonium-241, the number is 2.93.) When fission is caused by neutrons of sufficiently high energy, the number of fission neutrons may be three or four (or more) per fission.

Fissile species	Neutrons per fission (average)
Uranium-233	2.49
Uranium-235	2.42
Plutonium-239	2.87

Fission Products

When a heavy nucleus fissions, it splits in one or another of about 40 different possible ways (or modes). Hence, when a large number of nuclei undergo fission, as is usually the case, all 40 modes will be occurring simultaneously, although to widely different extents. Since each mode produces two fission fragments, roughly 80 such fragments are formed. Essentially all the fission fragments are radioactive, and they immediately start to decay by the emission of beta particles. On the average, each fragment is the parent of a four-stage decay chain; consequently, a complex mixture consisting of more than 300 different radioactive species is formed. The general term *fission products* is applied to this highly radioactive mixture of fission fragments and their decay products. A few of the fission products are gases or vapors, but most are solids.

The mass numbers of the fission fragments have been found to range from roughly 72 to 161; in the fission of

uranium-235, the pair with mass numbers 95 and 139 is the most common. Since the mass number remains unchanged in the emission of a beta particle, these numbers apply to the fission products as a whole. However, the atomic numbers, that is, the identities of the elements, change as the fission products decay. Isotopes of at least 37 different elements, with atomic numbers from 30 (zinc) to 66 (dysprosium), have been found among the decaying fission products.

Although each of the fission products has a definite half-life, the overall rate of decrease in activity of the complex mixture cannot be expressed in a simple manner.

FIXED BED

A bed of solid particles in intimate contact with a fluid (liquid or gas) flowing at a rate that is not sufficient to cause any significant movement of the solid. Fixed beds may be contrasted with **entrained** and **fluidized beds** in which the flow rates are higher. Fixed beds are used in some **coal gasification** processes (e.g., **Lurgi** and **Wellman-Galusha processes**) and catalytic hydrogenation reactions (see **Hydrotreating**).

FLASH HYDROGENATION (OR HYDROPYROLYSIS) PROCESS

A process being developed by Rockwell International Corporation and Cities Service Company for **coal gasification** or partial **coal liquefaction** and gasification by direct reaction of hydrogen gas at high temperature. The carbon in coal combines with hydrogen to form either **methane** gas as the main product or liquid **hydrocarbons** in addition to methane, depending on the operating conditions. The process can be used for caking as well as noncaking coals. The term "flash" in the title arises from the very short reaction time of the coal and hydrogen.

A special feature of the process is the injection device, based on the design of aerospace rocket fuel injectors, for introducing the coal and hydrogen gas into the reactor vessel. An injector unit consists of a central coal feed with a number of surrounding hydrogen jets directed to impinge on the coal. This arrangement permits very rapid mixing, heating, and reaction of the pulverized coal with high-temperature, pressurized hydrogen gas. In practice, several such injector units are located at the top of the cylindrical reactor vessel.

The hydrogen is preheated to about 1500°F (815°C) in a **heat exchanger** with hot gases and then to 2000°F (1095°C) by combustion with a small amount of injected oxygen gas. The temperature in the reactor is maintained at 1500 to 1900°F (815 to 1035°C) by the heat liberated in the formation of methane (see **Coal Gasification**). The reactor pressure can be in the range of 34 to 136 atm (3.4 to 13.8 MPa).

At the bottom of the reactor vessel, the gases are quenched to a temperature of 900°F (480°C) or less with water under pressure. As a result, the reaction with hydrogen virtually ceases. The gases (and vapors) pass to a condenser where liquid products, if any, are removed, leaving the gaseous products (and excess hydrogen). Unreacted **char** removed from the reactor vessel would be used to produce the hydrogen (or hydrogen-rich gas) required for the process (see **Hydrogen Production**).

The relative amounts of gaseous and liquid products of the process are determined by the residence time of the coal and hydrogen in the reactor. (The residence time is the time interval between injection and quenching; it depends on the rate of injection and the height of the reactor vessel.) For short residence times (e.g., one-tenth of a second or less) roughly equal amounts of gaseous and liquid products can be obtained. As the residence time is increased, further hydrogenation occurs, and at times of 1 or 2 sec (or more) only gaseous products result especially at higher temperatures.

After removal of excess hydrogen, the residual gas is about 90 volume percent methane and 10 percent carbon monoxide. The latter can be converted into methane by the **methanation** reaction with hydrogen, thereby producing a **high-Btu fuel gas** (or **substitute natural gas**).

The liquid product of the flash hydrogenation process can be divided into two main distillation fractions: a **naphtha** fraction boiling below 400°F (205°C) and a fraction with a higher boiling range. The relative amounts of these fractions can be changed by varying the proportions of hydrogen gas and coal in the reactor feed. The naphtha fraction is of special interest because it consists largely of benzene, a commercially valuable material usually made from petroleum (see **Petroleum Products**).

FLASH POINT

The lowest temperature at which the vapor from a liquid fuel will ignite (i.e., cause a flash of flame to occur) in air under specified test conditions. The flash point provides an indication of the potential fire hazard associated with the liquid and is generally applied to **hydrocarbon** mixtures, such as **kerosine, jet fuel, diesel fuel,** and **fuel (heating) oil.** The flash point is measured by heating the fuel at a controlled rate in air in a semiclosed vessel; a small flame or an electrically heated wire is applied from time to time. The temperature of the fuel when a flame flashes through the air–vapor mixture and is then extinguished is the flash point. The fire point is that (higher) temperature at which burning is sustained for more than 5 seconds. At the spontaneous (or autogenous) ignition temperature the air–vapor mixture will ignite in the absence of a flame or other hot source.

Approximate flash points of fuel oils are given below:

Kerosine and jet fuels	100 to 140°F	38 to 60°C
Diesel fuels	above 140°F	above 60°C
Fuel oils	above 150°F	above 66°C

FLEXICOKING PROCESS

A commercial process developed by the Exxon Research and Engineering Company for converting heavy (i.e., high density, viscous) oils, such as **heavy crude oil, tar sands** bitumen, and residues (or "bottoms") from the **distillation** of petroleum and **coal liquefaction** products (see, for example, **Exxon Donor Solvent Process**), into light **hydrocarbons.** The heavy oil is thermally cracked (see **Cracking**) to yield a mixture of simpler liquid products; the residual coke is reacted with a mixture of steam and air to produce a **low-Btu fuel gas.**

A Flexicoking unit consists of three main components: reactor (cracker), heater, and gasifier. The heavy oil feed is sprayed into the reactor vessel where it is cracked at a temperature above 1000°F (540°C). The heavier vapor products are condensed in a scrubber at the top of the reactor vessel and are returned for further cracking. The lighter vapors are cooled to form the liquid product of the process (see below).

The **coke** (or **char**) formed in the reactor vessel is carried by steam to the heater where its temperature is increased, as explained shortly. Part of the heated coke is recirculated to the reactor to provide the heat required for cracking, and the remainder is transferred to the gasifier (Fig. 49). Here the coke reacts with air and steam at a temperature of 1500 to 1800°F (815 to 980°C) to produce a gas, called coke gas, containing carbon monoxide and hydrogen as the main fuel components and nitrogen (from air) and carbon dioxide as inert diluents (see **Coal Gasification**).

The heat that maintains the cracking temperature is generated in the gasifier by the combustion of the coke in air. It is transferred to the reactor through the heater partly by the hot coke gas and partly by circulation of unburned coke between the gasifier and heater.

The coke gas leaving the heater vessel at about 1100 to 1200°F (590 to 650°C) is cooled by generating steam in a waste-heat (**heat exchanger**) boiler and is then treated for hydrogen sulfide removal (e.g., by the **Stretford process**). The product is a clean low-Btu fuel gas with a **heating value** of roughly 120 Btu/cu ft (4.4 MJ/cu m).

The liquid product, condensed from the vapor leaving the reactor, is a mixture of light and medium hydrocarbons. It may be used without further treatment as a low-sulfur **fuel (heating) oil** or it may be subjected to fractional **distillation,** catalytic cracking, and other conventional **petroleum refining** processes to produce **gasoline, diesel fuel,** etc.

FLUIDIZED BED

A bed of small solid particles with a fluid, usually a gas, flowing upward through it. The velocity of the gas flow exceeds that

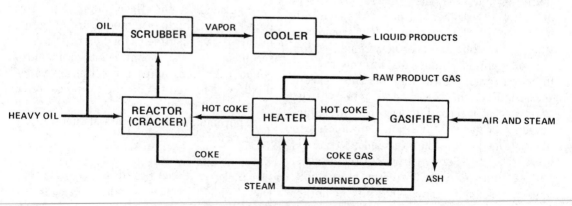

Fig. 49 Flexicoking process.

required to support the particles against the force of gravity but is not sufficient to carry them out of the container. The solid particles are then suspended in the gas in a state of turbulent motion, and there is intimate contact of the particles with the gas. As a result, chemical reaction (including catalysis) and heat transfer between the solid and the gas are greatly enhanced.

Fluidized-bed reactors are utilized extensively in the petroleum refining industry (e.g., in catalytic **cracking,** hydrocracking, and other processes). Several proposed methods for **coal gasification** are based on the use of fluidized beds. An important application under development is **fluidized-bed combustion** for the efficient burning of coal.

See also **Ebullated Bed; Entrained Bed; Fixed Bed.**

FLUIDIZED-BED COMBUSTION

A system under development for burning coal in a **fluidized bed** of an inert material (e.g., coal ash) with air as the fluidizing gas. Turbulence in the fluidized bed provides intimate mixing of the coal, air, and hot inert material; as a result combustion is rapid. The potential advantages of fluidized-bed combustion are (1) the ability to burn a wide variety of coals, including those with high sulfur and/or mineral (ash) contents; (2) a reduction in emission of the atmospheric pollutants, **nitrogen oxides** and **sulfur oxides**; (3) high combustion efficiency; and (4) the lower cost of coal crushing as compared with the widely used pulverized-coal firing (see **Steam Generation**).

Fluidized-bed combustors are being designed for producing steam (for **steam turbines**) or hot gases (for **gas turbines**). In each case, the inert material bed consists of coal ash and limestone (calcium carbonate) or dolomite (calcium and magnesium carbonates). The purpose of the limestone (or dolomite) is to absorb the sulfur oxides produced from sulfur in the coal.

The solid particles are fluidized by blowing preheated air up through nozzles at the bottom of the bed. Crushed coal, mixed with 15 to 30 percent by weight of crushed limestone (or dolomite), is fed continuously to the bed, and ash and spent sulfur oxide absorber are removed.

A special feature of the fluidized-bed combustor is that, in addition to heating by convection from the hot combustion gases, as in a conventional furnace, water, steam, or air is heated further by passage through tube bundles submerged within the fluidized bed. This design has several important consequences.

Heat transfer at the surfaces within the bed is several times more efficient than by gas convection; hence, the same heating capability can be achieved in a smaller furnace (or boiler). Furthermore, because of this efficient heat transfer, the main combustion temperature is well below the 2700°F (1480°C) or more in a conventional furnace. At the lower temperature, calcium oxide, formed from limestone or dolomite, readily absorbs sulfur oxides. Some 90 to 95 percent of the sulfur in a high-sulfur (3 to 4 percent) coal can be removed in this manner. The presence of a small amount of sodium chloride (common salt) appears to increase limestone utilization. Finally, because of the relatively low combustion temperature, nitrogen oxides in the stack (flue) gases are minimal, even with high-nitrogen (2 percent) coal.

In the multicell concept for steam generation, the furnace consists of two or more interconnected cells arranged horizontally or vertically. Coal is burned successively in these cells in fluidized beds at a temperature around 1600°F (870°C). Unburned carbon particles, carried over with the stack gases, are removed in a **cyclone separator** and transferred to a smaller carbon burnup cell (CBC). In this fluidized-bed cell, which operates at about 2000°F (1140°C), combustion is completed in excess of air. Feedwater is first heated to produce steam by convection from the combustion gases, initially in the CBC and subsequently in the other cells. Finally, the steam is superheated in the tube bundles submerged within the beds of the latter cells.

Fluidized-bed combustion of coal is also being considered for use in **cogeneration** and **combined-cycle systems.** In a proposed cogeneration system, the working fluid for a gas turbine generator is pressurized air heated to about 1500°F (815°C) by passage through a **heat exchanger** submerged in the fluidized bed. The turbine exhaust, at some 465°F (240°C), could be used prior to discharge as a source of industrial process heat or for space and water heating. Preheated combustion air at atmospheric pressure for the fluidized bed is provided by a completely independent circuit.

In the pressurized (as distinct from atmospheric) fluidized-bed combustion concept for a combined-cycle system, the gas turbine working fluid and the combustion air are parts of the same circuit. The airflow, at a pressure of 6 to 10 atm (0.6 to 1 MPa), is split, with about one-third providing the combustion air and the remainder passing through a heat exchanger within the fluidized bed. The air heated in this manner is then mixed with the hot combustion gases after removal of suspended particulate matter (ash) in cyclone separators to prevent turbine blade damage. The resulting hot gas is the working fluid for a gas-turbine generator. The turbine exhaust is passed through a waste-heat (heat exchanger) boiler to provide steam for a steam-turbine generator. Preliminary studies with a pressurized fluidized-bed combustor have given promising results, with greater sulfur oxide removal and less nitrogen oxide formation than at atmospheric pressure.

Regardless of their type or purpose, all fluidized-bed combustors discharge a mixture of coal ash and partially spent sulfur absorber. This material could possibly be treated to regenerate the absorber and recover the sulfur (see **Limestone Injection Desulfurization Process**). If regeneration is not practical, the discharged solid may find use in agriculture for treatment of acid soils or as a road-bed material in highway construction.

FLUIDIZED-BED HYDROGENATION (FBH) PROCESS

A process developed in the United Kingdom, utilizing hydrogenation (i.e., reaction with hydrogen gas) to convert a liquid petroleum product into a **high-Btu fuel gas** (or **substitute natural gas**). The method is especially applicable to a wide range of **hydrocarbon** feed materials that are heavier than **naphtha.**

The feed is reacted with a hydrogen-rich gas in a **fluidized bed** of coke particles at a temperature of about 1400°F (760°C) and a pressure of 50 atm (5 MPa) or more. Part of the feed is gasified and the remainder is converted into liquid hydrocarbons, largely aromatics, and a small amount of **char.** The hydrogen-rich gas can be made from a petroleum product or obtained in other ways (see **Hydrogen Production**).

The product gas contains hydrogen sulfide, from the sulfur in the feed, and it is removed in a conventional manner (see **Desulfurization of Fuel Gases**). The composition of the purified gas depends on the nature of the feed; a rough average is 50 volume percent of **methane** with most of the remainder hydrogen and ethane in varying proportions. The heating value is in the high-Btu gas range, but the substantial hydrogen content may result in a flame speed too high for general use. However, passage of the gas over a hydrogenation catalyst would convert most of the hydrogen and ethane into methane, thereby producing a substitute natural gas.

FLUOR ECONAMINE PROCESS

A process for removal of **acid gases** (hydrogen sulfide and carbon dioxide) from fuel gases by chemical absorption in a weakly alkaline solution. The absorber is an aqueous solution of the alkanolamine diglycolamine (DGA). The general principles are similar to those described for chemical absorption processes in the **desulfurization of fuel gases** (see Fig. 26).

FLUOR SOLVENT PROCESS

A process for the removal of **acid gases** (hydrogen sulfide and carbon dioxide) from fuel gases by physical absorption in the organic solvent propylene carbonate. The impure feed gas at a pressure of 5 atm (0.5 MPa) or more enters the bottom of an absorber column and meets a downflow of the solvent at about ambient temperature. The general principles are similar to those described for physical absorption processes in the **desulfurization of fuel gases** (see Fig. 25).

FLYWHEELS

Rotating wheels that can store energy in mechanical form (i.e., as kinetic energy) as their rate of rotation is increased; the energy can subsequently be released as the rate of rotation decreases.

Flywheels are used in piston (or reciprocating) engines of all types—gasoline, diesel, and steam. (In piston-engine aircraft the propeller serves as the flywheel.) Reciprocating engines produce energy in pulses, and the flywheel attached to the crankshaft makes possible an essentially uniform operating speed. The flywheel stores energy during the power stroke and releases it during the other stages of the operating cycle (see **Diesel Engine; Spark-Ignition Engine**). The speed of the engine's flywheel actually increases and decreases continuously as energy is stored and released, respectively, but the changes are small and not noticeable for a sufficiently heavy flywheel.

Energy Storage

In recent years attention has been directed at the development of flywheels that could store very much larger amounts of energy. In principle, flywheels could be used by electric utilities for load leveling, for storing solar and wind energy, and for the propulsion of electric vehicles (see **Energy Storage**). Flywheel-operated buses were used at one time in Switzerland, but their energy storage capacity (and travel range) between charges was small. For storing substantial amounts of energy, flywheels would need to be much larger, different in design, and made of different materials, as described later.

To store energy, the rate of rotation of the flywheel would be increased by means of an electric motor and a variable-speed transmission. Electrical energy supplied to the motor is then stored as mechanical energy in the flywheel. To recover the stored energy as electricity, the flywheel is connected to an **electric generator.** In practice, the motor and generator are usually the same (motor/generator) unit. When it is supplied with electricity, the unit serves as a motor that produces rotational motion to drive the flywheel. On the other hand, if its shaft (or armature) is rotated by the flywheel, the unit becomes an electric generator (see **Electric Motor**).

The amount of energy stored in a flywheel is related to its mass (or weight) and the square of the rate of rotation. Thus, energy storage can be increased by increasing the mass and especially the rate of rotation. However, as the flywheel rotates faster, the centrifugal force increases and this causes a stress within the wheel. If the stress should exceed the maximum (or ultimate) tensile strength of the material, the wheel would break apart. For a given rate of rotation, the magnitude of the stress decreases with the density of the material. Hence, for maximum energy storage, a flywheel should be made of material that has a low density and a high tensile strength. There are other requirements in flywheel design but these two are basic.

Flywheels for automobile and other engines are usually made of high-strength steel, but the relatively high density places a limit on the maximum safe rate of rotation. This is compensated to some extent by the mass of the flywheel. Since considerable damage could result from the disruption of a heavy steel flywheel, such wheels are operated well below their calculated maximum speeds.

Advanced Composite Flywheels

Because of the high material density, steel flywheels are not suitable for large-scale energy storage. Better flywheels should result from the development of low-density, fibrous materials with very high strength in the fiber direction. The fibers are lined up in a parallel manner and are embedded in a matrix, usually a plastic, of lower strength. The composite produced in this way has a low density but is very strong in the direction of fiber alignment. Flywheels made of such composites are called advanced composite flywheels. Because of their low mass, these flywheels would have to be run at high rotational speeds to achieve substantial energy storage.

A potential advantage claimed for flywheels made from fiber composites is that failures caused by overstress would lead to powdering and shredding rather than breaking into massive lumps. Damage would then be negligible in comparison with that which could result from the failure of a steel flywheel.

Possible materials for use in composite flywheels are graphite, glass, silica, or synthetic polymer fibers. Of these, fused silica fibers have the highest tensile strength and should be capable of providing the largest energy storage per unit flywheel mass. However, silica fibers suited to this purpose are not available commercially. A practical alternative is an organic (polyamide) polymer designated PRD-49 (trade named Kevlar). It appears to be superior to all fiber materials tested other than fused silica.

Composite flywheels would have to be specially designed to take advantage of the properties of unidirectional fiber composites. One design consists of concentric rings of composite materials separated by narrow gaps containing a resilient bonding material; Fig. 50, for example, shows such

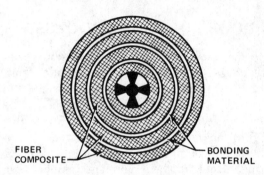

FIBER COMPOSITE — — BONDING MATERIAL

Fig. 50 Composite flywheel.

a flywheel with four rings. Several organizations are working under the sponsorship of the U. S. Department of Energy to develop advanced composite flywheels. The preferred fibers are presently Kevlar and graphite.

Losses in a flywheel energy-storage system arise mainly from bearing friction and atmospheric drag on the rotating wheel. Bearings of special design can reduce the friction losses, and drag can be decreased by running the flywheel in an evacuated containment. A low-pressure atmosphere of hydrogen will result in the least drag, but since this gas forms an explosive mixture with air, inert helium may be preferred. A well-designed flywheel system is estimated to have a possible efficiency of 90 (or more) percent; that is, 90 percent of the electrical energy supplied to drive the flywheel could be recovered. Such a low-loss flywheel is thought to be capable of storing energy for several months.

Vehicle Propulsion

A system for vehicle propulsion consists of a flywheel and attached motor/generator unit, as well as one or more drive motors. Energy is stored in the flywheel by connecting the motor/generator to an electrical outlet. The stored energy of the rotating flywheel would then be used as required to generate electricity for the drive motors to propel the vehicle.

A vehicle with a flywheel drive can take advantage of regenerative braking to increase the overall efficiency. In the braking process, the drive motor serves as a generator to supply power to the flywheel motor. The vehicle is slowed down and its kinetic (motion) energy is stored in the flywheel for subsequent use instead of being wasted as frictional heat in the brakes.

Flywheel-propulsion systems have features in common with electric vehicles operated by **storage batteries.** The major difference is that in the former energy provided by electricity is stored in the flywheel, whereas in the latter the energy is stored in batteries. Advantages claimed for the flywheel over the storage battery are faster energy release, which makes for faster acceleration, and more rapid storage. For example, storage of a given amount of energy might require a few minutes in a flywheel compared with a few hours for a battery. Furthermore, the efficiency of a flywheel system may be 80 to 90 percent compared with 75 percent, at best, for batteries.

Since an automotive vehicle carries its means of propulsion, the mass (or weight) of the propulsion unit is important. A meaningful comparison of flywheel and battery systems is, however, not yet possible, because advanced composite flywheels are still in the early experimental stages and lightweight storage batteries are also under development. Ultimately, the mass difference for a given travel range may prove to be small.

Hybrid Flywheel Vehicles

The travel range of a flywheel-driven vehicle is limited by the amount of energy that can be stored. One way of increasing the range would be to combine a storage flywheel with a conventional internal-combustion engine. Periodic (or steady) operation of the engine under near optimum conditions would supply energy to

be stored in the flywheel. The vehicle would be driven by electric motors, as described above, with advantage being taken of regenerative braking. The overall efficiency should be greater than that of a conventional vehicle driven by an internal-combustion engine alone.

Flywheels may be especially useful in battery-operated electric vehicles. Instead of supplying current to the drive motors, the batteries would be connected to the flywheel motor. The energy stored in the flyweel would then supply electric power to the drive motors in the usual manner. A substantial increase in efficiency could be achieved in this way. The amount of energy that can be drawn from a charged storage battery depends on the current demand (or discharge rate). When the vehicle is accelerated, the demand is large and this tends to decrease the energy available from the battery and hence the range of the vehicle. With a flywheel drive, the battery can be discharged at a fairly steady average rate; as a result, the available energy and vehicle range are increased. The steady discharge also increases the life of the battery. The use of a flywheel for regenerative braking helps to even out the discharge rate of the battery and to increase the overall efficiency of energy utilization in the vehicle.

Utility and Related Energy Storage

A flywheel coupled to a motor/generator unit could conceivably be used for electric-utility load leveling. Energy from excess power generated during periods of low demand (e.g., at night and over weekends) could be stored in flywheels for use at times of peak demand. Similarly, if first converted into electricity, energy from intermittent sources, such as the sun and the wind, could be stored for use when the direct sources are not available.

For utility or related storage, flywheels of much greater size and mass than those considered for vehicle propulsion would be required. In one concept, for example, the flywheel would have a diameter of 15 to 20 ft (4.5 to 6 m) and weigh roughly 100 to 200 tons (90,000 to 180,000 kg); the energy storage capacity would be 10,000 to 20,000 kilowatt-hours at a speed of 3500 rotations per minute. Individual flywheels would be located near several load centers throughout the utility's distribution system. Because flywheels would operate with little noise and no pollution, siting—possibly in a covered pit for safety—would not present a problem.

FOSSIL FUEL

A solid, liquid, or gaseous fuel material formed in the ground by chemical and physical changes in plant (mainly) and animal residues under high temperature and pressure over geological time periods. The major fossil fuels are **coal, petroleum,** and **natural gas.** Petroleum products, such as **fuel oils,** are often included among the fossil fuel. **Peat** is thought to be an intermediate stage in the formation of most coals.

Fossil fuels consist mainly of carbon and hydrogen; the proportion of carbon is largest in coal and smallest in natural gas. Carbon and hydrogen (and their compounds) can burn in air, combining with oxygen to form carbon dioxide (CO_2) and water (H_2O), respectively, and generating heat (see **Heating Value**). Hence, burning (or combustion) of fossil fuels is a major source of heat energy. In a **heat engine,** such as an **internal-combustion engine** or a **turbine,** heat can be converted into mechanical work; the work can then be used for generating electricity, vehicle propulsion, or operating machinery.

Although **oil shale** contains a combustible material derived from fossilized plant residues, the oil obtained by heating the shale is generally regarded as a **synthetic fuel** (or synfuel) rather than a fossil fuel.

FOSTER–WHEELER COAL GASIFICATION PROCESS

A Foster-Wheeler Energy Corporation process for **coal gasification** with steam and air; the product is a **low-Btu fuel gas.** The process is similar to the **BI-GAS process,** but differs in using air instead of oxygen gas and in operating at a lower pressure. Either caking or noncaking coal can be used without pretreatment.

The temperatures in the Foster Wheeler gasifier are much the same as in the BI-GAS process, and the gas pressure is 25 to 35 atm (2.5 to 3.5 MPa). The ash remaining after gasification of the coal is expected to form a molten slag, but if the melting point should exceed about 2700°F (1480°C), it may be necessary to add limestone as a flux.

The product gas, after removal of hydrogen sulfide (see **Desulfurization of Fuel Gases**), contains roughly 29 volume percent of carbon monoxide, 14 percent of hydrogen, and 3 percent of methane (dry basis) as fuel components; most of the remainder is inert nitrogen. The **heating value** is approximately 160 Btu/cu ft (5.9 MJ/cu m).

FUEL CELL

A cell (or combination of cells) capable of generating an electric current by converting the chemical energy of a fuel directly into electrical energy. The fuel cell is similar to other electric cells in the respect that it consists of positive and negative electrodes with an electrolyte between them. Fuel in a suitable form is supplied to the negative electrode and oxygen, often from air, to the positive electrode. When the cell operates, the fuel is oxidized and the resulting chemical reaction provides the energy that is converted into electricity. Fuel cells differ from conventional electric cells in the respect that the active materials (i.e., fuel and oxygen) are not generally contained within the cell but are supplied from outside.

But for its cost, pure (or fairly pure) hydrogen gas would be the preferred fuel for fuel cells. Alternatively, impure hydrogen obtained from **hydrocarbon** fuels, such as **natural gas** or **substitute natural gas** (methane), **liquefied petroleum gases** (propane and butane), or liquid **petroleum products,** can be used in fuel cells. Efforts are being made to develop cells that can use carbon monoxide as the fuel; if they are successful, it should be possible to utilize coal as the primary energy source.

The principle of the hydrogen-oxygen fuel cell was demonstrated by W. R. Grove in the United Kingdom in 1839, and about a hundred years later the British engineer F. T. Bacon initiated a series of studies that led to the construction of the first effective fuel cells. In the United States, the successful use of hydrogen-oxygen cells to provide electric power for the Gemini and Apollo spacecraft stimulated interest in fuel cells for other purposes. The practical nature of fuel cells has been demonstrated, but their general use will require the development of low-cost cells with a reasonably long life (at least 5 years) that can utilize fuels other than hydrogen gas for the negative electrode.

A first step in this direction was the TARGET (Team to Advance Research for Gas Energy Transformation) program started in 1967 by the Pratt and Whitney Division of United Technologies Corporation with the support of a group of gas and electric utilities. The TARGET program cell system had a design **power** of 12.5 kilowatts (see **Watt**) and used natural gas as the fuel with oxygen from the air. Some 60 such cells were installed in residences and other buildings in the United States. Experience gained with these cells led to the development of a similar cell with an electrical **capacity** of 40 kilowatts.

In 1976, the FCG (Fuel Cell Generator) program, sponsored by United Technologies Corporation and several utilities, resulted in the demonstration of a fuel cell with a capacity of 1 megawatt (MW), where 1 MW = 1000 kilowatts. The FCG cell differs from the TARGET cell in that it uses a less expensive fuel, namely, **naphtha** rather than natural gas. With the support of the U.S. Department of Energy and the Electric Power Research Institute, a cell of the same design with a capacity of 4.8 MW has been built for testing in conjunction with an electrical utility in New York City.

Advantages of Fuel Cells

The use of natural gas as the energy source in a fuel cell is justified by the potential for a higher efficiency for conversion into electrical energy than can be achieved by the usual combustion methods. The same should be true for other fuels, including hydrogen, petroleum products, or fuels derived from coal.

In the conventional thermal process for generating electricity, heat energy produced by combustion of the fuel is converted partially into mechanical energy in a **heat engine** (e.g., a **steam turbine**) and then into electricity by means of a generator (see **Electric Power Generation**). The efficiency of a heat engine is limited by the operating temperatures, and in the large modern steam-electric plants about 40 percent of the heat energy of the fuel is converted into electrical energy. For most existing steam (and nuclear) power plants, the efficiencies are lower.

Fuel cells, on the other hand, are not heat engines and are not subject to their temperature limitations. Moreover, the cells can be installed near the use point, thus reducing electrical transmission requirements and accompanying losses. Consequently, considerably higher efficiencies are possible. The reason for the superior efficiencies of fuel cells for converting thermal energy (heat) into electrical energy will be explained shortly.

Another advantage of fuel cells is that they have few mechanical components; hence, they operate fairly quietly and require little attention. Moreover, the associated atmospheric pollution is small. If the primary energy source is hydrogen, the only waste product is water; if the source is a hydrocarbon, carbon dioxide is also produced. **Nitrogen oxides,** such as accompany combustion of fossil fuels in air, are not formed in fuel cells. Some heat is generated by a fuel cell, but it can be dissipated to the atmosphere or possibly used locally. There is no requirement for large volumes of cooling water such as are necessary to condense exhaust steam from a turbine in a conventional power plant.

Fuel Cell Applications

The electromotive force (emf) or voltage of a fuel cell depends to some extent on the discharge current strength (described later), but a reasonable average would be 0.75 volt per cell. By joining a number of cells in series (i.e., the positive electrode of one cell connected to the negative of the adjacent cell), the voltage can be increased proportionately. The current that can be drawn from a cell is limited by the electrode area. The effective area, and hence the current strength, can be increased, however, by connecting cells in parallel (i.e., all the positives are connected together and so also are all the negatives). See **Storage Battery.**

A combination of cells in series and parallel can thus provide any reasonable desired voltage and current. The power output (in watts) is the product of the emf (in volts) and the current (in amperes). Fuel cells generate direct current which can be used for electric lamps and some small appliances. For larger appliances, heat pumps, motors, etc., conversion into alternating current by means of an inverter might be necessary.

Fuel cells can be made in modules of different size that are readily transportable. They can then be assembled at any location to provide a specified voltage and power output. The modular design should make it possible to construct plants of various capacities for different requirements. Additional units could be added to an installation if the power demand should increase.

If fuel cells of reasonably low cost and long life can be produced, a major use might be by electric utilities for load leveling, as explained below. A long-term possibility is a central-station power plant in which coal is gasified and the gas used to generate electricity directly by means of fuel cells. Such an installation is expected to have a higher efficiency for fuel utilization than a conventional steam-electric plant.

Large generating stations operate most efficiently at a steady (rated) power output, but the demand for power is variable (see **Energy Storage**). When the demand (or **load**) is less than the rated output, the excess would be used to generate hydrogen by electrolysis of water (see **Hydrogen Production**). At times when the load is greater than the power supply, the hydrogen would be used in fuel cells to satisfy the additional demand. By siting fuel cells near load centers where the demand exists, electrical transmission and distribution costs would be reduced, although there would be some cost for transporting the hydrogen.

Sometimes new load centers are formed as a result of housing and industrial developments. To satisfy the power demand, utilities will either build additional large plants, which require considerable capital expenditures, or utilize **diesel engines** or **gas turbines** operated by natural gas or a petroleum fuel. The same fuel might be utilized more economically in fuel cells located near the new load center.

Fuel cells have been proposed for remote or unattended locations, for mobile and emergency power sources, and for vehicle propulsion (see **Electric Vehicle**). The fuel would then be stored with the cells. The total electrical energy generated would be limited by the amount of fuel available, but it would be in a form that could be easily replaced. Possible fuels are hydrogen (as compressed gas, liquid, or solid hydride), methanol, and hydrazine.

Fuel Cell Fundamentals

In most fuel cells, hydrogen (pure or impure) is the active material at the negative electrode, and oxygen (from pure oxygen or air) is active at the positive electrode. Since hydrogen and oxygen are gases, a fuel cell requires a solid electrical conductor to serve as a current collector and to provide a terminal at each electrode. The solid electrode material is generally porous and has a large number of sites where the gas, electrolyte, and electrode are in contact; the electrochemical reactions (see below) occur at these sites. The reactions are normally very slow, and a catalyst is included in the electrode to expedite them.

The best electrochemical catalysts are finely divided platinum or platinum-like metal deposited on or incorporated with the porous electrode material. Since the platinum metals are expensive, other catalysts, such as nickel (for hydrogen) and silver (for oxygen), are used where possible. The very small catalyst particles provide a large number of active sites at which the electrochemical reactions can take place at a fairly rapid rate.

During the course of cell operation, however, the catalyst gradually loses its activity. This loss is often due to "poisoning" of the catalyst by impurities in the gas. But a more important cause is thought to be the spontaneous aggregation of the small catalyst particles into larger ones with fewer active sites. The development of electrodes of long life at a reasonable cost is one of the keys to the future of the fuel cell.

Although practical fuel cells differ in design details, the essential principles are the same, as indicated by the schematic illustration in Fig. 51. Hydrogen (or

hydrogen-rich) gas is supplied to one electrode and oxygen gas (or air) to the other. Between the electrodes is a layer of electrolyte. Most existing fuel cells operate at temperatures below about 390°F (200°C); the electrolyte is then usually an aqueous solution of an alkali or acid. The liquid electrolyte is generally retained in a porous membrane, but it may be free-flowing in some cells. Direct electric current is drawn from the cell in the usual manner by connecting a load between the electrode terminals. (High-temperature fuel cells, with a fused salt or solid oxide as electrolyte, have also been proposed but they are not intended to be simple hydrogen–oxygen cells.)

The electrochemical reactions occurring at the electrodes of a hydrogen–oxygen cell may vary with the nature of the electrolyte, but basically they are as follows. At the negative electrode, hydrogen gas (H_2) is converted into hydrogen (H^+) ions (i.e., hydrogen with a positive electric charge) plus an equivalent number of electrons (i.e., e^-); thus,

$$H_2 \rightarrow 2H^+ + 2e^-$$

When the cell is operating and producing current, the electrons flow through the external load to the positive electrode; here they interact with oxygen (O_2) and water (H_2O) from the electrolyte to form negatively charged hydroxyl ions (OH^-); thus,

$$\frac{1}{2}O_2 + H_2O + 2e^- \rightarrow 2OH^-$$

The hydrogen and hydroxyl ions then combine in the electrolyte to produce water:

$$H^+ + OH^- \rightarrow H_2O$$

Addition of the three foregoing reactions shows that when the cell is operating, the overall process is the chemical combination of hydrogen and oxygen (gases) to form water; that is,

$$H_2 + \frac{1}{2}O_2 \rightarrow H_2O$$

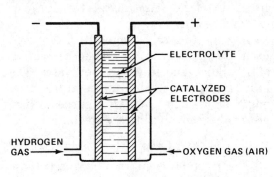

Fig. 51 Simple fuel cell.

This is always true in a hydrogen-oxygen cell, regardless of the nature of the electrolyte and the details of the individual electrode reactions.

Conversion Efficiency

The electrical energy generated by a fuel cell depends on what is called the "free" energy, rather than on the heat energy, of the overall cell reaction. The free energy of formation of 1 mole (18 grams) of liquid water from hydrogen and oxygen gases at atmospheric pressure is 225 **British thermal units** (Btu) or 237 kJ at 77°F (25°C). The heat energy (or enthalpy) of the reaction under the same conditions is 271 Btu (286 kJ). The theoretical efficiency for the conversion of heat energy into electrical energy in a hydrogen-oxygen fuel cell is thus (225/271) × 100 = 83 percent.

Efficiencies as high as 70 percent have been observed, but practical cells using pure hydrogen and oxygen generally have conversion efficiencies in the range of 50 to 60 percent. The efficiencies are somewhat lower when air is the source of oxygen. The overall thermal to electrical conversion efficiencies are also lower when the hydrogen is derived from hydrocarbon sources. Nevertheless, they should be higher than those obtainable from the same fuels in most steam-electric plants.

The theoretical emf (or voltage) of a fuel cell can be calculated from the reaction free energy. For the hydrogen-oxygen cell at 77°F (25°C), with the gases at atmospheric pressure, the ideal emf is 1.23 volts; at 390°F (200°C), it is about 1.15 volts. The discharge voltages observed in actual cells are always below the theoretical value, the difference increasing with increasing strength of the current drawn from the cell (Fig. 52). For the moderate currents at which fuel cells normally operate, the emf is 0.7 to 0.8 volt. This deviation from the theoretical emf accounts for the conversion efficiency of a fuel cell being below the ideal maximum value.

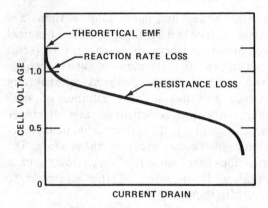

Fig. 52 Current and voltage in discharge of fuel cell.

The departure of a fuel cell from ideal behavior arises from several factors. One is the inherent slowness of the electrode reactions; this is dominant at low current drains. It can be reduced by an effective electrochemical catalyst and by increasing the operating temperature. At larger currents there is an additional contribution from the electrical resistance of the electrolyte (multiplied by the current strength). A low-resistance (i.e., high-conductivity) electrolyte is therefore desirable.

Even in an ideal hydrogen-oxygen cell, 100 − 83 = 17 percent of the chemical reaction energy (enthalpy) would be liberated as heat. The proportion is increased in an actual cell because the conversion efficiency is less than the maximum of 83 percent. In order to avoid an excessive temperature rise, heat is removed from the fuel cell during operation. Possible ways of doing this are by the flow of excess air past the positive electrode or by circulating the electrolyte through an external cooler.

In some fuel cell designs it is proposed to utilize the heat to provide space heating of a building and to supply hot water. The heat released in high-temperature cells might be used for industrial purposes (i.e., process heat) or to generate steam.

Hydrogen Fuel Cells

The most efficient fuel cells are those using hydrogen gas and oxygen (or air) as the active species. The electrolyte is usually an aqueous solution of potassium hydroxide. In smaller cells, the electrolyte is generally held in a porous membrane, but in larger cells a free-flowing liquid may be preferred to facilitate heat removal. Electrodes (i.e., current collectors) made of porous nickel are commonly used in cells of this type. Finely divided nickel or a platinum metal deposited on the outer surface serves as the hydrogen (negative) electrode catalyst; for the oxygen (positive) electrode the catalyst may be silver or a platinum metal. These cells may operate at temperatures up to 195°F (90°C). The hydrogen required for the cell can be stored either as the compressed gas, as liquid at low temperature, or as a solid metal hydride (see **Hydrogen Fuel**). The oxygen is usually obtained from the air, but in some cases it is stored as liquid (see below).

The gases in the hydrogen–oxygen (air) cell must be free from carbon dioxide, because this gas can combine with the potassium hydroxide electrolyte to form potassium carbonate. If this occurs, the electrical resistance of the cell is increased and its output voltage decreased. Consequently, when air is used to supply the required oxygen, carbon dioxide must first be removed by scrubbing with an alkaline medium (e.g., lime).

Hydrogen–oxygen (air) fuel cells have been proposed for propulsion of electric vehicles, with the hydrogen provided by a metal hydride. However, this use is limited by the heavy weight of the hydride and the cost of the relatively pure hydrogen required. It appears that for the present, at least, the use of hydrogen-oxygen cells will be restricted mainly to manned space vehicles. Such cells with porous nickel electrodes and potassium hydroxide electrolyte have been used to provide electric power for the Apollo and shuttle spacecraft. The hydrogen and oxygen for operating the cell are stored in liquid form to minimize the volume occupied. The fuel cells employed previously in the Gemini program were unusual in the respect that the electrolyte, a conductor of hydrogen ions, was a solid ion-exchange (sulfonic acid) resin membrane referred to as a *solid polymer electrolyte* (or SPE). A layer of finely divided platinum deposited on each surface of the membrane served as electrochemical catalyst and current collector. Somewhat similar cells, with an inorganic SPE, have been proposed for high-temperature operation.

Fossil-Fuel Cells

The most interesting fuel cells for the near future are modified hydrogen–air cells in which a gaseous or liquid hydrocarbon is the source of hydrogen. Eventually, coal may serve as the primary energy source for fuel cells. Cells based on **fossil fuels** have three main components: (1) the fuel processor which converts the fossil fuel into a hydrogen-rich gas, (2) the power section consisting of the actual fuel cell (or combination of cells), and (3) the inverter for changing the direct current generated by the fuel cell into alternating current to be transmitted to the user (Fig. 53). In the following descriptions of fuel cells, only the first two components will be considered; the third is referred to in the section on **Electric Power Transmission.**

Phosphoric Acid Cells. The most highly developed fossil-fuel cells in the early 1980s

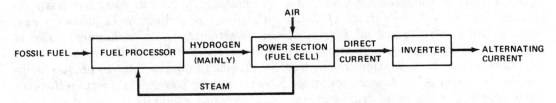

Fig. 53 **Main components of fuel cell systems.**

have a concentrated aqueous solution of phosphoric acid as the electrolyte. Cells of this type were used in the TARGET and FCG programs referred to earlier. The primary fuel may then be a light hydrocarbon, such as natural gas (in the TARGET program) or naphtha (in the FCG program).

In the processor, the fuel is subjected to high-temperature catalytic reactions with steam (i.e., **steam reforming** and **water-gas shift reaction**). In order to prevent poisoning (i.e., decrease in activity) of the catalysts, as well as of the cell electrodes, the fuel must be essentially free of sulfur compounds (see **Desulfurization**). The product of these reactions is a gas mixture containing roughly 80 volume percent of hydrogen; the rest is mainly carbon dioxide with a small proportion of carbon monoxide (see **Hydrogen Production**). The carbon dioxide has little or no effect on the phosphoric acid electrolyte and so can be tolerated.

In the fuel cell (or power section), the phosphoric acid solution is retained in a porous asbestos or carbon sheet. This is sandwiched between fibrous-carbon sheet electrodes that have a platinum-based electrochemical catalyst deposited on their outer surfaces. The hydrogen-rich gas from the processor is supplied to the negative electrode side of the sandwich, and air provides oxygen to the positive side. The operating temperature is 300 to 390°F (150 to 200°C) and the discharge voltage 0.7 to 0.8 volt. Each cell unit is only a few millimeters thick so that a large number can be stacked in a package of reasonable size to produce the desired voltage and power.

Molten-Carbonate Cells. High-temperature fuel cells, with a molten carbonate mixture as the electrolyte, offer the prospect for use with a variety of fossil fuels, including coal. A special feature of these cells is that during operation they can oxidize carbon monoxide to carbon dioxide as well as hydrogen to water. Hence, gaseous mixtures of hydrogen and carbon monoxide, which are relatively inex-pensive to manufacture, can be used in the cell; the presence of carbon dioxide would have only a minor effect. The theoretical emf of the carbon monoxide-oxygen cell is roughly 1 volt at 1290°F (700°C), which is approximately the same as for the hydrogen-oxygen cell at this temperature. However, the actual discharge emf is somewhat lower.

Several methods are available for fuel processing. These methods are essentially the same as those used for the commercial production of mixtures of hydrogen and carbon monoxide, known as **synthesis gas.** Synthesis gas can be made from coal by reactions with air and steam or from various petroleum products. In order to prevent poisoning of the electrodes in the fuel cell, the gas mixture must be desulfurized before being supplied to the power section.

The electrolyte in the high-temperature fuel cell under development is a molten mixture of alkali-metal (lithium, sodium, and potassium) carbonates at a temperature of 1110 to 1290°F (600 to 700°C). This is retained in an inert porous matrix sandwiched between two porous nickel electrodes. Because of the high temperature, electrochemical catalysts are apparently not necessary on the electrodes. The mixture of hydrogen and carbon monoxide is supplied to the negative electrode and oxygen (from the air) to the positive electrode. The discharge emf of the cell is about 0.8 volt.

An important aspect of molten-carbonate fuel cells is that the discharged gases, consisting mainly of the steam and carbon dioxide products and nitrogen from the air, are at a temperature exceeding 1000°F (540°C). The hot gases could be used to provide industrial process heat, to operate a gas turbine, or to produce steam in a waste-heat (**heat exchanger**) boiler to drive a steam turbine. The turbine would be attached to a generator to produce additional electric power. The overall efficiency for fuel use would thus be substantially increased.

Solid-Oxide Electrolyte Cells. Certain solid, ceramic oxides are able to conduct electricity at high temperatures and can serve as electrolytes for fuel cells. These cells could utilize the same fossil fuels as the molten-carbonate cells. The processing operation would then be the same as described above.

Solid-oxide cells are in the early stages of development. A possible electrolyte is zirconium dioxide containing a small amount of another oxide to stabilize the crystal structure; this material is able to conduct oxygen ions (O^{2-}) at high temperatures. The electrode material might be porous nickel and the operating temperature in the range of 1110 to 1830°F (600 to 1000°C). Electrochemical catalysts would not be required.

Miscellaneous Fuels

Alternative energy sources that can be conveniently stored and transported in liquid form, such as methanol (see **Alcohol Fuels**), ammonia, and hydrazine, have been proposed for fuel cells. Methanol can be catalytically reformed with steam at about 390°F (200°C) to yield a mixture of hydrogen (75 volume percent) and carbon dioxide. This gas can be supplied to the negative electrode of a fuel cell with air at the positive electrode. The cell, with aqueous phosphoric acid solution as the electrolyte, might be similar to those already described.

In a cell developed by the U. S. Army for military use, the methanol is not converted into hydrogen but is added directly to the potassium hydroxide electrolyte solution. The electrodes are silver screens with a special type of finely divided palladium-platinum electrochemical catalyst for the negative electrode and silver for the positive (air) electrode. When the cell operates, the methanol is oxidized to carbon dioxide which reacts with the electrolyte solution to form potassium carbonate. When the methanol is used up, the cell can be regenerated with fresh potassium hydroxide electrolyte and methanol.

In the ammonia (NH_3)-oxygen (air) fuel cell, ammonia gas, obtained from the stored liquid, is decomposed catalytically into hydrogen (75 volume percent) and nitrogen. Part of the hydrogen is burned in air to provide the heat required for the decomposition. The bulk of the hydrogen is then supplied to the negative electrode of a hydrogen-air fuel cell. The most suitable electrolyte would probably be potassium hydroxide solution. The nitrogen formed in the decomposition of ammonia is inert and plays no role in the cell.

A compact fuel cell for a mobile power source, possibly for vehicle propulsion, utilizes the liquids hydrazine (N_2H_4) and hydrogen peroxide (H_2O_2) or air as the energy source. Hydrazine is injected as required into the aqueous potassium hydroxide electrolyte to provide the active material at the negative electrode. The oxygen for the positive electrode is obtained either by the catalytic decomposition of hydrogen peroxide at ordinary temperatures or from the ambient air. Each electrode may consist of a nickel screen matrix with nickel (negative) or silver (positive) as the electrochemical catalyst. The overall cell reaction is the oxidation of hydrazine to water and nitrogen, but the discharge emf is similar to that of the hydrogen-oxygen cell.

Aluminum–Oxygen (Air) Cell

The aluminum-air cell (or battery), intended primarily for electric-vehicle propulsion, is being developed by the Lawrence Livermore National Laboratory in conjunction with industrial contractors. This cell is unusual in the respect that the metal aluminum is effectively the fuel which is consumed during operation and replaced as required. The aluminum (Al) forms the negative electrode of the cell, and oxygen (from air) is the positive electrode; the electrolyte is an aqueous solution of sodium hydroxide. The overall cell reaction is symbolically

$$Al + \frac{3}{4}O_2 \text{ (air)} + \frac{3}{2}H_2O \rightarrow Al(OH)_3$$
$$(-) \qquad (+)$$

so that aluminum, oxygen (from air), and water (from the electrolyte) combine to form aluminum hydroxide, $Al(OH)_3$.

The aluminum–air battery has two main compartments: a cell (or battery) compartment containing the electrodes and a crystallization compartment where solid aluminum hydroxide is deposited. The electrolyte, an approximately 20 weight percent aqueous solution of sodium hydroxide, is circulated between the two compartments.

The battery consists of a number of alternating plate-type aluminum and air electrodes separated by a space through which the electrolyte flows. The aluminum (negative) electrodes are made of the metal containing a small amount of gallium, and the air (positive) electrodes are carbon coated with an electrochemical catalyst, possibly silver. Before entering the battery, the air is scrubbed to remove carbon dioxide, for the reason given earlier. The operating temperature of the battery is about 120 to 140°F (50 to 60°C).

As the battery is discharged, the aluminum hydroxide formed dissolves in the electrolyte to form a sodium aluminate solution, which is effectively a supersaturated solution of aluminum hydroxide in sodium hydroxide. This supersaturated solution passes from the battery compartment to the crystallization compartment where it comes into contact with a solid aluminum hydroxide "seed." In the presence of the seed, the excess aluminum hydroxide in the supersaturated solution separates as a solid, leaving sodium hydroxide solution for reuse in the battery compartment.

In addition to aluminum metal, oxygen and water are consumed when the cell operates. Oxygen is always available from the air, but the water must be replaced

periodically, possibly every 250 to 375 miles (400 to 600 km) of travel in an electric vehicle. At the same time, accumulated solid aluminum hydroxide is removed and sent to a smelter for recovery of the aluminum metal.

The amount of energy that can be obtained from an aluminum–air battery, and hence the travel range of an electric vehicle, depends primarily on the aluminum electrodes. A practical cell might contain initially enough aluminum metal for at least 1000 miles (1600 km) of travel for an average five-passenger vehicle. Then, at intervals of about 1000 miles, the spent negative electrodes would be removed from the battery and fresh aluminum electrodes inserted. If it were desirable, the travel range could be increased by using more or larger aluminum plates, but the increased weight might reduce the overall efficiency.

FUEL (HEATING) OIL

A liquid mixture of **hydrocarbons,** commonly a petroleum product less volatile than **gasoline,** that is burned to generate heat. Heating (or furnace) oils are either **distillates** or **residual oils** (or a mixture). Oils of both these types are classified as light, medium (middle or intermediate), and heavy; the **density** and **viscosity** (resistance to flow) increase and the volatility decreases in this order. Heavy residual oils are sometimes made lighter (i.e., less viscous and more volatile) by blending with distillate oils.

The American Society for Testing and Materials (ASTM) divides fuel oils into the following six categories, based on the type of burner for which the oil is suitable:

No. 1 Heating Oil. A light distillate of relatively high volatility and low viscosity for use in burners in which the oil is vaporized by contact with a heated surface or by radiation. **Kerosine** used for small space heaters falls into this category.

No. 2 Heating Oil. A medium distillate used in atomizing burners which spray the oil in small droplets into a combustion chamber. This grade of oil is used in central-heating systems for residences and small commercial and industrial buildings.

No. 4 Fuel Oil. Either a heavy distillate or a light residual oil used in burners that can atomize oils of higher viscosity than No. 2 Heating Oil. It is used mainly for space heating and does not require preheating to decrease its viscosity, except in extremely cold weather.

No. 5 Fuel Oil (Light). A medium residual oil, somewhat more viscous than No. 4 Fuel Oil. It can be atomized in suitable burners without preheating except in cold weather. This oil is used for heating commercial and industrial buildings and large apartment houses.

No. 5 Fuel Oil (Heavy). A medium residual fuel oil more viscous than No. 5 Fuel Oil (Light) which is used for the same purposes. It often requires preheating, especially in colder climates.

No. 6 Fuel Oil. A high-viscosity, heavy residual oil, sometimes referred to as Bunker C oil, used mostly to produce steam for ship propulsion and for generating electricity. This heavy oil requires preheating in the storage tank to permit pumping and additional preheating prior to atomization at the burners.

The **heating value** of a fuel oil (per unit mass) generally decreases from about 20,000 Btu/lb (46.5 MJ/kg) for the lightest oils to roughly 18,500 Btu/lb (43.7 MJ/kg) for the heaviest. However, because the **specific gravity** increases at the same time, from roughly 0.82 to 0.95 (34 to 17° API), the heating value per unit volume at ordinary temperatures increases from approximately 136,000 Btu/gal (38.1 MJ/liter) for the lightest to about 148,000 Btu/gal (41.5 MJ/liter) for the heaviest oils.

FUSION ENERGY

A form of **nuclear energy** released by the fusion (or combination) of two light nuclei (i.e., nuclei of low mass number) to produce a heavier nucleus (see **Atom**). For two nuclei to fuse, they must come close enough to interact. However, similar electric charges repel one another, and since all nuclei carry a positive charge, an increasing force of repulsion develops as the two nuclei are brought closer together. Consequently, for the nuclei to fuse, they must have enough kinetic (or motion) energy to overcome the force of electrical repulsion that keeps them apart. (Actually more is involved in fusion than overcoming the force of repulsion, but this simple point of view is adequate for the present.)

For the force of repulsion to be small, and thus make fusion easier to achieve, the interacting nuclei should have small positive charges (i.e., low atomic numbers). The element hydrogen has the lowest atomic number, since the nuclei of its three isotopes all carry a single positive charge. Hence, hydrogen isotopes, which also have the lightest nuclei, should be particularly suitable for the production of energy by the fusion of two light nuclei (similar or dissimilar) to form a heavier nucleus.

The three hydrogen isotopes are: ordinary or light hydrogen (H), with a mass number of 1; heavy hydrogen or deuterium (D), mass number 2; and tritium (T), mass number 3. Of the several possible reactions between pairs of hydrogen isotope nuclei, the fusion of a deuterium nucleus (or deuteron) and a tritium nucleus (or triton) has been found to take place most readily. This is referred to as the D–T reaction and is expressed by

$$D + T \rightarrow {}^4He + n + 17.6 \text{ MeV}$$

where ${}^4He$ represents a common helium-4 nucleus and n is a neutron. The fusion of one deuteron and one triton releases 17.6 million electron volts (MeV) of energy. (For the definition of electron volt, see **Nuclear Energy.**)

In more familiar units, the D–T fusion energy would be 94 million kilowatt-hours (kW-hr) per kilogram of a mixture of deuterium (0.4 kg) and tritium (0.6 kg). Assuming the same proportions of deuterium and tritium, this is equivalent to 43 million kW-hr (or 145 billion Btu) per pound (see **British Thermal Unit; Watt**). The energy that would be released by fusion of a given mass of deuterium and tritium is thus more than four times that released by the fission of the same mass of a fissile material (see **Fission Energy**).

If nuclear fusion becomes a practical source of energy, it will undoubtedly be first by way of the D-T reaction. At a later stage, however, fusion involving deuterium nuclei only, which is more difficult to attain, may become important. There are two reactions between deuterons, namely,

$$D + D \rightarrow {}^3He + n + 3.2 \text{ MeV}$$

and

$$D + D \rightarrow T + H + 4.0 \text{ MeV}$$

which take place at about the same rate. The tritium nucleus (T) formed in the second of these reactions immediately fuses with a deuterium nucleus by the D-T reaction. The net result is the liberation of 66 million kW-hr of energy per kilogram of deuterium (or 102 billion Btu/lb).

The great advantage of D-D fusion would be that it would not be necessary to supply tritium from an outside source. Such tritium as is involved is produced in one of the D-D reactions. However, it is unlikely that the conditions for D-D fusion on a practical scale will be realized for many years. Consequently, the following discussion will refer primarily to D-T fusion.

The basic raw materials for the D-T reaction are water (for deuterium) and lithium minerals (for tritium). All natural waters contain a small proportion of HDO molecules in which deuterium (D) has replaced one hydrogen (H) atom in ordinary water (H_2O). Although there is only one deuterium atom for every 6500 ordinary hydrogen atoms, the amount of water is so large that more than 10 trillion (i.e., 10^{13}) tons of deuterium are present on earth. Deuterium, in the form of heavy water (D_2O), is extracted from water on a large scale at a moderate cost in both the United States and Canada. Its main use at present is in certain fission reactors (see **Heavy-Water Reactors**).

Tritium is made by reaction of neutrons with nuclei of the element lithium, especially of its lighter isotope lithium-6. The required neutrons are produced in the D-T reaction, as seen above, and lithium can be obtained from certain minerals and natural brines. If it should become necessary, this element could be extracted from seawater although at a higher cost. Because tritium is radioactive, decaying with a half-life of 12.3 years, it cannot be stockpiled for any length of time (see **Radioactivity**).

Some of the expected advantages of nuclear fusion as an energy source are:

1. An ample supply of raw materials for fuel (deuterium and tritium) at a moderate cost.

2. The amount of fuel in the fusion system at any time would be so small that a large accidental release of energy would be impossible.

3. Inherent safety with no danger of overheating if the cooling (heat removal) system should fail.

4. Minimum problems of radioactive release to the environment and of waste disposal.

5. High **thermal efficiency** for the conversion of fusion energy into electrical energy because of the high temperatures which should be attainable.

It is because of these potential advantages that much effort is being expended in the United States and in other countries in order to make fusion power a practical reality. Although the general requirements are well understood, many scientific and technological problems still remain to be solved.

Conditions for Nuclear Fusion

As already seen, an essential requirement for fusion is that the interacting nuclei have sufficient energy to overcome the force of repulsion. In fusion studies, the energy of the nuclei is increased by increasing the temperature. Nuclear fusion reactions that occur at high temperature are often called *thermonuclear reactions.* In the so-called hydrogen (or fusion) bomb, energy is released in an uncontrolled manner by the fusion of hydrogen isotope nuclei at very high temperatures. The problem is to bring about these reactions under conditions that permit the controlled release of fusion energy.

In a gas at high temperatures, the oppositely charged nuclei and electrons, which are normally held together in atoms by electrical attraction, become separated from each other. Such a gas, containing free positively charged nuclei (or *ions*) and free negatively charged electrons, is called a *plasma.* At fusion temperatures, and even well below, the hydrogen isotopes gases are actually plasmas. The unusual properties of plasmas have an important bearing on the efforts to realize controlled nuclear fusion.

Ignition Temperature. As a result of the electrical interactions between the nuclei and electrons, a plasma emits energy as radiation. In a fusion (D-T or D-D) plasma at moderately high temperatures, more energy is radiated than is produced by fusion. As the temperature is increased, however, fusion energy production increases more rapidly than radiation emission. At a sufficiently high temperature, "ignition" can occur; that is to say, the fusion reaction can be sustained in the plasma once it has been initiated. (The term *ignition* arises from the similarity to a fire; once it is started, the fire will continue as long as fuel is available.) In order to achieve ignition, the fusion energy remaining in the plasma must exceed the radiated energy.

The products of the D-T reaction are a helium-4 nucleus with a positive charge and an electrically neutral neutron. Under the conditions existing in controlled fusion reactions, only the charged particle energy is deposited in the plasma; this amounts to 20 percent of the total D-T fusion energy. (The remaining 80 percent carried by the neutrons can be absorbed and utilized outside the plasma.) Hence, the ideal ignition temperature for a D-T fusion plasma is the temperature at which the energy emitted as radiation is less than 20 percent of the fusion energy. Some energy will then remain in the system to sustain the reaction. As will be seen shortly, the actual ignition temperature is higher than the ideal value.

If the radiation energy arises solely from interactions of the electrons with singly charged ions (e.g., deuterium and tritium nuclei), the ideal ignition temperature for D-T fusion is calculated to be about 50 million °C. (In fusion studies, temperatures are expressed in terms of kilo-electron volts or keV, where 1 keV is equivalent to 11.6 million K or approximately 11.6 million °C.) In practice, there are additional radiation losses because of the presence of ions with more than a single charge (e.g., helium and impurity nuclei); furthermore, not all the charged particle energy remains in the plasma. The practical fusion ignition temperature for a D-T plasma is thus taken to be roughly 100 million °C. (For the D-D reactions, the corresponding ignition temperature is estimated to be 1000 million °C.)

Driven Systems. In one scheme (magnetic mirrors) for producing fusion energy, the loss of some charged particles is inevitable (see **Magnetic-Confinement Fusion**). The ignition temperature for D-T plasmas would then be much higher than 100 million °C. It is improbable that ignition could be realized in such systems, but a net production of fusion energy should nevertheless be possible. This is achieved by injecting energy continuously into the plasma and multiplying this energy by fusion reactions. Systems of this kind are said to be "driven," as distinct from ignition systems which are self-sustaining after initiation. In all cases, however, whether ignition occurs or not, the minimum practical temperature for D-T fusion is considered to be roughly 100 million °C.

Energy Break-Even Condition. An essential requirement for the net production of nuclear fusion energy is that the *break-even* condition be exceeded. This condition is that the plasma be confined for sufficient time to permit the total recoverable fusion energy to balance the energy required to heat the plasma and to compensate for radiation loss. By making a number of simplifying assumptions. J. D. Lawson in England showed in 1957 that the energy break-even condition could be expressed by the product $n\tau$, where n is the particle density and τ is the confinement time in seconds. The particle density is usually stated as the number of deuterium or tritium nuclei per cubic centimeter (cm^3) and the confinement time in seconds; hence, $n\tau$ is commonly given in sec/cm^3 units.

The quantity $n\tau$ is called the Lawson number (or *confinement parameter*) and the minimum value for fusion energy break-even is referred to as the *Lawson criterion.* The Lawson criterion depends on the nature of the fusion reaction and the plasma temperature; for a D-T plasma at 100 million °C, the break-even condition is calculated to be around 6×10^{13} sec/cm^3. The attainment of this value of $n\tau$ does not imply ignition. For ignition to occur, the fusion energy production must exceed the amount required merely to break even. The $n\tau$ product for D-T ignition at 100 million °C is thus higher at about 3×10^{14}. (For D-D fusion, the break-even $n\tau$ value would be larger.)

The calculation of the Lawson criterion is based on several approximations and assumptions which would not necessarily be valid in practice. Nevertheless, the confinement parameter $n\tau$ is commonly used in fusion studies as a general indication of the approach to break-even conditions. One of the objectives of current research is to attain a value close to 10^{14} sec/cm^3 at a temperature near 100 million °C.

Plasma Confinement

In a high-temperature plasma, the nuclei and electrons move randomly in all directions at average speeds of several kilometers (or miles) per second. Hence, unless confined in some way, all the particles would soon strike the walls of a containing vessel, thereby imparting some of their energy to the walls. The particles would then return to the plasma with less energy—that is, at a lower temperature. Thus, the plasma would be cooled, and the fusion temperature could not be realized or maintained. Two quite different confinement methods are currently being investigated: magnetic confinement and inertial confinement.

Note that a distinction is made between containment and confinement. *Containment* applies to the vessel or chamber which contains (or holds) the plasma, whereas *confinement* refers to the means for restraining the plasma from coming into contact with the walls of the containment vessel.

Magnetic confinement is based on the fact that electrically charged particles have difficulty escaping from a magnetic field. Since a fusion plasma consists almost entirely of charged particles, it should be possible to confine a plasma by a suitable magnetic field. Several different magnetic field arrangements have been (or are being) studied in the effort to confine a high-temperature plasma for a sufficient time to satisfy the Lawson criterion for energy break-even (see **Magnetic-Confinement Fusion**).

In *inertial confinement,* a small pellet or sphere of a deuterium–tritium mixture (solid or liquid) is heated by a very short burst of energy from either laser beams or beams of high-energy charged particles. Not only is the temperature increased, but as a result of various complex processes, the pellet material is compressed to a high density. The fusion reaction should then occur so rapidly, in about one-trillionth (10^{-12}) of a second, that inertia (i.e., reluctance to move from rest) would prevent the pellet from flying apart while fusion is in progress. No other confinement is required (see **Inertial-Confinement Fusion**).

Magnetic and inertial confinement represent two extremes of plasma conditions. In magnetic confinement, the plasma particle density, n nuclei/cm^3, would be in the vicinity of 10^{15}; a confinement time of 0.1 sec would then be required to satisfy the Lawson criterion for energy break-even. In inertial confinement, however, the particle density would be roughly a hundred billion (10^{11}) times greater at 10^{26} nuclei/cm^3; the break-even confinement time would then be as short as 10^{-12} sec.

Further details of the studies and problems of the two approaches to fusion plasma confinement are given in the sections on **Inertial-Confinement Fusion** and **Magnetic-Confinement Fusion.** The attainment of both a temperature of 100 million °C and a confinement parameter $n\tau$ of 10^{14} sec/cm^3 in a D-T plasma has proved to be difficult. Nevertheless, the steady approach to these conditions over the years has been encouraging. Worldwide efforts are thus being continued for the purpose of realizing a new, clean, and almost inexhaustible energy source. It does not seem probable, however, that a commercial fusion power plant will be operative before the early years of the 21st century.

See also **Fusion Hybrid.**

FUSION HYBRID

A concept for utilizing the neutrons produced by nuclear fusion (see **Fusion Energy**), especially of deuterium and tritium, to convert a fertile nuclear species (e.g., uranium-238 or thorium-232) into a fissile species (i.e., plutonium-239 or uranium-233, respectively). The fissile material would then be used to generate energy in a conventional nuclear fission reactor (see **Fission Energy**). By making use of fertile material in this manner, a fusion hybrid provides an alternative to the **breeder reactor.** Because of the additional energy that would be available from fission, the requirements of a fusion hybrid would be less severe than for a pure fusion system.

The earliest nuclear fusion reactors would be based on the deuterium-tritium (D-T) reaction in which 80 percent of the energy released is carried by the neutrons produced. In a pure fusion reactor, these neutrons would enter a lithium-containing blanket in which they would deposit their energy. Interaction of the neutrons with lithium nuclei would generate the tritium required for the D-T fusion fuel.

The main difference in a fusion-hybrid reactor would be that the blanket contains fertile material, say uranium-238, in addition to lithium. The high-energy fusion neutrons can interact with uranium-238 nuclei in two ways that result in an increase in the number of neutrons. First, fission of uranium-238 by the fusion neutrons is accompanied by the liberation of three or four neutrons to replace the one causing fission, and second, a high-energy neutron can eject one or two neutrons of lower energy from a uranium-238 nucleus. Some of the neutrons are captured by uranium-238 and converted into fissile plutonium-239 (see **Nuclear Fuel**), whereas others interact with lithium to produce tritium.

After a period of operation, the blanket would have to be removed and the fissile material extracted for use as a reactor fuel (see **Nuclear Fuel Reprocessing**). The net result would be the release of several times more energy than could be obtained by fusion alone. The only additional material required would be either normal (i.e., natural) uranium or the depleted uranium residues of the isotope enrichment process (see **Uranium Isotope Enrichment**).

The foregoing description has referred in particular to uranium-238, but it is equally applicable to thorium-232 as the fertile species.

G

GARRETT COAL GASIFICATION PROCESS

A process of the Garrett Corporation for partial **coal gasification** by **pyrolysis**; the products are an **intermediate-Btu fuel gas** and a residual **char** which can serve as a solid boiler fuel. Any type of coal can be used in the process.

The crushed coal is carbonized in a pyrolysis reactor by heating rapidly in the absence of air to a temperature of roughly 1700°F (930°C). The products are a fuel gas and a char. The char is transferred to a separate vessel (char heater), where part is burned in air to generate heat. The remaining hot char is recycled to the pyrolytic reactor, thereby providing the heat required for pyrolysis. (The scheme is similar to that used in the **COGAS Process**, as shown in Fig. 18.) The flue gas from the char heater is discharged, and the excess char is removed for use as a boiler fuel.

The gaseous pyrolysis product is somewhat similar to **coal gas** (i.e., with relatively large poportions of hydrogen and methane). The **heating value** on a dry basis, after removal of hydrogen sulfide (see **Desulfurization of Fuel Gases**), is roughly 550 Btu/cu ft (21 MJ/cu m).

GARRETT COAL PYROLYSIS PROCESS

A modification of the **Garrett Coal Gasification Process** designed to maximize **coal liquefaction**, rather than gasification, by rapid pyrolysis. For this purpose, the process is conducted at a lower temperature [about 1100°F (590°C)]. Tars removed from the vapors by washing constitute about 35 percent of the pyrolysis products. They are treated with hydrogen gas under pressure (hydrogenation) to yield a **synthetic crude oil** (syncrude); other products are **char** and an **intermediate-Btu fuel gas.**

GAS-COOLED FAST (BREEDER) REACTOR (GCFR)

A conceptual **nuclear power reactor** which has no moderator, so that most of the fissions are caused by fast (i.e., high-energy) neutrons, and uses helium gas as coolant; its purpose is to breed fissile material (i.e., to produce more than is consumed during operation) while generating useful power (see **Breeder Reactor; Fast Reactor**). The GCFR is designed primarily to breed fissile plutonium-239 (together with plutonium-241) from uranium-238, although it can be modified to produce fissile uranium-233 from thorium-232 (see **Nuclear Fuel**).

No GCFR has been constructed, and the following outline is based on design studies. The helium coolant circulation and steam generating systems are similar to those in the **High-Temperature Gas-Cooled Reactor** (HTGR), except that the helium gas pressure is higher at 90 atm (9 MPa) because of the smaller core. The core design, on the other hand, is essentially the same as in the **Liquid-Metal Fast Breeder Reactor** (LMFBR).

There is no graphite in the reactor, because it would slow down the neutrons, and the fuel rods are stainless steel tubes packed with pellets of mixed uranium and plutonium dioxides. The blanket rods (internal and external) contain uranium dioxide alone. In a modified version, the external blanket includes rods of thorium dioxide; uranium-233 generated in these rods could be used in the fuel for an HTGR.

The GCFR is considered as a possible alternative to the LMFBR. The claimed advantages of the former arise from the use of helium gas in place of liquid sodium as coolant. For example, helium gas causes less slowing down of neutrons than does the more dense liquid sodium; as a result of the faster neutron distribution, the breeding ratio in the GCFR is larger and the doubling time is shorter than in the LMFBR.

Since helium, unlike sodium, does not become radioactive when exposed to neutrons, an intermediate heat-exchanger is not required in the GCFR between the reactor core and the steam generators. Steam is then generated in a heat exchanger with helium gas coming directly from the core. Furthermore, in an LMFBR there is some danger of local boiling and the formation of sodium vapor voids in the reactor core; such voids can lead to operational instability. A potential advantage of the GCFR is that the coolant is a gas and the void effect cannot arise.

One drawback to the GCFR is that heat transfer from fuel to coolant and from coolant to water in the steam generator is less efficient for helium than for sodium. Consequently, core temperatures are higher in the GCFR than in the LMFBR for the same steam temperature. Another matter is that the energy required for pumping the helium coolant is greater than that for sodium, so that under equivalent conditions the net power delivered is less for a GCFR; this would be compensated by the increased breeding ratio.

Because of the close relationship between the GCFR and the HTGR, development of the former depends on the success or failure of the latter.

GAS OIL

A liquid **hydrocarbon** mixture, mostly less volatile than **kerosine,** obtained by direct or vacuum **distillation** of petroleum or petroleum products (see **Petroleum Refining**). Gas oil is characterized as a medium (middle or intermediate) **distillate.** Its boiling range at atmospheric pressure is variable, but it is usually from 500 to 700°F (260 to 370°C) or more, especially for vacuum distillates. The name gas oil originates from its early use for enriching **water gas**. Gas oil is now a sole or partial component of some **diesel fuels** and **fuel (heating) oils.** It can be cracked to yield more volatile products, including gasoline components (see **Cracking**).

GAS RECYCLE HYDROGENATION (GRH) PROCESS

A process developed in the United Kingdom, utilizing catalytic hydrogenation (i.e., reaction with hydrogen in the presence of a catalyst) to convert a liquid petroleum product, especially **naphtha,** into a **high-Btu fuel gas** (or **substitute natural gas**).

The naphtha is heated with a hydrogen-rich gas and a catalyst at a temperature in the vicinity of 1000°F (540°C); the pressure is commonly 20 to 30 atm (2 to 3 MPa), but higher pressures are being studied. (The term "recycle" in the title arises from the use of part of the product gas to generate the hydrogen-rich gas by catalytic **steam reforming**.) A typical composition of the product gas is about 47 volume percent of **methane,** 25 percent of ethane, 25 percent of hydrogen, and small amounts of carbon oxides. Although the heating value is close to 1000 Btu/cu ft (37 MJ/cu m), the large hydrogen content of the gas may result in a flame speed too high for general use. Passage of the gas over a hydrogenating catalyst would convert most of the ethane and hydrogen into methane and thus produce a substitute natural gas.

GAS (OR COMBUSTION) TURBINE

A machine, using a high-temperature, high-pressure gas as the working fluid, in which part of the heat supplied by the gas is converted directly into mechanical work of rotation. (For the general principles of turbine operation, see **Turbine**.)

The gas turbine is a **heat engine** operating on an approximation to the idealized **Brayton cycle**. As with other heat engines, the **thermal efficiency** (i.e., the proportion of the heat energy supplied to the gas that is converted into net mechanical work) depends on the upper and lower temperatures between which the machine operates. High gas intake temperatures and low discharge (or exhaust) temperatures favor high thermal efficiencies.

In most cases, the hot gases for operating a gas turbine are obtained by the combustion of a fuel in air; hence, gas turbines are often referred to as combustion turbines. Exceptions are the **expansion turbine** and a proposed closed-cycle gas turbine in which the gas is heated by nuclear fission (see **High-Temperature Gas-Cooled Reactor**).

The fuel for a combustion turbine may be natural gas or other combustible gas or a liquid distillate fuel (e.g., **diesel fuel**). Of special interest are the relatively cheap low-Btu gases made from coal (see **Low-Btu Fuel Gas**). Since the hot combustion gases come into direct contact with the turbine blades, an essential requirement is that the fuel be free from impurities (e.g., sulfur and metal compounds) that might cause corrosion of or ash deposition on the blade surfaces.

Because they are compact, lightweight, and simple to operate, gas turbines have found many applications; the most important uses are probably in jet aircraft and in the generation of electricity, especially for intermediate and peaking demands (see **Electric Power Generation**). Gas turbines are also utilized to operate pumps and gas compressors, to propel ships and trains and, to some extent, freight trucks and buses. The possibility of using gas turbines in automobiles is being studied.

Open-Cycle Turbines

The simplest combustion gas turbine engine is the open-cycle system. In such a system, the cycle, in which the working gas would be restored to its initial state, is not completed. The working gas is discharged in the exhaust and fresh gas (air) is taken in from the atmosphere, just as in **diesel** and **spark-ignition** (gasoline) **engines**.

An open-cycle gas turbine engine has three main components: the compressor, the combustion chamber (or combustor), and the turbine itself. The compressor and turbine are usually on the same shaft, and part of the mechanical energy produced in the turbine drives the compressor (Fig. 54). The remaining energy is available for doing useful work. In some gas-turbine sys-

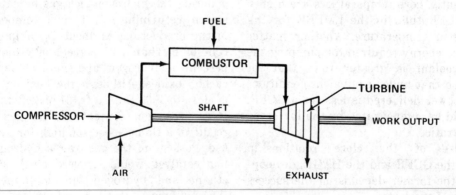

Fig. 54   Open-cycle gas turbine.

tems (e.g., turboprop aircraft and possibly automobile engines) separate turbines on independent shafts are used to drive the compressor and to do useful work, respectively.

Combustor. The compressor takes in air from the atmosphere, compresses it, and forces it into the combustor, which commonly surrounds the shaft connecting the compressor to the turbine. The fuel, as a gas under pressure or as a fine spray of liquid droplets, is injected continuously into the combustor where it burns in the compressed air. Once combustion has been initiated by means of an electrical (or other) igniter, it can continue smoothly as long as fuel and air are supplied. Combustion takes place at almost constant pressure; hence, the pressure in the combustor is much the same as that of the air leaving the compressor, namely, 10 to 20 atm (1 to 2 MPa).

The mass (or weight) of air supplied to the combustor is three to four times the amount required theoretically for complete combustion (about 50 to 60 parts by weight of air to one part of fuel). The excess air mixes with the very hot combustion products and moderates the temperature of the gas somewhat, thus protecting both the combustor and the turbine blades from damage.

The hot, pressurized gas leaving the combustor enters the turbine where it expands (i.e., the pressure decreases) as it passes between the fixed blades. The internal energy of the gas is thus partially converted into energy of motion (kinetic energy). Conversion into rotational mechanical energy then occurs in passage between the adjacent moving (rotor) turbine blades. After expansion in the open-cycle turbine, the exhaust gas, with its temperature and pressure decreased, is discharged to the atmosphere. (Utilization of the heat in the exhaust gas is considered later.)

Compressor. Compressors are commonly of the axial-flow type, although centrifugal compressors are often used in small gas turbines. An axial-flow compressor consists of sets of moving and fixed blades, resembling a turbine in reverse. In traversing the passages between the blades, the kinetic (motion) energy of the gas imparted by the rotation is changed into pressure (internal) energy (i.e., the pressure of the gas is increased). In the centrifugal compressor, air taken in near the shaft of a rotating impeller blade is accelerated outward by centrifugal force. At the periphery, the high-speed air enters a diffuser, that is, a nozzle designed to convert kinetic energy into pressure energy.

Turbine. Gas turbines generally utilize both impulse and reaction principles to provide optimum performance (see **Turbine**). In some cases, the rotor blades are shaped so that impulse predominates at the inner (shaft) end and reaction predominates at the outer end. A gas turbine may consist of one or more stages of fixed and moving (rotor) blades, with the gas pressure decreasing from one stage to the next. The overall pressure drop, from intake to exhaust, in a gas turbine is usually much less than in a **steam turbine;** consequently, there are fewer stages in a gas turbine and a smaller overall increase in turbine diameter from intake to exhaust.

Thermal Efficiency

Gas Temperature. The thermal efficiency of a gas turbine, as defined earlier, depends in the first place on the intake gas temperature, which should be as high as possible. In practice, this temperature is limited by the potential for blade damage. Gas temperatures are commonly in the range from 1470 to 1650°F (800 to 900°C). By the use of special alloys and protective refractory coatings for the blades, the temperature can be increased to about 2280°F (1250°C) or so. For still higher temperatures, it would probably be necessary to use special means of cooling the blades. However, the increase in thermal efficiency resulting from an increase in gas temperature must be balanced against the greater cost of the turbine.

Regeneration. A high thermal efficiency also requires a low temperature of the turbine exhaust gas. Since this gas is quite hot when it is discharged, there is a loss of energy. A substantial increase in efficiency can be achieved by means of a regenerative cycle. Before it is discharged to the atmosphere, the exhaust gas is passed through a **heat exchanger,** called a *regenerator*; here, heat is transferred to the compressed air before it enters the combustor (Fig. 55). Less fuel is then required to provide turbine intake gas at a specified temperature.

Pressure Ratio. The thermal efficiency of a gas turbine is related to the pressure ratio (i.e., the pressure in the combustor relative to the exhaust gas pressure). Up to a point, an increase in the pressure ratio, to about 10 at moderate gas intake temperatures or to 20 at high temperatures, is accompanied by an increase in efficiency. Once again, however, the increased cost of the equipment must be taken into account.

Intercooling. During compression of the air, the temperature rises; this reduces the compressor efficiency because compression of the heated air requires more work than does cold air. An overall efficiency improvement can thus be realized by compressing the air in stages and cooling the air between the stages. The air is cooled by transfer of heat to water in a heat exchanger (or intercooler), as shown in Fig. 55; for this purpose, a supply of cold water would be required.

Reheating. In steam turbines, the thermal efficiency is improved by reheating the steam between stages. Reheating is also possible in a gas turbine with separate high-pressure and low-pressure sections. The exhaust gas from the high-pressure section is reheated in a separate combustor and then used to drive the low-pressure section. Since the exhaust gas contains a large amount of air, the reheating combustor requires only a fuel supply.

Combined Cycle and Cogeneration. Another approach to increasing the efficiency of fuel utilization would be in a combined-cycle or cogeneration system. The still-hot exhaust gas from the turbine provides the heat for generating steam in a waste-heat boiler. The steam is then used to operate a steam turbine (see **Combined-Cycle Generation**). Alternatively, the hot gas might be used to produce process heat (see **Cogeneration**).

Net Efficiency. The efficiency, in terms of the useful mechanical work output relative to the heat supplied by the fuel, of an open-cycle gas turbine is usually not more than about 25 percent. This could be

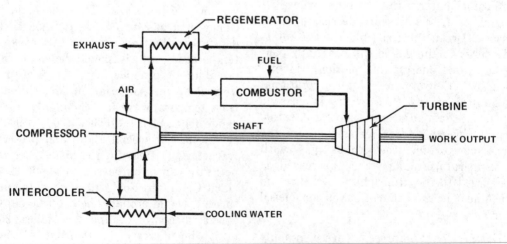

Fig. 55   Gas turbine with regenerator (and intercooler).

increased to perhaps 35 to 45 percent by regeneration, intercooling, and reheating. A further increase in overall efficiency might be achieved in a combined-cycle cogeneration system. As in all heat engines, a fraction of the heat supplied must eventually be rejected at the lower operating (exhaust) temperature and cannot be converted into work. Other losses in efficiency arise from mechanical effects (e.g., friction) in the compressor and turbine.

Closed-Cycle Turbines

In an open-cycle gas turbine, the exhaust gas is discharged, either directly or by way of a regenerator (or in a combined cycle), and fresh air is continuously drawn into the compressor. In a closed-cycle turbine, on the other hand, the working gas (e.g., air) leaving the turbine is restored to the same state (or condition) in which it entered the compressor; thus, the same working gas is used over and over again. A closed-cycle gas turbine is more complex and more expensive than an open-cycle system, but it can operate at a higher thermal efficiency and use a wider variety of fuels.

A closed-cycle gas turbine arrangement with a precooler is shown in Fig. 56. Cooling the working gas in the precooler before it enters the compressor increases its density and so increases the mass of gas that can be compressed in a given machine. The working gas forms a closed circuit, with heat added by way of a heat exchanger in an independent heater. The heat is provided by hot gases obtained by burning any fossil fuel, including pulverized coal (or possibly by utilizing nuclear heat). Since the combustion gas does not come into contact with the turbine blades, the choice of fuel is less restricted than in an open-cycle system. If a fossil fuel is burned in the heater, the required air may be preheated by the turbine exhaust gas in a heat-exchange regenerator located between the turbine outlet and the precooler.

Environmental Aspects

Since open-cycle gas (combustion) turbines must use clean fuels, the exhaust emissions are low in **sulfur oxides** and particulate matter. Furthermore, proper design of the combustor and the use of excess air permit essentially complete combustion of the fuel. Consequently, the levels of the atmospheric pollutants hydrocarbons and carbon monoxide are also low.

High flame temperatures in the combustor could result in the formation of substantial amounts of **nitrogen oxides** from the nitrogen and oxygen in the air. However, the combustion gas temperatures and hence the nitrogen oxide emissions can be controlled by the excess air invariably used to burn the fuel.

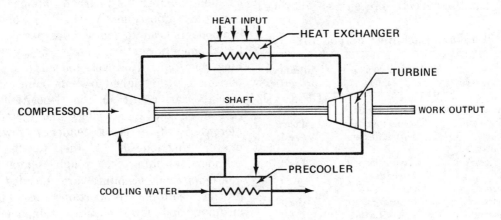

Fig. 56 Closed-cycle gas turbine.

A fossil fuel that is burned in the heater of a closed-cycle system may contain impurities that are prohibited in an open-cycle gas turbine. In this event, it would be necessary to remove pollutants from the exhaust before they are discharged to th atmosphere.

GASOLINE

The liquid most widely used as fuel for internal-combustion engines with electric spark ignition (see **Spark-Ignition Engine**). It consists of a mixture of **hydrocarbons** usually containing 5 to 12 carbon atoms per molecule and boiling almost entirely in the temperature range of about 90 to 400°F (32 to 205°C). The **specific gravity** is around 0.74 (60° API). Commercial gasolines are usually blends of petroleum refinery products (see **Petroleum Refining**) which provide the characteristics required for different engines under various conditions. However, gasolines of the same quality (e.g., boiling range, antiknock behavior, etc.) often vary widely in composition, since the desired properties can be obtained by blending different components.

The sources, main hydrocarbons, and antiknock properties (see below) of the more common gasoline components are given in the table. **Natural gasoline** is the liquid condensed from "wet" natural gas and from the **casinghead gas** collected at the top (or head) of oil-well casing.

Common Gasoline Components

Source	Main hydro-carbon	Antiknock properties
Natural gasoline	Normal paraffins	Poor
Straight-run	Normal paraffins	Poor
Cracking (catalytic)	Isoparaffins and aromatics	Good
Hydrocracking	Isoparaffins and aromatics	Good
Reforming	Aromatics	Good
Alkylation	Isoparaffins	Good
Isomerization	Isoparaffins	Good

Straight-run gasoline is obtained by direct distillation of petroleum (crude oil). The other components are products of oil refinery treatment. (The various sources are described more fully in separate articles.)

Gasoline Properties

Volatility. Two important characteristics of gasoline are volatility and antiknock properties. As a general rule, the most volatile hydrocarbons (i.e., with the lowest boiling points) are those with the smallest number of carbon atoms in the molecule. Readily volatile components are desirable to facilitate engine starting in cold weather. On the other hand, if the volatility is too high, the fuel–air mixture entering a cylinder may contain more fuel than can be burned completely in the available air ("rich" mixture). Rapid vaporization of the gasoline can also lead to blockage of the fuel line either by a bubble of vapor (vapor lock) or by ice, as a result of the cooling accompanying vapor formation. Components of medium volatility are required to achieve fast warm-up.

Once the engine has become warm, the less volatile (higher boiling point) components provide a suitable "lean" mixture (i.e., with less fuel than can be burned completely). These less volatile components, especially the aromatic hydrocarbons, also have a larger energy content by volume than do the volatile hydrocarbons with a smaller number of carbon atoms per molecule. Consequently, the less volatile components lead to more miles per gallon (or kilometers per liter) under steady running conditions. However, too large a proportion of the higher boiling point components can result in the deposition of carbon on the spark plugs and cylinder. Moreover, there is a tendency toward uneven distribution of fuel among the cylinders, leading to rough operation.

Commercial gasolines are blended to provide suitable proportions of high, medium, and low volatility components. These proportions are adjusted according to the local climatic conditions and altitude. At high (summer) temperatures and high altitudes, volatility increases and smaller proportions of the more volatile components are required. In cold climates increased volatility is necessary, but the danger of icing of fuel lines is also increased; this problem is overcome by the addition of an anti-icing agent (see below).

Aviation gasolines have some special volatility requirements. At flight altitudes, the ambient temperature and pressure are lower than on the ground. Hence, the more volatile components tend to favor icing and vapor lock. In addition, the least volatile components have the drawback of causing nonuniform distribution of the fuel-air mixture to the cylinders. In order to minimize these problems, the boiling point range of aviation gasoline is restricted to about 120 to 340°F (50 to 170°C). The proportions of both the least and most volatile hydrocarbons are less than in automobile gasoline.

Antiknock. The efficiency of a gasoline engine can be increased by increasing the compression ratio (i.e., the extent to which the fuel-air mixture is compressed before it is ignited by a spark). At high compressions, however, "knocking" can occur, leading to loss of power and engine overheating. Hence, good antiknock characteristics, as expressed by the **octane number,** are desirable. The so-called "regular" gasolines have posted octane numbers around 90 and "special" gasolines about 95. Aviation gasolines often have performance numbers, equivalent to octane numbers, exceeding 100.

The antiknock properties of gasoline can be improved by decreasing the proportion of normal paraffins (e.g., natural and straight-run gasolines) and increasing the amounts of isoparaffins and aromatic hydrocarbons. The octane number of gasoline can be increased significantly by the addition of a small amount of an antiknock agent; tetraethyl lead (TEL) has been used extensively for this purpose in both automobile and aviation gasolines. However,

the lead in TEL poisons (i.e., renders inactive) the catalyst in the catalytic converters attached to automobile exhausts to reduce hydrocarbon and carbon monoxide emissions. Hence, lead-free gasolines with sufficiently high octane numbers are made by increasing the proportion of isoparaffin and/or aromatic hydrocarbons. An alternative possibility is to add up to 7 percent of the octane "booster" methyl tertiary butyl ether (MTBE) to a low-octane fuel.

Additives

Apart from antiknock agents, small amounts of several materials are added to gasoline to serve a variety of purposes. Olefins (unsaturated) compounds are susceptible to oxidation by air, especially in the presence of traces of certain metals (e.g., copper), leading to the formation of gum-like deposits in the carburetor and other engine components. Consequently, antioxidants and metal-deactivating agents are added to the gasoline to inhibit gum formation. Detergents are added to prevent deposition of dirt and other matter in the carburetor and elsewhere in the fuel system. Other additives include anticorrosion and anti-icing agents to reduce formation of rust and ice, respectively.

GEOTHERMAL ENERGY

Energy present as heat (i.e., thermal energy) in the earth's crust; the more readily accessible heat in the uppermost 6 miles (10 km) or so of the crust constitutes a potentially useful and almost inexhaustible source of energy. This heat is apparent from the increase in temperature of the earth with increasing depth below the surface. Although higher and lower temperatures occur, the average temperature at a depth of 6 miles (10 km) in the United States is about 390°F (200°C).

Hot molten (or partially molten) rock, called "magma," is commonly present at depths greater than 15 to 25 miles (24 to 40 km). In some places, however, anomalous geologic conditions cause the magma to be pushed up toward the surface. In an active volcano, the magma actually reaches the surface, but more often "hot spots" occur at moderate depths, where the heat of the magma is being conducted upward through an overlying rock layer.

If surface (or ground) water has access to the high-temperature rock, a geothermal reservoir of steam and/or pressurized hot water can be formed. (A *geothermal reservoir* is defined as a region where there is a concentration of extractable heat.) The presence of such wet reservoirs may be apparent at the surface as hot springs, geysers, and steam vents (or fumaroles). On the other hand, where water does not penetrate, there may be dry geothermal reservoirs; their existence is then inferred from other observations (see below).

The total amount of heat energy in the outer 6 miles (10 km) of the earth's crust exceeds greatly that obtainable by the combustion of coal, oil, and natural gas. At present, however, only the relatively small proportion of the geothermal energy in wet reservoirs may be regarded as economically useful. Nevertheless, this amount is large enough to make a significant contribution to the United States' energy resources. With advances in technology, a portion of the much larger dry geothermal energy resources may also become available.

There are three main applications of the steam and hot water from wet geothermal reservoirs: (1) generation of electricity, (2) space heating for various kinds of buildings, and (3) industrial process heat.

1. Electricity was first generated from geothermal steam, although on a very small scale, at Larderello, Italy, in 1904; a commercial plant started operation in the same area in 1912. This was followed by commercial generation of electricity from geothermal steam in the Wairakei field, New Zealand, in 1958 and in The Geysers area, north of San Francisco, CA, in 1960. Electric power is now also being produced from geothermal energy in several other countries, including Iceland, Japan, the Philippines, Mexico, and the U.S.S.R.

2. Since 1928, geothermal energy has been used in Iceland for space and greenhouse heating. Almost all the buildings in the capital city of Reykjavik, population about 80,000, are heated by natural hot water and steam. Among other countries where geothermal energy is used for space heating are France, Hungary, New Zealand, and the U.S.S.R. In the United States, the cities of Boise, ID, and Klamath Falls, OR, have utilized geothermal heat for several years. District heating systems are also being developed in California and Colorado.

3. In Iceland and New Zealand in particular, and to a lesser extent in Italy, Japan, the United States and the U.S.S.R., steam and hot water from geothermal sources are used to supply heat for industrial operations. These include pulp and paper and brewing industries, drying timber, wool washing, crop drying, and processing of diatomaceous earth.

A stimulus to the wider application of geothermal energy in the United States was provided by the passage of the Geothermal Steam Act of 1970 and especially of the Geothermal Energy Research, Development and Demonstration Act of 1974. The latter act requires that the Federal Government encourage and assist private industry in the development of practicable means for producing useful energy from geothermal sources. Coordination and management of the Federal Geothermal Energy Program is the responsibility of the U. S. Department of Energy.

U. S. Geothermal Resources

Potential geothermal sources have been identified mainly in areas associated with relatively recent volcanic activity and mountain building, and also to some extent with deep sedimentary formations. The major sources are in the western United States, and their locations are shown on the map in Fig. 57. In addition, large reservoirs of water (or brine) at high tempera-

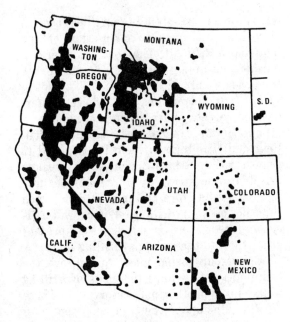

Fig. 57   Geothermal areas in the western United States.

logical Survey has identified almost 2900 sq miles (7500 sq km) of KGRAs in the Western States. In addition, there are regions with a total area some 50 times larger that are believed to have potential for development.

Geothermal Exploration

The existence of a geothermal field is often apparent from steam or hot water at the surface. Because of the high costs of geothermal drilling, however, something needs to be known about the energy potential of a reservoir before drilling is undertaken. In some geothermal fields, the surface indications provide meager (or misleading) information as to the reservoir capacity.

At present, relatively little is known about methods for predicting geothermal reservoir characteristics. Procedures under study include the following: rate of upward heat flow in the ground; chemical composition of surface and groundwaters; electrical resistivity of the ground at depth; and passive and active seismic measurements. These procedures are discussed briefly below. In addition, gravity and magnetic surveys can sometimes provide useful information.

Heat Flow Rate. The heat flow (or flux) in the ground is measured in boreholes to a depth of 330 ft (100 m) or more. It is commonly expressed in *heat flow units* (HFU), where 1 HFU is equal to 10^{-6} calorie per sq cm per sec (0.0418 watt/sq m). The average heat flux is about 1.3 HFU in the eastern United States and 1.5 HFU in the west. The observation of a substantially higher value would indicate a possible geothermal reservoir.

Water Chemistry. Analysis of hot well or spring water can often indicate the probable origin of the water and the reservoir temperature. Concentration of mineral salts in the water may be used to distinguish between geothermal reservoirs of different types.

ture and unusually high pressure, containing significant amounts of dissolved **methane** gas, exist along the north coast of the Gulf of Mexico. There are also promising geothermal sites in Alaska and Hawaii.

In the eastern United States, Arkansas, Georgia, Virginia, and West Virginia each have a number of hot springs indicating the presence of wet geothermal reservoirs, although it is not yet known if they are extensive enough to merit exploitation. There are a few hot springs in other eastern states, but they do not appear to indicate significant geothermal resources. However, a potential low-to-moderate temperature resource may be related to radioactive elements in rocks underlying parts of the Atlantic Coastal Plain.

In the Geothermal Steam Act of 1970, which stated the conditions for leasing public lands for geothermal exploitation, *known geothermal resource areas* (or KGRAs) were defined as areas where "the prospects for the extraction of geothermal steam or associated thermal resources are good enough to warrant expenditure of money for that purpose." The U. S. Geo-

Electrical Resistivity. Measurements are made of the electrical resistivity of the ground at a specific depth at many points and also at several depths at a fixed location over a potential geothermal reservoir. The electrical resistivity depends largely on the salinity and temperature of the groundwater and the porosity of the rocks. A low resistivity may indicate the presence of hot and/or saline water. Hot water containing substantial amounts of dissolved mineral salts is often present in wet geothermal reservoirs.

Seismic Monitoring. Geothermal reservoirs may be associated with persistent microseismic (minor earthquake) activity which can be readily detected by a seismometer. This is referred to as a passive seismic technique. There are some indications that the microseismic signals may be related to reservoir depth and temperature gradient.

Active seismic surveys are similar to those used in exploring for oil (see **Petroleum: Occurrence**). A shock wave is generated in the ground by means of an explosion or impact with a heavy mass. The signals detected at a distance can provide information about the subsurface structure. More needs to be known, however, concerning the response of geothermal reservoirs in order to interpret the results.

Geothermal Well Drilling

Drilling into geothermal reservoirs is usually conducted with rotary bits in a manner similar to that used in drilling for oil or gas. Geothermal drilling is, however, more difficult and expensive. The temperatures, up to 660°F (350°C), encountered in geothermal drilling are higher than in oil- (or gas-) well drilling; moreover, it may be necessary to penetrate hard, abrasive volcanic rocks. Even the best tungsten carbide bits suffer excessive wear and must be renewed frequently.

The drilling muds used to lubricate and cool the drill bit deteriorate rapidly at temperatures above about 350°F (175°C). The deterioration results in an increase in viscosity, thus making circulation of the mud more difficult. The mud may form a hard, caked lining in the well bore and ruin the production.

In dry formations or where water-bearing strata have been cased off, a stream of air can be used instead of drilling mud. The high air velocity required to remove rock cuttings can, however, weaken the drill-pipe string by abrasion; the drill pipe must therefore be replaced more often than in oil-well drilling. Where water is encountered, there is no choice at present other than to use ordinary drilling mud.

Efforts are being made to improve the techniques and reduce the costs of drilling wells into geothermal reservoirs. These efforts include the development of (1) improved drill bits including special synthetic diamond bits, (2) drilling techniques that do not require rotary bits or which permit changing the cutting face while the bit is down-hole, (3) high-temperature drilling fluids and well-cementing materials, and (4) well-logging instruments that can operate at high temperatures. (Well logging involves the lowering of various instruments down a well hole to provide indications of the characteristics of the formations being penetrated.)

Geothermal Resources

Five general categories of geothermal resources have been identified:

1. *Hydrothermal convective systems* are wet reservoirs at moderate depths containing steam and/or hot water under pressure at temperatures up to about 660°F (350°C). These systems are further subdivided, depending upon whether steam or hot water is the dominant product. Hydrothermal resources represent only a small fraction of the potential geothermal resources, but they are the only ones that have been utilized commercially so far. If the temperature is high enough, the water or steam can be used to generate electricity;

otherwise the geothermal energy is best applied to process and space heating.

2. *Geopressured resources* occur in large, deep sedimentary basins. The reservoirs contain moderately high-temperature water (or brine) under very high pressure. They are of special interest because substantial amounts of methane (natural gas) are dissolved in the pressurized water (or brine) and are released when the pressure is reduced. The geopressured resources are quite large; they could be used for the generation of electric power and the recovery of natural gas if a suitable technology could be developed and if individual reservoir productivity and longevity prove to be adequate.

3. *Hot dry rocks* are very hot solid rocks occurring at moderate depths but to which water does not have access, either because of the absence of groundwater or the low permeability of the rock (or both). In order to utilize this resource, means must be found for breaking up impermeable rock at depth, introducing cold water, and recovering the resulting hot water (or steam) for use at the surface. Efforts in this direction are in progress.

4. *Magma resources* consist of partially or completely molten rock, with temperatures in excess of 1200°F (650°C), which may be encountered at moderate depths, especially in recently active volcanic regions. These resources have a large geothermal energy content, but they are restricted to a relatively few locations. Furthermore, the very high temperatures will make extraction of the energy a difficult technological problem.

5. *Normal (or near-normal) thermal gradient resources* are those in which the heat flux is close to the world average value of 1.5 HFU. In the United States, temperatures at a depth of 6 miles (10 km) mostly range from roughly 340°F (170°C) in the east to 555°F (290°C) in the west. The total geothermal resources which might be utilized are very large, but the cost of extraction may not yet be economic. In a few locations, the decay of natural radioactive elements may result in higher temperatures which might be exploited.

The geothermal resources outlined above and the means used or proposed for their development are described more fully in the section on **Geothermal Resources Development.**

GEOTHERMAL RESOURCES DEVELOPMENT

Processes for exploiting the energy content of geothermal resources. These resources are described briefly in the section on **Geothermal Energy.**

Hydrothermal (Convective) Resources

Hydrothermal resources arise when water has access to high-temperature rocks; this accounts for the description as "hydrothermal." The heat is transported from the hot rocks by circulatory movement (i.e., by convection) of the water in a porous medium. The general geological structure of a hydrothermal convective region is shown in simplified form in Fig. 58. The molten rock (magma), raised by internal earth forces, is overlaid by an impervious rock formation, through which heat is conducted upward. Above this is a permeable layer into which water has penetrated, often from a considerable dis-

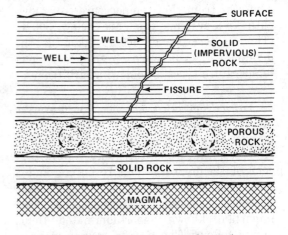

Fig. 58 Hydrothermal convective region.

tance. The permeability could result from fractures or intergranular pores. The heat taken up by the water from the rocks below is transferred by convection to a layer of impervious rocks above. (In convection, the heated water rises, because of its lower density, and then descends when it is cooled by transferring heat to the colder rocks above.)

Hot water or steam often escapes through fissures in the rock, thus forming hot springs, geysers, fumaroles, etc. In order to utilize the hydrothermal energy, wells are drilled either to intercept a fissure or, more commonly, into the formation containing the water (i.e., the hydrothermal reservoir). Most hydrothermal wells range in depth from about 2000 to 7000 ft (600 to 2100 m), although there are some shallower and deeper production wells.

For practical purposes, hydrothermal resources are further subdivided into vapor-dominated and liquid-dominated types. In *vapor-dominated* systems, the wells deliver steam, with little or no liquid water, usually at temperatures of about 300 to 480°F (150 to 250°C) but occasionally higher. *Liquid-dominated* systems, on the other hand, produce a mixture of steam and hot water or hot water only. Two categories of liquid-dominated resouces are distinguished, according to the temperature of the delivered fluid; from high-temperature (or high-enthalpy) systems the temperatures are above 300°F (150°C), whereas from low-temperature (or low-enthalpy) systems they are below 300°F. (The temperature of 300°F taken as the dividing line between high- and low-temperature types is somewhat arbitrary.)

Vapor-Dominated Systems. These are the most attractive geothermal resources because they are the most easily developed. However, they constitute only a few percent of the hydrothermal resources and a much smaller proportion of the accessible geothermal energy resources. The most important known vapor-dominated fields are as follows: The Geysers region in Cali-

fornia, which may be the largest, the Larderello and some smaller areas in Italy, and a small field (or fields) at Matsukawa, Japan.

Conversion of the heat in geothermal steam into electricity is technologically simple. The dry steam from a well is supplied to a **steam turbine** which drives an **electric generator** in the usual manner (see **Electric Power Generation**). The essential difference between this system and a conventional steam turbine-generator system, using fossil or nuclear fuel, is that geothermal steam is supplied at a much lower temperature and pressure.

At The Geysers installation, the more recent deep [roughly 8200 ft (2500 m)] wells supply steam to the turbines at a temperature of about 355°F (180°C) and a gauge pressure of 7.8 atm (0.78 MPa). (The temperatures and pressures in the reservoir are higher.) For comparison, the steam temperature in a modern fossil-fuel plant may be 1000°F (540°C) and the pressure 240 atm (24 MPa). Hence, the geothermal-steam electric plants require special low-pressure turbines. Furthermore, because of the low steam temperature, the **thermal efficiency** (i.e., the proportion of heat supplied to the turbine that is converted into electrical energy) is only about 15 percent, compared with 40 percent (or so) in a modern fossil-fuel plant.

As with steam (or any vapor) turbines in general, the efficiency is improved if the back (or exhaust) pressure is maintained at a low level by condensing the steam. In a conventional plant, the pure water produced in the condenser is required as feedwater for the steam boilers; consequently, the steam and impure condenser cooling water are kept separate (see, for example, Fig. 158). In a hydrothermal plant, however, there is no need for feedwater. Hence, a direct-contact system is used at The Geyers in which the turbine exhaust steam is condensed by direct contact with cooling water. The resulting warm water is circulated through a mechanical-draft cooling tower and returned to the condenser (Fig. 59).

Condensation of the steam continuously increases the volume of the cooling water. Part of this is lost by evaporation in the cooling towers, and the remainder is injected deep into the ground for disposal. The ability to produce their own cooling water is a unique feature of many hydrothermal plants; conventional steam-turbine generating plants require an external source of condenser (makeup) cooling water.

Since The Geysers installation started commercial operation in 1960, its electrical **capacity** has been steadily increased and is expected to reach 1000 megawatts (MW), where 1 MW = 1000 kW = 1 million watts (see **Watt**), by the early 1980s. This is about the same capacity as a single modern fossil-fuel or nuclear **steam-electric** power plant. The eventual capacity of The Geysers field, determined by the maximum rate at which steam can be withdrawn without substantially affecting the reservoir conditions, has been estimated to be about 4000 MW. But this is largely a guess. The capacity is not likely to be limited by the geothermal heat supply but rather by the rate of access of groundwater to the hot rocks. The

reinjection of excess condensed water into the ground may help in this respect.

A number of environmental effects are characteristic of geothermal-steam electric facilities. The steam may contain 0.5 to 5 percent by weight of noncondensable gases which appear in the turbine exhaust. (At The Geysers, the proportion varies but averages around 1 percent of the steam.) These gases consist mainly of carbon dioxide with small amounts of methane and ammonia, which are largely innocuous in the quantities present. In addition, the gases may contain up to 4 or 5 percent of hydrogen sulfide; not only does this gas have an unpleasant (rotten-eggs) odor, it can also be harmful to plant and animal life if it should accumulate. At The Geysers, hydrogen sulfide is considered to be a nuisance rather than a hazard.

In the past the noncondensable gases have been released to the atmosphere where the hydrogen sulfide is gradually destroyed by oxidation. The products, however, are oxides of sulfur which can themselves be harmful at appreciable concentrations (see **Sulfur Oxides**). Plans are therefore under way to remove most of the hydrogen sulfide from the gases before they are discharged.

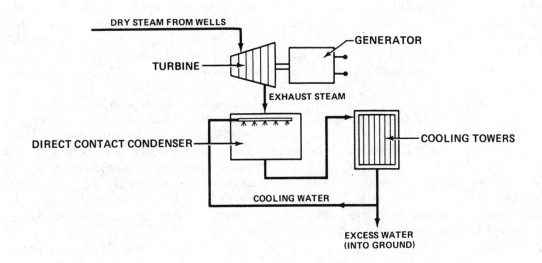

Fig. 59 Hydrothermal power plant.

The geothermal steam may also contain boron, arsenic, mercury, and other potentially poisonous elements which, together with some of the ammonia, are found in the turbine condensate. These substances must be disposed of in a safe manner. This is achieved at The Geysers by the reinjection of excess condensate into the ground at a considerable depth.

The withdrawal of large amounts of steam (or water) from a hydrothermal reservoir may result in surface subsidence. Such subsidences have sometimes occurred in oil fields and are dealt with by injecting water into the ground. The reinjection of excess water from a geothermal-steam electric plant may serve the same purpose. Care must be taken, however, that the water is reinjected at some distance from a ground fault. Experience in Colorado has shown that water injected near a fault can cause minor earthquakes; however, it may prevent larger ones.

Steam-electric plants of all types discharge to the atmosphere much of the heat present in the turbine exhaust steam. In The Geysers facilities this discharge occurs in the cooling towers. Because of the low thermal efficiency, a larger proportion of the heat supplied is discharged than from fossil-fuel or nuclear plants. Vaporization of water in the cooling towers also results in the addition of large amounts of moisture to the atmosphere. Hitherto, geothermal steam has been used on a relatively moderate scale and the environmental effects of the heat and moisture have not been significant.

Liquid-Dominated (High Temperature) Systems. Liquid-dominated hydrothermal resources in the United States are estimated to be from ten to 20 times as large as the vapor-dominated type. But for certain reasons, as will be seen shortly, their development has been more difficult. Nevertheless, the future prospects appear promising.

In a liquid-dominated reservoir, the water temperature is above the normal boiling point 212°F (100°C). However, because the water in the reservoir is under pressure, it does not boil but remains in the liquid state. When the water comes to the surface, the pressure is reduced; rapid boiling then occurs and the liquid water "flashes" into a mixture of hot water and steam. The steam can be separated and used to generate electric power in the usual manner. The remaining hot water can be utilized to generate electric power or to provide space and process heat, or it may be distilled to yield purified water.

Steam from liquid-dominated, high-temperature reservoirs is being used in several countries to generate electric power. The most extensive development has been in the volcanic Wairakei field in New Zealand. The water in the hydrothermal reservoir is at a temperature of about 445°F (230°C) and a pressure of 40 atm (4 MPa). The liquid originates mainly from depths of 2000 to 4600 ft (600 to 1400 m) and is flashed into a mixture of steam and water at the surface.

After passage through a **cyclone separator** to remove the water, the steam is supplied to the turbines connected to electric generators. The maximum initial steam temperature is about 350°F (175°C) and the gauge pressure is 3.5 atm (0.35 MPa). The exhaust steam is condensed by direct contact with cold water from the nearby Waikato River; the warm condenser cooling water is then discharged to the river. This simple procedure, which does not require cooling towers, is possible only because of the ample flow of river water.

The hot water withdrawn from the steam–water separator after flashing is also discharged to the river. Because heat present in the water is discarded in this manner, the overall thermal efficiency for electric power generation at Wairakei is only about 8 percent. The hot water could, however, be used for space heating before discharge.

The largest presently known high-temperature, liquid-dominated hydrothermal resource in the United States is in the

Salton Sea in southern California. It differs from the New Zealand reservoir in being in a sedimentary, rather than a volcanic formation, in having few obvious surface manifestations, and in the high salinity (i.e., high mineral salt content) of the water. The Salton Sea resource was discovered in the course of drilling for oil. Its total electric power potential has been estimated to be about 100,000 MW with a lifetime of some 50 years if it could be fully exploited. The major problem has been related to the very high salinity of the water.

There are apparently two separate hydrothermal systems at different depths underlying the Salton Sea region; the upper system is less saline but it also has a lower temperature than the lower one. Because of the high water temperature of more than 570°F (300°C) and a pressure of about 120 atm (12 MPa) at a depth of 4000 ft (1220 m), the lower (and larger) reservoir is of special interest as a source of energy. However, the mineral content (mainly silica and sodium, potassium, and calcium chlorides) of the water ranges up to 20 or 30 percent by weight; this may be compared with the 3.3 percent (average) of salts in seawater. The hot, highly saline water, referred to as *geothermal brine*, is very corrosive. Moreover, when the temperature is lowered, as it is when the heat is utilized as an energy source, dissolved silica and salts and corrosion products form solid deposits on pipes, pump surfaces, etc.

A number of methods are being studied for converting the heat content of geothermal brines into electrical energy. Some of these techniques, which can also be used with low-salinity water, are described below.

1. The simplest procedure under development is similar to the one used at Wairakei (and elsewhere). As the high-temperature brine reaches the surface, it is flashed into a mixture of steam and more concentrated brine. The steam is separated, scrubbed to remove suspended saline drop-

lets, and used to drive a turbine and generator. The exhaust steam from the turbine is condensed, possibly in a direct-contact condenser as in The Geysers operation. The residual concentrated brine from the flash tank and excess condensate are reinjected deep into the ground.

2. Another technique for generating electric power from the Salton Sea geothermal brines, called the *binary-fluid* (or *two-fluid*) *system*, is being investigated. The hot geothermal fluid, either as unflashed liquid or as steam produced by flashing, is circulated through a primary **heat exchanger.** In the latter, a volatile liquid, such as isobutane, is vaporized and heated under pressure to provide the working fluid for a turbine-generator. The exhaust vapor from the turbine is cooled in the regenerative heat exchanger and then condensed, using either an air-cooled condenser or a water-cooled condenser and cooling tower. The condensed liquid isobutane is returned to the primary heat exchanger by way of the regenerative heat exchanger (Fig. 60). The hot geothermal fluid and the isobutane constitute the two fluids of the binary-fluid system.

If the temperature and salinity of the geothermal brine are not high, the tendency for solids (scale) to deposit on surfaces in the heat exchanger is not too great. The liquid brine under pressure may then be pumped directly through the heat exchanger and reinjected into the geothermal reservoir.

However, where the temperature and salinity of the brine are high, this procedure may not be practical because the heat exchangers may soon be rendered ineffective by scale deposition. Methods for scale control and removal are being developed, but in the meantime the high-temperature brine is flashed into steam, which is scrubbed and passed on to the heat exchanger. Nearly all the salt content remains in the residual brine in the flash tank. To achieve maximum utilization of the heat content of the brine, it may be flashed in two more stages, at successively

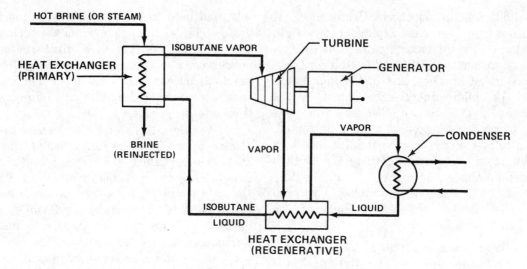

Fig. 60   Binary-fluid geothermal power system.

lower temperatures and pressure, before it is reinjected into the ground.

3. A third approach, called the *total-flow* concept, would utilize both the kinetic energy and the heat energy of the steam–liquid mixture produced by flashing the geothermal brine. The overall efficiency for conversion into electrical energy should be greater than in the methods described above in which only the heat content of the brine is utilized.

One proposed total-flow system utilizes the principle of the Lysholm machine known in this connection as the helical rotary (or screw) expander. Gas compressors, based on the invention of A. Lysholm in Sweden, have been in wide use for some 20 years. The expander is simply a compressor in reverse.

Liquid-Dominated (Low-Temperature) Systems. The low-temperature [below 300°F (150°C)], liquid-dominated reservoirs are the most numerous hydrothermal resource in the United States. The main use would be to provide heat for homes, commercial, industrial, and agricultural buildings, including greenhouses and animal shelters, and for food and industrial processes. This is already being done in several other countries and to a smaller extent in the United States (see **Geothermal Energy**). Hot

water can also be utilized for air conditioning and refrigeration. The general principles of possible applications are the same as those considered in connection with solar energy (see **Solar Energy: Direct Thermal Applications**).

Because the water temperature is not very high, little mineral matter is extracted from the rock medium; hence, the mineral salt content of the water is usually small. The geothermal water is then not very corrosive, and it can often be circulated directly through a heat distribution system and reinjected into the ground. If there is danger of corrosion, however, a heat exchanger would be used to transfer heat from the natural hot water to ordinary service water which is then distributed. The relatively small geothermal water circuit could then be made of more expensive corrosion-resistant materials.

Economic use of the hot water referred to above depends on the proximity of utilization centers. It does not appear to be feasible at present to transport hot water for distances greater than about 15 miles (24 km). This is so partly because of the cost of the pumps, piping, etc., and partly because heat losses increase with distance. The cost of installing a distribution system in an existing city would be prohibitive, but in a new housing development the hot-water piping could be laid at the same time as that for service water.

If the hot water cannot be used directly in the vicinity of its source, the heat may be converted into electrical energy which can be transmitted to more distant points. A binary cycle could be used in which the hot water serves to vaporize and heat a liquid of low boiling point, as described earlier (see Fig. 60). Whether or not such a system for generating electricity will prove to be economic depends on circumstances (e.g., cost of drilling wells, temperature of the hot water, availability of condenser cooling water, etc.).

Geopressured Resources

Drilling for oil and gas has revealed the existence of reservoirs containing salt water at moderately high temperatures and very high pressures in a belt some 750 miles (1200 km) in length along the Texas and Louisiana coasts of the Gulf of Mexico. The belt, which extends up to about 100 miles (160 km) inland and 150 miles (240 km) offshore, is divided into several independent volumes separated by faults running roughly parallel to the coastline. Because of the abnormally high pressure of the water, up to 1350 atm (137 MPa) in the deepest layers, the reservoirs are referred to as *geopressured*. Other deep sedimentary basins are known to contain geopressured waters, but the Gulf Coast reservoirs are by far the largest in the United States. (Note that the pressures of the Salton Sea geothermal brines are much lower than in geopressured zones.)

The geopressured hot water (or brine) reservoirs were apparently formed by the accumulation of geothermal heat, stored over several million years, in water trapped in a porous sedimentary medium by the overlying impervious layers. The upward loss of heat is relatively small and there are no obvious surface indications of the deep, high-temperature reservoirs. Were it not for the drilling operations, their pres-

ence might not have been suspected. In typical geopressured systems in Texas, the pressures are from 680 to 950 atm (68 to 95 MPa) and the temperatures from 320 to 390°F (160 to 200°C) at depths from 2.5 to 3 miles (4 to 5 km). Higher pressures and temperatures have been measured at greater depths. The amount of dissolved salt in the water varies with the location and depth of the reservoir, ranging from very small to about three times that in seawater.

A special feature of geopressured waters (or brines) is their content of methane (**natural gas**). The energy value of the brines thus depends on the recoverable gas as well as on their temperature. The solubility of methane in water at normal pressure is quite low, but it is increased at the high pressures of the geopressured reservoirs. When the water is brought to the surface and its pressure reduced, the methane gas is released from solution. The gas content of geopressured brines is usually about 10 to 20 cu ft (measured at normal temperature and pressure) per **barrel** (42 gal) of water (1.9 to 3.8 cu m gas per cu m water), but higher values have been reported in brief tests. However, the amount of natural gas recoverable economically from geopressured reservoirs is presently unknown.

Drilling wells into deep reservoirs containing brine at high temperature and pressure, with proper precautions against blowout, is difficult and expensive. Hence, information must be available about the potential of a geopressured reservoir before drilling is initiated. Studies are under way to identify reservoirs with sufficient volume and permeability to sustain large water flows over long time periods. The amount of methane gas that can be recovered must also be determined, since this is a primary factor in the economics of geopressured water resources.

Hot Dry Rock Resources

There are many places, especially in the western United States, where intrusions of hot rocks have produced regions with temperatures above average at a given depth but which are not associated with hydrothermal activity. These are the *hot dry rock* (HDR) geothermal resources. Such resources, with rock temperatures exceeding about 390°F (200°C) at depths up to 16,000 ft (5 km), are estimated to be significant and worthy of development as a source of energy.

The presence of HDR formations may be indicated by larger than normal heat flow rates and absence of hydrothermal manifestations at the surface. Such indications can, however, be misleading and other possible exploration methods are being studied. At present, reliable information is obtainable only by drilling a test well.

Hot dry rocks exist because they are impermeable to water or because water does not have access to them. Quite often both conditions occur; that is to say, the rocks are impermeable, and there is little or no surface water in the vicinity. In principle, the recovery of heat from such hot dry rocks involves breaking up or cracking the rock to make it permeable and then introducing water from the surface. The water is heated by the rock and is returned to the surface where the heat is utilized.

A possible method for achieving this objective is to detonate a high explosive at the bottom of a well drilled into the rock. Water would be injected into this well, circulated through the rubble, and withdrawn as hot water or as a water–steam mixture through another well. It appears, however, that the production of a useful amount of broken rock would require an uneconomically large quantity of explosive. Nuclear explosives have been suggested but discarded as an alternative to conventional high explosive, because of the associated environmental and institutional problems.

A more promising approach, under investigation by the Los Alamos National Laboratory for the U. S. Department of Energy, is to use hydraulic fracturing to produce the heat transfer surface and permeability required to extract energy at a high rate from hot dry rock. Hydraulic fracturing, which is performed by pumping water at high pressure into the rock formation, is commonly used in oil and gas fields to improve the flow. The crack produced in this manner is roughly in the form of a large, thin vertical disk, resembling a pancake, possibly a few hundred meters (yards) in diameter but less than a centimeter (half inch) thick. Heat can then be extracted from the hot rock by circulating water through the crack (Fig. 61).

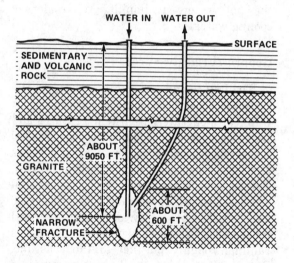

Fig. 61   Heat extraction from hot dry rocks.

In a test of the foregoing technique, a number of hydraulic fractures were made in a granite rock at a location adjacent to the Valles Caldera, to depths up to about 9800 ft (3.0 km). (A caldera is a large, bowl-like depression formed by the collapse of a volcano.) Water from the surface was then pumped into the cracks through a cased hole at a depth of roughly 9050 ft (2.76 km) where the temperature was 365°F (185°C).

To recover the heated water, another hole, drilled at a distance of about 250 ft (76 m), was angled toward the cracks. At a depth of some 8850 ft (2.7 km) or so, circulation was established and a steady flow of high-temperature water was realized. Electricity was generated with a binary liquid system using Freon (R-114) as the turbine working fluid (see Fig. 60). Only about 5 percent of the water introduced was lost in the ground in the initial tests and so a relatively small proportion of makeup water was required.

One of the uncertainties associated with the hydraulic fracturing and heat-extraction procedure is the useful lifetime of the cracked region. This can be determined only by continued testing. However, computer modeling suggests that, as heat is extracted, the cracking should increase with time and make more hot rock accessible to water. The effective lifetime of the fractured system should thus be extended.

Magma Resources

In some cases, especially in the vicinity of relatively recent volcanic activity (e.g., in Hawaii), molten or partially molten rock (i.e., magma) occurs at moderate depths [e.g., less than 3 miles (5 km)]. The very high temperatures, above 1200°F (650°C), and the large volume make magma a substantial geothermal resource. However, extraction of the heat from the molten rock will be difficult and may not be feasible for some time.

A concept being studied for the U. S. Department of Energy by the Sandia National Laboratories is to place a heat exchanger within the magma. Heat would be transferred to a suitable liquid and brought to the surface. The hot liquid could be used to produce a working fluid, possibly steam, to operate a turbine and electric generator. The liquid would then be recirculated through the heat exchanger in the magma. One problem would be to construct a heat exchanger that will withstand the high temperatures of the magma. Another is to maintain flow of the viscous magma around the heat exchanger to provide a steady supply of heat and to prevent deposition of solid.

Normal (or Near-Normal) Thermal Gradient Resources

Hot rocks and near-surface magmas occupy a relatively small area. In most parts of the United States, the increase of temperature with depth (i.e., the thermal gradient) is normal or near normal. Consequently, rocks at a moderately high temperature are invariably encountered at sufficient depth. Heat could presumably be extracted by a method similar to that used for hot dry rocks. Thermal gradients are, on the whole, greater to the west of the Rocky Mountains than to the east, as shown by the rough estimates of average temperatures at different depths given in the table. [The thermal gradient is equal to the heat flux divided by the thermal conductivity of the rocks; hence, the local value of the heat flux (in HFU) is commonly taken as a measure of the thermal gradient; see **Geothermal Energy**.]

Average Temperatures in Areas of Near-Normal Thermal Gradients

Depth, ft	Temperature	
	West	East and Midwest
9,800 (3 km)	194°F (90°C)	131°F (55°C)
19,700 (6 km)	356°F (180°C)	230°F (110°C)
32,800 (10 km)	554°F (290°C)	338°F (170°C)

Water at a temperature around 195°F (90°C), which should be obtainable from rocks at a depth of about 2 miles (3 km) in the west and 3 miles (5 km) in the east and midwest of the United States, could be used for space heating and similar purposes. For the generation of electric power, the depths would need to be roughly 3.75 miles (6 km) and 6.2 miles (10 km), respectively, in order to obtain water at a sufficiently high temperature.

Because of the high costs of drilling to great depths, especially in hard rock, it may be several years before the normal thermal gradient geothermal resources become economically useful. Nevertheless, they represent a very large energy source that may ultimately prove to be of great value if the technological problems of heat extraction can be solved.

In addition to the potential geothermal resources from normal (or near-normal) thermal gradients, there are several areas in the eastern United States (e.g., the Atlantic Coastal Plain) where the thermal gradients at modest depths of less than a mile (1.6 km) are somewhat greater than expected. This is apparently due to heat generated over long periods of time by the decay of the radioactive elements uranium, thorium, and potassium in the ground. Both warm dry rocks and aquifers have been detected. The temperatures are not high enough for economic electric power generation, but they would be adequate to provide space and process heat.

GIAMMARCO VETROCOKE PROCESSES

Two processes for the removal of **acid gases** from fuel gases: one form is designed for removing hydrogen sulfide and the other for carbon dioxide. In both cases the absorber is a solution of a carbonate containing an arsenic compound. The hydrogen sulfide is removed by oxidation whereas the carbon dioxide is absorbed by a weakly alkaline solution. If both hydrogen sulfide and carbon dioxide are to be removed, the former must be removed first.

In the Giammarco Vetrocoke Sulfur process, hydrogen sulfide is removed from the fuel gas by chemical oxidation with sodium arsenate (Na_3AsO_4) in a weakly alkaline (sodium carbonate) solution. The arsenate is made by the action of oxygen (in air) on a solution of arsenite (Na_3AsO_3) obtained by dissolving arsenic trioxide (As_2O_3) in the carbonate. The impure fuel gas at just above atmospheric pressure enters at the bottom of the absorber column and meets a downflow of absorber

solution at a temperature up to about 300°F (150°C). The hydrogen sulfide is oxidized to elemental sulfur and the arsenate is reduced to arsenite. The sulfur is precipitated as a solid and is removed by froth flotation. The spent absorber solution is heated and sent to the regeneration column through which air is blown. The regenerated arsenate solution is recycled to the absorber (see **Desulfurization of Fuel Gases.**)

For removing carbon dioxide the absorber is an aqueous solution of a carbonate, usually potassium carbonate (K_2CO_3), containing potassium arsenite (K_3AsO_3) produced by the action of arsenic trioxide on the carbonate. The impure feed gas enters the bottom of the absorber column at a pressure up to 75 atm (7.5 MPa) and meets a downflow of absorber solution at 120 to 210°F (about 50 to 100°C). The carbon dioxide is removed by chemical reaction with the carbonate and arsenite, forming potassium bicarbonate ($KHCO_3$) and arsenic trioxide, respectively. The clean gas leaves at the top of the column. The spent absorber solution is removed and its pressure reduced to atmospheric; it then passes to the regenerator column where it is heated. At the higher temperature, the bicarbonate is decomposed giving off carbon dioxide gas and leaving potassium carbonate, together with arsenite. The regenerated solution is recycled to the absorber and the carbon dioxide is discharged.

GILSONITE

A black, asphalt-like solid, consisting mainly of a mixture of complex, high molecular weight **hydrocarbons**; it is found in eastern Utah and western Colorado. Gilsonite has been mined hydraulically with high-pressure water jets, pumped to the surface, and transferred as a slurry by pipeline to a refinery. After separation from the water by filtration, the solid Gilsonite is heated and distilled in much the same way as in **petroleum refining.** The products are gaseous, liquid, and solid (coke) fuels.

GRAPHITE-MODERATED REACTOR (GMR)

A thermal nuclear fission reactor in which the moderator is graphite (i.e., a form of the element carbon); the coolant is usually, although not always, a gas, either carbon dioxide or helium (see **Thermal Reactor**). An important aspect of GMRs is that they can be designed to operate with normal uranium metal as fuel.

Graphite-moderated **nuclear power reactors,** with carbon dioxide as coolant, are used extensively in the United Kingdom for generating electricity. Most of those operating in the early 1980s are known as Magnox reactors, because the normal uranium metal fuel is clad with a magnesium–aluminum alloy called Magnox. The coolant gas leaves the reactor core at a temperature approaching 750°F (400°C), passes to a **heat-exchanger** steam generator, and returns to the reactor at a lower temperature. The steam operates a turbine which drives an electrical generator in the usual manner (see **Electric Power Generation**).

In the United Kingdom's newer advanced gas-cooled reactors (AGRs), the fuel is uranium dioxide, enriched to about 1.2 percent in uranium-235, clad with stainless steel. Operation with this fuel is possible at higher temperatures than in Magnox reactors and the carbon dioxide coolant leaves the core at near 1110°F (600°C). The higher temperature of the steam generator results in a better **thermal efficiency** for power generation. A problem with the AGR is corrosion of the graphite moderator by the hot carbon dioxide gas, especially in the presence of nuclear radiation in the reactor. Special measures are therefore taken to minimize the corrosion.

The earliest experimental and research reactors in the United States used graphite as moderator and were air-cooled; these reactors operated at very low heat (power) levels. The first plutonium production reactors (for weapons) were also graphite-moderated, but the coolant was water because of its superior heat-removal properties. The graphite-moderated, water-cooled Hanford-N reactor, which is the only power reactor of its kind in the United States, developed from these production reactors. The core consists of a graphite matrix with a large number of horizontal tubes containing the clad uranium metal fuel. Water under high pressure, to prevent boiling, is pumped through the tubes to remove the fission heat. The high-temperature, high-pressure water leaving the reactor then produces steam from water at a lower pressure in a heat-exchanger steam generator (see **Pressurized-Water Reactor**).

The major, although limited, interest in graphite-moderated reactors in the United States (and in West Germany) is in those of the **high-temperature gas-cooled reactor** (HTGR) type. The coolant is the inert gas helium which can be used at high temperatures but does not require high pressure. The special feature of the HTGRs is that the fertile component of the fuel is thorium-232, whereas it is uranium-238 in nearly all other graphite reactors; during operation the thorium-232 is converted into fissile uranium-233 (see **Nuclear Fuel**). The fissile material in the fuel would initially be uranium-235, but this would be replaced as far as possible by uranium-233 as it became available.

Another graphite-moderated reactor concept is the **molten-salt breeder reactor** (or MSBR). It is also based on thorium-232 as the fissile species and is designed to breed uranium-233, that is, to produce more of this fissile species than is consumed during reactor operation (see **Breeder Reactor**). The MSBR is unusual in several respects; one is that molten salt mixtures serve as both fuel and coolant.

H

H-COAL PROCESS

A Hydrocarbon Research, Inc. (HRI), process primarily for **coal liquefaction** by direct hydrogenation in the presence of a catalyst. The process, which can be used with any type of coal, including those with a high sulfur content, is related to HRI's H-Oil process for converting heavy, residual petroleum oils into light oils. By changing the ratio of coal to hydrogen, the liquid products of the H-Coal process can range from heavy boiler fuel to a **synthetic crude oil** (syncrude) suitable for refining (see **Petroleum Refining**). A **high-Btu fuel gas** is also produced.

Pulverized coal, slurried with a heavy oil generated in the process, is first heated with hydrogen (or hydrogen-rich gas) to a temperature of 650 to 700°F (345 to 370°C) in an externally fired preheater. The mix-ture is then fed to the bottom of an **ebullated-bed** catalytic hydrogenation reactor (Fig. 62). The catalyst is cobalt molybdate (CoO–MoO_3) on a predominantly alumina base. The temperature of the bed is maintained at about 850°F (455°C) by the hot feed and by the heat released in the hydrogenation reactions. The pressure in the reactor is around 200 atm (20 MPa).

The vapor leaving the reactor is cooled to condense the less volatile components, leaving a gas containing mainly methane and hydrogen, with hydrogen sulfide and ammonia as impurities. The impurities are removed, and the hydrogen is separated and recycled for combination with the input slurry. The remainder is a clean high-Btu fuel gas.

The liquid product from the reactor, with suspended solid particles, is fed to a

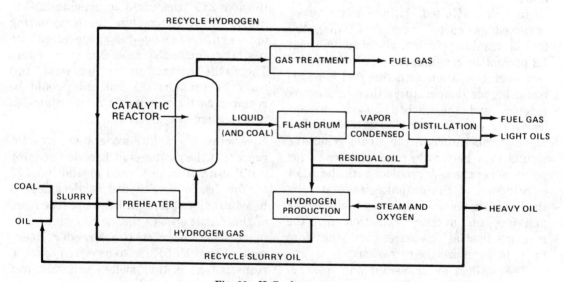

Fig. 62   H-Coal process.

flash drum where the pressure is decreased and the more volatile components are vaporized. The vapor is cooled and fractionally distilled at atmospheric pressure. The products are a high-Btu fuel gas and various **hydrocarbon** oils. The heavy fraction provides the oil for slurrying the coal feed.

The residue from the flash drum consists of heavy oils, which are not readily vaporized, and unreacted coal (and ash). The components are separated, and the oil is fed to the atmospheric fractional **distillation** unit; the coal can be reacted with steam and oxygen to produce the hydrogen (or hydrogen-rich) gas required for the process (see **Hydrogen Production**).

The characteristics of the products of the H-Coal process depend on the conditions in the reactor. An increase in the ratio of coal to hydrogen in the feed leads to preferential formation of fuel gas and heavier distillates suitable as boiler fuel. On the other hand, an increase in the hydrogen content favors a syncrude, which can be treated to yield lighter fuels, as in petroleum refining.

HEAT ENGINE

A machine or device for converting heat energy into mechanical energy (or work). Examples of heat engines are conventional **steam engines, steam** and **gas turbines, spark-ignition** and **diesel engines** and the **Stirling engine.** The basic thermodynamic principles of all such engines are the same: a working fluid, such as water (liquid and vapor) or a gas, takes in heat from a "source" at an upper temperature (or temperatures), converts part of this heat into an equivalent amount of mechanical work, and rejects the remainder to a "sink" at a lower temperature (or temperatures) (Fig. 63).

Heat engines function in cycles, that is, in a repeated sequence of operations. At the beginning of each cycle (or sequence) the initial condition of the working fluid is restored in preparation for that sequence.

In at least one stage of the cycle, the working fluid at a high pressure expands and as a result mechanical work is done *by* the fluid. However, during expansion the pressure falls and hence in a later stage (or stages) the working fluid must be compressed in order to restore the initial high pressure. Work is then done *on* the fluid. The difference between the work done by the fluid (during expansion) and that done on it (during compression) is the net (or useful) work output in each cycle of the heat engine. This net work is equivalent in energy to the difference between the heat taken up from the source and that rejected to the sink.

Since part of the heat taken up from the source must be rejected to the sink, a heat engine operating in cycles cannot convert heat completely into mechanical work. (A conceivable, but not realizable, exception would occur if the sink were at the absolute zero of temperature.) The fraction of the

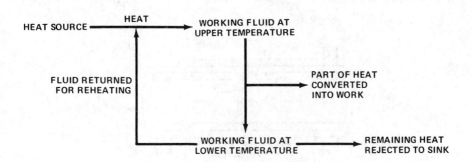

Fig. 63   Principle of heat engine.

heat taken up that is converted into net mechanical work is called the **thermal efficiency** of the heat engine. The maximum possible efficiency of a heat engine is that of a hypothetical (ideal) cycle called the **Carnot Cycle.**

Practical heat engines operate on less efficient cycles (see **Brayton, Diesel, Otto, Rankine,** and **Stirling Cycles**), but a general conclusion is applicable to all cycles: the thermal efficiency is increased by having the upper (or source) temperature as high as possible and the lower (sink) temperature as low as possible.

See also **Prime Mover.**

HEAT EXCHANGER

A device for transferring heat from a hotter fluid (gas or liquid) to a colder fluid (or vice versa). The purpose may be to heat the colder fluid (e.g., using the exhaust gases to heat the combustion air in a **gas turbine** or to heat water in a waste-heat boiler) or to cool the hotter fluid (e.g., in an automobile radiator or the condenser of a **steam turbine**).

The shell-and-tube heat exchanger is the most common type. It consists of a bundle of parallel tubes (or pipes) contained within a surrounding, usually cylindrical, structure called the shell. One fluid flows through the pipes and the other flows outside, generally in the opposite direction (countercurrent flow). The tubes may have exterior fins to increase the heat transfer area and thus facilitate heat exchange between the hot and cold fluids.

HEAT PIPE

Usually, a means for transporting heat efficiently from a source to a cooler receiver by utilizing the latent heat of vaporization of a liquid. (A different kind of heat pipe is described at the end of this section.) In essence, a heat pipe is a closed space containing a suitable working liquid and its vapor. One part of the space is in contact with the heat source and another with the cooler material to which the heat is to be transported. The interior wall of the space is lined with a porous material called a *wick*. For high-temperature operation, the wick can be made of several layers of a woven metal wire mesh. A simple form of heat pipe is shown in Fig. 64, but more complex arrangements which satisfy the basic conditions outlined above are possible.

In the hotter (i.e., heat source) part of the heat pipe, the working liquid is vaporized, thereby taking up the latent heat of vaporization. The vapor diffuses toward the cooler region because the pressure is lower there and condenses to liquid. In doing so, it deposits the heat of vaporization taken up from the source. The liquid is returned to the heat source region by capillary action of the wick. There is thus a continuous movement of vapor from the heat source to the receiver and of condensed liquid back to the source, accompanied by the transfer of heat.

Since the working liquid is returned to the heated region by capillary action, the heat source may be located above the heat receiver. However, if the receiver is at a

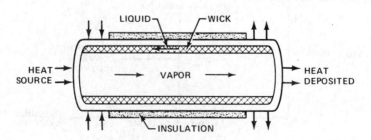

Fig. 64   Simple heat pipe.

higher level than the heat source, the return of the liquid is assisted by gravity. The vapor always moves from the higher pressure (i.e., the heat source) region to the lower pressure (i.e., heat receiver) region regardless of their relative positions.

A special feature of the heat pipe is its ability to transport heat at a high rate even when there is only a small temperature difference between the source and receiver. For other methods of heat transport (i.e., conduction, convection, and radiation), the rate is highly dependent on the temperature difference, but in a heat pipe, the important consideration is the heat of vaporization of the liquid. Another advantage is that a heat pipe can serve as a concentrator; heat can be taken up by vaporization from a large area and deposited during condensation on a small area.

For operation of a heat pipe at high temperatures, sodium, melting at 207°F (97°C), has been used as the working liquid. Its large heat of vaporization and good thermal conductivity make it particularly suitable for this purpose. A sodium heat pipe might be used to transfer heat from a source, possibly stored heat, to a **Stirling engine.**

Chemical Heat Pipe

The purpose of the chemical heat pipe concept under development is to transfer heat from a source to a distant receiver, up to 100 miles (160 km) away. The operating principle is quite different from that of the physical heat pipe described above.

The chemical heat pipe is based on a reversible chemical reaction, that is, a reaction that proceeds predominantly in one direction or in the opposite direction according to the conditions (e.g., temperature). Heat is supplied from a source (fossil fuel, nuclear, or solar) to substances that absorb the heat and undergo a chemical reaction. The products are then pumped through a pipeline to a distant receiver where the conditions are changed so that the chemical reaction is reversed and heat

is released. The original interacting materials are recovered and returned by another pipeline to the heat source for reuse. The principles are the same as described under Reversible Chemical Reactions in the section on **Energy Storage.**

HEAT PUMP

In general, a device operating in a manner opposite to a **heat engine;** it takes up heat at a lower temperature and, by utilizing mechanical work, discharges heat at a higher temperature. More specifically, the term heat pump is applied to a machine that can be used for space heating in the winter and for cooling (air conditioning) in the summer. The most common type of heat pump for domestic use is the air-to-air system in which heat is both taken up from and transferred to air. Water-to-air and water-to-water heat pumps are also available, and an air-to-water heat pump has been designed for domestic hot-water heating.

Operating Principles

In essence, a heat pump is a vapor-compression type refrigerator (see **Refrigeration**), consisting of an evaporator unit (where latent heat is taken up) and a condenser unit (where heat is discharged), with a mechanical compressor between them. One of the units is installed in the interior of the building and is connected to the air-distribution system; the other unit is outdoors. In the winter, the interior unit becomes the condenser and discharges heat into the building; heat is taken up from the outside air by the evaporator. In the summer, the situation is reversed. The indoor unit becomes the evaporator and removes heat from the interior air, thereby cooling it; heat is discharged to the surroundings from the outdoor (condenser) unit.

The operation of a heat pump for space heating and cooling may be understood from Figs. 65A and 65B, which show heat-

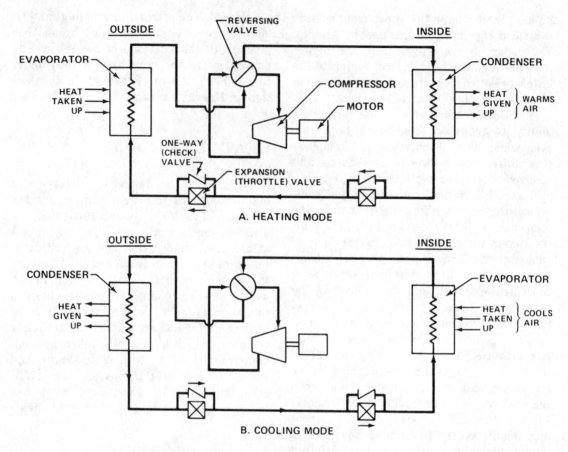

Fig. 65   Heat pump. A: Heating mode; B: Cooling mode.

ing and cooling modes, respectively. The components are the same in both cases, but the refrigerant (working fluid) flows in opposite directions. The change in operational mode, from winter to summer (and vice versa), is achieved by a reversing valve and a combination of expansion and one-way (check) valves. A blower (or fan) draws room air across the coils in the inside (right-hand) unit to provide space heating in winter or cooling in summer.

Rating of Heat Pumps

Heat pumps are rated according to their heating or cooling capacity and their operating efficiency. The capacity of a heat pump for either heating or cooling is commonly expressed in terms of **British thermal units** (Btu) per hour of heat delivery or removal. The cooling capacity is some-

times stated in tons, where 1 ton is equivalent to a heat removal rate of 12,000 Btu/hr (see **Refrigeration**).

The efficiency of a heat pump, as distinct from its capacity, is given by the **coefficient of performance** (or COP). This is the rate of heat delivery or removal relative to the total rate of energy input (or power) required to operate the heat pump and associated air blower (or blowers). The cooling efficiency is sometimes expressed by the **energy efficiency ratio** (or EER), where the EER in Btu/watt-hr is 3.41 times the COP. The COP is a pure ratio and has no units.

The average COP of a heat pump in the heating mode (i.e., in winter) may be in the range of 2 to 3. In other words, the heat delivered to a building may be two or three times as great as the energy input required to operate the heat pump and air blower. The excess energy delivered (as heat) over the input (as electricity) is derived from the outdoor air. Even when the ambient air temperature is low, it contains heat which is taken up by the evaporator unit of the heat pump. The heat in the air is derived from the sun; hence, in a sense, an air-to-air heat pump makes use of solar energy for heating.

The COP of a heat pump is generally not the same in the heating and cooling modes. In any case, the COP varies with the conditions, in particular with the temperatures at which the machine takes up heat (in the evaporator unit) and gives it up (in the condenser unit). This is important for indoor heating in the winter because the efficiency decreases as the temperature of the outdoor air, from which heat is taken up, decreases. When the outdoor temperature falls below about 32°F (0°C), the efficiency of a heat pump is so low that auxiliary, electrical resistance heating is required to maintain comfortable indoor conditions. Consequently, heat pumps are not recommended for areas where low winter temperatures are common.

There are two main reasons for the low COP of a heat pump in the winter when the outside temperature is low. The most important is the low vapor pressure of the refrigerant and consequent decrease in mass passing through the compressor. The result is a lower rate of heat transfer from outside (evaporator) to inside (condenser). Another reason for the decrease in efficiency is that, as the temperature difference between outside and inside becomes larger, as it does in very cold weather, more energy is required to deliver a certain amount of heat.

See also **Annual Cycle Energy System.**

HEAT RATE

As applied to an **electric power generation** plant, the amount of heat in **British thermal units** (Btu) supplied to generate 1 kilowatt-hour (kW-hr) of electrical energy in the given plant (see **Watt**). The heat rate, expressed in Btu per kW-hr, is a measure of the **thermal efficiency** of the plant. The gross thermal efficiency is equal to the amount of electrical energy generated, expressed in Btu (or kW-hr) divided by the heat supplied, also in Btu (or kW-hr). Since the heat rate is the heat supplied to generate 1 kW-hr, which is equivalent to 3413 Btu, the thermal efficiency is equal to 3413 divided by the heat rate in Btu/kW-hr.

For modern **fossil-fuel** power plants, the heat rate may be as low as 8100 Btu per kW-hr of electricity generated. The gross thermal efficiency, expressed in percent, is then $(3413/8100) \times 100 = 42$ percent. However, older fossil-fuel plants have higher heat rates, up to 11,000 Btu/kW-hr (or so), and lower efficiencies, down to about 31 percent. For nuclear power plants utilizing **light-water reactors,** the heat rate is around 10,300 Btu/kW-hr, and the gross thermal efficiency is $(3413/10,300) \times 100 = 33$ percent.

In metric (SI) units, the heat rate might be defined as the amount of heat in **joules** (J) supplied to generate 1 J (i.e., 1 watt-sec) of electrical energy. The heat rate would then be the reciprocal of the thermal efficiency expressed as a fraction (rather than as a percentage).

HEATING VALUE

The amount of heat energy produced by the complete combustion of a unit quantity (mass or volume) of a fuel, usually a **fossil fuel** (or derived from a fossil fuel); it is also called the heat of combustion or the calorific value. For solid fuels, the heating value is quoted for a unit of mass whereas for gaseous fuels it is given for unit volume. The heating values of liquid fuels may be stated for either unit mass or unit volume, depending on circumstances.

In the common engineering units, heat is expressed in **British thermal units** (Btu) and mass in pounds (lb); hence, the heating values of solid fuels (e.g., coal) and some liquids are given in Btu/lb. In metric (SI) units, heat is in **joules** (J), kilojoules (kJ), or megajoules (MJ), where 1 MJ = 1000 kJ and 1 kJ = 1000 J, and mass in kilograms (kg); thus, heating values of solids are commonly stated in MJ/kg. Heating values in Btu/lb are converted into MJ/kg upon multiplying by 2.32×10^{-3}.

For gases and some liquids the common unit of volume is the standard cubic foot (scf), that is, 1 cu ft measured under standard conditions, usually 30 in. of mercury pressure and a temperature of 60°F. Heating values are then in Btu/scf. In metric units, the corresponding values are in MJ per standard cubic meter (cu m). For conversion of Btu/scf into MJ/cu m, multiply by 3.72×10^{-2}.

Heating values of **fuel oils** are often quoted in Btu/gallon. The equivalent in metric units, MJ/cu m (or MJ/liter), is obtained upon multiplying Btu/gal by 0.279 (or 2.79×10^{-4}).

The heating value of a fuel is usually stated as the gross or *higher heating value* (HHV), as distinct from the net or *lower heating value* (LHV). The HHV is the amount of heat produced in combustion, assuming the products (carbon dioxide and water) to be cooled to the initial (room) temperature, so that the water is condensed to liquid. The LHV is less than the HHV by the latent heat of vaporization of the water; that is, the water is assumed to be entirely in the vapor state.

Heating values of gases given in this book in terms of Btu/cu ft (or MJ/cu m) are the HHVs, with volumes (cu ft or cu m) measured under the standard conditions given above.

HEAVY (CRUDE) OIL

A high-density, viscous (i.e., thick), dark-colored **petroleum** found in several parts of the world; in the United States, the main reservoirs of heavy crudes are in California. Because of the very high viscosity (i.e., flow resistance) of the oil, less than about 10 percent can be recovered by conventional pumping. Further recovery is achieved by heating the oil in the reservoir so as to decrease its viscosity and hence increase the rate of flow to the well.

Heating by steam, called *steam flooding*, has been used to improve recovery from heavy-oil reservoirs at relatively shallow depths. In the discontinuous (or cyclic) method, a period of steam injection is followed by pumping from the same well; this procedure is repeated as long as it is effective. In continuous steam flooding, steam is injected into one well and the oil is pumped simultaneously from an adjacent production well (or wells).

In the methods just described, the steam is produced at the surface; a procedure is being tested, primarily for use in deep heavy-oil deposits, of generating the steam within the reservoir. The steam generator, located at the bottom of an injection well, burns propane gas in compressed air; the fuel gas and air, together with water, are introduced through pipes from the surface. The steam and hot combustion gases produced cause the oil to flow so that it can be recovered from nearby production wells.

For deep reservoirs, tests are also being made of heating by in-situ (or in-place) combustion of part of the heavy oil; this procedure is also referred to as *fire flooding*. The oil is ignited at the bottom of an injection well, and combustion is sustained by the continuous injection of air. A slow-burning front moves through the reservoir and heats the oil which can then be pumped from a production well. In one form of in-situ combustion, water is introduced ahead of the burning front; the resulting hot water and steam increase the mobility of the oil.

In the U.S.S.R., substantial amounts of heavy oil are recovered by a form of mining, but the procedure has been used to only a small extent in the United States. The details of oil mining depend on the local geological formation, but the general principles are as follows. A number of horizontal tunnels are excavated from a vertical shaft and many relatively short holes are drilled into the heavy-oil reservoir. The oil, preferably heated by steam to increase its mobility, drains through the holes into production galleries where it is collected and pumped to the surface.

Refining of heavy crudes presents special problems because of the large proportion of heavy **hydrocarbon** molecules, with a low ratio of hydrogen to carbon, and usually high sulfur content. Straight **distillation** (see also **Petroleum Refining**) yields mostly low-priced, high-boiling-range products (e.g., heavy **distillates** and **residual oils**). More valuable lighter products, such as **gasoline** and **diesel fuel,** can be obtained from heavy crudes by severe **hydrotreating,** to remove sulfur and increase the ratio of hydrogen to carbon, or **thermal cracking,** followed by conventional refining (see **Flexicoking Process**).

HEAVY (FUEL) OILS

Fuel oils of high density and viscosity (i.e., flow resistance) boiling in a high temperature range. As a general rule, **fuel (heating) oils** in ASTM categories No. 5 and No. 6 (Bunker C), which are **residual oils** of petroleum distillation, are called heavy oils (see **Petroleum Refining**).

HEAVY-WATER REACTOR (HWR)

A thermal **nuclear power reactor** in which heavy water (deuterium oxide), D_2O, is both moderator and coolant (see **Thermal Reactor**). Deuterium nuclei have a much smaller tendency than ordinary (light) hydrogen to capture neutrons; as a result, HWRs can operate with normal (unenriched) uranium dioxide as fuel whereas **light-water reactors** (LWRs) require fuel enriched in uranium-235. This is the main reason for the development of

the HWR for power generation in Canada where there are ample supplies of uranium but no enrichment facilities (see **CANDU Reactor**). However, the lower fuel costs for the HWR, as compared with the LWR, are compensated by the higher cost of the heavy water.

Another difference between HWRs and LWRs is that, because of the neutronic characteristics of deuterium, the fuel rods in the HWR are spaced farther apart. The extra spacing makes possible a pressure-tube design in which the heavy-water moderator is kept separate from the heavy-water coolant.

The coolant, under high pressure to prevent boiling, flows through tubes within which the fuel rods are supported. The heated heavy water passes to a heat-exchanger steam generator, where heat is transferred to ordinary water to produce steam and then returns to be reheated by the fuel (see **Pressurized-Water Reactor**). The heavy-water moderator surrounds the pressure tubes; it remains at a relatively low temperature and does not need to be pressurized.

In an LWR, the same water serves as both moderator and coolant, and a large, strong pressure vessel is required to contain the water. In an HWR, on the other hand, only the coolant tubes are required to withstand a high pressure. The vessel which contains the moderator (and the reactor core) does not have to support a significant pressure, and so it does not have to be especially strong.

HIGH-BTU FUEL GAS

A fuel gas consisting mainly of **methane,** with a **heating value** of approximately 950 to 1050 Btu/cu ft (35 to 39 MJ/cu m). **Natural gas** is a high-Btu fuel of major importance which is described elsewhere. The treatment here is concerned with the manufacture of **substitute natural gas** (or SNG) from coal and petroleum products. In the United States, major interest lies in the conversion of coal into a methane-rich high-Btu gas of **pipeline quality.**

Production from Coal

A high-Btu fuel gas may be produced from coal either (1) by using **synthesis gas** or (2) by direct reaction of the carbon in coal with hydrogen to produce a methane-rich gas. Although the direct (or hydrogasification) process has potential advantages, it has proved difficult to implement.

1. Synthesis gas, consisting largely of carbon monoxide (CO) and hydrogen (H_2), is made by the high-temperature reaction of coal with steam. Essentially all **intermediate-Btu fuel gases** are also forms of synthesis gas; hence, the process based on synthesis gas is one for the conversion of an intermediate-Btu to a high-Btu gas.

The production of methane requires a volume (or molecular) ratio of 1 CO to 3 H_2, but synthesis gas generally contains a smaller proportion of hydrogen. Hence, the first step is to increase the proportion to the desired value by subjecting the gas to the **water-gas shift reaction** with steam. The final composition is adjusted by varying the proportions of synthesis gas and steam. A small excess of hydrogen is desirable to limit residual poisonous carbon monoxide in the final product gas.

The catalyst used in the next (**methanation**) stage is very sensitive to sulfur compounds, so these must be removed. The purified gas is passed over the catalyst at a controlled temperature. Reaction of the hydrogen (H_2) and carbon monoxide (CO) then leads to the formation of methane (CH_4) and water (H_2O) vapor. After cooling to condense the water, the gas is dried in the same way as natural gas. The final product will generally contain about 90 volume percent of methane and 5 percent or so of hydrogen and should be a high-Btu gas (see Fig. 171).

2. The formation of methane from coal by direct hydrogasification with hydrogen, that is,

$$C + 2H_2 \rightarrow CH_4$$

is possible in principle but difficult in practice. One reason is that a high temperature is required for the reaction to occur at a satisfactory rate, but the high temperature favors the decomposition of the methane formed. A short residence time at high temperature in the hydrogasification reactor may provide a partial solution to this problem.

Direct reaction of carbon in coal with hydrogen gas occurs to a substantial extent in the **Hydrane** and **Flash Hydrogenation processes.** In the **HYGAS process** there is also some reaction between carbon and hydrogen formed in the carbon–steam reaction. In both cases, a moderate degree of methanation, possibly without water-gas shift, yields a high-Btu fuel gas.

The **Exxon Catalytic Coal Gasification process** is unusual in the respect that methane is formed by direct reaction of coal with steam as the immediate source of the hydrogen. The separate generation of hydrogen is therefore not necessary.

Production from Petroleum Products

The average ratio of hydrogen to carbon atoms in liquid petroleum (**hydrocarbon**) products is almost 2 to 1; in methane, however, the ratio is 4 to 1. Hence, conversion of a liquid hydrocarbon to a high-Btu gas requires the addition of hydrogen. This is done in two different ways: (1) reaction of the hydrocarbon with steam, generally referred to as **steam reforming**, and (2) direct addition of hydrogen gas.

1. The **Catalytic Rich Gas,** Lurgi Synthan, and Methane Rich Gas processes, which are similar in principle, use low-temperature [around 930°F (500°C)] catalytic steam reforming to convert a liquid petroleum product, especially **naphtha,** into a high-Btu gas. Several reactions occur leading to the formation of methane,

hydrogen, and carbon oxides. Subsequent methanation produces a gas consisting mainly of methane.

2. Direct reaction of hydrogen gas with a liquid petroleum hydrocarbon at high temperature yields methane and ethane; the latter can be converted into methane if desired. The two major hydrogenation processes for making a high-Btu fuel gas from a liquid petroleum product are the **Fluidized Bed Hydrogenation** (FBH) **process** for treating heavy oils and the **Gas Recycle Hydrogenation process** for naphtha.

HIGH-TEMPERATURE GAS-COOLED REACTOR (HTGR)

A thermal **nuclear power reactor** with graphite as moderator and helium gas as coolant (see **Thermal Reactor**); the unusual property of graphite in becoming stronger with increasing temperature, up to about 4500°F (2480°C), and the chemically inert nature of helium gas make reactor operation possible at very high temperatures. Superheated steam, obtained by heat transfer from the helium coolant to water in a heat exchanger, is used to drive a turbine and generate electric power with a high **thermal efficiency.** Although other fuel materials can be used, the HTGR is designed to take advantage of the effective conversion of fertile thorium-232 into fissile uranium-233 in a thermal reactor (see **Nuclear Fuel**).

The HTGR developed in the United States is related to the experimental Dragon reactor which commenced operation in the United Kingdom in 1964. A demonstration plant, with an electrical **capacity** of 40 megawatts (MW), built by the General Atomic Co. at Peach Bottom, PA, attained full power in 1967 and operated for some seven years. Construction of a larger commercial installation of the same type at Fort St. Vrain, CO, designed to produce 330 MW (electric), was completed in 1974. A low-power HTGR of a different type, called the Pebble Bed Reactor, has operated successfully in West Germany since 1967, and the 300-MW (electric) Thorium High-Temperature Reactor should be completed in 1982. In the United States, however, the HTGR project is in abeyance, but the concept is of interest because of its potential advantages in efficient power generation and the utilization of nuclear fuel.

General Description

Fuel. Eventually, the HTGR is expected to operate with fuel consisting of uranium-233 as the fissile species and thorium-232 as the fertile species; the reactor might then be (or approach) a thermal breeder (see **Breeder Reactor**). But since uranium-233 does not occur naturally, it must be produced by neutron capture in a reactor with uranium-235, which can be obtained from natural uranium, as the fissile material. Consequently, in the early stages, the HTGR fuel consists essentially of uranium-235 and thorium-232 (i.e., natural thorium).

In order to simplify recovery of the uranium-233 formed from thorium-232, the fissile and fertile components of the fuel are in separate particles. The fissile particles consist of almost pure (more than 90 percent) uranium-235 as the dicarbide (UC_2) or as a mixture of uranium dioxide (mainly) and the dicarbide, referred to as "oxycarbide." The fertile particles are thorium dioxide (ThO_2). The fissile particles are coated with three layers of carbon and the fertile particles with two layers to retain fission products. The fissile particles also have an intermediate layer of silicon monocarbide (SiC) to facilitate separation for reprocessing (see below).

The fissile and fertile particles are mixed in such proportions that the uranium-235 constitutes 5 percent by weight of the total. The HTGR fuel is made by incorporating this mixture into graphite cylinders, roughly 0.6 in. (1.6 cm) in diameter and 31 in. (0.8 m) long. Eight of these cylinders placed vertically end to end constitute a fuel rod. The carbon coatings on the fuel particles make additional cladding unnecessary.

Core. The following description refers in particular to the conceptual design for a 1000-MW (electric) system; however, no such plant has yet been built. The core consists of almost 4000 hexagonal graphite blocks, stacked in several hundred closely packed columns in a roughly cylindrical form. The overall core dimensions are 21 ft (6.4 m) high and 27 ft (8.2 m) across. The graphite reflector, which surrounds the core, is 47 in. (1.2 m) thick at top and bottom and a little over 40 in. (1 m) at the sides.

Three types of holes are drilled vertically in the core blocks: fuel holes, into which the cylindrical fuel rods fit tightly, somewhat wider holes for coolant flow, and still wider ones for control rods. There are more than 35,000 fuel holes and roughly one coolant hole for two fuel holes. The 130 or so control rods, made of stainless steel tubes packed with a mixture of boron carbide (B_4C) and graphite, are distributed throughout the core. (The control material is the strong neutron absorber boron.) The control rods, operated from above the core, are withdrawn to start the reactor operating and inserted to shut it down.

Steam System. Helium, at a pressure of 50 atm (5 MPa), is forced downward through the coolant holes in the core (and reflector) by gas circulators and the heated gas leaving the core enters the steam generators. A large HTGR may have four to six main circulators and steam generators located around the reactor (Fig. 66). The gas circulators draw the high-temperature helium upward through the steam generators and return the somewhat cooler gas for reheating in the reactor core. The reactor, steam generators, and helium circulators are enclosed in a steel-lined, prestressed concrete reactor vessel (or PCRV).

Helium gas enters the steam generator at a temperature around 1300°F (705°C)

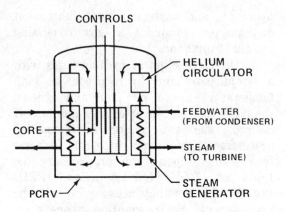

Fig. 66   High-temperature gas-cooled reactor.

and leaves at 610°F (320°C) for return to the reactor. Superheated steam (i.e., steam at a temperature well above the boiling point of water at the existing pressure) is produced at a temperature of some 1000°F (540°C) and a pressure of 170 atm (17 MPa). The steam conditions would thus approach those of a large modern fossil-fuel steam plant. The superheated steam operates a turbine system for generating electrical power; the thermal efficiency should be 38 to 40 percent.

Gas Turbine. As an alternative to producing steam to operate a turbine, the high-temperature helium gas leaving the reactor core might be used as the working gas in a closed-cycle **gas turbine.** Instead of heating the gas by combustion of a fossil fuel, as in a conventional gas turbine, the heat would be supplied by nuclear fission. Direct use of helium gas in a turbine would thus eliminate the steam generator. Furthermore, the overall thermal efficiency might be increased.

Safety Features

In an emergency requiring rapid reactor shutdown, the control rods would drop into the core under the influence of gravity. An independent backup (or reserve) shutdown system consists of pellets of graphite containing boron carbide. These pellets can be released into the core from hoppers by reactor operator action.

If the main coolant circulation system should fail for any reason, the reactor would shut down at once, but the generation of heat in the fuel would continue from the radioactive decay of fission products (see **Radioactivity**). The large mass of graphite in the core and reflector would then act as a heat sink and prevent the fuel temperature from rising too rapidly. This would allow time for some or all of the three auxiliary helium circulators to commence operation. The gas would be drawn downward through the core in the usual way, then through heat exchangers where heat is transferred to water, and finally back to the core by way of the circulators.

Fuel Reprocessing

Since an objective of the HTGR is to recover the uranium-233 formed during its operation for eventual use as the fissile species, the fuel must be reprocessed from time to time (see **Nuclear Fuel Reprocessing**). About one-fourth of the fuel rods would be removed annually and treated to separate the fuel particles from the graphite matrix and then the fissile particles from the fertile ones.

Substantial amounts of uranium-235 remaining in the fissile particles must be recovered for further use by removal of accompanying fission products. The fertile particles must be treated for recovery of both uranium-233 and thorium-232. However, radioactive decay of uranium-232 present in the uranium product and of thorium-228 in the thorium product leads to the formation of strong gamma-ray emitters. Consequently, special protective measures have to be taken in handling the recovered materials (see **Nuclear Fuel**).

HYDRANE PROCESS

A Pittsburgh Energy Technology Center (U. S. Department of Energy) process for **coal gasification** by direct reaction of coal with hydrogen gas. The carbon in the coal combines with hydrogen to form **methane**, which is the main component of the product gas. The latter is readily converted into a **high-Btu fuel gas** (or **substitute natural gas**).

The Hydrane process is conducted in two stages. In the upper stage the coal reacts at a temperature of about 1650°F (900°C) with hydrogen coming from below; the pressure is around 70 atm (7 MPa). The **char** from this stage falls into the lower **fluidized-bed** region where further reaction with hydrogen takes place. The unconsumed char from this stage is used to generate hydrogen for the process by reaction with steam and oxygen (see **Hydrogen Production**).

The two-stage Hydrane process proved to be too complex for commercial application. Consequently, development studies have been terminated in favor of the one-stage **flash hydrogenation process** for the direct production of methane from coal.

HYDRAULIC TURBINE

A **turbine,** located at a lower level, driven by water originating at a higher level; the kinetic energy present in the flowing water is converted into mechanical work of rotation. Hydraulic turbines operate on the same general principles as other turbines; jets (or a jet) of fluid, in this case water, interact with blades or vanes attached to a rotor, called the *runner*; rotational motion then occurs as a result of impulse and/or reaction.

Unlike gas and steam turbines in which the working fluid is a gas or vapor, hydraulic turbines are not heat engines. Consequently, their efficiency is not determined by the operating temperatures (see **Heat Engine**). In theory, all the energy removed from the water could be recovered as rotational energy in a hydraulic turbine. However, because of various mechanical losses, the practical recovery efficiency is commonly from 80 to 90 percent at moderate and high loads. (The load here refers to the actual power output of the turbine relative to its design power.) At low loads (e.g., about 40 percent or less), the efficiency falls off substantially.

The old-fashioned water wheel, once widely used for flour milling, is a simple hydraulic turbine of the impulse type. The wheel is caused to rotate by the weight and energy of water falling onto paddles (or buckets) attached to the periphery of the rotor wheel.   Hydraulic turbines of improved design are now used mainly to generate electricity from water power (see **Hydroelectric Power; Tidal Power**). The three major classes of hydraulic turbines are (1) the Pelton wheel (impulse), (2) the Francis turbine (mainly reaction), and (3) the Kaplan or propeller turbine (mainly reaction).

Pelton Wheel Turbine

The Pelton wheel turbine consists of a wheel-like runner (rotor) with spoon-shaped buckets around the circumference of the wheel. Water from a high level is discharged through a stationary nozzle (or nozzles) and emerges as a high-speed jet. The initial potential (or gravitational) energy of the water is thus converted into kinetic (or motion) energy of the jet. The water jet is directed at the buckets, and its impulse causes the runner to rotate about its axis (or shaft). The buckets have a central ridge which improves the operation by splitting the jet into two streams. The speed of the turbine can be changed by varying the size of the water jet.

In the earlier Pelton wheel designs, the runner is vertical and it rotates about a horizontal shaft (Fig. 67). In such designs

only one or two jets can be accommodated. More recently, large Pelton wheel turbines have been developed with a vertical shaft and horizontal runner. Four to six jets can then be located around the circumference to increase the turbine power.

The Pelton wheel turbine is preferred for use with high waterheads (i.e., large vertical distance between upper and lower water levels). It is generally employed in hydroelectric power plants where the waterhead exceeds roughly 1000 ft (300 m).

Francis Turbine

Although there are exceptions, the Francis type (mainly reaction) turbine is used for waterheads from about 100 to 1000 ft (30 to 300 m). The runner is horizontal, and it rotates about a vertical shaft. Some 12 to 16 slightly curved vertical blades are attached to the hub of the runner. The blades are farther apart at the circumference than at the hub, and so they serve as rotating nozzles. Water enters the turbine in a radial (inward) direction through an arrangement of movable wicket gates that control the water supply and act as stationary nozzles (Fig. 68). The water flows between the runner blades initially in a radial direction and is then discharged downward (axial direction); this is referred to as mixed flow.

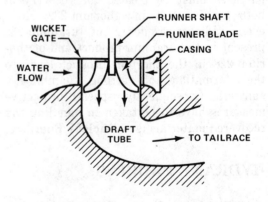

Fig. 68   Francis-type hydraulic turbine.

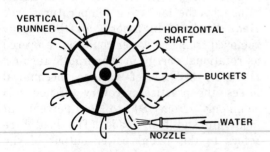

Fig. 67   Pelton wheel hydraulic turbine.

Kaplan (Propeller) Turbine

Turbines with propeller-type runner blades are most efficient for lower waterheads, generally less than 100 ft (30 m) or so. The blades are fairly broad relative to their length and resemble ships' propellers. From three to eight blades are attached to the vertical shaft to form the runner. Water enters in a radial direction through wicket gates above the runner and then changes direction and flows axially between the blades (Fig. 69). The turbine motion results mainly from reaction. The blade angle (or pitch) may be fixed, but in the widely used Kaplan turbine the blades are pivoted at the hub and the angle is changed automatically to suit the operating conditions.

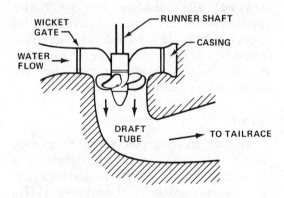

Fig. 69   Kaplan-type hydraulic turbine.

Draft Tube

In a Pelton wheel (impulse) turbine, the water is discharged at atmospheric pressure into an open space above the *tailrace* (i.e., the channel through which the water flows from the turbine to the surroundings). If this mode of discharge were used in a reaction turbine, useful power would be lost because of the kinetic energy (or velocity) remaining in the water. This loss is largely eliminated by discharging the water into a *draft tube* below the turbine (see Fig. 69). The draft tube provides a continuous column of water through the turbine to the tailrace. The turbine runner can then be located above the tailrace (for safety) and still permit full use of the available waterhead.

Very Low Waterheads

Modifications of the Kaplan-type turbine have been developed for waterheads below some 50 ft (15 m). In one such type, called the *tube turbine*, the turbine with a horizontal (or almost horizontal) shaft is located in a tubular channel; the water flows in an axial direction through the channel and the runner. The water is directed between the propeller blades by fixed guide vanes, and the flow is controlled by wicket (or similar) gates.

If the shaft is horizontal, the channel slopes downward beyond the turbine so that the attached generator is above the water level in the tailrace (Fig. 70). Alternatively, the turbine shaft may be sloped upward to achieve the same objective. An advantage claimed for the tube turbine is that the water does not change direction and hence lose energy before it enters the turbine.

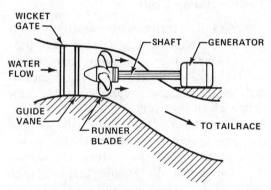

Fig. 70   Tube turbine.

For very low waterheads, about 20 ft (6 m) or less, the compact Kaplan-type *bulb turbine* is recommended. It is used, for example, in the Rance River tidal power plant in France where the maximum operating waterhead is only 27 ft (8.2 m). Bulb-type turbines are also utilized to generate electricity from low-head dams in Europe and the U.S.S.R. and may find application for the same purpose in the United States.

In the bulb turbine, the generator is enclosed in a watertight housing (or bulb) which is supported centrally within a horizontal water channel. The turbine, located at the downstream end, is connected to the generator by way of a sealed shaft. Water flows in the axial direction around the bulb and through guide vanes and wicket gates to the turbine runner with propeller-type blades (Fig. 71).

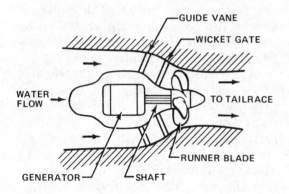

Fig. 71   Bulb turbine.

Turbine/Pump Units

Hydraulic turbines generally use a waterhead to drive a generator that produces electricity; it is possible, however, to reverse the operation with reaction turbines. Electricity is supplied to the generator which then functions as a motor (see **Electric Motor**); the motor drives the reaction turbine in reverse so that it serves as a pump to lift the water to a higher level (or head). Reversible turbine/pump (with generator/motor) units are used in several **pumped-storage** plants and in tidal-power installations. The waterheads in pumped-storage systems are mostly in the range of 200 to 1200 ft (60 to 360 m), and the turbine/pump units are of a modified Francis type. At the Rance River tidal-power plant, where the waterhead is low, the bulb type turbine-generator units are designed to operate also in the pump–motor (or reverse) mode.

HYDROCARBONS

A large group of chemical compounds containing only hydrogen and carbon. They are the major (or almost sole) constituents of **petroleum** and **natural gas** and are also produced by heating **coal, oil shale,** and **tar sands.** The simplest hydrocarbons, generally with one to four carbon atoms per molecule (i.e., C_1 to C_4), are gases at normal temperature and pressure; the more complex C_5 to very roughly C_{20} hydrocarbons are usually liquids, whereas those with larger numbers of carbon atoms are solids.

The importance of hydrocarbons from the energy standpoint is that they generate large amounts of heat per unit mass (or volume) upon combustion in air. For this reason, hydrocarbon mixtures are used extensively as fuels (e.g., **diesel fuels, fuel (heating) oils, gasoline,** and **jet fuels**). The hydrocarbons of special interest in this respect may be considered in four categories, as described below. Other types of hydrocarbons are known, but they are not important as fuel materials.

Paraffins

Paraffin hydrocarbons (or alkanes) have the general formula C_nH_{2n+2}, where n is the number of carbon atoms per molecule. The lightest paraffins, **methane** (CH_4), ethane (C_2H_6), propane (C_3H_8), and butane (C_4H_{10}), are present in natural gas and in petroleum gases. Paraffins with up to 60 carbon atoms (i.e., $C_{60}H_{122}$) and possibly more have been found in crude petroleum.

Paraffins with straight (or unbranched) chains of carbon atoms are called *normal paraffins* (or *normal alkanes*), whereas those with branched chains are known as *isoparaffins* (or *isoalkanes*). Typical normal paraffin and isoparaffin carbon chains are depicted below; the open-ended lines indicate where hydrogen atoms are attached to the carbon atoms.

Normal paraffin (straight chain)   **Isoparaffin** (branched chain)

Olefins

Like the paraffins, the *olefin* (or *alkene*) hydrocarbons consist of straight or branched chains of carbon atoms; however, in the olefins, two adjacent carbon atoms are joined together by double (rather than single) bonds, as shown. As a result, the number of hydrogen atoms in the hydrocarbon molecule is less than in a paraffin with the same number of carbon atoms. Olefins are thus described as unsaturated hydrocarbons. Strictly speaking, the term olefin refers to hydrocarbons with a single double bond, having the general formula C_nH_{2n}, but it is often used more generally to include straight- or branched-chain hydrocarbons with two or more double bonds.

Normal olefin   **Iso-olefin**

Olefins occur only rarely (and in very small proportions) in crude petroleum. They are produced, however, in large quantities in petroleum refining and are used in other refinery processes as well as in petrochemical raw materials (feedstocks).

Cycloparaffins

Cycloparaffins (or *cycloalkanes*), commonly called *naphthenes* in the petroleum industry, are saturated hydrocarbons in which the chains are closed to form one or more rings. The most common (and stable) cycloparaffins have rings of five or six carbon atoms and are known as cyclopentanes and cyclohexanes, respectively. Single-ring compounds with side chains of carbon atoms are shown. More complex cycloparaffins, in which several rings are joined or fused together (by sharing carbon atoms), occur in some petroleum crudes.

Cyclopentane with side chains   **Cyclohexane with side chains**

Aromatics

These are cyclic, unsaturated hydrocarbons with six-membered carbon rings in which alternate pairs of carbon atoms are joined by double bonds. (The actual situation is more complex, but this brief description is adequate here.) The simplest *aromatic hydrocarbon* is benzene, with the formula C_6H_6; the arrangement of carbon atoms in a benzene ring with two side chains is shown below. Aromatic hydrocarbons present in petroleum or obtained from coal may contain two or more fused rings in which carbon atoms are shared by adjacent rings. Polycyclic aromatic hydrocarbons (or PAH), consisting of several attached benzene rings, are carcinogenic; traces are sometimes found in coal and petroleum products.

Benzene ring with side chains   **Fused rings**

Three important aromatic hydrocarbons, namely, benzene, toluene (one benzene ring with a single $-CH_3$ side chain), and xylene (one benzene ring with two $-CH_3$ side chains) are often present in refined petroleum and **coal liquefaction** products. They are collectively referred to as BTX, from the initial letters of the components, and are largely used as petrochemical feedstocks (see **Petroleum Products**).

HYDROELECTRIC POWER (HYDROPOWER)

Electric power generated by flowing water in descending from a higher to a lower level. The difference in level can arise either from differences in ground surface elevations or from the rise and fall of the tides. The term hydroelectric power (or hydropower) generally refers to the former situation; power derived from tidal motion is described elsewhere (see **Tidal Power**).

General Principles

The energy utilized in the generation of hydropower is actually a form of solar energy. Ground-level water is vaporized by the sun's heat; the vapor is subsequently condensed and falls as rain or snow. Water from the rain and melted snow deposited at higher altitudes contains potential energy which can be converted into useful energy (or mechanical work) when it descends to a lower altitude. Potential energy, in general, is the capacity of a body or system for doing work as a consequence of its state or position; in the present case, the potential energy of the water at the higher level is due to its position (or altitude) (see **Energy**).

Hydroelectric plants for the conversion of the available energy in water into electrical energy vary in detail, but they all have certain common features. Water at a higher level from a forebay, commonly a reservoir or sometimes a canal, is delivered by way of a wide pipe, called a penstock, to a **hydraulic turbine** at a lower level where the energy of the flowing water is converted into mechanical energy of rotation. The water is finally discharged from the turbine by way of a tailrace (Fig. 72).

Electrical energy is generated in the conventional manner by connecting the turbine to an electrical generator (see **Electric Power Generation**). In some

cases, the turbine has a horizontal shaft and the generator is located alongside, as it commonly is in steam-turbine generator plants. In large hydroelectric power installations, the shaft is usually vertical and the generator is mounted above the turbine, as in Fig. 72.

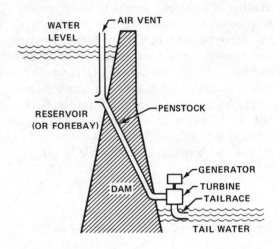

Fig. 72   Hydroelectric power system.

The **power** (i.e., energy per unit time) that can be obtained from water flowing from a higher (or upstream) level to a lower (downstream) level depends on the *head* (i.e., the vertical distance between the two levels) and the mass flow rate of the water. Suppose the waterhead at a particular location is H and the mass flow rate at a given instant is Q; then the theoretical amount of power P available at that time is given in metric units by

$$P(\text{kilowatts}) = 9.81\ H(\text{meters}) \times Q(\text{kilograms/sec}) \times 10^{-3}$$

or in common units by

$$P(\text{kilowatts}) = 1.36\ H(\text{ft}) \times Q(\text{lb-mass/sec}) \times 10^{-3}$$

The flow rates of water are commonly expressed in terms of volume rather than mass; the corresponding expressions for the available power are then

$$P\text{(kilowatts)} = 9.81\ H\text{(meters)}$$
$$\times\ Q\text{(cu m/sec)}$$

and

$$P\text{(kilowatts)} = 0.085\ H\text{(ft)}$$
$$\times\ Q\text{(cu ft/sec)}$$

Because of unavoidable losses, mainly due to friction, and to energy remaining in the discharged water, not all the power available in the water is converted into electric power. A well-designed hydroelectric plant might have an overall conversion efficiency of about 80 percent; that is to say, the maximum electrical capacity of the plant (in kilowatts) would be 0.8 of the theoretical available power, as calculated from the equations given above. For example, the electric power that could be generated by water with a head of 328 ft (100 m) flowing through a turbine at a rate of 7060 cu ft/sec (200 cu m/sec) would be $0.085 \times 328 \times 7060 \times 0.8 = 157 \times 10^3$ kilowatts (kW) or 157 megawatts (MW), where 1 MW = 1000 kW = 1 million watts (see **Watt**).

Run-of-the-River and Storage Plants

Hydroelectric plants fall into two main categories, namely, run-of-the-river and storage types. *Run-of-the-river* plants are preferably located where there is both a fairly sharp drop in the ground level and an ample year-round flow of water. The hydroelectric plant at Niagara Falls is an outstanding example of a run-of-the-river system. Water from the upper level of the Niagara River is diverted through conduits and penstocks to turbines at the base of the Falls.

Run-of-the-river plants usually do not have facilities for storing substantial quantities of water. The power generated may thus be variable, since it depends almost entirely on the stream flow which changes with the seasons. The Niagara Falls plant is somewhat exceptional in this respect because the Great Lakes system serves as a storage reservoir that permits an essentially constant water flow rate.

In *storage* hydropower plants, water is deliberately stored in a reservoir or lake created by building a dam across the river at a suitable location. The height of the dam and the volume of water impounded will depend on the local topography. Except in special circumstances (see below), storage of a large volume of water in this manner permits the plant to operate in accordance with the power demand rather than the stream flow. Reservoirs often perform useful functions in flood control and irrigation, in addition to the generation of electric power.

If water collects in the reservoir at a greater rate than it can be utilized for power generation, the level will rise. When the water reaches the maximum safe level, the excess is allowed to flow over a spillway. On the other hand, in periods of drought, the falling water level in the reservoir will decrease the plant's power capacity. This would be aggravated if the reservoir is also required to supply irrigation water to farms. Thus, even hydroelectric plants with large storage volumes, such as those in the Pacific Northwest, have been forced to curtail their electric power output during a sustained drought.

Hydropower plants associated with large reservoirs contained by high dams can generate electricity cheaply. Although construction costs are high, there is no expenditure for fuel. Furthermore, the plants are simpler to operate and maintain than fossil-fuel or nuclear steam plants. As will be seen later, however, there are some environmental and related problems that may limit the siting of large hydroelectric plants in the United States.

Waterhead and Turbines

Although there is no clear line of demarcation between them, hydropower plants are commonly distinguished by being low-head, medium-head, or high-head types. One reason for making this distinction is that it has an important bearing on the choice of hydraulic turbine design. Roughly speaking, in a low-head system, the waterhead is less than about 100 ft (30 m), whereas in a medium-head system the head is commonly in the range of 100 to 1000 ft (30 to 300 m). If the head is more than 1000 ft (300 m) or so, the hydropower system is of the high-head type. It should be emphasized, however, that these numbers are by no means exact and are given only to provide a general indication.

Since the power available in water is related to the product of the head and the flow rate, low-head plants require large flow rates in order to generate a substantial amount of power. Fixed- or variable-pitch propeller turbines (e.g., Kaplan type) are best suited to low-head operations. For very low heads, less than about 50 ft (15 m), "tube" or "bulb" type turbines are recommended (see **Tidal Power**). Because large volumes of water must be carried, flow pipes must be wide. Furthermore, the relatively slow rate of rotation of the turbines means that large-diameter generators may be necessary.

Medium-head plants operate most efficiently with Francis-type reaction turbines; these are often employed even for heads above 1000 ft (300 m). However, for high heads, impulse turbines of the Pelton-wheel type are commonly used. They can operate at moderately high speeds with a relatively low water flow rate and generators of small diameter.

Hydropower in the United States

Electricity from water power was first produced commercially in the United States, although on a small scale, at Appleton, Wisconsin, in 1882. The first storage dam for hydroelectric power generation was completed on the Willamette River in Oregon in 1894. Since that time, the generating capacity of hydropower plants in the United States has increased steadily and almost doubled between 1960 and 1980.

In the early 1980s, the hydroelectric generating capacity is about 65,000 MW and the annual electrical energy output is roughly 300 billion kW-hours, representing an average use factor of some 53 percent. Hydroelectric plants generate 13 to 14 percent of the total electrical energy presently used in the United States.

From stream flow records over many years, the U. S. Geological Survey concluded that the potential hydroelectric capacity of the United States is about 190,000 MW. The developed capacity is thus roughly 34 percent of the total. It appears, therefore, that considerable potential hydropower resources are undeveloped; however, about one-fourth of the undeveloped capacity is in Alaska and is not readily accessible.

Large hydroelectric plants with large reservoirs generate electricity most economically, but there are now relatively few sites remaining for such plants. Furthermore, the development of these sites is faced with many problems.

There is increasing public resistance to the building of dams and the flooding of large areas, often of valuable agricultural land. Frequently, interference with the original stream flow has adverse ecological consequences. In Oregon and Washington, dams have been provided with "fish ladders" to permit salmon to migrate upstream to spawn. But such remedial measures are not always possible or effective. Furthermore, the best hydroelectric sites are usually far from areas where the demand for power exists, and long transmission lines are required.

A possible future development of hydropower lies in the use of existing small reservoirs. There are thousands of low-head dams in the United States built for flood control, irrigation, or recreational purposes. A small proportion of these dams have turbine–generators to produce electric power, but most do not. According to estimates made by the U. S. Army Corps of

Engineers in 1977, small dam sites have a potential electrical generating capacity of about 54,000 MW. This is almost equal to the capacity of existing large hydroelectric plants.

HYDROGEN (ENERGY) ECONOMY

A concept in which a primary energy form (e.g., **fossil-fuel, nuclear,** or **solar energy**) would be used to produce hydrogen gas from water, possibly at remote locations. The hydrogen would then be stored and transmitted by pipeline and in other ways to load centers where the energy would be recovered locally as heat, electricity, or mechanical energy, with little or no atmospheric pollution. Water produced in the energy recovery process would balance that used to obtain the hydrogen in the first place. Proponents of the hydrogen economy concept claim that the use of hydrogen, rather than electricity, as an energy carrier would have economic as well as environmental advantages.

See **Hydrogen Fuel.**

HYDROGEN FUEL

The use of hydrogen as a means for storing, transmitting, and utilizing energy. At ordinary temperatures and pressures, hydrogen is a colorless and odorless gas. The combination of hydrogen with oxygen (e.g., from air) results in the liberation of energy, with water as the sole material product; thus,

$$H_2 + \frac{1}{2}O_2 \rightarrow H_2O + energy$$

Hydrogen   Oxygen   Water

The reaction can be carried out and the energy made available in several different ways, so that hydrogen is a versatile fuel material. Some possibilities are burning with or without flame to produce heat, use as fuel for a **gas turbine** or for a **spark-ignition engine,** and generation of electricity in a **fuel cell.**

Except for small quantities in the upper atmosphere, hydrogen does not exist on earth in the uncombined (or free) state. However, large amounts of combined hydrogen are present as water, which contains 11 percent by weight of hydrogen. By applying energy in a suitable form, water can be split into its constituent hydrogen and oxygen; the hydrogen can then be used as a fuel.

Unlike **fossil** (coal, natural gas, and petroleum) and **nuclear fuels** and solar radiation (see **Solar Energy**), which are primary energy sources, hydrogen is a secondary fuel that is produced by utilizing energy from a primary source. Much of this energy can be recovered by recombination of the hydrogen with oxygen in the ways indicated above.

One of the potential advantages of hydrogen as a secondary fuel is that it can be transmitted and distributed by pipeline in much the same way as natural gas. Alternatively, the hydrogen gas could be converted into liquid form by cooling to a very low temperature and transported in insulated tanks by highway or railroad. On a smaller scale, hydrogen gas could be compressed into transportable cylinders. Another possibility is to form a solid compound of a metal with hydrogen from which the gas can be recovered, when required, by heating. Thus, hydrogen can serve as a means of carrying energy from the place where a primary source is available to a distant load center where the energy is used.

An important aspect of hydrogen utilization is that it is accompanied by little or no atmospheric pollution. In many cases, such as when hydrogen is burned in pure oxygen, in a flameless (catalytic) burner in air, or in a fuel cell with air or oxygen, water is the sole product. However, if hydrogen is burned normally in air or in a spark-ignition engine or a gas turbine, **nitrogen oxides** may be produced by combination at high temperature of the nitrogen and oxygen in the air. By using an excess of air relative to the hydrogen consumed or by introducing water vapor, the flame temperature can be reduced to the point where nitrogen oxide formation is small.

The simplest practical way to obtain hydrogen from water is by means of electrical energy (see **Hydrogen Production**). Hence, electrical energy can be conveniently stored and transmitted by way of hydrogen (see **Energy Storage**). In some situations, electricity is generated from distributed primary sources, such as **geothermal energy, wind energy** and some other forms of solar energy, often at a distance from a load center or an electric utility grid. Conversion into hydrogen fuel would then be a practical means of energy transmission.

Properties

Hydrogen at ordinary temperature and pressure is a light gas with a density only $\frac{1}{14}th$ that of air and $\frac{1}{9}th$ that of natural gas under the same conditions. By cooling to the extremely low temperature of $-423°F$ $(-253°C)$ at atmospheric pressure, the gas is condensed to a liquid with a specific gravity of 0.07, roughly $\frac{1}{10}th$ that of gasoline.

The standard **heating value** of hydrogen gas is 325 Btu/cu ft (12.1 MJ/cu m) compared with an average of 1030 Btu/cu ft (38.3 MJ/cu m) for natural gas. The heating value of liquid hydrogen is 52,000 Btu/lb (120 MJ/kg) or 30,200 Btu/gal (8400 MJ/cu m); the corresponding value for gasoline (or approximately for jet fuel) is 19,000 Btu/lb (44 MJ/kg) or 115,000 Btu/gal (32,000 MJ/cu m). Hence, for producing a specific amount of energy, liquid hydrogen is superior to gasoline (or jet fuel) on a weight basis but inferior on a volume basis.

The flame speed of hydrogen burning in air is much greater than for natural gas, and the energy required to initiate combustion (i.e., the ignition energy) is less. One consequence of the low ignition energy is that flameless combustion on a catalytic (finely divided metal) surface is possible with hydrogen at much lower temperatures than flame burning.

Mixtures of hydrogen and air are combustible over an exceptionally wide range of compositions; thus, the flammability limits at ordinary temperatures extend from 4 to 74 percent by volume of hydrogen in air. (Detonation can occur between 18 and 59 percent.) This wide range has an important bearing on the use of hydrogen fuel in internal combustion engines. The engine will operate, although not necessarily with the same efficiency, from very rich (excess fuel) to very lean (excess air) mixtures. The adjustment of air-to-fuel ratio is thus much less critical than in a gasoline engine.

Storage

Hydrogen Gas. The cheapest way to store large amounts of hydrogen for subsequent distribution would probably be in underground facilities similar to those used for **natural gas**; these facilities would include depleted oil and gas reservoirs and aquifers. More expensive alternatives would be caverns produced by conventional mining or by dissolving out salt with water. Since hydrogen gas tends to escape readily through a porous material, some geologic formations that may be suitable for storing natural gas may not be suitable for hydrogen.

Liquid Hydrogen. On a small or moderate scale, hydrogen is frequently stored under high pressure in strong steel cylinders; this type of storage would be too costly for large-scale applications. A more practical approach is to store the hydrogen as liquid at a low temperature. For example, the liquid hydrogen fuel used as rocket propellant in the space program is stored in large tanks. The largest such tank in the United States is at the Kennedy Space Center, Florida; it can hold about 900,000 gal (3400 cu m) of liquid hydrogen, with a total heating value of some 28 billion Btu (29 million MJ), which is equivalent to 8 million kW-hr.

Because of the low temperatures that must be maintained for liquid hydrogen, the tanks utilize the principle of the Dewar vessel, more commonly known as the Thermos bottle. The storage tanks are double-walled steel spheres with a vacuum space between the walls to minimize access of heat to the interior. The space is filled with perlite, a granular volcanic glass, to provide additional thermal insulation.

One of the problems of liquid hydrogen storage arises from the boil-off, that is, hydrogen gas that continuously escapes from the tank as a result of the slow vaporization of the liquid. Because of the flammable nature of the gas, special precautions must be taken in disposing of the boil-off. Another drawback to liquid hydrogen storage, to which reference will be made later, is associated with the possible liquefaction of oxygen from the air at the low storage temperatures. Finally, some 25 to 30 percent of the energy content of the hydrogen is required to condense it to a liquid; the overall energy efficiency of hydrogen as a fuel is thus substantially decreased.

Metal Hydrides. A number of metals and alloys form solid compounds, called metal hydrides, by direct reaction with hydrogen gas. When the hydride is heated, the hydrogen is released and the original metal (or alloy) is recovered for further use. Thus, metal hydrides provide a possible means for hydrogen storage. An important property of metal hydrides is that the pressure of the gas released by heating a particular hydride depends mainly on the temperature and not the composition. At a fixed temperature, the gas pressure remains essentially constant until the hydrogen content is almost exhausted.

Several studies are being made to find a metal hydride that would satisfy the requirements for hydrogen storage. These requirements include the following: (1) the metal (or alloy) should be fairly inexpensive, (2) the hydride should contain a large amount of hydrogen per unit volume and per unit mass, (3) the hydride should be formed without difficulty by reaction of the metal with hydrogen gas, and it should be stable at room temperatures, and (4) the gas should be released at a significant pressure from the hydride at a moderately high temperature [preferably below 212°F (100°C)].

Three of the more promising hydrides are those of lanthanum-nickel ($LaNi_5$), iron-titanium (FeTi), and magnesium-nickel (Mg_2Ni) alloys. The maximum hydrogen contents are represented approximately by the formulas $(LaNi_5)H_6$, $(FeTi)H_2$ and $(Mg_2Ni)H_4$, respectively. These hydrides contain somewhat more hydrogen than an equal volume, but much less than an equal weight, of liquid. Thus, in theory $(LaNi_5)H_6$ contains 1.35 percent by weight, $(FeTi)H_2$ contains 1.9 percent, and $(Mg_2Ni)H_4$ contains 3.6 percent of hydrogen. It appears that hydrides would be of interest for stationary storage of hydrogen, when the small volume is advantageous. They might be less useful, however, for mobile storage on a light vehicle, where their weight would be a drawback (see later).

Transmission (or Transportation)

Pipelines. Hydrogen gas for fuel could be transmitted and distributed from storage by pipelines similar to (or the same as) those used for natural gas. In comparing hydrogen with natural gas, it should be recalled that the heating value of hydrogen, as given earlier, is less than one-third that of an equal volume of natural gas. On the other hand, because of its lower density (and viscosity), hydrogen flows more readily and the carrying capacity of a given pipeline is three times as great by volume as for natural gas.

Since the larger capacity roughly balances the reduced heating value, the total energy available from the gas carried by the pipeline should be almost the same for hydrogen as for natural gas. This is true at atmospheric pressure, but at the higher pressure of a conventional pipeline [average

50 atm (5 MPa)], the energy content is about 25 percent smaller for hydrogen.

Because of the larger volume of hydrogen gas that must be pumped for a specific energy content, there is a three- to four-fold increase in the pumping power compared with natural gas at the same pressure. The transmission costs per unit of heat would then be substantially greater for hydrogen. The cost difference could be decreased by operating the hydrogen pipeline at a higher pressure [e.g., in the range of 100 to 140 atm (10 to 14 MPa)], but this may introduce a problem.

A number of metals lose their mechanical strength on exposure to hydrogen; the phenomenon, called hydrogen embrittlement, is especially significant for steel in hydrogen under pressure. It is probable that ordinary pipeline steel would not suffer embrittlement at normal pipeline pressures (up to 50 atm), but the behavior at higher pressures is uncertain.

Short and moderately long pipelines are used by the petroleum industry for carrying hydrogen gas. For example, a 50-mile (80-km) network is reported to be in operation in Texas. The most extensive pipeline system, now 127 miles (204 km) in length, has been in continuous use in the industrial Ruhr district, West Germany, since 1938. The line is made of seamless steel pipe and carries gas at about 15 atm (1.5 MPa) pressure.

Liquid Hydrogen Transportation. Hydrogen in bulk can be transported and distributed as the liquid. Double-walled, insulated tanks of liquid hydrogen with capacities of 7000 gal (26.5 cu m) or more are carried by road vehicles and up to 34,000 gal (129 cu m) by railroad cars. Distribution of liquid hydrogen by pipelines, jacketed with liquid nitrogen, has been proposed. The costs would be substantially greater than for gas pipelines, but it might be justifiable for certain fuel applications where the liquid is required (shown later).

Metal Hydride Transportation. Hydrogen can also be transported as a solid metal hydride. The main drawback, as noted earlier, is the weight of the hydride relative to its hydrogen content.

Applications

Residential Uses. The burners of domestic appliances (e.g., stoves) would have to be modified if hydrogen were to replace natural gas. In the first place, since the heating value per unit volume of hydrogen gas is less than for natural gas, a larger volume would have to reach the burners to achieve the same heating effect. Furthermore, the greater flame speed when hydrogen burns in air would facilitate backward propagation of the flame (i.e., backfiring) in a natural-gas burner. These problems can be overcome by changing the design of the burner, including the hole size and the air supply system.

Water (vapor) is the only direct combustion product of hydrogen in air. In addition, however, nitrogen oxides may be formed from the nitrogen and oxygen in air at the flame temperature. The amounts of these oxides should be quite small and, unless the humidity level from the water vapor were objectionable, the stove might not require venting to the outside atmosphere.

Because of the possibility of flameless combustion on a catalytic surface, hydrogen would be especially useful in radiant space heaters. Such devices would operate spontaneously when the gas was turned on and no pilot light or other ignition system would be required. Because of the low combustion temperature, nitrogen oxide formation would be negligible and venting would be unnecessary.

Electricity for lighting and for operating domestic appliances (e.g., refrigerators) could be generated by means of fuel cells, with hydrogen gas at one electrode and air at the other.

Industrial Uses. There are many potential uses for hydrogen in industry, either as a fuel or a chemical reducing (i.e., oxygen removal) agent, if the economics were favorable. For example, in several industrial processes natural gas has been the most satisfactory source of heat. In a hydrogen energy economy, hydrogen could replace natural gas in these operations. Hydrogen gas could also be used with advantage, instead of coal or coal-derived gases, to reduce oxide ores (e.g., iron ore) to the metal (iron).

Land Vehicles. The use of hydrogen fuel in internal-combustion engines for automobiles, buses, trucks, and farm machinery has attracted interest as a means of conserving petroleum products and of reducing atmospheric pollution. Because the fuel is a gas, the conventional carburetor of a spark-ignition engine, in which liquid gasoline is vaporized in air, must be modified for use with hydrogen. Three different approaches to this problem are described below. Another modification arises from the high speed of the hydrogen flame in air; this requires that the ignition time be retarded (i.e., less spark advance) compared with a gasoline engine (see **Spark-Ignition Engine**).

The main types of hydrogen fueled engine designs are based on the following principles:

1. A mixture of the fuel gas and air, with an approximately constant fuel-to-air ratio, is introduced into the cylinder intake manifold. The engine power (i.e., vehicle speed) is controlled by varying the quantity of mixture entering the cylinder by means of a throttle valve. (The same procedures were used in the Otto and other early internal-combustion engines with coal gas as the fuel.) Stable operation, especially at higher speeds, may require addition of water vapor to the fuel–air mixture; this can be achieved by returning part of the exhaust gas to the manifold.

2. The hydrogen gas under pressure is injected through a valve directly into the engine cylinder, and the air is admitted through another intake valve. Since the hydrogen and air are supplied separately, an explosive mixture does not occur except in the cylinder. This scheme is considered

to be safer than the one outlined above in which such a mixture is formed in the manifold. The engine power output is controlled by varying the hydrogen gas pressure from about 14 atm (1.4 MPa) at low power to 70 atm (7 MPa) at high power. The ability to achieve such a variation would appear to require that the hydrogen be stored as a compressed gas.

3. The hydrogen gas at normal or moderate pressure is drawn through a throttle valve into the engine cylinder during the intake stroke. At the same time, unthrottled air is drawn in through the intake port. Here, also, there is no explosive mixture except in the cylinders. The engine power is varied by adjusting the hydrogen inlet throttle. Since the air supply is unthrottled, there is a change in the proportion of fuel in the cylinder and consequently a change in the power developed. This scheme of power variation is possible because of the wide composition range over which hydrogen–air mixtures can be ignited.

One of the advantages claimed for hydrogen-fuel engines is that they can have higher efficiencies (i.e., utilize a higher proportion of the energy of the fuel) than gasoline engines. Another is that carbon monoxide and hydrocarbons, if any, in the exhaust would be very small since they would originate only from the cylinder lubricating oil. However, because of the high combustion temperature, the nitrogen oxide levels may be high. One way to reduce the combustion temperature, and hence the nitrogen oxide emissions, is to inject water vapor into the cylinder from the exhaust, as described earlier.

An entirely different manner in which hydrogen could serve as a vehicle fuel is by way of fuel cells. The electricity generated would operate electric motors to propel the vehicle (see **Electric Vehicle**). Atmospheric pollution would be essentially zero.

Hydrogen Fuel Storage on Vehicles. In all cases, regardless of whether a spark-ignition engine or an electric motor produces the motive power, the main problem is storage of hydrogen on the vehicle; the possibilities are (1) compressed gas, (2) liquid, and (3) metal hydride.

1. Compressed hydrogen gas contained in heavy steel cylinders has been used in most experimental hydrogen-fuel engines. The weight of the cylinders is so great, however, that this may not be a practical means for storing hydrogen on a vehicle. The fuel tank of an automobile holds, on the average, some 20 gallons (76 liters) of gasoline weighing roughly 122 lb (55 kg); this is equivalent in energy content to 45 lb (20 kg) of hydrogen. Allowing for the greater energy utilization efficiency of hydrogen, the actual weight equivalent would be somewhat less. Nevertheless, the weight of the compressed-gas containers might be as much as 2200 lb (1000 kg).

2. Storage of hydrogen as liquid would solve the weight problem, but there would be several disadvantages. The volume of liquid hydrogen would be three to four times that of the equivalent gasoline. The weight of the double-walled storage tank would be acceptable, but the volume would be large and the cost high. Other drawbacks are the loss of hydrogen by boil-off and the need for safe disposal of the gas. Refueling with liquid hydrogen would not be simple; it might even require complete replacement of the empty fuel tank by a full one.

3. Solid metal hydrides would be very convenient for storing hydrogen, but the weight would be considerable. Of the three hydrides mentioned earlier, magnesium–nickel hydride initially contains the largest proportion by weight of hydrogen and is also the least expensive. However, even this hydride contains only $1/27\,th$ its weight of hydrogen; the energy equivalent of a 20-gallon tank of gasoline would weigh roughly 1100 lb (500 kg). Moreover, magnesium–nickel hydride must be heated to above 480°F (250°C) before the hydrogen gas pressure is sufficient for operating an internal-combustion engine. This temperature can be obtained by using the exhaust gases for heating, but the engine could not be started when cold. One proposal, which is being tested on buses in West Germany, is to utilize the iron-titanium hydride for starting; this hydride contains less hydrogen by weight than magnesium–nickel hydride but it gives off hydrogen gas at a lower temperature.

In spite of the problems, hydrogen in some form may eventually play a role as an automotive fuel, especially in heavier road vehicles (e.g., trucks and buses).

Aircraft. The earliest application of liquid hydrogen fuel is expected to be in jet aircraft; this possibility was demonstrated in a subsonic aircraft in 1957. The principal advantage is the much lower overall weight of the fuel and storage tank than for ordinary jet fuel. (The comparisons given earlier for hydrogen and gasoline also apply reasonably well to jet fuel.) The volume of liquid hydrogen would be greater than for regular fuel, but this could be accommodated on a large aircraft. The cold liquid hydrogen could be used directly or indirectly to cool the engine and the airframe surfaces of a high-speed aircraft. If a hypersonic aircraft (i.e., speed more than five times the local speed of sound) is ever developed, liquid hydrogen may be the only practical fuel.

Utilities. It is unlikely that hydrogen would serve as a major fuel for electric power generation by a utility. However, its substitution for natural gas in peak-shaving turbines is possible (see **Electric Power Generation**). Hydrogen could also be used as a means for storing and distributing electrical energy, as already noted.

Safety Considerations

Objection has been raised to the extensive use of hydrogen on the grounds that it would be dangerous. This fear of hydrogen, described as the "Hindenburg syndrome," has resulted from the spectacular fire on the airship *Hindenburg* in 1937. Since hydrogen is a highly flammable gas, it must be handled with care in special equipment designed for safety. Natural gas and gasoline are also hazardous materials, yet they are in common daily use. By taking proper precautions with hydrogen, the danger of fire or explosion can be minimized. Large quantities of the gas and liquid have been carried by pipeline, road, and rail for use in industry and in space vehicles. Manufactured (or town) gas made from coal was distributed by pipeline for many years to residences and industries; this gas contained about 50 volume percent (on a dry basis) of hydrogen as well as 30 percent of methane and 7 percent of poisonous carbon monoxide as the flammable constituents (see **Coal Gas**).

In many respects, hydrogen gas is no more, and possibly less, hazardous than natural gas. The lower flammability limit of hydrogen is 4 percent (by volume) in air, whereas that of methane in natural gas is 5 percent. However, hydrogen will escape from a leak of fixed size about three times as fast as natural gas; hence, in a closed space, the flammability will be reached sooner with hydrogen. On the other hand, the energy contained in the space at the lower flammability limit is about one-fourth that for natural gas. Because of its low density, hydrogen disperses more readily in an open (or ventilated) space, and it would therefore take a much longer time for the flammability limit to be reached.

The chief danger from hydrogen gas is associated with the very low ignition energy, which is less than a tenth that of natural gas. Consequently, a spark that is too weak to ignite a flammable air-natural gas mixture may ignite an air-hydrogen mixture. In handling hydrogen, special care is taken to avoid flames and to prevent spark formation.

Liquid hydrogen presents another safety problem. The low temperature of the liquid may cause air in the vicinity to liquefy; since oxygen is more readily liquefied than nitrogen, an oxygen-rich liquid may form. This would greatly increase the flammability danger. Such a situation could arise if a liquid-hydrogen tank or pipeline were not properly insulated so that the exterior temperature fell low enough to liquefy oxygen from the air. The special double-walled insulation described earlier is thus desirable.

Liquid oxygen could also be formed from the air present in an "empty" tank or pipe when it is being filled with liquid hydrogen. The air must therefore be previously removed by purging with hydrogen gas or preferably helium. If hydrogen is used, care must be taken in disposing of the purged gas mixture.

HYDROGEN PRODUCTION

Processes for obtaining hydrogen gas from water (and sometimes also from **hydrocarbons**). The decomposition of water (H_2O) into its components hydrogen and oxygen requires the addition of energy. The methods for producing hydrogen may be classified according to the immediate source of this energy, thus, electrical energy (in electrolysis), heat energy (in thermochemical methods), fossil fuels, and solar energy.

Electrolysis

The process of splitting water into hydrogen and oxygen by means of a direct electric current is known as *electrolysis*; this is the simplest method of hydrogen production. In principle, an electrolysis cell consists of two electrodes, commonly flat metal or carbon plates, immersed in an aqueous conducting solution called the electrolyte. A source of direct current voltage is connected to the electrodes so that an electric current flows through the electrolyte from the positive electrode (or anode) to the negative electrode (or cathode). As a result, the water in the electrolyte solution is decomposed into hydrogen gas (H_2) which is released at the cathode, and oxygen gas (O_2), released at the anode (Fig. 73). Although only the water is split,

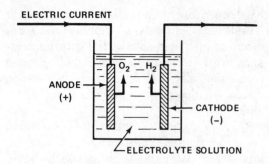

Fig. 73 Simple electrolytic cell.

an electrolyte (e.g., potassium hydroxide solution) is required because water itself is a very poor conductor of electricity.

Ideally, a voltage of 1.23 volts should be sufficient for the electrolysis of water at normal temperature and pressure. For various reasons, especially the slowness of the electrode processes that lead to the liberation of hydrogen and oxygen gases, higher voltages are required to decompose water. The decomposition voltage increases with the current density (i.e., the current per unit area of electrode). Since the rate of hydrogen production is proportional to the current strength, a high operating current density is necessary for economic reasons. Hence, in practice, the decomposition voltage (per cell) is usually around 2 volts.

Theoretically, 79 kilowatt-hours (kW-hr) of electrical energy (see **Watt**) should produce 1000 cu ft of hydrogen gas (or 2.8 kW-hr per cu m). Because of the higher than ideal decomposition voltage, however, the actual electrical energy requirement is generally from 110 to 130 kW-hr per 1000 cu ft (3.9 to 4.6 kW-hr per cu m). This means that the efficiency of electrolysis (i.e., the proportion of the energy supplied that is used in electrolysis) is roughly 60 to 70 percent.

The electrolysis efficiency can be increased by decreasing the decomposition voltage for a given current density. To achieve this, the electrode surface must be able to catalyze (i.e., expedite) the electrode processes. One of the best catalysts is platinum in a finely divided form deposited on a metal base. However, because of the high cost of platinum, other electrode surface materials are used commercially.

For practical water electrolysis, the electrodes are generally of nickel-plated steel. The effective electrode surface area (and hence the rate of the electrode process) is increased by depositing porous nickel on a wire gauze or a highly corrugated steel base. Research is being directed at the development of improved electrodes that will give better electrolysis efficiency at a reasonable cost.

Two types of electrode arrangements are used by industry for the electrolysis of water.

Tank Type Electrolyzer. In the *tank type* electrolyzer, a number of electrode plates, alternately anodes (+) and cathodes (−), are suspended vertically in a tank containing a 20 to 30 percent solution of potassium hydroxide in water. The adjacent (+ and −) electrodes are separated from each other by a porous (e.g., asbestos) diaphragm which permits passage of the electrolyte but prevents mixing of the hydrogen and oxygen gases. All the anodes in the tank are connected to the same positive terminal of the direct-current voltage source, and all the cathodes are connected to the same negative terminal (Fig. 74).

By connecting the electrodes in parallel in this manner the voltage required for a tank of several electrode pairs, regardless of the number, is little more than for a single pair (or cell), that is, about 2 volts. As a rule, a number of tanks are connected in series; the operating voltage is then roughly $2T$ volts, where T is the number of tanks so connected.

Filter-Press Electrolyzer. The alternative and more widely used water electrolysis system is called the *filter-press* electrolyzer because of its superficial resemblance to a filter press. Except at the ends of the cell, the electrodes are bipolar; that is, one face of each plate electrode is an anode and the other face is a cathode (Fig. 75). As in the tank system, porous diaphragms between

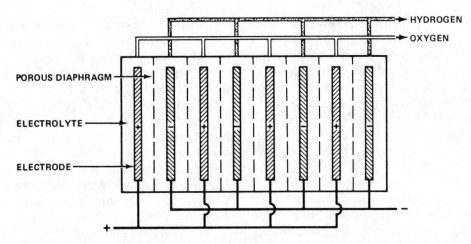

Fig. 74   Tank-type electrolytic cell.

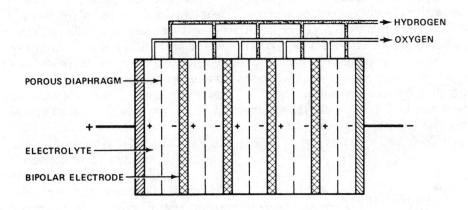

Fig. 75   Filter-press type electrolytic cell.

adjacent electrodes prevent mixing of the hydrogen and oxygen gases.

In the filter-press electrolyzer the cells are connected in series with an anode at one end and a cathode at the other end of the series. The total voltage required is then approximately $2C$ volts, where C is the number of cells in the series.

A comparison of the two electrolyzer types shows that the tank type has the advantage that it can be operated with a lower voltage source for the same total number of cells. Moreover, because the cells are in parallel, failure of one cell will not affect the others. In the filter-press type, on the other hand, failure of a single cell would make the whole series inoperative.

The construction is also more complex than for the tank type because the individual cells must be sealed in such a way as to prevent the hydrogen produced at one face of an electrode from mixing with the oxygen from the other face.

The filter-press electrolyzer is generally preferred, however, because it occupies less space and can be operated at a higher current density than the tank type. The economics are thus more favorable, since a larger hydrogen production is possible in a plant of a given size.

Electrolyzer Modifications. In addition to the search for improved electrodes, efforts are being made to increase electrolyzer efficiencies. For example, operation at temperatures and pressures above normal is reported to decrease the decomposition voltage. In a commercial German design, for which a high efficiency is claimed, the pressure in the cells is maintained at some 50 atm (5 MPa) and the electrolyte temperature is 194°F (90°C).

A novel and compact type of electrolyzer is being developed in the United States by the General Electric Company; it is, in principle, a reversal of the **fuel cell** used in the spacecraft *Gemini* in the mid 1960s. The electrolyte is an ion-exchange (sulfonic acid) polymer, referred to as a solid polymer electrolyte (SPE), which conducts hydrogen ions. The SPE is kept saturated with pure water, and an electrolyte solution is not required. A cell consists of a thin sheet of SPE with a finely divided precious metal (e.g., platinum) coated on each side to catalyze the electrode processes. Metal plates pressed against the surfaces of the SPE provide electrical contact with the coatings and serve as electrodes for connection with the voltage source.

Thermochemical Methods

The overall efficiency for the conversion of primary energy from fossil and nuclear fuels into hydrogen by electrolysis is dependent, in the first place, on the net efficiency of generating electricity. This efficiency may be up to about 38 percent for modern fossil-fuel plants and 32 percent for nuclear installations. Assuming that electrolyzer efficiency can be increased to 80 percent, the overall efficiency for hydrogen production would be only 25 to 30 percent.

A higher conversion efficiency might be possible if the heat produced by the primary fuel could be used directly to decompose water, without the intermediary of electrical energy. Such direct decomposition into hydrogen and oxygen is possible, but it requires a temperature of at least 4530°F (2500°C). It is conceivable, however, at least in theory, that water can be split into its constituent elements in an indirect manner by means of a series of chemical reactions at substantially lower temperatures.

In the reaction series, water is taken up at one stage, and hydrogen and oxygen are produced separately in different stages. The net result is the decomposition of water into hydrogen and oxygen. The operation is called a *thermochemical cycle* because energy is supplied as heat at one or more of the chemical stages. Apart from the decomposition of water, all other materials are recovered when a cycle is complete.

In a thermochemical cycle, heat is taken up in a high-temperature stage (or stages) and part is released at lower temperatures. Thermochemical cycles thus have some features in common with thermal cycles (see **Heat Engine**). The efficiency with which heat energy is utilized in the decomposition of water is therefore limited by the upper temperature at which heat is taken up; hence, this temperature should be as high as feasible (see **Thermal Efficiency**).

For practical reasons, primarily the availability of structural and containment materials, the maximum temperature in a thermochemical cycle is considered to be about 1740°F (950°C). Heat energy should then be convertible into hydrogen energy with an efficiency around 50 percent; this is a marked improvement over what is possible by electrolysis. Unless the upper temperature of the thermochemical cycle is above 1290°F (700°C), the efficiency is little better than for electrolysis.

A very large number of thermochemical cycles have been proposed, based on chemical and thermodynamic considerations. Several of the more promising cycles are being investigated in the United States and other countries. Although the various stages in a cycle may be thermodynamically possible, there is no assurance, without experimental studies, that they will take place at a useful rate. Also, several cycles involve highly corrosive substances, such as sulfur dioxide and trioxide, chlorine, and hydrogen bromide; containment problems may then be dominant. On the whole, the thermochemical production of hydrogen is regarded as a long-term project.

It does not appear possible to summarize the characteristics of the proposed thermodynamic cycles; hence, two of different types have been chosen for purposes of illustration. It should be emphasized, however, that these are not necessarily the best, nor are they typical of the many proposed thermochemical cycles. If the reactions in the stages of each of these cycles are added and substances that appear on both sides are eliminated, the net result is seen to be

$$H_2O \rightarrow H_2 + \frac{1}{2}O_2$$

(or twice this), that is, the decomposition of water into hydrogen and oxygen.

Fossil Fuel Methods

In most cases, the first stage in the production of hydrogen by using a fossil fuel (e.g., natural gas, petroleum product, or coal) is the formation of a gaseous mixture of carbon monoxide and hydrogen. Such a mixture can be made by any method used

for an **intermediate-Btu fuel gas, synthesis gas,** or **water gas.** The procedures in common use are **steam reforming** of **methane** or other hydrocarbon gas (see **Liquefied Petroleum Gas**) or a light liquid hydrocarbon (e.g., **naphtha**) and partial oxidation of a heavier hydrocarbon in the presence of steam at a high temperature. In all these cases, part of the hydrogen produced originates in the hydrocarbon.

To remove the carbon monoxide, the mixture is submitted to the **water-gas shift reaction** with steam. The carbon monoxide is thereby converted into carbon dioxide with the formation of additional hydrogen. The carbon dioxide is an **acid gas** that can be absorbed in an alkaline medium. If the small amounts of carbon monoxide and dioxide remaining are undesirable, they can be converted into methane (see **Methanation**) which can be separated as a liquid by cooling to a moderately low temperature.

Several processes proposed for converting coal into gaseous and liquid hydrocarbon fuels require a hydrogen-rich gas (see **Coal Gasification; Coal Liquefaction**). The hydrogen would then be made by reacting coal or **char** obtained in the early

Possible Cycles for Hydrogen Production

Chemical Reaction	Temperature	
	°F	°C

I

Chemical Reaction	°F	°C
$2CrCl_2 + 2HCl \rightarrow 2CrCl_3 + \underline{H_2}$	620	325
$2CrCl_3 \rightarrow 2CrCl_2 + \underline{Cl_2}$	1610	875
$\underline{H_2O} + Cl_2 \rightarrow 2HCl + \frac{1}{2}O_2$	1560	850

II

Chemical Reaction	°F	°C
$Fe_3O_4 + \underline{2H_2O} + 3SO_2 \rightarrow 3FeSO_4 + \underline{2H_2}$	260	125
$3FeSO_4 \rightarrow \frac{3}{2}Fe_2O_3 + \frac{3}{2}SO_2 + \frac{3}{2}SO_3$	1340	725
$\frac{3}{2}Fe_2O_3 + \frac{1}{2}SO_2 \rightarrow Fe_3O_4 + \frac{1}{2}SO_3$	1700	925
$2SO_3 \rightarrow SO_2 + \underline{O_2}$	1700	925

stages of coal treatment with steam and a limited amount of oxygen. The heat generated when the carbon in the coal (or char) reacts with oxygen produces the high temperature required for the carbon-steam reaction. The product is a mixture of hydrogen and carbon monoxide. The carbon monoxide may be removed in the manner described above.

Air can be used instead of oxygen to supply the heat for the carbon-steam reaction, but about half (by volume) of the product gas is inert nitrogen from the air. Although this procedure is more economical, the high proportion of nitrogen, which cannot be removed easily, is often a drawback. However, the iron-steam process, as developed in the United States by the Institute of Gas Technology (IGT), is designed to use air, steam, and coal char to make hydrogen essentially free from nitrogen and also from carbon monoxide without the need for water-gas shift.

The IGT method, which is a modification of an obsolete process, depends on the reaction of steam with iron at a temperature of about 1500°F (815°C) at a pressure of some 70 atm (7 MPa). The products are fairly pure hydrogen gas and a solid iron oxide; thus

Iron + Steam → Iron oxide + Hydrogen

The iron is recovered from the oxide in a separate vessel and returned for further reaction with steam. The conversion (reduction) of iron oxide to iron is achieved by means of a reducing-gas mixture of carbon monoxide, hydrogen, and nitrogen at a temperature of 2000°F (1095°C) made by the air-steam-char process (Fig. 76). Because the iron-steam reaction occurs in another vessel, carbon monoxide and nitrogen are essentially absent from the product hydrogen.

Solar Energy Methods

The use of solar energy to produce hydrogen from water is a long-term pros-

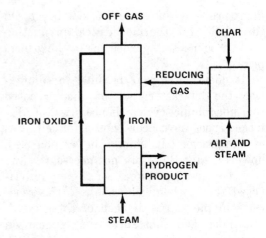

Fig. 76   Hydrogen production by the iron process.

pect. Some possible methods have been demonstrated on a small scale for short time periods in the laboratory, but much research will be required before they have practical value. The processes under consideration fall into two categories: biophotolysis and photoelectrolysis.

Biophotolysis. In biophotolytic methods, use is made of living systems (or materials derived from such systems) to split water into hydrogen and oxygen. In normal photosynthesis in green plants (see **Biomass Fuels**), the green pigment chlorophyll takes up energy from sunlight and in a complex series of reactions breaks up water molecules into oxygen gas, hydrogen ions (i.e., hydrogen with a positive electric charge), and electrons (i.e., particles with a negative charge). The oxygen is evolved from the green plant, but the hydrogen ions and electrons are removed by interaction with carbon dioxide (from the air) to produce simple sugars.

Certain single-celled green algae are able to make the enzyme (i.e., biological catalyst) hydrogenase. In these algae, the second stage of photosynthesis can be circumvented by eliminating carbon dioxide. The hydrogen ions and electrons then combine in the presence of hydrogenase to form hydrogen gas. Thus exposure of these algae to sunlight and water (plus essential mineral salts) yields a mixture of oxygen and hydrogen gases that can be separated in various ways. The formation and activity of hydrogenase are inhibited by accumulated oxygen gas; consequently, the gases produced are removed by a flow of nitrogen.

Blue-green algae differ from green algae in several respects. In particular, in addition to normal photosynthesis cells in which reaction with carbon dioxide occurs, they contain some larger cells (heterocysts) where hydrogen can be formed. In the presence of nitrogen (e.g., from the atmosphere), however, the nitrogen combines with the hydrogen ions and electrons to produce ammonia. By preventing access of nitrogen (e.g., in an inert argon atmosphere), blue-green algae decompose water in sunlight to yield hydrogen and oxygen.

Instead of using living algae to obtain hydrogen from water, a more convenient approach is to utilize biological materials obtained from plants or bacteria. One advantage is the ability to vary the conditions to optimize hydrogen production.

Chloroplasts, the small bodies containing the chlorophyll in green plants, retain their photosynthetic activity when extracted from the plant. Hydrogen and oxygen can then be obtained from water by exposing chloroplasts to sunlight together with the enzyme hydrogenase and ferredoxin, an electron carrier, also of biological origin. It is possible, although less efficient, to replace the ferredoxin by a synthetic electron carrier and the hydrogenase by an inorganic (platinum) catalyst. An ultimate objective of research on the decomposition of water by sunlight is the efficient simulation of biological processes without using biological materials.

Photoelectrolysis. In ordinary electrolysis, as outlined at the beginning of this section, water is decomposed into hydrogen and oxygen by passing an electric current, from an outside source, between two electrodes in an electrolyte solution. In photoelectrolysis, a current is generated by exposing one or both electrodes to sunlight. Hydrogen and oxygen gases are liberated at the respective electrodes by the decomposition of water, just as in ordinary electrolysis. At least one of the electrodes in photoelectrolysis is usually a **semiconductor**; a catalyst may be included to facilitate the electrode process. In the cells studied so far, the efficiency for the conversion of solar energy into hydrogen-oxygen energy has been very low. Research is being directed at increasing this efficiency by selection of electrode materials, electrolyte solutions, and electrode catalysts.

HYDROTREATING

A general name for a variety of processes used extensively in **petroleum refining** involving reaction of natural or synthetic crude oils and petroleum products with hydrogen gas at high (or moderately high) temperature and pressure in the presence of a catalyst. The reaction conditions and nature of the catalyst depend on the objective of the hydrotreating process. A common catalyst is derived from cobalt oxide (CoO) and molybdenum trioxide (MoO_3), often referred to as cobalt molybdate, dispersed on a predominantly alumina base.

Among the applications of hydrotreating are: **desulfurization of liquid fuels;** removal of substances (e.g., nitrogen compounds) that deactivate catalysts used in subsequent treatment; conversion of olefin (unsaturated) to paraffin (saturated) **hydrocarbons** to reduce gum formation in gasoline; improvement of color, odor, and other characteristics of various fuels; and upgrading of synthetic crude oils from **coal liquefaction, oil shale,** and **tar sands** to facilitate refining.

HYGAS PROCESS

An Institute of Gas Technology process for **coal gasification** partly by direct reaction of the carbon in coal with hydrogen gas to form **methane** and partly by reaction with steam to produce hydrogen and carbon monoxide. The methane-rich product can serve as an **intermediate-Btu fuel gas,** but it can be readily converted into a **high-Btu fuel gas** (or **substitute natural gas**). If caking coal is to be used in the process, it must be pretreated by heating in air at a temperature of about 750°F (400°C).

The HYGAS gasifier consists of four regions, one below the other. The uppermost region is the feed (slurry) dryer section which is followed by first and second hydrogasification stages (Fig. 77). In the bottom (gasifier) region, unreacted **char** from the upper sections interacts with steam and oxygen to produce hydrogen and carbon oxides (see **Hydrogen Production**). The hydrogen-rich gas and excess steam rise through the gasifier and react with the carbon in coal (and char) in the two hydrogasifier stages.

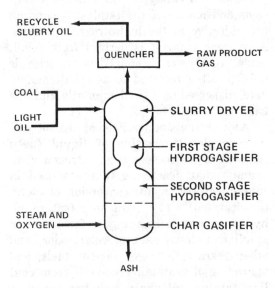

Fig. 77   HYGAS process.

Crushed coal, pretreated if necessary to prevent caking, is slurried with a light oil and fed under pressure into the dryer section. Here the oil is vaporized at a temperature around 600°F (315°C). The vapors, together with the hydrogasification products from below, leave the top of the vessel; the oil, recovered by quenching the vapors, is recycled to the coal slurrying operation.

By means of a special design, the coal leaving the dryer descends through a pipe and is then carried upward by the hot gases rising from the stage below. This constitutes the first hydrogasification stage where reaction with hydrogen occurs at a temperature of 1250°F (675°C). The pressure is about 68 atm (6.8 MPa), as it is throughout the gasification vessel. The unreacted coal (char) from the first stage descends to the second, **fluidized-bed** hydrogasification stage where the carbon reacts with hydrogen and steam from the lowest section (i.e., the char gasifier). The temperature is maintained around 1700°F (925°C) by the opposing effects of these reactions: heat is liberated by the reaction with hydrogen but is absorbed in the reaction with steam. The latter reaction also produces hydrogen which is used in the hydrogasification process.

The char leaving the second hydrogasification stage is finally consumed in the bottom section. Reaction with steam and oxygen at a temperature of 1850°F (1010°C) results in the formation of hydrogen (and carbon oxides). The residual coal ash is discharged.

After leaving the quencher, where the slurry oil is recovered, the raw gas from the gasifier is treated for hydrogen sulfide removal (see **Desulfurization of Fuel Gases**). The purified gas contains about 25 volume percent of methane, 25 percent of carbon monoxide, and 30 percent of hydrogen (dry basis) as fuel components; the **heating value** is about 370 Btu/cu ft (14 MJ/cu m). The gas could be used as a fuel, but the main objective of the HYGAS process is to produce a high-Btu gas. Hence, after adjusting the ratio of hydrogen to carbon monoxide, if necessary, by the **water-gas shift reaction,** the gas would be subjected to **methanation**.

I

IFP (INSTITUT FRANÇAIS DU PÉTROLE) PROCESS

A process for removing small amounts of hydrogen sulfide remaining in the off-gas from the **Claus process**. Basically, it involves the interaction of hydrogen sulfide with sulfur dioxide in the presence of a solvent containing a catalyst at a temperature of 250 to 280°F (120 to 140°C). The chemical process is represented by

$$2H_2S \ + \ SO_2 \ \rightarrow \ 2H_2O \ + \ 3S$$

Hydrogen   Sulfur   Water   Sulfur
sulfide   dioxide   vapor   (liquid)

and the molten sulfur formed is removed from the bottom of the catalytic reactor. The sulfur dioxide required can be obtained by oxidation of the hydrogen sulfide in the Claus off-gas.

INERTIAL-CONFINEMENT FUSION

Controlled nuclear fusion in a small sphere of a deuterium-tritium (D-T) fuel mixture under such conditions that the disassembly (or confinement) time of the sphere at the high fusion temperature (about 100 million °C) is longer than the fusion reaction time (see **Fusion Energy**). That is to say, as a result of inertia (i.e., resistance to motion), the fuel remains confined (or holds together) long enough to permit nuclear fusion to occur to a substantial extent.

For a sphere with a diameter of 1 mm, the confinement time at the accepted D-T fusion temperature of 100 million °C is estimated to be about 10^{-10} sec (i.e., one ten-billionth of a second). If the sphere were smaller, the disassembly time would be decreased roughly in proportion to the diameter. For significant D-T fusion, the reaction time would have to be even shorter than this short confinement time.

In a small sphere (or pellet) of a liquid (or solid) D-T mixture at normal density, the reaction time would be much longer than the confinement time. By compressing the sphere, the diameter would be decreased and the density of the D-T mixture would be increased. The decrease in diameter would result in a decrease in the confinement time, but the increase in density would produce a much greater decrease in the fusion reaction time. Therefore, at a sufficiently high compression, the reaction time could become less than the confinement (or disassembly) time and inertial-confinement fusion should be possible.

For example, in a D-T mixture compressed to 10,000 times normal density, the confinement time at 100 million °C of a sphere initially 1 mm in diameter would be roughly 10^{-11} sec. The reaction time, however, might be about 10^{-12} sec, thus complying with the condition for inertial confinement. Furthermore, at 10,000 times normal density, the particle density n would be close to 4×10^{26} nuclei/cubic centimeter (cm^3). The energy confinement time τ is approximately the same as the fusion reaction time, 10^{-12} sec. The product $n\tau$ would then be 4×10^{14} sec/cm^3 so that, by the Lawson criterion for D-T, ignition and energy break-even should be realized. (For an explanation of these terms, see the article on **Fusion Energy**).

Ablation-Implosion Compression

The enormous pressure, estimated to be 10^{12} times atmospheric pressure, required to compress a liquid (or solid) D-T mixture to 10,000 times normal density should be attainable by using the ablation-implosion principle. A thin spherical shell containing the fusion fuel is heated uniformly in an extremely short time by means of a high-energy pulse. The shell surface is vaporized and ablated (or blown off) at a high speed in the outward direction. Then, in accordance with Newton's Third Law of Motion (i.e., every action has an equal and opposite reaction), the remainder of the material is imploded—that is, it is forced inward. In other words, the D-T fuel material is compressed. Under appropriate conditions, the compression necessary for inertial-confinement fusion should be realized.

As the sphere of D-T fusion fuel is imploded, the increase in the interior pressure gradually slows down the inward motion until a brief stationary state is reached. A strong shock (or pressure) wave then moves into the compressed fuel and increases the temperature at the center to at least 100 million °C, at which D-T fusion occurs at a substantial rate.

The neutrons produced in the D-T reaction escape, but as a result of the high density of the compressed fuel mixture, the accompanying helium nuclei do not. The energy carried by these nuclei, representing 20 percent of the total fusion energy, is deposited as heat in the bulk of the fuel. With the additional heat energy available, the fusion reaction, initiated at the center by shock wave heating, spreads through the fuel sphere in a time that is shorter than the confinement (disassembly) time.

Two general ways have been proposed for supplying the initial energy pulse required to cause surface ablation followed by compression and heating of a fusion fuel sphere (or pellet). One is by means of intense laser beams, and the other is to use beams of high-energy, electrically charged particles, such as negative electrons and positive ions; these are described later. The energy available for compression can be optimized by pellet design. One concept is based on the use of two shells: an outer (ablator) shell of lighter elements (e.g., plastic) and an inner (pusher or compresser) shell of a heavier element (e.g., gold).

Laser-Beam Fusion

The initiation of nuclear fusion by laser beams was reported first from the U.S.S.R. in 1968 and was subsequently confirmed in the United States. However, the extent of fusion, as indicated by the neutrons produced, was very small. Laser beams of very much greater energy than are available even in the early 1980s would be required to realize a practical laser fusion reactor for generating useful power.

Theory had indicated that laser beams of short-wavelength radiation would provide the most efficient compression of a D-T fuel pellet. Consequently, neodymium (Nd)-glass (i.e., glass doped with a small amount of neodymium) lasers have been (and are being) used in basic studies of laser-beam fusion. Because these lasers have a low efficiency (i.e., a small fraction of the "pumping" energy is converted into radiation energy), a large proportion of the energy supplied is ineffective. This energy is converted into heat, and the need for cooling the solid lasing material may make Nd-glass lasers impractical for reactor use.

An alternative is the carbon dioxide (gas) laser; this has a higher efficiency than the Nd-glass laser and, furthermore, heat can be removed from the gas by circulation through an external cooling system. The radiation from the carbon dioxide laser has a longer wavelength than from a Nd-glass laser, but it appears that this may not be such a drawback as had been thought.

Existing lasers, no matter what the lasing material, are not powerful enough and do not have the other requirements for a fusion reactor. Efforts are being made to develop such lasers, possibly with novel lasing materials (e.g., krypton fluoride). According to present estimates, a fusion power reactor would require a laser system with an efficiency of at least 10 percent, capable of delivering 1 million joules (0.28 kW-hr) of energy per pulse in pulses of less than 10^{-9} sec duration at a rate of one pulse per second (see **Joule; Watt**).

Laser Fusion Reactor. A laser-beam fusion reactor is conceived as a strong, spherical chamber (or cavity) of some 6.6 ft (2 m) radius. A small [e.g., 0.04 in. (1 mm) diameter] pellet (or sphere) of solid or liquid D-T fuel is injected into the cavity. When it reaches the center, several symmetrically located high-energy laser beams are focused simultaneously on the fuel for about 10^{-9} sec. An ablation-implosion occurs, and fusion energy is released. After an interval of a second or so, another fuel pellet is injected, and the process is repeated.

The helium nuclei formed in the fusion reaction are absorbed in the fuel, but the neutrons, carrying 80 percent of the fusion energy, will escape. They are captured in a lithium blanket where they deposit their energy as heat and also produce the tritium required as fuel material. The heat could be utilized to generate electricity in a more-or-less conventional manner by producing steam to drive a turbine–generator (see **Electric Power Generation**).

Apart from the actual fusion process, a laser-beam cavity reactor would be similar in principle to (but smaller than) other nuclear fusion reactors (see **Magnetic-Confinement Fusion**). A problem is expected to arise from deterioration of the inner wall of a laser-fusion cavity as a result of the intense neutron bombardment. Proposed ways for overcoming this problem are to keep the inner wall continuously covered with a thin layer of liquid lithium or to maintain a swirling vortex of this liquid within the cavity.

Although the fusion process initiated by laser beams would be self-sustaining in a single D-T fuel pellet, a continuous succession of fresh pellets and laser-beam pulses would be required to maintain operation of the reactor. A laser-fusion reactor is thus a driven energy (or power) amplifier. The energy supplied in the short pulse of laser radiation would be amplified by a factor of 100 or more in the resulting fusion reaction.

If the laser energy is 1 million joules per pulse, as suggested earlier, the fusion energy produced would be about 100 million joules (or 28 kW-hr) per pulse. For a repetition rate of one pellet (or pulse) per second, the fusion **power** output per cavity would be 100 megawatts (MW). Because only 80 percent of this energy, which is carried by the neutrons, is available outside the cavity and there are inevitable losses in the laser system and elsewhere, it is expected (optimistically) that a single cavity could generate useful power at a rate of some 20 MW.

The 100 million joules per pulse of fusion energy is equivalent to the explosion energy of 53 lb (24 kg) of TNT. However, the fusion process is expected to have a minor impact on the cavity because of the small mass (less than 1 milligram) of the D-T fuel. The proposed cavity radius of 6.6 ft (2 m) should thus provide ample protection against the blast effect of the repeated fusion energy pulses.

Charged-Particle Beam Fusion. A pulse of energy supplied by a beam of high-energy, electrically charged particles should, like a laser beam, be able to ablate and compress a fusion fuel pellet (or sphere). This capability was demonstrated with electron beams and deuterium pellets in the U.S.S.R. in 1976 and in the United States in 1977. The production of neutrons showed that nuclear (D-D) fusion had occurred. There is no reason to doubt that fusion could also be realized with beams of heavier charged particles.

Proponents of charged-particle beam fusion claim that it has advantages over laser-beam fusion. Pulses of the required energy should be easier to generate, and the generation efficiency is expected to be higher than for laser beams. The coupling of the beam energy to the fuel pellet should also be better. The pellets may be larger and less spherical and, consequently, cheaper to fabricate.

By the early 1980s, most studies of charged-particle beam fusion had been made with electron beams. High-energy electron beams can be readily obtained with accelerators; but for fusion, pulses of unusually high current (in amperes) and short duration (less than 10^{-8} sec) are required. Furthermore, the beams must be transported from the accelerator without spreading and focused onto the small fuel pellet.

Beams of positively charged light ions, especially protons (i.e., ordinary hydrogen nuclei), are also being considered for inertial-confinement fusion. With such beams, fusion may be achieved with beams of higher voltage but smaller currents than for electron beams. The required pulsed proton accelerators may then be easier to build than the electron accelerators. For this and other reasons, proton beams are now preferred to electron beams for charged-particle fusion experiments.

Since 1976, interest has grown in the use of beams of accelerated heavy ions, with mass numbers (see **Atom**) from 124 to 238, to produce inertial-confinement fusion. It is possible that these charged-particle beams may prove more effective than electron, proton, or laser beams. Because the heavy ions can travel only a very short distance in matter, essentially all their energy would be absorbed in the outer layers of the pellet. Ablation and compression of the fusion fuel pellet should then be more effective than in the other proposed methods for achieving inertial-confinement fusion.

As far as can be seen at present, a nuclear fusion reactor utilizing high-energy particle beams would be similar in principle to the laser-beam reactor described earlier.

INTERMEDIATE-BTU (OR MEDIUM-BTU) FUEL GAS

A fuel gas with a **heating value** generally in the range of roughly 280 to 400 BTU/cu ft (10 to 15 MJ/cu m). The compositions of intermediate-Btu gases vary according to the method of manufacture, but they usually consist of roughly 20 to 40 volume percent of carbon monoxide, 25 to 40 percent of hydrogen, and 5 to 15 percent of methane (dry basis); the remainder is largely inert carbon dioxide and a small amount of nitrogen. The heating value of both carbon monoxide and hydrogen is close to 320 Btu/cu ft (12 MJ/cu m); hence, any mixture of these gases has essentially the same heating value. In an actual intermediate-Btu gas, the heating value is increased by the presence of **methane**, with a heating value of 1030 Btu/cu ft (38 MJ/cu m), and decreased by the inert gases.

Intermediate-Btu fuel gases are most commonly produced by the action of steam on carbon in coal at a temperature of 1500°F (815°C) or more. This reaction is accompanied by the absorption of heat which is supplied by the combustion of coal in oxygen gas or air (see **Coal Gasification**). Most processes for manufacturing intermediate-Btu gases are based on the simultaneous reactions of coal with steam and oxygen gas (e.g., **BI-GAS; COED; Kellogg Molten Salt Gasification; Koppers–Totzek; Lurgi; Synthane; Texaco Coal Gasification; Wellman–Galusha; and Winkler Processes**). In a few cases (e.g., **Agglomerating Burner; Carbon Dioxide Acceptor; and COGAS Process**), coal or char is burned in air (rather than oxygen gas) in one vessel, and the heat generated is transferred to a separate vessel where the carbon (coal)–steam reaction takes place. In this way, the inert nitrogen gas from air is largely kept out of the product gas.

Because of its moderately low heating value, distribution of intermediate-Btu gas would not be economic; hence, the gas would be used locally. Possible applications are as a clean boiler fuel or as the fuel for a **gas turbine** that might be the first stage of a **combined-cycle generation** system.

Two other potential applications for intermediate-Btu gas are of interest. One is to use the gas as a **synthesis gas** for the production of various liquid fuels (see **Fischer–Tropsch Process**) or of methanol (see **Alcohol Fuels**). The other is conversion into a **high-Btu fuel gas** comprised mainly of methane. The intermediate-Btu gas is first subjected to the **water-gas shift reaction**, to adjust the volume ratio of carbon monoxide to hydrogen to 1 to 3, and then to **methanation**. The resulting gas would be a **substitute natural gas**.

INTERNAL-COMBUSTION ENGINE

Usually a **heat engine** in which fuel is burned in air within a cylinder enclosed by a movable piston; expansion of the hot gas produced by the combustion pushes back the piston and, as a result, mechanical work is done.

The open-cycle gas (or combustion) turbine is sometimes considered to be an internal-combustion (IC) engine because the fuel is burned in a combustion chamber from which the hot gases pass directly to the turbine. The operating principles of the gas turbine are so different, however, from those of other IC engines that it is described in a separate section (see **Gas Turbine**).

Internal-combustion engines are distinguished from external-combustion engines, such as the conventional **steam engine, steam turbine,** and the **Stirling engine,** in which heat is generated (e.g., by combustion or in other ways) outside the region where the mechanical work is done.

Internal-combustion engines are heat engines which convert part of the heat generated by combustion of the fuel into mechanical work. Like other heat engines, IC engines operate in a succession of cycles, each consisting of four stages: (1) compression of air which may be mixed with fuel vapor or into which the fuel is injected, (2) ignition of the fuel in the air, resulting in the production of hot gases under pressure, (3) expansion of the hot, high-pressure gases, thus pushing back a piston and doing mechanical work, and (4) discharge of the gases at a lower pressure and temperature as exhaust.

The two main types of IC engines differ primarily in the mode of ignition of the compressed air–fuel mixture in the second stage. In the **spark-ignition engine**, the mixture is ignited by an electric spark, whereas in the compression-ignition engine, the fuel is ignited by injecting it into air that has been heated by rapid (adiabatic) compression (see **Carnot Cycle**). Spark ignition is commonly used in automobile and other gasoline engines, whereas the compression-ignition engine is better known as the **diesel engine**.

Thermodynamically, the spark-ignition engine operates (ideally) on the **Otto cycle**, and the compression-ignition (diesel) engine is based on the **Diesel cycle**. However, in neither case is the cycle closed; the working gas is not restored to its initial state but is discharged to the atmosphere at the end of the cycle. Fresh air and fuel are then taken in at the start of the next cycle.

One of the differences between the two types of IC engines is in the nature of the fuel. For a spark-ignition engine, the fuel must be a gas or a volatile liquid (i.e., one which is easily converted, at least in part, into vapor), usually **gasoline**. Diesel-type engines, on the other hand, can use a wide variety of fuels, ranging from natural gas to petroleum distillates that are less volatile and less expensive than gasoline (see **Diesel Fuel**).

In most spark-ignition engines, a mixture of air and fuel is drawn into the cylinder by way of a carburetor; the mixture is then compressed and ignited. The air-to-fuel ratio in the mixture depends on the carburetor design and is almost constant during normal operation. The speed (or power) of the engine is controlled by a throttle valve which changes the amount of the air–fuel mixture that enters the cylinder. In a diesel engine, however, a constant amount of air is taken into the cylinder and the speed is controlled by adjusting the quantity of fuel that is injected. In other words, the speed is changed by varying the air-to-fuel ratio. (This method of control has been used in an experimental spark-ignition engine using hydrogen gas as fuel, see **Hydrogen Fuel**.)

For a given compression ratio (i.e., volume of air prior to compression divided by the volume after compression in stage 1 above), a spark-ignition engine has a higher **thermal efficiency** (i.e., it converts a larger proportion of the heat supplied by the fuel into useful mechanical work) than a diesel engine. However, diesel engines can be (and are) operated at higher compression ratios and with higher thermal efficiencies than spark-ignition engines. Consequently, in practice, fuel utilization is usually better in a diesel engine.

The pressures developed in a diesel engine are very high, so that strong and heavy construction is required. Hence, in the past, diesel engines have been used mainly as stationary power sources (e.g., for electrical generators and pumps) and as mobile power sources for relatively heavy vehicles (e.g., ships, locomotives, earth-moving machinery, freight trucks, and buses). However, because they can make efficient use of fuels that are cheaper than gasoline, diesel engines are now being developed for automobiles.

Both spark-ignition and diesel engines have cooling systems with water (or sometimes air) as the coolant. The cooling system plays no part in the operation; its sole purpose is to prevent mechanical weakening of the cylinder walls by overheating.

IRON OXIDE DESULFURIZATION PROCESS

A high-temperature process for the **desulfurization of fuel gases** by chemical absorption of hydrogen sulfide. Iron (ferric) oxide was used extensively at one time to remove hydrogen sulfide from **coal gas**. Efforts are being made at the Morgantown Energy Technology Center (U. S. Department of Energy) to improve the efficiency of the iron oxide absorber by forming it into pellets with fly ash (from coal combustion) or powdered silica. The operating temperatures are in the range of 1000 to 1700°F (540 to 925°C). The iron oxide absorber becomes iron sulfide which is regenerated by heating in air. The sulfur dioxide gas formed can be converted into sulfuric acid or elemental (solid) sulfur (see **Desulfurization: Waste Products**).

ISOMERIZATION

In general, the conversion of one molecule into another with the same atomic composition but with the atoms arranged differently; the two molecules related in this manner are called *isomers*. In **petroleum refining**, isomerization usually refers to the conversion of a normal (straight-chain) paraffin **hydrocarbon** into a branched-chain isoparaffin with the same atomic composition; the isomerization of normal butane, for example, to isobutane is represented below.

Normal butane

Isobutane

Catalytic isomerization of butane is used extensively in the petroleum industry to obtain the isobutane required for the **alkylation** process. Another important application is the conversion of normal pentane and hexane in **straight-run gasoline** into the isomeric branched-chain compounds, which have substantially higher **octane numbers**. In the catalytic **reforming** of certain naphthenes (cycloparaffins) to aromatic hydrocarbons, a five-membered carbon ring with a side chain is first isomerized to a six-membered ring; the latter is then dehydrogenated to yield an aromatic hydrocarbon.

The isomerization catalyst is either aluminum chloride (with hydrogen chloride) or platinum. The operating temperature is usually in the range of about 300 to 400°F (150 to 205°C), and the pressure is from 20 to 100 atm (2 to 10 MPa), depending on the feed and the catalyst. When the latter is platinum, hydrogen gas may be included to increase the effective life of the catalyst.

ISOTOPE POWER SYSTEMS

Devices for using the energy deposited by the absorption of alpha or beta particles from a radioactive material (or radioisotope) as the heat source for a turbine (see **Radioactivity**). Isotope power systems have been designed especially to provide electric power for space applications. The favored radioactive material is the alpha-particle emitter plutonium-238, as the dioxide (PuO_2); it is expensive but has a long useful lifetime (see **Radioisotope Thermoelectric Generator**).

The Brayton Isotope Power System (or BIPS) utilizes a closed-cycle **gas turbine** operating on the **Brayton cycle**. The working fluid is a mixture of helium and xenon gases. Another scheme is the Kilowatt Isotope Power System (KIPS) which is a **Rankine-cycle** (condensing) turbine with Dowtherm A, a high-boiling-point hydrocarbon mixture, as the working fluid (see **Vapor Turbine**). In each case, heat is supplied at a high temperature by the radioactive material; part of the heat is converted into useful work in the turbine, and the remainder is discharged to the environment (i.e., space). An electric generator connected to the turbine produces electric power (see **Electric Power Generation**).

J

JET FUEL

A mixture of liquid hydrocarbons, consisting either of the **kerosine** fraction of petroleum **distillation** (see **Petroleum Refining**) or a blend of kerosine and **gasoline** (or **naphtha**), for use as fuel in jet-propelled aircraft. Desirable properties of a jet fuel are volatility, to facilitate starting and relighting, especially at the low temperatures at high altitudes, and a high **heating value** per unit mass, to minimize the weight of fuel to be carried. The average heating value for jet fuel is about 19,000 Btu/lb (44 MJ/kg). Various inhibitors are added to jet fuel to minimize gum formation, icing in the fuel lines, and corrosion (see **Gasoline**).

Aromatic hydrocarbons in jet fuels can cause carbon deposition in the engine and a smoky exhaust. These hydrocarbons must not exceed 20 volume percent in the fuel for civilian jet aircraft or 25 percent for military aircraft. The proportion of aromatic hydrocarbons in the fuel can be reduced by extraction with various solvents (e.g., liquid sulfur dioxide, furfural, dimethyl sulfoxide, and others).

The American Society for Testing and Materials (ASTM) has developed specifications for the following jet fuels for civilian aircraft:

Jet A: A kerosine-type distillate with a moderately high **flash point** [above 100°F (38°C)] for safety against fire. Used mainly for domestic flights.

Jet A-1: Similar to Jet A but with a lower freezing point. Used for aircraft flying long distances and at high altitudes.

Jet B: A blend of gasoline components and kerosine having a wide boiling range and a low freezing point (see JP-4 below).

Jet fuels for military aircraft are indicated by the symbol JP, as follows:

JP-1: Similar to Jet A for civilian aircraft; little used at present.

JP-4: Substantially identical with Jet B. It is used extensively in subsonic military aircraft and (with oxygen) as the propellant in large, liquid-fueled (booster) rocket engines.

JP-5: A kerosine-type fuel with a high flash point [above 140°F (60°C)] for naval carrier operations.

JP-6: A kerosine–gasoline blend for supersonic aircraft.

JOULE

The international (SI) unit of work or **energy**, including heat. It is defined as the work done (or its equivalent) when the point of application of a force of 1 newton moves a distance of 1 meter in the direction of the force. A newton is the force that gives a mass of 1 kilogram an acceleration of 1 meter per second per second. The joule (J) is equivalent to 1 watt-second (i.e., a **power** of 1 **watt** operating for 1 second) or 0.239 calorie or 9.48×10^{-4} **British thermal unit**. A direct current of 1 ampere flowing for 1 second under an electromotive force (voltage) of 1 volt develops 1 J of energy.

K

KELLOGG MOLTEN-SALT GASIFICATION PROCESS

An M. W. Kellogg Company process for **coal gasification** with steam and oxygen gas (or air) in the presence of a molten salt (sodium carbonate). The sodium carbonate acts as a catalyst for the carbon (coal)-steam reaction and also serves to remove sulfur from the coal. The product is either an **intermediate-Btu fuel gas** if oxygen is used for combustion of the coal or a **low-Btu fuel gas** if air is used. The process is applicable to both caking and noncaking coals without pretreatment.

Crushed coal and sodium carbonate, in a stream of preheated oxygen (or air) and steam, are fed to the gasifier containing the molten carbonate. Combustion of the coal in oxygen (or air) generates the heat required to maintain the temperature at about 1800°F (980°C). At this temperature, the coal reacts with steam to produce mainly hydrogen and carbon monoxide. The pressure in the gasifier may be around 30 atm (3 MPa).

With oxygen as the combustion medium, the product gas contains, typically, 45 volume percent of hydrogen, 33 percent of carbon monoxide, and a few percent of methane (dry basis) as the fuel constituents; the remainder is mostly inert carbon dioxide. The **heating value** is around 300 Btu/cu ft (11 MJ/cu m). If air is used in place of oxygen gas, the product gas is about 50 percent inert nitrogen (from the air) and the heating value is roughly 150 Btu/cu ft (5.6 MJ/cu m). Since the sulfur in the coal is retained by the sodium car-

bonate as sodium sulfide, the product gas does not require treatment for sulfur removal.

A bleed stream of molten sodium carbonate, containing coal ash and sodium sulfide, is withdrawn from the bottom of the gasifier. The sodium carbonate and sulfide are extracted with water, and the ash residue is discarded. Passage of carbon dioxide gas through the solution results in the precipitation of sodium bicarbonate; this is separated and heated to regenerate sodium carbonate. At the same time, the carbon dioxide acts on the sodium sulfide to produce hydrogen sulfide gas which can be converted into elemental sulfur by the **Claus process**.

See also **Rockwell Coal Gasification Process.**

KEROSINE

A mixture of liquid **hydrocarbons**, less volatile than **gasoline** but more volatile than **gas oil**, obtained by the distillation of petroleum (see **Petroleum Refining**). The boiling range is approximately 300 to 550°F (150 to 290°C), although brands of kerosine for specific uses generally have a narrower boiling range. Originally kerosine served as a fuel in oil lamps and for domestic space heating and cooking. The main current uses are as the sole or partial component of **diesel fuels, jet fuels,** and light **fuel (heating) oils.**

KOPPERS–TOTZEK PROCESS

A commercial process, developed by Heinrich Koppers (now Krupp–Koppers)

GmbH in West Germany and Koppers Company, Inc., in the United States, for **coal gasification** with steam and oxygen gas. The product is an **intermediate-Btu fuel gas** which can also serve as a **synthesis gas** or it can be converted into a **high-Btu fuel gas** (or **substitute natural gas**). Any kind of coal, caking or noncaking, can be used in the process without pre-treatment. Oxygen cannot be replaced by air in the Koppers–Totzek process, but operation is possible with oxygen-enriched air. However, as a general rule, oxygen alone is used.

Some 40 Koppers–Totzek gasification units have been constructed in Europe, Asia, and Africa; they are intended mostly for the manufacture of hydrogen or for a synthesis gas. A small demonstration plant was built in the United States in 1948, and a larger installation is due for operation in the early 1980s.

Pulverized coal, entrained in a stream of oxygen gas and low-pressure steam in a mixing nozzle, is introduced into the gasifier (Fig. 78). The original Koppers–Totzek units had two nozzle heads, each with two adjacent burners. Later designs have four heads and a total of eight burners; units with six heads and 12 burners are under consideration to increase the gasifier capacity.

As it leaves the burners, the entrained coal reacts very rapidly with the oxygen and steam at a temperature of about 3500°F (1930°C). The pressure in the gasifier is usually just above atmospheric (see, however, **Shell–Koppers Process**). Oxidation of the carbon in the coal provides the heat required for the reaction of carbon with steam to produce carbon monoxide

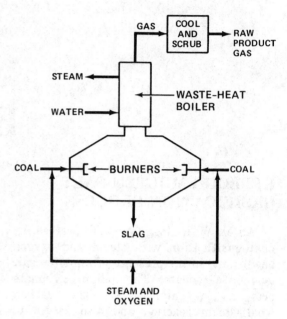

Fig. 78   **Koppers–Totzek process.**

and hydrogen. The coal ash melts at the high temperature of the gasifier, and liquid slag is withdrawn from the bottom.

The product gas leaving the gasifier at a temperature of 2700°F (1480°C) is passed through a waste-heat boiler (**heat exchanger**) to generate steam. It is subsequently cooled and scrubbed with water and treated for hydrogen sulfide removal (see **Desulfurization of Fuel Gases**). The resulting clean gas contains, typically, 55 volume percent of carbon monoxide and 34 percent of hydrogen (dry basis), as the fuel components; most of the remainder is inert carbon dioxide. The **heating value** of the gas is about 290 Btu/cu ft (11 MJ/cu m). It can be converted into a high-Btu fuel gas by the **water-gas shift reaction** followed by **methanation**.

L

LIGHT-WATER BREEDER REACTOR (LWBR)

A **pressurized-water reactor** (PWR), with light (ordinary) water as the moderator and coolant (see **Thermal Reactor**), in which the core is modified to breed fissile uranium-233 (i.e., produce more than is consumed during operation) from fertile thorium-232 while generating useful power (see **Breeder Reactor**). The LWBR concept is being tested by installing a demonstration core, designed by the Division of Naval Reactors, U. S. Department of Energy, in the 90-MW (electric) Shippingport reactor. Although the breeding ratio is not expected to be much greater than unity, the uranium-233 recovered, after allowing for inevitable reprocessing losses, should ultimately be sufficient (in principle) to refuel the reactor. Hence, in the long term, thorium would be the only fuel material required.

The core of the LWBR is made up of hexagonal assemblies each consisting of several hundred closely spaced vertical fuel rods of two types. The rods in the center of an assembly, called the "seed" region, contain fertile thorium dioxide with zero to 6 percent of fissile uranium-233 dioxide; the outer rods, in the internal blanket region, contain thorium with zero to 3 percent of uranium-233 as the dioxides (see **Fission Energy**). Twelve such assemblies form the reactor core. The outer blanket consists of extensions of thorium dioxide alone at the top and bottom of the core rods, with rods of the same fertile material around the core. The flow of coolant water through the core (and blanket) and steam generator is similar to that in the PWR (see Fig. 104).

A special feature of the LWBR is the method of control which avoids the loss of neutrons by absorption in a neutron poison, such as is used in other power reactors (see **Nuclear Power Reactor**). Control is achieved by moving the whole central (seed) region in or out of each assembly. If the seed is lowered out of an assembly, neutrons escape from the core and the fission rate (and hence the thermal power) decreases. Restoration of the seed to the core causes the power to increase. The essential point is that the neutrons escaping from the core are not lost; they are largely captured by thorium-232 in the blanket regions to produce uranium-233.

LIGHT-WATER REACTOR (LWR)

A thermal **nuclear power reactor** in which light (ordinary) water is both moderator and coolant (see **Thermal Reactor**). An advantage of the LWR is that, even when specially purified, water is a cheap and readily available material, but this must be weighed against the cost of the uranium fuel which must be enriched to 2.5 to 3 percent in uranium-235 (see **Nuclear Fuel; Uranium Isotope Enrichment**). Furthermore, because water boils at the low temperature of 212°F (100°C) at atmospheric pressure, LWRs must operate at high pressures to prevent boiling altogether or to permit boiling at a high temperature. With very few exceptions, all commercial nuclear power reactors operating in the United States are LWRs.

There are two types of LWRs: **pressurized-water reactors** (PWRs) and **boiling-water reactors** (BWRs). In the PWR, the system pressure is high enough to prevent boiling at the maximum water temperature of about 626°F (330°C). The water heated in the reactor core passes to a separate **heat-exchanger** steam generator. Heat is transferred from the high-pressure water to water at a lower pressure; the latter boils and produces steam at close to 555°F (290°C).

In the BWR, the system pressure is such as to permit the water to boil (and produce steam) at a temperature of about 550°F (288°C). Thus, by allowing the water in a BWR to boil in the reactor core, the temperature and pressure of the steam produced are similar to those from a PWR steam generator.

LIME/LIMESTONE DESUL-FURIZATION PROCESS

The most widely used method for the **desulfurization of stack gases**. It is a wet, throwaway (nonregenerative) process with lime (CaO) or limestone ($CaCO_3$) as the chemical absorber of **sulfur oxides**, mainly sulfur dioxide (SO_2). In each case, the absorption of sulfur dioxide results in the formation of calcium sulfite ($CaSO_3$), some of which is oxidized by oxygen in the air to calcium sulfate ($CaSO_4$). Calcium sulfite and sulfate are not soluble to any great extent in water, and so they are present in solid form.

In the lime/limestone process, the stack (flue) gas is scrubbed with a water slurry containing 5 to 15 weight percent of cal-

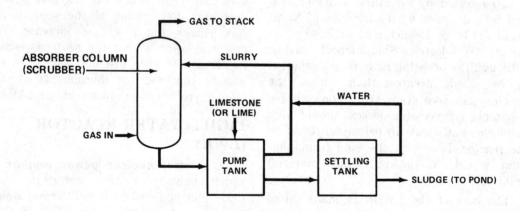

Fig. 79   Lime/limestone desulfurization process.

The **light-water breeder reactor** (or LWBR) is a modified PWR designed to breed fissile uranium-233 from fertile thorium-232—that is, to produce more of the fissile species than is consumed during reactor operation (see **Breeder Reactor**). All other LWRs are converters with uranium-238 as the fertile species; they produce less fissile material, in this case plutonium-239, than they consume (mainly uranium-235).

cium sulfite and sulfate to which small amounts of lime or limestone are added continuously. Part of the slurry is drawn off and pumped to a settling tank where some of the water is recovered. The remaining thick slurry is sent to a disposal pond where a sludge of calcium sulfite and sulfate containing roughly 50 percent of water accumulates (Fig. 79). The disposal of this sludge is described under **Desulfurization: Waste Products**.

Several factors have to be considered in choosing between limestone and lime as the absorber. Limestone is less expensive than lime and generally results in the deposition of less scale in the scrubbers. On the other hand, limestone does not absorb sulfur dioxide as readily as does lime. The lower absorption rate is compensated by using excess of limestone and providing for larger stack gas–slurry contact times in the scrubbers. Because of the additional absorber required, the quantity of waste sludge to be disposed of is greater with limestone than lime.

LIMESTONE INJECTION (FLUIDIZED-BED) DESULFURIZATION PROCESS

A process for removal of **sulfur oxides** from the combustion gases of a coal-fired furnace by injecting limestone (or lime) with the coal feed (see **Steam Generation**). The limestone ($CaCO_3$) then absorbs sulfur dioxide (SO_2) from the combustion gases before they leave the boiler furnace. The absorption results in the formation of calcium sulfite ($CaSO_3$) which is oxidized by air to calcium sulfate ($CaSO_4$). The latter, removed with the ash, could be treated for sulfur recovery and absorber regeneration, as explained below. Studies with conventional boilers revealed several problems. At best, only about 50 percent of the sulfur dioxide was absorbed. Furthermore, boiler fouling occurred, and the performance of **electrostatic precipitators** for fly ash removal was degraded.

The **fluidized-bed combustion** system shows promise for decreasing the sulfur content of stack gases. Crushed coal and limestone, which may contain magnesium carbonate as dolomite, are fluidized in the combustion bed with air. If the combustion temperature is maintained in the vicinity of 1600°F (870°C), some 90 to 95 percent of the sulfur in a high-sulfur (3 percent) coal may be removed as calcium sulfate. At lower temperatures, the absorption of sulfur dioxide by limestone is too slow to be useful; at higher temperatures, the calcium sulfate tends to decompose and release sulfur oxides. The activity of the limestone in the furnace can be extended by adding a small quantity of sodium chloride (common salt).

The overflow from the fluidized bed, consisting of coal ash and calcium sulfate (plus salt), can be treated for regeneration of the absorber. One approach is to reduce the spent material with carbon (coal) or a **hydrocarbon** gas (e.g., **natural gas**) at high temperature. The calcium sulfate is converted into calcium oxide (lime), which can be reused as absorber, and a gas rich in sulfur dioxide is produced. The recovered lime is less reactive as an absorber than fresh lime or limestone, apparently because of the high regeneration temperature, approaching 1950°F (1065°C). Hence, an appreciable amount of replacement is necessary.

A possible alternative regeneration procedure is to reduce the calcium sulfate to calcium sulfide (CaS) by means of hydrogen or carbon monoxide (or a mixture made from coal). Reaction of the calcium sulfide with steam and carbon dioxide results in the formation of calcium carbonate in a reactive form, for use as a sulfur dioxide absorber, and the liberation of hydrogen sulfide gas. The latter can be converted into elemental (solid) sulfur by the **Claus process**.

See also **Desulfurization: Waste Products**.

LIQUEFIED NATURAL GAS (LNG)

A liquid formed by cooling commercial **natural** gas to a low temperature. To liquefy **methane**, the main constituent of natural gas, the temperature must not exceed −116°F (−82°C); the required pressure is then 46 atm (4.6 MPa). Methane can be liquefied at normal atmospheric pressure, but the temperature would have to be lowered to $-259°F$ ($-162°C$). In practice, natural gas is liquefied at a moderate pressure at a temperature between the extremes given above.

The volume of LNG is smaller than that of the initial gas by a factor of about 600, thus offering the possibility for convenient storage as well as overseas transportation by special tankers. Controlled vaporization of the liquid produces a gas that can be pressurized and distributed by pipeline and used in exactly the same way as commercial natural gas.

The use of LNG as a peak-shaving fuel, which is stored in the summer for use in the winter when the demand is greatest, was initiated in the United States in 1941. However, as a result of a severe accident in 1944, about 20 years elapsed before this type of storage was resumed. Several local gas-distributing facilities now use LNG for peak shaving. Since natural gas cannot exist in liquid form at normal temperature, LNG is stored at low temperature in insulated tanks holding up to 315,000 **barrels** or 1.7 million cu ft (50,000 cu m) of liquid. This is equivalent to more than a billion (10^9) cu ft (30 million cu m) of natural gas measured at standard (atmospheric) temperature and pressure.

Most steels become brittle at the low temperature of LNG, and storage tanks may be constructed from high-nickel steels or certain aluminum alloys. The tanks are double-walled with insulation between the walls. Underground insulated tanks of prestressed concrete have also been used successfully. A concept for using unlined holes in the ground for storing LNG has been tested but cracks developed in the frozen soil leading to considerable losses.

A major potential source of LNG is from places outside the United States where abundant supplies of natural gas are available. Since about 1965, shipments of LNG by tanker ships have steadily increased. The LNG is carried in a number of separate, well-insulated containers. In one design, the containers are double-walled spheres, about 120 ft (39 m)

interior diameter, with approximately 3 ft (1 m) of insulation between the walls. A large tanker, with five such containers, has a total capacity of 780,000 barrels (125,000 cu m), equivalent to 2.6 billion (i.e., 2.6×10^9) cu ft (75 million cu m) of natural gas under standard conditions. Because of the low density of the liquid (specific gravity about 0.55), the deadweight (and draft) of an LNG tanker is less than that of a petroleum tanker of equal volume capacity.

The potential fire hazard associated with the transportation and storage of LNG requires careful attention. However, with the safety regulations now in effect, there seems to be no reason why LNG cannot be handled as safely as **gasoline, liquefied petroleum gas** (LPG), or liquid hydrogen (see **Hydrogen Fuel**).

LIQUEFIED PETROLEUM GAS (LPG)

A fuel product obtained from **natural gas** and **casinghead gas,** consisting mainly of the light paraffin **hydrocarbons** propane (C_3H_8) and/or butane (C_4H_{10}). These substances are gases at normal temperature and pressure, but they can be liquefied by moderate compression even at normal atmospheric temperature. The resulting liquid (LPG) can be readily stored and transported. Upon opening a valve to reduce the pressure in the container, the liquid vaporizes and gas is released for use; the volume of gas produced is about 250 times the liquid volume. Liquefied petroleum gas is often supplied in portable steel cylinders or tanks and is then referred to as "bottled" gas.

Much of the LPG used as a fuel in the United States is separated from "wet" natural gas prior to its distribution by pipeline. When the raw natural gas is cooled and compressed at ordinary temperatures, the propane and butane present are condensed; butane gas liquefies at a pressure of 2 atm (0.2 MPa), whereas propane requires a pressure of 7 atm

(0.7 MPa). Considerable quantities of propane and butane gases are produced in petroleum refining processes, but only a portion is sold as LPG. The remainder is used in other refinery and petrochemical operations.

Liquefied petroleum gas is widely utilized as a fuel, especially in mobile homes and in rural areas where natural gas is not available. Even where natural gas is normally used, LPG frequently supplies the additional demand in exceptionally cold weather. The major fuel applications of LPG are for domestic heating and cooking, for numerous farming operations, and as a substitute for gasoline in modified **internal-combustion engines.** Although longer chain normal paraffins have poor antiknock properties (see **Gasoline**), propane and butane are good antiknock fuels with high **octane numbers.** Both propane and butane are odorless gases; consequently, traces of a sulfur compound with an offensive smell (e.g., a mercaptan) are added to LPG in order to facilitate leak detection.

Because liquid butane does not vaporize (i.e., revert to gaseous form) readily at low temperatures, the LPG used in cold climates consists mainly of propane. This form of LPG is generally preferred for domestic use because of its more rapid vaporization when the pressure is reduced.

The approximate **heating values** of liquid and gaseous commercial propane and butane are given below. The heating values

Heating Values of LPG

	Liquid				Gas	
	Btu/lb	MJ/kg	Btu/gal	MJ/liter	Btu/cu ft	MJ/cu m
Propane	20,000	46.5	87,000	24	2500	92
Butane	19,000	44	98,000	27	3300	123

of the gases at atmospheric pressure may be compared with approximately 1030 Btu/cu ft (38 MJ/cu m) for natural gas. Because of these differences in heating value per unit volume, and also because LP gases have heavier molecules and escape

more slowly from an orifice, different burner adjustments are required for LPG than for natural gas. When propane is used to supplement natural gas for local distribution, a small amount of air is included to reduce the heating value per unit volume.

LIQUID-METAL FAST BREEDER REACTOR (LMFBR)

A **nuclear power reactor** which has no moderator, so that most of the fissions are caused by fast (i.e., high-energy) neutrons, and uses a liquid metal (sodium) as coolant; its purpose is to breed fissile material (i.e., to produce more than is consumed during operation) while generating useful power (see **Breeder Reactor; Fast Reactor**). As a general rule, the fertile material in an LMFBR is uranium-238 which captures neutrons to produce plutonium-239 and some plutonium-241 (see **Nuclear Fuel**), both of which are fissile.

Breeding of plutonium was first demonstrated in the United States in the Experimental Breeder Reactor No. I (EBR-I), completed in 1951 under the direction of the Argonne National Laboratory. Subsequently, other experimental fast breeder reactors were built, including the EBR-II which started in 1962 and is still operating in the early 1980s. A commercial demonstration LMFBR, the Enrico Fermi reactor, near Monroe, Michigan, commenced operation in 1964, but difficulties experienced indicated the need for further development. Another demonstration power reactor, the Clinch River Breeder Reactor, with an electrical capacity of about 350 megawatts (MW), is to be constructed at Oak Ridge, Tennessee (see **Watt**).

Because of the availability in the United States of uranium for nuclear fuel in nonbreeder reactors (see **Light-Water Reactor**), there has not been an urgent need for breeding. But if nuclear power should be required into the next century, breeding, which uses uranium more efficiently, will be necessary. In other countries, however, where uranium resources are more limited, the LMFBR is being developed more extensively. In France, a power plant with an electrical capacity of 230 MW has been in operation since the end of 1973, and a 1200-MW (electric) project is scheduled for completion in 1983 or 1984. Prototype commercial LMFBRs have been (or are being) built in Japan, the United Kingdom, the U.S.S.R., and West Germany.

General Principles

Core and Blanket. In outline, an LMFBR consists of a central core containing the nuclear fuel, surrounded by a blanket of fertile material (uranium-238). The core is made up of a large number of vertical fuel rods (or "pins") containing both fissile (e.g., plutonium-239) and fertile species. The fissile material undergoes fission, producing heat energy and releasing neutrons (see **Fission Energy**). Some of the neutrons in excess of those required to maintain the fission chain are captured in the fertile material present in the fuel rods; most of those escaping from the fuel are captured either in the external blanket surrounding the core or in rods of fertile material, called the internal blanket, within the core. Fissile plutonium is produced from the fertile uranium-238 in the fuel rods and in both external and internal blankets.

In the interest of safety, the core of an LMFBR has a flattened cylindrical ("pancake") form, with the diameter two or three times the height. If the reactor has an internal blanket, which is also a safety feature, the core is built up of a series of rings. The central region of the core consists of blanket (uranium-238) rods, and this is followed by two to four rings of fuel rods alternating with blanket rods. The external (radial) blanket rods follow the last ring of fuel rods.

Controls. Control of an LMFBR is achieved by means of neutron absorber (poison) rods which are inserted into the core from above to shut the reactor down or removed gradually to start it operating.

The common poison material is boron, as boron carbide (B_4C). Fast neutrons are not readily absorbed, but boron carbide, which can withstand the high temperature in an LMFBR, is the best available control material. Two separate control systems are provided; one is for normal startup, operation, and shutdown, and another is for backup shutdown in an emergency.

Coolant System. Liquid sodium coolant is pumped upward through the core, where most of the fission heat is liberated, and also (at a lower rate) through the blanket where some fissions occur. As a result of neutron capture, the sodium becomes radioactive and emits beta particles and gamma rays (see **Radioactivity**). Consequently, as a protective measure, heat is transferred in an intermediate **heat exchanger** (IHX) from the radioactive reactor (or primary) sodium to an independent, nonradioactive secondary sodium circuit. The primary sodium is pumped back to the reactor, and the hot secondary sodium proceeds to the steam generator. Here superheated steam, to drive a turbine and electrical generator, is produced by transfer of heat from the secondary sodium to pressurized water. The secondary sodium is returned to the IHX for reheating (Fig. 80).

Two different configurations have been used in LMFBRs. In the "pot" or "pool" type, the reactor (core and blanket), IHXs (three or four), and primary coolant pumps are contained in a large tank (or pot) filled with liquid sodium. In the alternative "loop" type, only the reactor and liquid sodium coolant are contained in the reactor vessel as in Fig. 80; each pump and each IHX is in a separate vessel (or cell).

Because sodium has such a high boiling point [1616°F (880°C)] at atmospheric pressure, the sodium tank or other containing vessel does not have to withstand high pressure, thereby simplifying the construction. However, sodium surfaces must have an inert gas (e.g., nitrogen) cover to

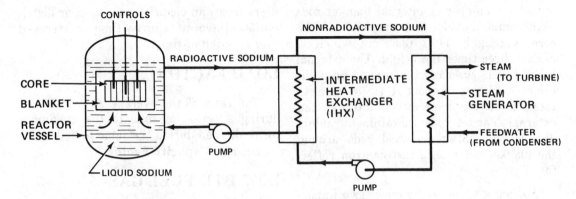

Fig. 80 Liquid-metal fast breeder reactor.

prevent access of air (oxygen) which reacts chemically with sodium.

The main advantage of the pool type of LMFBR is that the large amount of sodium in the tank provides a heat sink to absorb heat from the fuel and thus prevent damage in case of a primary coolant pump failure. Since there is no primary coolant outside the tank, a loss-of-coolant accident, such as is possible in reactors of other types, cannot occur (see **Boiling-Water Reactor**). On the other hand, maintenance and repair of such components as IHXs and pumps are difficult because they are immersed in liquid sodium. This difficulty does not arise in the loop system, but a loss-of-coolant accident could occur.

The pool configuration has been preferred for large LMFBRs in Europe, whereas U. S. designers have tended to favor the loop type. In a proposed modification of the loop system, the reactor vessel would be larger than necessary to accommodate the core and blanket. The vessel would contain enough sodium to provide an adequate heat sink if there should be a break in the primary coolant system.

Fuel Reprocessing. From time to time, the fuel and blanket rods must be removed from the reactor for reprocessing; that is, for the recovery of both plutonium and residual uranium (see **Nuclear Fuel Reprocessing**). As a result of breeding, the fissile material recovered would eventually be sufficient to refuel the breeder reactor

and also provide part of the fuel requirement of another reactor of the same or a different type. Fuel reprocessing is thus an essential aspect of LMFBR operation; without it, the LMFBR would not be economically justifiable.

Design Specifications

Core and Blanket. The following information does not apply to any specific reactor, but it indicates the general characteristics of a large, loop-type LMFBR with an electrical capacity of about 1000 MW. The fuel is a mixture of 20 percent plutonium dioxide and 80 percent uranium dioxide. The latter may be obtained from normal (i.e., natural) uranium or from the depleted "tails" of a uranium-235 enrichment plant (see **Uranium Isotope Enrichment**).

The fuel rods (or pins) consist of small, cylindrical pellets of the mixed oxides packed into long, thin stainless steel tubes, 0.23 to 0.3 in. (6 to 8 mm) in external diameter. The length of the fuel region of the rods, which is also the height of the active core when the rods are assembled vertically, is only 4 ft (1.2 m). The core may include 60,000 closely packed fuel rods with narrow spaces in between for coolant flow.

Above and below the fuel region, the stainless steel tubes contain pellets of normal or depleted uranium dioxide to form 14-in. (36-cm) lengths of upper and lower external blankets. The core also includes a number of internal blanket rods of uranium dioxide. The diameter of the core is roughly 11 ft (3.35 m), which is almost three times the height. The external blanket rods surround the core to an additional thickness of about 1 ft (30 cm). In total, there are some 30,000 internal and external blanket rods containing uranium dioxide only. Stainless steel rods around the blanket serve as a fast-neutron reflector.

Sodium and Steam Systems. Two important characteristics of sodium as a coolant are utilized in the LMFBR. First, because of its excellent heat-transfer properties, liquid sodium can remove fission heat from the closely spaced fuel rods; moreover, there is only a minor temperature penalty in transferring heat from the primary to the secondary sodium in the IHX. Second, the high boiling point of liquid sodium at ordinary pressure permits the attainment of high coolant temperatures without pressurization. The high temperature of the sodium makes possible the production of superheated steam and hence a high **thermal efficiency** for power generation.

In the large LMFBR design being considered, the sodium leaves the reactor and enters the IHX at close to 930°F (500°C). The temperature of the secondary sodium leaving the IHX and entering the steam generator is only about 30°F (17°C) lower at 900°F (483°C). In the steam generator preheated water at a pressure of 150 atm (15 MPa) is converted into superheated steam (i.e., at a temperature above the normal boiling point at the existing pressure) at roughly 850°F (455°C). The expected thermal efficiency for electric power generation by means of a steam turbine is about 38 percent, which is comparable with that of the most efficient fossil-fuel plants (see **Electric Power Generation**).

LOAD

As applied to electrical **power**, the power, usually expressed in kilowatts (see **Watt**), that is delivered by (or consumed by users from) an electric generation or distribution system at a given time or averaged over a period of time.

LOAD FACTOR

The ratio of the average electrical **load** during a period of time to the maximum (or peak) load during the same period.

See also **Capacity Factor**.

LOW-BTU FUEL GAS

A fuel gas with a **heating value** generally in the range of about 120 to 200 Btu/cu ft (4.5 to 7.5 MJ/cu m). Low-Btu gases are related to **producer gas** and are commonly made by the reaction of coal with steam and air at temperatures of 1200 to 1800°F (650 to 980°C) and moderately high pressures. Combustion of the coal in air provides the heat required for the carbon (coal)–steam reaction (see **Coal Gasification**). In a few processes (e.g., **Rockwell Gasification Process**), low-Btu gas is obtained by **pyrolysis** (i.e., heating) and partial oxidation of the coal in air without steam. In all cases, the product gas contains some 40 to 50 volume percent (dry basis) of carbon monoxide and hydrogen, in various proportions, plus a few percent of methane as the fuel constituents; the remaining 50 percent or so is mainly inert nitrogen (from air).

Processes designed primarily (or solely) for the manufacture of low-Btu fuel gas by the action of steam and air on coal are described under the following headings: **BCR TRI-GAS; Combustion Engineering; Foster–Wheeler Coal Gasification;** and **Westinghouse Coal Gasification Processes.** Several of the processes for producing **intermediate-Btu fuel gases** by the interaction of coal with steam and oxygen gas will yield low-Btu gases if the oxygen is replaced by air. Examples are **BI-GAS; HYGAS; Kellogg Molten-Salt; Lurgi; Synthane; U-GAS; Wellman–Galusha;** and **Winkler Processes**.

Most of the sulfur originally present in the coal appears in the raw product gas as hydrogen sulfide which can be readily removed (see **Desulfurization of Fuel Gases**). Because its production does not require oxygen gas, the clean low-Btu gas is cheaper (per heat unit) than gases with higher heating values. Hence, conversion into a low-Btu fuel gas is a possible method for utilizing high-sulfur coal in an economic and environmentally acceptable manner (see **Sulfur Oxides**). However, because of the high transmission costs (per heat unit), low-Btu gas would be used near the point of production, preferably not far from a coal mine.

Low-Btu gas could be burned as a boiler fuel to produce steam (e.g., for an electric power plant). A more attractive alternative is to use the gas as the fuel for a **gas turbine** forming the first stage of a **combined-cycle generation** system.

LURGI DRY-ASH PROCESS

A commercial process developed by Lurgi Mineralöltechnik GmbH in Germany (associated with the American Lurgi Corporation) for **coal gasification** with steam and oxygen gas (or air). When oxygen is used, the product is an **intermediate-Btu fuel gas** which can be converted into a **high-Btu fuel gas** (or **substitute natural gas**), or it can serve as a **synthesis gas**. If the oxygen is replaced by air, the product contains some 50 volume percent of nitrogen and is a **low-Btu fuel gas**. The dry-ash Lurgi gasifier is designed for noncaking coals, but it can be modified for use with caking coals.

The Lurgi process is one of the most highly developed coal gasification methods. More than 50 units are in operation outside the United States. A considerable number of these are used in the Republic of South Africa to produce synthesis gas for conversion of coal into liquid fuels by the **Fischer–Tropsch process**. Lurgi gasifiers are attracting interest in the United States, and some are expected to be in operation in the early 1980s.

The coal, entering the top of the **fixed-bed** gasifier, travels slowly downward interacting with the hot gases rising from below (Fig. 81). Oxygen gas (or air) and steam, introduced at the bottom, flow upward through spaces in a rotating grate. Rotation of the grate facilitates ash removal and favors uniform distribution of oxygen (or air) and steam throughout the coal bed. The ratio of oxygen (or air) to steam fed to the gasifier determines the operating temperature; this is controlled to be high enough to ensure complete gasification of

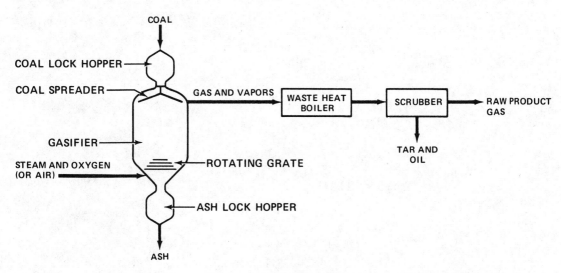

Fig. 81 Lurgi Dry Ash Process.

the coal while not so high as to permit damage to the rotating grate and fusion of the residual coal ash. The dry" ash is then discharged.

Crushed and sized coal is fed to the gasifier by way of a lock hopper and is spread out by a rotating distributor. At the top of the gasifier the entering coal is first preheated and dried by the rising hot gases, and then, as it descends, it is devolatilized (i.e., volatile matter is driven off). Lower in the gasifier, part of the residual **char** interacts with steam and oxygen gas (or air) at a temperature of 1150 to 1500°F (620 to 815°C) and a pressure of 25 to 30 atm (2.5 to 3.0 MPa) to produce carbon monoxide and hydrogen. The remaining char descends into the combustion (lowest) zone of the gasifier where the contained carbon burns in oxygen (or air) at a temperature of 1800 to 2500°F (980 to 1370°C) to generate the heat required to dry and devolatilize the coal and for the carbon–steam reaction.

The hot product gas leaving the gasifier passes through a waste-heat boiler (**heat exchanger**) to generate steam. It is then scrubbed to separate tar, oil, and other impurities and treated for sulfur removal (see **Desulfurization of Fuel Gases**). The composition of the resulting clean gas depends on the nature of the coal and on the process conditions. When oxygen gas is used for combustion, the product might contain roughly 40 volume percent of hydrogen, 20 percent of carbon monoxide, and 10 percent of methane (dry basis) as the fuel constituents; the rest is mainly inert carbon dioxide. The heating value is about 300 Btu/cu ft (11 MJ/cu m). This intermediate-Btu gas can be subjected to the **water-gas shift reaction** and **methanation** to yield a high-Btu fuel gas.

If air is used instead of oxygen in the gasifier, the product contains some 40 percent of inert nitrogen from the air, and the heating value is around 180 Btu/cu ft (6.7 MJ/cu m).

The original Lurgi gasifier was designed for noncaking coals and would not be satisfactory for the caking bituminous coals commonly found in the eastern United States. However, a modification, with rotating arms to break up the coal mass, has been operated successfully with highly caking coal.

See also **British Gas/Lurgi Slagging Gasifier Process.**

M

MAGNETIC-CONFINEMENT FUSION

The use of various magnetic field arrangements for confining high-temperature plasmas to achieve controlled nuclear fusion (see **Fusion Energy**). In a uniform magnetic field, electrically charged particles are forced to follow helical (i.e., corkscrew) paths along the field lines. (The method used for generating the magnetic fields is described below.) In a sense, charged particles are "tied" to the magnetic field lines. Since a high-temperature plasma consists almost entirely of charged particles (atomic nuclei and electrons), it should be confined by a magnetic field.

The maximum density of a plasma that can be confined at a given temperature depends on the magnetic field strength. There is a practical limit to the field strength and hence a corresponding maximum for the confined plasma density. At a temperature of 100 million °C, required for significant fusion of deuterium and tritium (D–T) nuclei, this maximum, n, would be in the range of 10^{14} to 10^{15} nuclei/cm^3. Hence, in order to satisfy the Lawson criterion for energy break-even (i.e., $n\tau$ greater than about 10^{14} sec/cm^3), the confinement time, τ, must be at least 1 to 0.1 sec, respectively. (For a discussion of D–T fusion, the confinement parameter, and the Lawson criterion, see the article on **Fusion Energy**.)

Although the confinement time appears short, it has been difficult to attain because several factors contribute to the escape of plasmas from magnetic fields. For example, collisions between charged particles moving along the field lines can cause the particles to move across the lines and escape from confinement. Furthermore, a plasma in a magnetic field has a tendency to become unstable and leak out of the confining field. Plasma instabilities are generally classified as large-scale (or macro) instabilities and small-scale (or micro) instabilities.

Macroinstabilities can cause the plasma as a whole to break up and escape from its confining magnetic field. *Microinstabilities*, on the other hand, lead to gradual leakage of the plasma across the field lines. Research on the magnetic confinement of high-temperature plasmas was initiated independently in the United States, the United Kingdom, and the U.S.S.R. around 1951, and much of the work done since has been concerned with understanding and overcoming plasma instabilities.

Plasma Heating

Ohmic Heating. Four main methods are available for heating a magnetically confined plasma to the high temperatures required for nuclear fusion. The simplest, called *ohmic* (or *resistance*) *heating*, is to pass an electric current through the plasma; the same principle is used in an ordinary electric heater. To heat a plasma, the current is generally induced from outside, as explained later, to avoid the need for inserting electrical connections in the containing vessel. The heating rate depends on the resistance of the plasma and the square of the current strength. Since the resistance decreases with increasing temperature, and because instabilities may develop, the maximum plasma temperature that can be attained by ohmic heating is commonly around 20 million °C.

Neutral-Beam Injection. Further heating of a magnetically confined plasma can be achieved by *neutral-beam injection.* This depends on two factors: (1) positively charged atomic nuclei (or ions) can be accelerated to high velocities (i.e., high energies) by means of a high voltage, but they cannot easily penetrate a magnetic field, and (2) neutral (i.e., uncharged) atoms, on the other hand, can readily penetrate a magnetic field but they cannot be accelerated directly. Plasma heating by neutral-beam injection, used specifically for deuterium, is conducted in the following manner.

Deuterium ions are generated from deuterium gas by an electrical discharge in an ion source; the ions are then accelerated by a high voltage (e.g., more than 100,000 volts). The accelerated, high-energy deuterium ions pass through a chamber containing neutral deuterium gas; here, electrons transfer from low-energy deuterium (D^0) atoms to high-energy ions (D^+) with the resulting formation of low-energy deuterium ions and high-energy atoms; thus,

$$D^0 \text{ (low energy)} + D^+ \text{ (high energy)} \rightarrow$$

electron

$$D^+ \text{ (low energy)} + D^0 \text{ (high energy)}$$

The low-energy D^+ ions are diverted and removed by a subsidiary magnetic field, while the high-energy D^0 atoms are injected into a magnetically confined plasma.

In the plasma, electrons are rapidly removed from the atoms to form high-energy ions which are trapped by the magnetic field and cannot escape. The high-energy ions then transfer part of their energy to the plasma particles in repeated collisions, thereby increasing the plasma temperature. An ion temperature of 75 million °C was reached in a deuterium plasma

in 1978 by neutral-beam injection, and it is expected that temperatures of 100 million °C will be attainable.

Radiofrequency Heating. Radiofrequency waves in the microwave and radar ranges, generated outside a magnetically confined plasma, can heat the plasma to very high temperatures. If the waves have particular frequencies (or wavelengths), part of their energy is transferred to the ions or electrons (or both) in the plasma. These high-energy particles then collide with other particles and thus increase the plasma temperature.

Compression Heating. A plasma, like a gas, can be heated by compression. If the plasma is confined by a magnetic field, the compression can be achieved readily by a sudden increase in the confining field strength. In addition to increasing the temperature, compression increases the plasma density and thus contributes to the confinement parameter $n\tau$.

Magnetic Fields

An electric current is always associated with a magnetic field such that the field lines (or lines of force) are perpendicular to the direction of current flow. The gases (e.g., deuterium and tritium) required for fusion are often contained in a hollow, doughnut-shaped chamber called a *torus.* Magnetic fields can be generated within a torus in two general ways.

In one case, an electric current is passed through a number of rings (or coils) surrounding the torus, as in Fig. 82A. The associated magnetic field, called a *toroidal field,* has its lines of force running around the long way of the torus. Alternatively, the current may be made to flow through the plasma, as in Fig. 82B; the lines of the resulting magnetic field, called a *poloidal field,* encircle the plasma in the polar (shorter) direction. (Incidentally, whenever a plasma is heated by passing a current through it, as described above, a poloidal field is produced.)

To generate a toroidal magnetic field, the rings, which are outside the toroidal chamber, are actually coils of many turns of insulated wire. Direct current of the same strength is then passed through all the coils. In order to avoid inserting electrical connections into the plasma, current within a plasma, as is required for a poloidal field (and ohmic heating), is produced by induction. In simple terms, an electric current which *varies with time* is passed through a conductor outside and parallel to the torus; this is referred to as the primary current. A similar secondary current is then induced in the plasma (see **Transformer**).

The time-varying primary current is usually a pulse which increases rapidly to a maximum and then decreases slowly to zero; the secondary current changes correspondingly. An important consequence is that a poloidal field consists of a succession of pulses. A toroidal field, on the other hand, can be a steady field.

In some nuclear fusion studies, the plasma is contained in a straight tube. An *axial field*, equivalent to a toroidal field with its lines parallel to the tube axis, is then produced by electric currents in coils that encircle the tube in a manner similar to Fig. 82A. The equivalent of a poloidal field in a straight tube, sometimes referred to as an *azimuthal field*, is now rarely used.

Plasma Studies

Several magnetic field arrangements proposed for plasma confinement have been discarded, mainly because of difficulty in controlling plasma instabilities. Others are still being investigated, and a few that appear promising in the early 1980s will be described. There is no assurance, however, that any will provide the basis for a practical nuclear fusion reactor. (The term "reactor" is commonly applied to a device capable of generating significant amounts of power by nuclear fission or fusion reactions.)

Because tritium is radioactive and requires special handling facilities, experiments with plasmas are made with nonradioactive hydrogen isotopes, either ordinary (light) hydrogen or deuterium. Apart from fusion reactions, which occur less readily or not at all, the general behavior is expected to be the same as when tritium is present. At a later stage of development, tritium will be introduced in order to study deuterium–tritium (D–T) fusion. This process will be used in the first controlled fusion reactors.

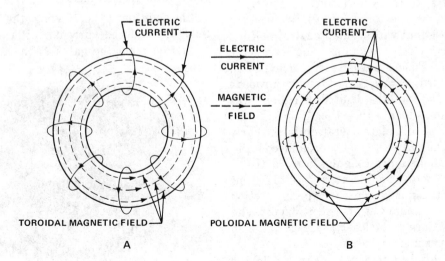

Fig. 82   Magnetic field. A: Toroidal; B: Poloidal.

The behavior of a magnetically confined, high-temperature plasma is followed by determining such properties as temperature, pressure, electron and ion densities and energies, magnetic field and current distributions, and extent of fusion reactions. These determinations are based on measurements of various radiations and particles emitted from the plasma and of magnetic and electric fields. In particular, the number and energy of neutrons produced provide an indication of the nature and extent of fusion reactions, since both D–D and D–T reactions are accompanied by the emission of neutrons.

Tokamak Systems

The tokamak principle of magnetic confinement in a torus was developed in the U.S.S.R. in the 1960s. (The name "tokamak" is derived from the initial letters of the Russian words meaning "toroidal," "chamber," and "magnetic," respectively.) After 1968, when convincing evidence of the tokamak confinement was reported, many experimental devices were built in the United States and elsewhere. As experience has been gained with tokamak operation, larger and larger machines have been constructed to increase the plasma confinement time. In particular, plasma stability is improved if the height of the torus is large relative to its horizontal diameter. In other words the torus has the shape of a "fat" doughnut with a small central hole.

Four large tokamak devices, the Tokamak Fusion Test Reactor (TFTR) in the United States, the Joint European Torus (JET) in the United Kingdom, the JT-60 in Japan, and the T-15 in the U.S.S.R., should start operating in the early 1980s. The stainless steel toroidal containment chambers are roughly 25 to 33 ft (7.5 to 10 m) across and 6.5 to 8 ft (2 to 2.5 m) high. A tokamak reactor for the production of useful amounts of energy is expected to be much larger, at least twice as large in each dimension.

Basically, the tokamak magnetic field is a combination of a strong, steady (for a time) toroidal field and a much weaker, pulsed poloidal field. In a purely toroidal magnetic field, as in Fig. 82A, the field lines are closed circles with different radii; as a consequence, the plasma drifts outward in the torus. By superimposing a weak poloidal field, the toroidal field lines acquire a slight twist so that they do not close on themselves. The twist in the field largely prevents the drift, and there is a marked increase in the plasma stability and confinement time.

The toroidal field component is produced by an electric current passing through coils which encircle the torus in a vertical direction. Eventually, a constant direct current will probably be used to maintain a steady toroidal field, but in experimental studies a short-duration field is adequate.

There are various ways of inducing the plasma current for generating the poloidal field. One procedure is to apply a pulse of primary current to horizontal coils located within the central hole of the "doughnut" (torus) and around it, as shown in the cutaway illustration in Fig. 83. (The torus itself is not shown in the figure.) The increase and decrease in the primary currents induce secondary currents within the torus, and these generate the desired poloidal magnetic field.

In addition to producing the poloidal field, the induced currents initiate plasma formation from the gases in the torus and

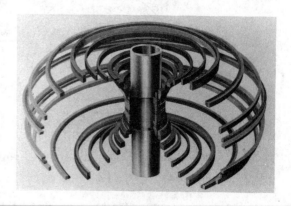

Fig. 83   Poloidal field coils in a tokamak.

increase the temperature by ohmic heating. The attainable temperature is limited by the development of large-scale instabilities to about 20 million °C. Neutral-beam injection is favored for further heating, but radiofrequency heating is also being considered.

Tokamak Reactor. By the mid-1980s it should be known whether tokamak confinement can form the basis of a reactor for the net generation of nuclear fusion energy. The essential requirements are the attainment of a temperature of about 100 million °C in a D–T plasma and a confinement parameter $n\tau$ exceeding the 3×10^{14} sec/cm^3 necessary for energy break-even and ignition. If these conditions can be realized, a tokamak fusion reactor would probably operate as follows.

A D–T plasma, formed and heated by a pulse of induced current, is confined in a torus by a combination of toroidal and poloidal fields. High-energy neutral deuterium beams are then injected until D–T ignition is initiated. The 20 percent of the fusion energy carried by the helium nuclei remains in the plasma and serves to maintain the fusion reaction temperature without further neutral-beam heating. During this stage, when fusion energy is being produced, the deuterium and tritium are replenished by injecting solid pellets of these substances into the plasma.

After some 10 to 20 minutes of operation, the pulse of induced plasma current will have dropped to zero; the poloidal field will then cease to exist, and stable plasma confinement is no longer possible. Residual helium gas and impurities are removed from the tokamak chamber, and the cycle is repeated, starting with the formation and heating of the D–T plasma by an induced current pulse followed by neutral-beam injection.

The neutrons produced in the D–T reaction are electrically neutral and are not confined by the magnetic field; they consequently escape, carrying 80 percent of the fusion energy. The toroidal plasma chamber is surrounded by a "blanket" containing lithium, either as liquid metal or a molten salt, and the escaping high-energy neutrons enter the blanket. Here they react with lithium nuclei to produce the tritium required for the fusion process.

At the same time, the energy of the neutrons, together with the energy released in their reaction with lithium, is deposited in the blanket as heat. The heat can then be used to produce high-temperature steam (or other vapor) to drive a turbine for generating electric power (see **Electric Power Generation**). Part of this power is required to produce the tokamak magnetic fields and to operate the neutral-beam injectors and other equipment; the remainder is then available for sale.

A practical fusion reactor will undoubtedly use superconducting coils to generate the magnetic fields. When certain metals and alloys are cooled to extremely low temperatures, usually below −441°F (−263°C), they become superconductors. Once a flow of current has been started in a superconductor, it will continue indefinitely in a closed circuit after the current source is removed. Although superconducting magnets, cooled by liquid helium, have been built, they are smaller than those that will be required for fusion reactors (see **Superconductivity**).

The Toroidal (or Z) Reversed-Field Pinch

The toroidal pinch, commonly called the Z pinch, was proposed as a means of plasma confinement and heating as far back as 1951. In simple terms, the idea was that a strong poloidal magnetic field, produced by a powerful current induced in a plasma, as shown in Fig. 82B, would compress (or pinch) and confine the plasma in a torus. (The name Z pinch arises from the flow of current in the z direction—that is, the long way around the torus.)

Pinching does occur but the plasma is highly unstable and confinement is extremely brief. Various efforts to improve stability were only partially successful and

Z-pinch projects were largely abandoned. However, with increased understanding of plasma behavior, studies of a modified toroidal Z pinch were revived in the 1970s, and the results have been encouraging.

The modified (reversed-field) Z pinch is similar to the tokamak in using combined toroidal and poloidal magnetic fields for plasma confinement. However, in the tokamak the toroidal field is much stronger than the poloidal field, whereas in the Z pinch the field strengths are approximately equal. Another difference is that in the Z pinch, the induced plasma current, and hence the poloidal magnetic field, is increased very rapidly. The fast-rising field causes a shock wave to move inward through the plasma.

The Z-pinch plasma is heated almost simultaneously in three ways: (1) by the shock wave produced by the rapidly increasing poloidal field, (2) by compression (or pinching) caused by the strong poloidal field, and (3) by ohmic heating due to the induced plasma current. Supplementary heating, by neutral-beam injection or radiofrequency waves, as required in the tokamak, may then not be necessary.

Soon after the induced plasma current reaches its maximum value, the current in the toroidal field coils is reversed. A toroidal field is thus produced outside the pinched plasma in the opposite direction to the initial toroidal field trapped within the plasma. (Under suitable conditions, the outer axial magnetic field can reverse spontaneously without reversing the current.) The combination of reversed toroidal fields and strong poloidal field confers large-scale stability on the pinched plasma. The presence of a metal conducting wall around the plasma also contributes to the stability.

The toroidal chamber was initially made of a ceramic insulating material (e.g., alumina), surrounded by a metal (e.g., aluminum) torus to serve as the stabilizing conducting wall. Subsequently it appeared that longer confinement times could be achieved if the plasma were contained in a metal toroidal chamber.

A nuclear fusion reactor based on the Z pinch would have a lithium-containing blanket and superconducting coils and would operate in pulses like a tokamak reactor, as described earlier. But the Z-pinch system has some potential advantages. For example, it is expected that a given toroidal magnetic field will be able to confine a plasma of higher density. The same power could then be generated in a smaller device. Moreover, operation should be more convenient with the Z pinch because the toroidal chamber can have a larger central hole than in an equivalent tokamak.

Open-Ended Magnetic Mirror Systems

Plasma confinement by magnetic mirrors is based on the principle that charged particles spiraling along magnetic field lines tend to be repelled, and their direction of motion reversed, when they enter a stronger field. This reversal of direction (or reflection) of the particles leads to the name *magnetic mirror* for the region where the field is stronger.

A simple magnetic mirror system for plasma confinement would consist of closely spaced coils at each end of a straight open-ended tube (Fig. 84A). Passage of an electric current through the coils results in a magnetic field which is

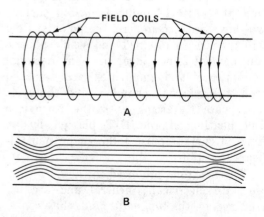

Fig. 84   Magnetic mirrors. A: Field coils; B: Magnetic field.

stronger at the ends. The ends, where the lines of force are closer together, then constitute the mirror regions (Fig. 84B). If a plasma is present between the mirrors, the charged particles may be reflected back and forth many times from one mirror to the other. If this occurs, the plasma is confined.

However, some loss of plasma cannot be avoided. If a particle has a sufficient component of motion parallel to the field lines, it will pass through the mirrors and escape from confinement. Since the directions of motion of the particles are continually changing as a result of collisions, the escape is continuous. Nevertheless, confinement of plasma between two magnetic mirrors for short periods has been demonstrated. Means for improving confinement are described below.

Plasma Heating. Ohmic (or resistance) heating of a plasma in a mirror field does not seem to be practical, and so other means are used. One promising method is to inject or form a low-density, low-temperature "seed" plasma between the mirrors and then to increase the density and temperature by the injection of first intermediate-energy and then high-energy neutral beams.

The seed plasma can be made outside the confinement region by means of an electrical discharge in a device called a plasma "gun." The low-temperature plasma is then introduced through the mirror ends by injection along the direction of the field lines. Alternatively, a seed plasma may be formed between the mirrors by heating a pellet of solid deuterium by means of a laser beam.

Magnetic Mirror Problems. Two major problems in magnetic mirror confinement are plasma instabilities and loss of plasma by escape through the ends. Theoretical studies have indicated that large-scale instabilities can be controlled if the confining field lines always bulge into the plasma. Examination of Fig. 84B shows that in the region between the mirrors the

field bulges outward; the confined plasma is then not stable. For stable confinement, the magnetic field must be a minimum in the inner region where the plasma is confined and increase outward in all directions. This is often referred to as a *magnetic-well configuration.*

One way to produce such a configuration is by the so-called yin-yang system. (The name arises from the symbols for the Yin and Yang principles of Confucianism.) This consists of two almost semicircular coils at right angles (Fig. 85). An electric

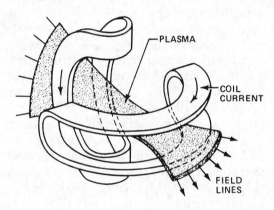

Fig. 85   Yin-yang coils.

current passing through the coils produces a magnetic field having fan-shaped ends which, like the coils, are at right angles. These ends constitute the two mirror regions. There is a central region between the mirrors where the field is a minimum and where stable plasma confinement is possible. The magnetic field strength increases outward in all directions from this region, as required for a stable (magnetic-well) configuration.

Although a plasma confined between magnetic mirrors may be stabilized in this manner, escape through the ends is unaffected. The losses may be decreased by increasing the field strength in the mirror region relative to that between the mirrors. There is, however, a practical limit to this increase. Losses may also be reduced by an increase in the plasma temperature so that collisions, which can lead to the escape of particles along the field lines, are decreased. This is one reason why operating temperatures in a nuclear fusion reactor using open-ended magnetic mirror confinement would probably be higher than in a toroidal system (e.g., tokamak or Z pinch).

Tandem Mirror System. The purpose of the tandem mirror system is to reduce the end losses from a confined plasma. In a simple, open-ended arrangement, there is a single mirror at each end with plasma in between. In the tandem mirror scheme, however, each single mirror is replaced by a fairly closely spaced mirror pair (or tandem mirror), such as could be generated by yin-yang (or similar) coils or in other ways.

An experimental tandem mirror system (of the early 1980s) consists of a wide, open-ended cylinder with a relatively long [18 ft (5.5 m)] central region where the fusion plasma is confined by a uniform, axial magnetic field. At each end of the uniform field region is a magnetic-well mirror pair, called an end cell, with 5 ft (1.5 m) between the mirrors. High-density, high-temperature plasmas are built up in the end cells by neutral-beam injection, possibly in conjunction with radiofrequency heating. A less dense plasma in the central region is heated by high-energy particles escaping from the end cells, so that fusion occurs.

It is improbable that the ignition condition, when the fusion reaction once initiated will continue without further heating, will be realized in any open-ended mirror arrangement. The tandem mirror is thus expected to be a driven system (see **Fusion Energy**). Energy supplied continuously (e.g., by high-energy, neutral-beam injection) to the end cells is amplified (or driven) by the fusion reaction. As a result, more energy should be produced than is required to sustain the magnetic fields, beam injectors, etc. Since ignition is not required, the confinement parameter $n\tau$ for energy break-even can be smaller than in a tokamak system at the same temperature.

Unlike tokamak and Z-pinch fusion reactors, which would operate in pulses, a tandem mirror reactor would operate continuously. There are no induced currents and the magnetic fields would be generated by steady currents, probably in superconducting coils. The reaction region between the end cells would be surrounded by a blanket containing lithium. Neutrons, carrying 80 percent of the fusion energy, would deposit their energy as heat in the blanket and also produce the tritium required for the D–T reaction.

Additional energy could be recovered from the high-energy charged particles that inevitably escape from the outer ends of the tandem mirrors. By means of a combination of magnetic and electrical fields, the motion energy of these particles could be converted directly into electrical energy.

Toroidally Linked Mirrors (Bumpy Torus)

If a chain of simple magnetic mirrors of the type shown in Fig. 84 are linked together so as to form a closed torus, particles escaping from any mirror field would simply enter an adjacent mirror field. Loss of plasma by motion of the particles along the field lines would thus be eliminated. Such an arrangement is called a "bumpy torus," because the field lines form a series of bumps (or swellings) between the mirrors, as shown in Fig. 86. However, plasma confined in a bumpy torus field is subject to large-scale instability.

A marked improvement in stability is achieved by surrounding each bump region with a microwave cavity. If the cavities are supplied with microwave energy at an appropriate frequency, a series of high-energy electron rings are formed around the plasma. The electron rings modify the bumpy field so as to produce magnetic wells in which the plasma has large-scale stability. Furthermore, collisions of the high-energy electrons with the plasma particles serve to heat the plasma. If additional heating is required, it could be provided by microwaves of higher frequency

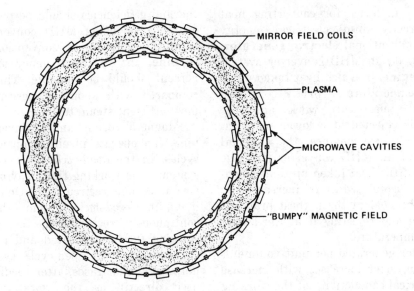

Fig. 86   Linked magnetic fields in the "bumpy torus."

(preferably) or by high-energy, neutral-beam injection.

A nuclear fusion reactor utilizing the bumpy torus principle would be similar to other toroidal reactors in having a lithium-containing blanket and super-conducting field coils. Although ignition conditions, in which the fusion reaction is sustained without additional heating, should be attainable, microwave energy would have to be supplied to maintain plasma stability. An important difference between the bumpy torus reactor and other toroidal (tokamak and Z-pinch) systems is that, since there are no induced currents in the former, it could operate continuously rather than in pulses.

In some respects a bumpy torus reactor would be more like a Z-pinch than a tokamak reactor. For the same power output, the magnetic field strength and plasma volume would be smaller than for a tokamak, and the central hole of the torus would be larger in proportion to the overall size.

MAGNETOHYDRODYNAMIC (MHD) CONVERSION

The direct conversion of kinetic (or motion) energy into electrical energy by the flow of an electrically conducting fluid, usually a gas or a gas–liquid combination, through a stationary magnetic field. If the flow direction is at right angles to the magnetic field direction, an electromotive force (or electrical voltage) is induced in the direction at right angles to both flow and field directions, as depicted in Fig. 87. (Note that the 90° angles appear to be distorted because a three-dimensional situation is represented in a two-dimensional figure.) This is the basic principle of magnetohydrodynamic (or MHD) conversion.

In a practical MHD converter (or generator), the energy of motion of the conducting fluid is derived from heat obtained by burning a **fossil fuel.** Hence, an MHD

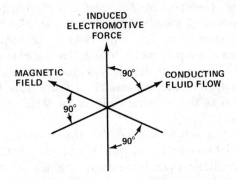

Fig. 87   Principle of magnetohydrodynamic conversion.

generator is a device for converting heat energy directly into electrical energy without a conventional **electric generator.** In this respect, an MHD converter system is a **heat engine,** in which heat taken up at a higher temperature is partly converted into useful (electrical) work and the remainder is rejected at a lower temperature. Like all heat engines, the **thermal efficiency** of an MHD converter (i.e., the proportion of the heat taken up that is converted into useful work) is increased by supplying the heat at the highest practical temperature and rejecting it at the lowest practical temperature.

The power generated per unit volume of an MHD converter increases with increase in the electrical conductivity of the working fluid, the velocity of flow of the fluid, and the magnetic field strength. The conditions are usually such that the conductivity is relatively low; consequently, MHD generators are designed to operate with high flow velocities and magnetic field strengths.

General Principles

In MHD schemes under study, the working fluid usually consists of a carrier gas which is itself a nonconductor; it is rendered electrically conducting in two different ways. One way is to inject a solid "seed" material into the gas, up to about 1 percent of the total flow rate. The seed contains an element, commonly potassium, which ionizes when heated; that is to say, the atoms of the seed element split off electrons (see **Atom; Thermionic Conversion**). The presence of the negatively charged electrons makes the carrier gas an electrical conductor. The other way is to incorporate a liquid metal into a flowing carrier gas. Since the metal is a good electrical conductor, the gas–metal mixture can be used as the working fluid in an MHD generator.

The major reason for the interest in MHD conversion is that it can take better advantage than other heat engines of the high temperatures attained in the combustion of a fossil fuel. As a result, higher thermal efficiencies should be possible. For example, by using MHD conversion as a **topping cycle** for a conventional **steam turbine,** an overall efficiency of 50 to 60 percent should be possible. This may be compared with about 40 percent for the most efficient steam turbines.

Magnetohydrodynamic conversion systems can operate in either open or closed cycles. In the *open-cycle* (or once-through) systems, the working fluid leaving the converter is not recirculated. On the other hand, in *closed-cycle* systems the fluid is continuously recirculated; the discharged working fluid is reheated and returned to the converter. In open-cycle systems, the hot combustion gases, after seeding, can be used directly as the working fluid; in closed-cycle systems, however, heat is transferred from the combustion gases to the working fluid by means of a **heat exchanger.** A higher working temperature and a better thermal efficiency are thus possible in open cycles, provided suitable construction materials are available (see later).

Open-Cycle Systems

Most MHD studies have been made with open-cycle systems. The carrier gas is obtained by burning a fossil fuel, preferably coal, in a combustion chamber (e.g., by **fluidized-bed combustion**). The seed material, generally potassium carbonate, is injected into the chamber; the potassium is then ionized by the hot combustion gases at temperatures of roughly 4200 to 4900°F (2300 to 2700°C).

To attain such high temperatures, the compressed air used to burn the coal (or other fuel) in the combustion chamber must be preheated to at least 2000°F (1100°C). A lower preheat temperature would be adequate if the air were enriched in oxygen. An alternative is to use compressed oxygen alone for combustion of the fuel; little or no preheating is then required. The additional cost of the oxygen might be balanced by the savings on the preheater.

The hot, pressurized working fluid leaving the combustor flows through a convergent–divergent (de Laval) nozzle similar to a rocket nozzle. In passing through the nozzle, the random motion energy of the molecules in the hot gas is largely converted into directed, mass motion energy. Thus, the gas emerges from the nozzle and enters the MHD generator unit at a high velocity.

The MHD generator is a divergent channel (or duct) made of a heat-resistant alloy (e.g., Inconel) with external water cooling. The magnetic field direction, which is at right angles to the fluid flow, would be perpendicular to the plane of the paper in Fig. 88. A number of oppositely located electrode pairs are inserted in the channel to conduct the electric current generated to an external load. The electrode pairs may be connected in various ways (see below), one of which is shown in Fig. 88. An MHD generator, unlike a conventional generator, produces direct current; this can be converted into the commonly used alternating current by means of an inverter.

The arrangement of the electrode connections is determined by the need to reduce losses arising from the *Hall effect.* By this effect, the magnetic field acts on the MHD-generated (or Faraday) current and produces a voltage in the flow direction of the working fluid, rather than at right angles to it. The resulting current in an external load is then called the Hall current.

Various electrode connection schemes have been proposed to utilize the Faraday current while minimizing the Hall current. A simple way, although not the best, is shown in Fig. 88. A better, but more complicated, alternative is to connect each electrode pair across a separate load, as in Fig. 89. Another possibility is to utilize the Hall current only; each electrode pair is short-circuited outside the generator, and the load is connected between the electrodes at the two ends of the MHD generator (Fig. 90).

As the working fluid travels along the MHD generator and its energy is converted into electricity, its temperature falls. When the gas temperature reaches about 3450°F (1900°C), the extent of ionization of the potassium is insufficient to maintain an adequate electrical conductivity. This places a lower limit on the useful operating temperature of the MHD system. The large residual heat available from the hot discharged working gas can then be utilized in several ways. For example, it can serve to preheat the combustion air by way of a heat exchanger similar to the regenerator in a gas turbine (see Fig. 55).

At this stage, some 25 to 30 percent of the heat energy in the working gas should have been converted into electrical energy. The still hot gas leaving the air preheater

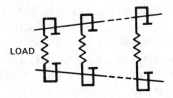

Fig. 89 MHD electrode connections to minimize Hall current.

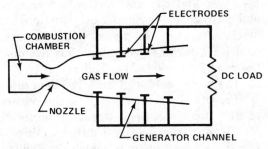

Fig. 88 Simple MHD generator. (Magnetic field, perpendicular to the plane of the page, is not shown.)

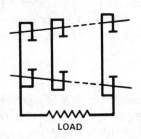

Fig. 90 Use of Hall current in MHD generator.

would be used in a waste-heat (**heat exchanger**) boiler to produce steam for operating a turbine–generator. In this way, another 25 to 30 percent of the initial heat energy should be recovered as electrical energy in a **combined-cycle system.**

Prior to the discharge of the working gas (as flue gas) from the steam boiler to the atmosphere, the fly ash from the coal fuel must be removed (see **Steam Generation**). However, instead of discharging the fly ash, as is usually done, it may have to be treated for recovery of the seed material which is mixed with the ash. Unless the sulfur in the coal has been removed (e.g., in a fluidized-bed combustor), the original potassium carbonate seed will have been converted into potassium sulfate. This must be extracted from the fly ash and reconverted by chemical reactions into potassium carbonate.

The removal of residual sulfur by the potassium carbonate seed eliminates the need for desulfurization of the flue gas, but **nitrogen oxides** are not removed. When oxygen alone is used for combustion of the coal (or other fossil fuel), the problem of nitrogen oxide formation does not arise. However, if nitrogen (from air) is present, the nitrogen oxide content of the combustion gases will be high because of the required high temperature of the working fluid. Consequently, a controlled combustion procedure is used to reduce the nitrogen oxide level in the discharged flue gas. The air supplied to the combustion chamber is not sufficient to permit complete fuel burning; combustion of the unburned fuel gases is then completed by introducing additional air at a later stage, beyond the MHD generator. The lower combustion temperature is accompanied by a decrease in the nitrogen oxide concentration.

MHD Developments. Since the 1950s, several experimental MHD conversion systems have been operated in the United States and in other countries, especially the U.S.S.R. The experiments have demon-strated the scientific feasibility of MHD generation, but the efficiencies attained so far have been relatively low and the life of the equipment has been short. Some of the matters requiring resolution before MHD generation can become economically practical are outlined below.

The combustor, MHD-generator channel, electrodes, and air preheater are exposed to corrosive combustion gases at very high temperatures; materials must be developed to permit an adequate operating life for the components. The ash (or slag) residue from the burning coal is carried over with the combustion gases and tends to cause erosion of exposed surfaces. However, deposition of the slag on such surfaces may provide some protection. Another problem is separation of the seed material (as potassium sulfate) from the fly ash and its reconversion into its original (carbonate) form.

The difficulties associated with slag and seed recovery can be eliminated by using a fuel gas derived from coal, rather than coal itself, in the combustor. An ash-free, **low-Btu fuel gas,** made from coal at a moderate cost and treated for sulfur removal, would make a suitable fuel for an MHD conversion combustor. Burning of the gas in preheated air should provide adequate working fluid temperatures.

A more advanced concept is to use hydrogen gas made from coal and water (see **Hydrogen Production.**). When this is burned in (compressed) oxygen, the product would be high-temperature steam. (Water, chemical formula H_2O, is a compound of hydrogen and oxygen only.) After seeding and passage as working fluid through the MHD generator, the steam would be used to drive a turbine–generator, thus avoiding the need for a waste-heat boiler.

The power output of an MHD generator is theoretically proportional to the square of the magnetic field strength; hence, a strong magnetic field is desirable. Conventional electromagnets, in which the field is generated by direct current passing through water-cooled copper coils, have been used in MHD studies. Such magnets would not be practical for large-scale MHD power generation because they would require very large currents. Consequently, superconducting magnets, which use very little current, will be necessary (see **Superconductivity**). Magnets of this type are being developed for use in MHD generators.

Closed-Cycle Systems

Two general types of closed-cycle MHD generators are being investigated. In one type, electrical conductivity is maintained in the working fluid by ionization of a seed material, as in open-cycle systems; and, in the other, a liquid metal provides the conductivity. The carrier is usually a chemically inert gas, although a liquid carrier has been used with a liquid metal conductor. The working fluid is circulated in a closed loop and is heated by the combustion gases using a heat exchanger. Hence, the heat source and the working fluid are independent.

In a closed-cycle system, the carrier gas operates in a form of **Brayton cycle.** The gas is compressed and heat is supplied by the source, at essentially constant pressure; the compressed gas then expands in the MHD generator, and its pressure and temperature fall. After leaving the generator, heat is removed from the gas by a cooler; this is the heat rejection stage of the cycle. Finally, the gas is recompressed and returned for reheating.

Nonequilibrium Ionization Systems. In these systems, the carrier is a monatomic gas (i.e., a gas with only one atom in the molecule); this is usually the chemically inert gas argon, although helium is a possible alternative. The seed is cesium metal, which volatilizes and ionizes more readily than potassium in potassium carbonate. Cesium is more expensive than potassium, but in a closed system the losses should be small. In the monatomic gas carrier, nonequilibrium ionization can occur; the extent of ionization then corresponds to that which would normally exist at a much

higher temperature (e.g., as in combustion gases). Adequate electrical conductivity for MHD applications can thus be attained in a monatomic gas with a cesium seed at temperatures below 3600°F (2000°C).

In a nonequilibrium ionization MHD converter, the carrier gas is compressed and then heated by combustion gases in a (primary) heat exchanger; the seed is injected into the hot gas. The resulting working fluid is introduced into the generator at high speed in the manner already described. Upon leaving the MHD generator, the still hot fluid enters a (secondary) heat exchanger, which serves as a waste-heat boiler to generate steam (Fig. 91). The steam is utilized to drive a turbine–generator. The working fluid leaving the heat exchanger is cooled, and the gas is recompressed to complete the cycle. The seed is condensed and separated for reuse.

Because the combustion system is separate from the working fluid, so also are the ash and flue gases. Hence, the problem of extracting the seed material from fly ash does not arise. The flue gases are used to preheat the incoming combustion air and then treated for fly ash and sulfur dioxide removal, if necessary, prior to discharge through a stack to the atmosphere (see **Steam Generation**).

The somewhat lower operating temperatures of a closed-cycle MHD converter than of an open-cycle system have an advantage in permitting a wider choice of materials. On the other hand, the lower temperature of the working fluid also means a lower thermal efficiency. Furthermore, temperatures in the combustion chamber are still high, and special construction materials are required for the primary heat exchanger.

Liquid-Metal Systems. When a liquid metal provides the electrical conductivity, an inert gas (e.g., argon or helium) is a convenient carrier. The carrier gas is pressurized and heated by passage through a (primary) heat exchanger within the combustion chamber. The hot gas is then incorporated into the liquid metal, usually

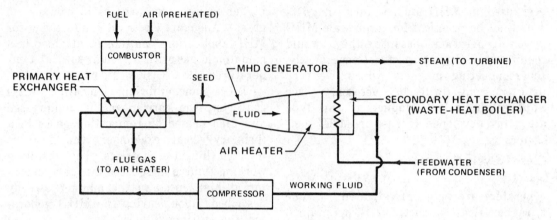

Fig. 91   Closed-cycle MHD generator.

hot sodium, to form the working fluid. The latter then consists of gas bubbles uniformly dispersed in an approximately equal volume of liquid sodium.

The working fluid is introduced into the MHD generator through a nozzle in the usual way; the carrier gas then provides the required high directed velocity of the electrical conductor (i.e., the liquid metal). After passage through the generator, the liquid metal is separated from the carrier gas. Part of the heat remaining in the gas is transferred to water in a (secondary) heat exchanger to produce steam for operating a turbine-generator. Finally, the carrier gas is cooled, compressed, and returned to the combustion chamber for reheating and mixing with the recovered liquid metal.

The electrical conductivity of the liquid metal (sodium) is not greatly affected by temperature; the initial temperature of the working fluid is then determined only by the need to avoid boiling of the liquid sodium. The boiling point of sodium, even under moderate pressure, is below 1650°F (900°C); hence, the working fluid temperature is usually around 1500°F (815°C).

The lower operating temperature than in other MHD conversion systems may be advantageous from the materials standpoint, but the maximum thermal efficiency is lower. A possible compromise might be to use liquid lithium, with a boiling point near 2370°F (1300°C), as the electrical con-ductor. Lithium is much more expensive than sodium, but losses in a closed system should be small.

MDEA PROCESS

A process for removing hydrogen sulfide (and to a lesser extent carbon dioxide) from fuel gases by chemical absorption in a weakly alkaline medium. The absorber is an aqueous solution of methyl diethanolamine (MDEA) at a temperature of 80 to 120°F (27 to 49°C); the process is not highly pressure sensitive. The general principles are the same as in other processes based on chemical absorption (see **Desulfurization of Fuel Gases**). The MDEA process differs from the **SNPA-DEA** and **Sulfiban processes,** which use similar alkanolamine solvents, in removing hydrogen sulfide preferentially to carbon dioxide.

MEROX PROCESS

A commercial process for reducing the sulfur content of and/or sweetening (i.e., removing the unpleasant odor from) fuel gases, **gasoline,** or other light fuels (see **Sweetening Processes**). The Merox process has two phases: (1) extraction of the malodorous mercaptans (sulfur compounds) and (2) sweetening by oxidation of the mercaptans to innocuous disulfides with oxygen in air. The name Merox is derived from the initial letters of "mercaptan" and "oxidation." Either one or both phases of the process may be required, according to circumstances.

In the treatment of gasoline, for example, the mercaptans are first extracted with the Merox solution, consisting of an aqueous alkali (caustic) and a proprietary catalyst. After extraction, the gasoline passes to the sweetening unit where it is brought into contact with more Merox solution and air. The treated gasoline is separated and stored, whereas the Merox solution is recycled to the sweetener.

The spent Merox solution from the extraction phase is regenerated with air. The dissolved mercaptans are oxidized to disulfides, which form a separate liquid layer that can be removed. The Merox solution is thus regenerated for further use.

METHANATION

A process for converting a mixture of carbon monoxide (CO) and hydrogen (H_2) gases into **methane** (CH_4) and water (H_2O), using a nickel catalyst. For complete conversion, the volume (or molecular) ratio of carbon monoxide to hydrogen in the gas should be 1 to 3 (see **Water-Gas Shift Reaction**). The methanation process

$$CO + 3H_2 = CH_4 + H_2O$$
$$+ 3400 \text{ Btu/lb CO } (7900 \text{ kJ/kg})$$

is accompanied by considerable evolution of heat. If carbon dioxide (CO_2) is present, it can also be methanated by the reaction

$$CO_2 + 4H_2 = CH_4 + 2H_2O$$
$$+ 1800 \text{ Btu/lb CO}_2 (4200 \text{ kJ/kg})$$

Methanation is used to some extent to remove carbon monoxide and dioxide from hydrogen gas, but its main interest here lies in the production of a **high-Btu fuel gas** to serve as a **substitute natural gas.** After removal of the water produced in the methanation process, the resulting gas consists mainly (about 95 volume percent) of methane (see **Natural Gas**).

Methanation is a catalytic process, and, in order to prevent inactivation (poisoning) of the nickel catalyst, the feed gas must be exceptionally free from sulfur compounds. Removal of most of the hydrogen sulfide (see **Desulfurization of Fuel Gases**) is followed by absorption of residual traces by zinc oxide. Organic sulfur compounds are removed by passing the gas over "activated" carbon. Processes for absorbing hydrogen sulfide generally also remove carbon dioxide (see **Acid Gases**); hence, the gas to be methanated is usually a mixture of carbon monoxide and hydrogen, with little carbon dioxide.

A major problem in methanation is the heat generated in the reaction. If, as a result, the temperature becomes too high, carbon monoxide is decomposed into carbon dioxide and solid carbon; thus,

$$2CO = CO_2 + C$$

The solid carbon deposits on the nickel catalyst and reduces its activity. On the other hand, if the temperature is below about 450°F (230°C), nickel is lost as nickel carbonyl, a gaseous compound of nickel and carbon monoxide. Hence, proper temperature control is an essential aspect of the methanation process.

In the conventional procedure, as used in hydrogen purification, the gas under a high pressure is passed over a fixed bed of nickel catalyst deposited on an alumina base. The feed enters the catalytic reactor at a temperature near 500°F (260°C), and the exit temperature should not exceed 850°F (455°C). This may be controlled by adjusting the gas feed rate.

For large-scale methanation, as required for the production of high-Btu fuel gases, more efficient techniques are being developed. In the hot-gas recycle (HGR) system, the carbon monoxide–hydrogen feed gas at high pressure is passed between a number of parallel steel plates that are flame-sprayed with a finely divided nickel preparation called Raney nickel. Temperature control is achieved by recycling some of the product gas after cooling in an external **heat exchanger.**

Another methanation system is the tube-wall reactor (TWR) in which the Raney nickel catalyst is flame-sprayed on the interior walls of 2-in. (5.1-cm) diameter steel tubes. The pressurized feed gas passes through the tubes, and the heat generated is removed by circulating a high-boiling-point organic liquid (Dowtherm) around the outside. In each case, water is condensed from the product. Final methanation, to minimize residual carbon monoxide, is conducted in the conventional manner, as described above.

Another approach, using liquid-phase (as distinct from gas-phase) methanation, is being studied. Small particles of solid nickel catalyst are suspended in an inert liquid (**hydrocarbon**) medium by feed gas entering from the bottom. The operating temperature is roughly 570°F (300°C), and the gas pressure is up to 68 atm (6.8 MPa). The heat generated in the methanation reaction is removed by circulating the liquid medium through an external heat exchanger before it is recycled to the reactor. If final methanation of the product gas is required, it can be carried out in the conventional manner after water removal.

METHANE

The simplest paraffin **hydrocarbon**; its molecular formula is CH_4. Methane is a combustible gas with a heating value of 1030 Btu/cu ft (38.3 MJ/cu m). It is the main constituent of **high-Btu fuel gases,** including **natural gas** and **substitute natural gas.** It is formed in nature by the decomposition of vegetable matter and occurs in coal mines (as firedamp), marshes (marsh gas), and sewage gas. A gas rich in methane can be obtained by the action of bacteria on animal manure in the absence of air (see **Biomass Fuels**).

MOBIL M-GASOLINE PROCESS

A Mobil Research and Development Corporation process for producing **gasoline** and **high-Btu fuel gas** from methanol.

The methanol (see **Alcohol Fuels**) can be made from **synthesis gas** obtained by the gasification of coal. Hence, the Mobil process provides an indirect method of producing liquid (and gaseous) fuels from coal (see **Coal Liquefaction**).

Vapor of crude methanol (e.g., from synthesis gas) is first dehydrated and then passed over a catalyst consisting of a special type of artificial zeolite (aluminosilicate). The special catalyst is an essential feature of the process. The pressure in the reaction chamber is about 20 atm (2 MPa), and the temperature increases from roughly 680°F (360°C) at the inlet to 780°F (415°C) at the outlet. The product, after condensation, consists of water, which is separated, and a mixture of light and moderately light (mostly C_3 to C_9) **hydrocarbons.** A good yield of gasoline (C_5 to C_9) can be obtained by fractional **distillation.** The C_3 and C_4 gases contain olefins and isobutane for conversion by **alkylation** into high-octane gasoline. **Substitute natural gas** and **liquefied petroleum gas** are obtained as by-products.

MOLTEN-SALT BREEDER REACTOR (MSBR)

A proposed **nuclear power reactor** which is unique in using a liquid (molten-salt) fuel; its purpose is to breed fissile uranium-233 (i.e., produce more than is consumed during operation) from fertile thorium-232 while generating useful power. Unlike plutonium-239 breeders, which are **fast reactors,** the MSBR is a **thermal reactor;** that is to say, nearly all the fissions are caused by slow neutrons (see **Breeder Reactor**).

The MSBR design is based on the Molten Salt Reactor Experiment (MSRE), a low-power reactor at the Oak Ridge National Laboratory. The MSRE started operation in 1965 and continued into late 1969 when the project was terminated for lack of funding.

In the MSBR concept, the reactor consists of an array of vertical graphite bars, with the molten-salt fuel flowing upward through channels between the bars. The fuel is a mixture of fluorides of lithium-7 (LiF), beryllium (BeF_2), thorium (ThF_4), and uranium-233 (UF_4), melting at 930°F (500°C). (The reason for using lithium-7 rather than normal lithium is that the isotope lithium-6, present to the extent of 7.4 percent, is a strong absorber of thermal neutrons and so could not be tolerated in a reactor.) The graphite (i.e., carbon) is the main moderator for slowing down the neutrons, but the lithium, beryllium, and fluorine in the fuel also make a substantial contribution.

The MSBR does not have a coolant in the conventional sense. To remove the heat generated in the core by uranium-233 fission, the liquid fuel is circulated continuously from the reactor through an external heat exchanger and back to the reactor. In the heat exchanger heat is transferred from the fuel to another molten salt mixture of sodium fluoborate ($NaBF_4$) and sodium fluoride (NaF), called the coolant salt. The hot coolant salt then passes to a steam generator, where steam is produced from highly pressurized water, and returns to the heat exchanger (Fig. 92).

According to the MSBR design, the molten fuel salt would leave the reactor and enter the heat exchanger at 1300°F (705°C). The hot coolant salt would leave the heat exchanger at 1150°F (620°C) and produce superheated steam at a temperature of 1000°F (540°C) and a pressure of 240 atm (24 MPa). The steam would drive a turbine

to generate electricity in the usual manner (see **Electric Power Generation**). The **thermal efficiency** of an MSBR plant for electricity generation is estimated to be an exceptionally high 44 percent.

A special feature of the MSBR is the continuous removal of fission products and other undesirable species from the fuel while the reactor is operating. Other breeders have solid fuel elements which are removed at intervals when the reactor is shut down. After removal of undesirable products, the recovered materials must be refabricated into fuel elements (see **Nuclear Fuel Reprocessing**). In the MSBR, however, the reprocessing would be simpler.

A small side stream of fuel enters a chemical processing unit at a rate that would permit the entire inventory to pass through every 10 days. Impurities are removed by a special technique developed for the treatment of molten salts, and the fuel returns to the reactor. Any excess of uranium-233 over that required to operate the MSBR could be used to fuel another reactor.

The fuel and coolant salts have low vapor pressures even at the high operating temperatures. Consequently none of the installation, except the steam side of the steam generator, has to be designed to withstand any significant pressure. On the other hand, the corrosive nature of the molten fluorides has presented a problem. This appears to have been solved by the development of a special nickel-base alloy

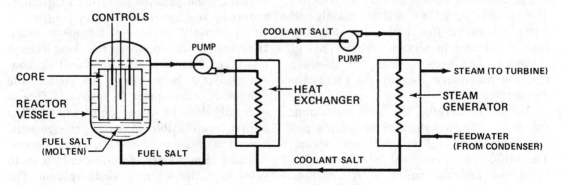

Fig. 92   Molten-salt breeder reactor.

(titanium modified Hastelloy N). All equipment coming into contact with molten fluoride would have to be made of this expensive material.

MOLTEN-SALT DESULFURIZATION PROCESS

A regenerative process of Rockwell International Corporation for **desulfurization of stack gases.** The process is unique in the respect that the sulfur dioxide (and trioxide) absorber is a molten salt mixture, consisting of 32 weight percent of lithium carbonate (Li_2CO_3), 33 percent of sodium carbonate (Na_2CO_3), and 35 percent of potassium carbonate (K_2CO_3). This (eutectic) mixture, represented by the general formula M_2CO_3, melts at 750°F (400°C) to form a clear liquid which can be pumped and sprayed like any other liquid. Stack (flue) gases are scrubbed with the molten carbonate mixture at about 850°F (450°C). Hence, the gases must be adjusted to this temperature before entering the scrubber. High-temperature **electrostatic precipitators** are used to remove particulate matter from the gas.

The carbonates react with sulfur dioxide to form sulfites (and some sulfates by oxidation), which dissolve in the melt, and carbon dioxide gas is evolved. The melt then enters the reducer where it is heated with carbon (**petroleum coke**) and a limited amount of air at about 1600°F (870°C). The mixed sulfites (M_2SO_3) and sulfates (M_2SO_4) are thereby reduced to sulfides (M_2S), and more carbon dioxide is given off. The liquid melt, consisting mainly of unreacted carbonates with dissolved sulfides, is filtered to remove carbon. Part of the clear melt is recycled to the absorber and the remainder goes to the carbonate regenerator.

In the regenerator, the melt containing sulfide is reacted with carbon dioxide gas (from scrubber and reducer) and steam. The sulfides are converted into carbonates for reuse, and the sulfur is removed as hydrogen sulfide gas. If sufficiently concentrated, this gas can be converted into sulfur by the **Claus process.**

See also **Desulfurization: Waste Products.**

MUNICIPAL WASTE FUELS

Energy sources derived from municipal wastes. Approximately 154 million short tons (140 million metric tons) of solid municipal wastes are generated annually in the United States. The composition of this waste is variable, but on the whole at least half the weight consists of combustible material, such as paper, cardboard, textiles, etc., that could serve as an energy source. The average **heating value** of the solid waste is roughly 5000 Btu per lb or 10 million Btu per short ton (116 MJ/kg).

Assuming that 50 percent of this heating value could be utilized, the solid municipal wastes produced each year in the United States would have the heat equivalent of almost 33 million short tons (30 million metric tons) of average bituminous coal or 130 million barrels of crude oil. Municipal waste thus constitutes a significant potential source of energy. It is of special interest as a fuel because of its low sulfur content of 0.1 to 0.2 weight percent (see **Sulfur Oxides**).

Solid municipal wastes in the United States have been disposed of mainly in so-called sanitary landfills where alternate layers of refuse and soil are compacted. The wastes are gradually decomposed by microorganisms over the course of time. However, some concern has been expressed regarding the possible pollution of groundwaters by seepage from sanitary landfills.

In densely populated European countries and in Japan, there has long been a shortage of suitable sites for landfills; consequently, it is necessary to reduce the volume of waste prior to disposal. This is generally done by direct incineration of the lighter combustible material; the residue, consisting mainly of metals, glass, ceramics, and mineral ash, occupies only 5 to 10 percent of the original waste volume. The heat released during the incineration process is used in boilers to produce steam for district heating, for industrial purposes, and for generating electricity in the conventional manner (see **Electric Power Generation**).

Since the late 1960s, in particular, interest has increased in the United States in the treatment of solid wastes, partly because of the unavailability of acceptable landfill sites and partly for the purpose of utilizing the energy content and recovering glass and metals. In the early 1980s, some 30 such solid waste treatment plants, referred to as *resource recovery systems*, were in operation, under construction, or being planned.

In the development stages, the cost of collecting, delivering, and processing the wastes is likely to exceed the value of the energy produced and the glass and metals recovered. The deficit then represents the cost to the local government of providing the essential service of waste disposal. In time, however, the situation may change and a profit made from the operation.

The methods for utilizing the energy potential of solid municipal waste fall into three general categories:

Incineration (or direct combustion)

Pyrolysis

Biochemical conversion

Some of the more highly developed processes in each category are described below; others, based on the same general principles, have been proposed.

Incineration Processes

1. In the most common incineration procedure, of which there are several variations, the waste is collected in a storage pit from which it is transferred, as required, by means of a crane onto a sloping grate of a stoker-type furnace (see **Steam Generation**). Grates of a number of different types have been developed, but they are all designed to keep the waste material in motion with fresh surfaces being continuously exposed for incineration. On the grate, the combustible portion of the waste is first dried and then burned in air supplied by blowers. If bulky objects are included in the waste, they are broken into smaller pieces by a shredder before being deposited on the grate.

The hot combustion gases pass from the furnace, where the waste is incinerated, to a steam boiler, usually of the common water-tube wall design. Such a system is often called a *water-wall incinerator*. The noncombustible residue left after the paper, cardboard, etc., have been burned collects in the ash pit at the furnace bottom. It is cooled and passed over a magnet to remove ferrous metals (mainly iron) which are compacted and sold to a steel mill. The remainder can be disposed of as an inert landfill or it may be ground and used in road construction or in land reclamation.

The flue gases leaving the boiler carry substantial amounts of fly-ash particles in suspension. These must be removed before the gases are discharged through a stack to the atmosphere. **Electrostatic precipitators** or **cyclone separators** are commonly employed for this purpose. Because of the low sulfur content of the wastes, scrubbing of the stack gases for sulfur oxides removal is unnecessary. However, attention may have to be paid to heavy metal emissions (e.g., mercury and antimony).

Extensive experience in Germany with direct incineration of wastes has shown that corrosion of the boiler tubes by the combustion gases can be a serious problem. Various procedures have been devised to minimize the corrosion. For example, the tubes in the lower part of the furnace (i.e., closest to the grate) are covered with a protective, refractory material. An excess of combustion air also helps to inhibit corrosion. By maintaining the temperature of the gases in the range of about 1600 to 1800°F (870 to 980°C), both corrosion and nitrogen oxide formation are reduced (see **Nitrogen Oxides**).

2. In an alternative incineration process, the combustible matter is first separated from the wastes. The mass of material to be incinerated is reduced to about half, and the heating value (per unit mass) is

roughly twice that of the original wastes. In addition, much less ash is formed in the furnace.

The collected solid waste is fed to a hammer mill where it is shredded into pieces generally less than 1 inch (2.5 cm) across. The shredded waste passes to an air "classifier" in which an upward flow of air takes with it nearly all the lighter, combustible material (i.e., paper, etc.) while the heavier, noncombustible material (i.e., metals, glass, ceramics, etc.) falls to the bottom.

At the power plant, a pneumatic feeder carries the shredded combustible material suspended in air to a furnace of the type normally used for pulverized coal or oil. To compensate for the variability in amount and heating value, the wastes are used as a supplementary heat source in boilers with coal as the main fuel. Pulverized coal and the combustible waste fraction are fed separately to the furnace and burned simultaneously. A control system adjusts the coal supply automatically to maintain constant combustion conditions. Up to 20 percent of the heat may be supplied by the wastes without affecting the operation of the boilers.

3. A process developed by Combustion Equipment Associates and Arthur D. Little, Inc., converts the wastes into a clean, transportable fuel, trade named Eco-Fuel, which can be used in a pulverized-coal furnace. After shredding in a flail mill and magnetic separation of ferrous metals, the wastes are screened to remove the larger pieces, which are returned to the shredder, as well as glass, dirt, etc. The remaining fraction, consisting largely of combustible matter, is treated by a patented chemical process for reducing the particle size. The resulting material is dried and then screened and air-classified to decrease the heavier inorganic (ash) content. The prodduct is a free-flowing combustible powder; it has a fairly high heating value and leaves relatively little ash when burned.

4. The CPU system of the Combustion Power Corporation is a modified waste incineration process in which the hot combustion gases, at high temperature and pressure, are used to drive a **gas turbine** for generating electricity. The compressor attached to the turbine provides compressed air for combustion of the wastes. In principle, the CPU system is equivalent to a conventional gas turbine, except that the fuel supplied to the combustor is derived from solid municipal waste (see Fig. 54).

The waste is shredded and the lighter, mainly combustible fraction is separated by air classification. The material is dried by hot gases from the turbine exhaust and is then fed with compressed air into a **fluidized-bed combustion** furnace, where the wastes are rapidly incinerated. After passage through cyclone separators to remove fly ash and granular filters to remove aluminum oxide which can deposit on the turbine blades, the combustion gases enter and drive the turbine in the usual manner. Heat remaining in the turbine exhaust is used to produce steam in a waste-heat boiler as well as to dry the wastes.

Pyrolysis Processes

In **pyrolysis** processes, several of which have been proposed, the waste is heated to a high temperature, usually in the range of roughly 950 to 1850°F (510 to 1010°C), in the essential absence of oxygen (from air). The carbonaceous material, mainly cellulose in paper, cardboard, and cotton rags, is decomposed, and gaseous and liquid fuel products can be obtained from the resulting vapors. The relative amounts of these products depend on the pyrolysis conditions; as a general rule, higher temperatures lead to an increased proportion of gaseous components. Possible advantages of pyrolysis over incineration are that the fuels produced by pyrolysis can be burned readily and leave little ash.

1. In the Landgard Resource Recovery System, developed by Monsanto Enviro-Chem Systems, Inc., the shredded waste, without prior treatment, is fed continu-

ously into one end of a refractory-lined, rotary kiln while fuel oil is injected at the other end together with a limited supply of air. Combustion of the fuel oil provides the heat required for pyrolysis of the waste. In the kiln, the solid wastes and the pyrolysis (and combustion) gases travel in opposite directions; the gases exit where the wastes enter, and solid residue leaves at the other end where the oil is burned. Temperatures in the kiln range from about 1200°F (650°C) where the waste enters to 1800°F (980°C) where it leaves.

On a dry basis, the pyrolytic gases contain roughly 80 percent by volume of inert nitrogen and carbon dioxide, arising chiefly from combustion of the fuel oil in air. The remainder is a mixture of the combustible gases carbon monoxide, hydrogen, methane, and ethylene. The overall heating value of the gas leaving the kiln is roughly 120 Btu/cu ft or 4.5 MJ/cu m (see **Low-Btu Fuel Gas**). The gas is used either alone or as an auxiliary fuel in a steam-generating boiler.

2. Another pyrolysis method for producing a fuel gas from solid municipal wastes is Union Carbide Corporation's Purox system. The untreated waste enters at the top of a vertical furnace, and oxygen gas is introduced near the bottom. Combustion of the waste char in the lower part of the furnace with oxygen produces gases at temperatures of 2600 to 3000°F (1425 to 1650°C). The hot gases travel upward through the waste and cause pyrolysis.

Because oxygen is used instead of air for combustion, the gas leaving at the top of the furnace contains essentially no nitrogen and the only inert component is carbon dioxide. There is about 50 volume percent (on a dry basis) of carbon monoxide, 30 percent of hydrogen, and some 4 percent of **hydrocarbons**; the heating value is roughly 300 Btu/cu ft or 11.3 MJ/cu m (see **Intermediate-Btu Fuel Gas**). The gas is cleaned by passage through an electrostatic precipitator and an acid absorber and is used as a boiler fuel. Because the gas consists mainly of hydrogen and carbon monoxide, it could possibly be used as a **synthesis gas** in the production of methane (see **Substitute Natural Gas**), methanol (see **Alcohol Fuels**), or other fuels (see **Fischer–Tropsch Process**).

As a result of the high temperature at the bottom of the Purox pyrolysis furnace, the waste residues form a liquid slag. This is drawn off and quenched to yield a granular material consisting mainly of glass and metal. The product, occupying not more than 3 percent of the original waste volume, can be used as constructional fill or in other ways.

3. The high-temperature process of Torrax Systems, Inc. (Carborundum Company), is similar to that of the Union Carbide Corporation, except that air preheated to about 2000°F (1090°C) is used instead of oxygen to burn the char and thus provide the heat required for pyrolysis. Before entering the pyrolysis furnace (or gasifier), the air is heated by burning a subsidiary fuel (e.g., natural gas or fuel oil) in a separate ceramic heat exchanger. Temperatures at the bottom of the gasifier are in the range of roughly 2600 to 3000°F (1430 to 1650°C). The furnace off-gas has a heating value of only 170 Btu/cu ft (6.4 MJ/cu m) since it contains a high proportion of nitrogen from the air. The waste residue is a liquid slag which is withdrawn from the bottom of the gasifier and quenched, as described above.

4. A proposed **fluidized-bed** pyrolysis process is of interest because it yields an intermediate-Btu fuel gas, as in the Purox process, but without using oxygen gas. The system consists of two separate fluidized-bed reactors, one for combustion and the other for pyrolysis. Sand particles, which form the fluidized beds, are circulated continuously from the combustor, where they are heated, to the pyrolyzer, where they supply the heat required for pyrolysis, and back to the combustor for reheating. The operation is of a type that is often used in petroleum refining.

The shredded, light fraction of municipal wastes is fed to the pyrolyzer where the hot fluidized-bed sand causes pyrolysis at a temperature of about 1500°F (815°C). The products are separated into a solid char and pyrolysis gas. The char is transferred to the combustor where it is burned in compressed air to generate heat; the air also fluidizes the bed. In the pyrolyzer, the fluidized bed is maintained by circulation of part of the pyrolysis gas.

Since air is restricted to the combustor, the pyrolysis gas contains essentially no nitrogen and consists mainly of carbon monoxide and hydrogen. Like the Purox gas, it can be used directly as an intermediate-Btu fuel gas or as a synthesis gas for conversion into other fuels.

5. In the American Thermogen pyrolytic process, the waste is heated by combustion of a subsidiary fuel (gas or oil) introduced at the bottom of the furnace. Temperatures up to 3000°F (1650°F) are attained. An unusual feature of the process is that the pyrolysis gases are not removed, as in other pyrolytic methods, but are burned in air in a combustion zone at the top of the furnace. The resulting hot combustion gases pass to a boiler for generating steam. The final residue is drawn off as a slag from the furnace bottom.

6. In the flash pyrolysis system of the Garrett Research and Development Company, the temperatures are substantially lower than in other pyrolysis processes. The main product is consequently a liquid fuel that can be transported by tank truck and used at a distance from the waste treatment plant. Pyrolytic gases, on the other hand, must be used nearby because transmission of intermediate- or low-Btu fuel gas over a significant distance is not economic.

The solid waste is shredded and subjected to air classification in the usual manner. The lighter fraction is dried and screened to remove most of the remaining inorganic matter. It is then ground to a fine powder and injected into a reactor which is heated externally to about 1000°F (540°C). Rapid (flash) pyrolysis occurs in the complete absence of oxygen (air). The products pass to a cyclone separator where solid char is removed while the vapors proceed to a wet scrubber. Here the pyrolytic oil is condensed and separated. Combustion of the accompanying gas together with part of the char in air provides heat for the pyrolysis reactor.

The oil product can be burned as fuel in a boiler, but its characteristics differ from those of common petroleum fuel oil. Because of the high (about 50 weight percent) oxygen content of cellulose, from which it is derived, the pyrolytic oil contains some 33 percent of oxygen, compared with less than 1 percent for most petroleum oils. The heating value of 10,500 Btu/lb (25 MJ/kg) is little more than half that of common fuel oil and it is more viscous. On the other hand, the sulfur and mineral (ash) contents are very low. A disadvantage of the pyrolytic oil is its corrosive nature, so that special materials are required for containment and transportation.

Biochemical Processes

1. Paper and paper products, the main components of the light fraction of solid municipal wastes, consist largely of cellulose which can be degraded into the fuel gas methane (and carbon dioxide) by the action of bacteria in the absence of air (see **Biomass Fuels**). The shredded waste material is made into a slurry with water or, preferably, sewage sludge and digested for several days in a closed tank. Nutrients and bacteria are added if necessary. Methane is also produced by the decay of refuse in landfills commonly used for municipal waste disposal. The possibility has been considered of collecting the gas for fuel by sinking wells in the landfills.

2. An alternative biochemical process for utilizing municipal wastes is by conversion of the cellulose into ethanol for use as a supplement to gasoline in automobiles. The cellulose (lighter) fraction of the wastes, separated in the usual way, could first be broken down into sugars which would then be fermented to produce ethanol (see **Alcohol Fuels**).

N

NAPHTHA

A general term applied, especially in **petroleum refining,** to various mixtures of liquid **hydrocarbons** with boiling points partially overlapping the ranges of **gasoline** and **kerosine.** The boiling range of naphtha is roughly 200 to 400°F (93 to 205°C). Products with increasing boiling ranges are often categorized as light, intermediate (or medium), and heavy naphthas. In petroleum refining the naphtha fraction may be converted into more desirable fuels by **reforming.** Naphthas are used in industry mainly as solvents and cleaning agents. Light distillation products of coal tar (mainly benzene and toluene) and of wood (mainly methanol) have also been called naphthas.

NATURAL GAS

A combustible (fuel) gas found in nature in underground reservoirs of porous rocks, either alone or in association with **petroleum** (i.e., crude oil). (Unconventional sources of natural gas are described later.) Natural gas consists mainly of a mixture of simple paraffin **hydrocarbons,** of which **methane** (CH_4) is by far the major constituent (generally 70 to 90 volume percent); ethane (C_2H_6), propane (C_3H_8), butanes (C_4H_{10}), and some higher paraffins are also present in decreasing proportions.

Impurities often found in raw (or crude) natural gas include carbon dioxide, nitrogen, and hydrogen sulfide gases. Natural gas containing appreciable amounts of sulfur compounds, especially hydrogen sulfide and mercaptans (thiols), is called "sour" gas; it has an unpleasant odor and corrodes metal. The sulfur compounds are therefore removed if the gas is to be distributed for use as a fuel. "Sweet" natural gas is gas essentially free of sulfur compounds.

Crude natural gas, as it is obtained from a gas well, may be categorized as "dry" or "wet" gas. Dry gas consists mainly of methane and ethane; only very small amounts of higher paraffin hydrocarbons (propane, butanes, etc.) that can be readily condensed to a liquid are present. Wet gas, on the other hand, contains substantial amounts of the higher hydrocarbons that can be condensed to form what are called **natural gas liquids;** they are a source of **liquefied petroleum gas** (LPG) and **natural gasoline.**

In some locations, natural gas contains up to about 2 volume percent (occasionally more) of helium formed by the radioactive decay of uranium and thorium minerals over long periods of time. Since helium is a valuable commercial product, it is separated from the crude natural gas by liquefying the other constituents. A considerable proportion of the helium produced in the United States is obtained from natural gas fields in Kansas, New Mexico, and Texas.

Occurrence and Production

Like petroleum, natural gas probably originated from the decomposition of microscopic living organisms which were deposited in natural waters, together with particles of clay, sand, or limestone (see

Petroleum: Occurrence). Natural gas occurs in structural traps, consisting of a reservoir and a cap rock, similar to those in which petroleum is found. Such reservoirs may contain gas alone or gas trapped over oil (see Fig. 100). Natural gas found alone is called *nonassociated gas* whereas *associated gas* occurs together with petroleum. Most of the commercial natural gas in the United States is nonassociated gas. Natural gas dissolved in petroleum under pressure in a reservoir is released when the oil is pumped to the surface; this is commonly referred to as **casinghead gas.**

Nonassociated natural gas is produced from wells which are drilled and cased in a manner similar to oil wells (see **Petroleum Production**). Pumping is, however, unnecessary because the gas in the reservoir is under a high natural pressure. As the gas is released, the pressure drops but it can be maintained by pumping water into the gas reservoir. Return of part of the natural gas already removed is sometimes used to maintain the pressure and increase the total gas recovery. As a general rule, from 50 to 80 percent of the gas in the reservoir can be recovered.

Treatment

Before distribution, natural gas must be treated for sulfur removal by one of the several available processes (see **Desulfurization of Fuel Gases**); when an alkaline absorber is used, most of the carbon dioxide is removed at the same time. If the crude natural gas is a wet gas, the natural gas liquids are separated by compressing the gas to about 7 atm (0.7 MPa) at normal temperature. A common procedure, however, is to scrub the wet natural gas with a hydrocarbon oil; the oil absorbs the heavier hydrocarbons but leaves the methane and ethane. The residual gas is thus equivalent to dry natural gas.

The oil containing the propane, butanes, etc., is heated to release the dissolved gases and vapors, which are then cooled and subjected to a form of low-temperature fractional **distillation.** The natural gasoline fraction condenses to a liquid at normal temperature [e.g., 60°F (16°C)] and atmospheric pressure (0.1 MPa); butane forms a liquid at a pressure of 2 atm (0.2 MPa) and propane at 7 atm (0.7 MPa) at this temperature. Liquid propane and butane produced from wet natural gas are the main commercial sources of LPG in the United States.

Dry natural gas or the gas remaining after removal of natural gas liquids may contain nitrogen and possibly helium. These inert (noncombustible) impurities are harmless, but they dilute the gas and decrease the heating value per unit volume. If helium is present, it is separated as described earlier; the nitrogen can then be removed at the same time by liquefaction followed by fractional distillation under pressure. In the absence of helium in the natural gas, the nitrogen is generally allowed to remain.

Finally, it may be necessary to remove moisture from the natural gas before distribution. If water vapor is present, solid compounds of methane (or ethane) and water may deposit in and block the distribution pipes in cold weather. The moisture is commonly removed by absorption of the water in an ethylene glycol, particularly triethylene glycol. The water is subsequently removed from the glycol by heating. An alternative method for preventing solid hydrate formation is to add some methanol (see **Alcohol Fuels**) to the natural gas. Because methane and ethane are odorless, a trace of a sulfur compound (e.g., a mercaptan) with a strong odor is added to facilitate leak detection.

Composition and Heating Value

The composition of commercial natural gas is variable; a range of typical values for gas from several (but not all) sources is given in the table. The standard **heating**

Approximate Composition Ranges of Commercial Natural Gas

Constituent	Volume percent
Methane	80 to 95
Ethane	2.5 to 7
Propane and butane	1 to 3
Carbon dioxide	Less than 1
Nitrogen	Less than 10

value of methane is 1030 Btu/cu ft (38.3 MJ/cu m) whereas that of ethane is 1760 Btu/cu ft (65.4 MJ/cu m). Because of the presence of ethane, the heating value of commercial natural gas is often higher than that of methane. On the other hand, inert gases decrease the heating value. The range of heating values of commercial natural gas is from about 960 to 1130 Btu/cu ft (36 to 42 MJ/cu m), with a rough average of 1000 Btu/cu ft (38 MJ/cu m). One of the requirements of **pipeline quality** gas is a minimum heating value of 900 Btu/cu ft (33.5 MJ/cu m).

Uses

Natural gas is a clean burning fuel, leaving no ash and producing essentially no **sulfur oxides;** the combustion products, carbon dioxide and water vapor, are readily assimilated by the atmosphere. Consequently, natural gas is in wide use in the United States as a heat source in residences, industry, and commerce, as well as for generating electricity. With modification of the fuel intake, natural gas can be used in place of gasoline in an **internal-combustion engine.** It can also serve as the fuel in a **fuel cell.**

There are several uses for natural gas in industry other than as an energy source. One of the most important is the production of hydrogen by **steam reforming** of methane for the manufacture of ammonia fertilizers. Natural gas is also used to make carbon black, acetylene, and other chemical products.

Distribution and Storage

Main pipelines, up to 4 ft (1.2 m) in diameter, carry natural gas from the sources, mostly in Kansas, Louisiana, New Mexico, Oklahoma, and Texas, to essentially all other parts of the United States. There are more than 250,000 miles (400,000 km) of main pipelines and at least twice this length of minor lines. Pumps located 50 to 100 miles (80 to 160 km) apart maintain a pressure up to 70 atm (7 MPa), with an average around 50 atm (5 MPa), in the main pipelines to provide a steady and optimum gas flow rate.

With liquid petroleum fuels, the decreased demand for heating oils in the summer is largely offset by the increased demand for gasoline, but there is nothing to offset the decreased use of natural gas in the summer. To keep the gas fields and pipelines in operation throughout the year, storage facilities have been established in many locations near large consumption centers. Gas is stored in the summer to provide a reserve supply for the winter. (The use of a reserve to supplement the normal supply at times of high demand is commonly called *peak shaving.*)

Natural gas for local distribution is often stored in steel tanks, but larger facilities have been developed. Most of these are depleted oil or gas reservoirs, into which the natural gas is pumped and stored under pressure. More than 300 such reservoirs exist in the United States. Where these facilities are not available, aquifers (i.e., water-bearing strata) in porous rock are being used increasingly to store natural gas. To be suitable for this purpose, the aquifer must have an impervious cap rock so as to form a trap, similar to the formations in which gas and crude oil normally occur (see Fig. 100). For storage, the natural gas is pumped into the aquifer to form a gas cap above the water. Natural gas has also been stored in salt caverns and in abandoned coal mines.

With proper precautions, natural gas can be stored safely in liquid form in insulated tanks at a very low temperature (see **Liquefied Natural Gas**). An advantage of liquid storage is that a given tank can hold about 600 times its volume of gas as measured at normal temperature and pressure.

A limited form of storage, called *line packing*, is often used to deal with variations in demand for natural gas in the course of the day. When the demand is low, the pressure in the distribution lines is allowed to increase. The lines can then store a larger mass of gas for use at times when the demand increases.

Unconventional Sources

The so-called unconventional sources of natural gas are coal beds, "tight" gas formations, Devonian (eastern) shales, and geopressured aquifers. These potential sources are being studied to determine if the natural gas, which is undoubtedly present in substantial quantities, can be extracted economically.

Coal Beds. All coal deposits release methane, the major constituent of natural gas, commonly known as "firedamp." Because it is an explosion hazard in coal mines, it is discharged to the surroundings with large volumes of ventilation air. However, if the methane were collected, it would be an important source of natural gas. Recovery of gas from an operating mine might be possible in some cases but, as a general rule, the extraction would be conducted before mining started. Gas could also be obtained from coal beds that cannot be mined (e.g., because of their location in a populated area, at a great depth, or in sloping beds).

Release of methane gas from coal occurs at external (exposed) and internal (fractured) surfaces. Methods for recovering the gas at a useful rate are thus based on extending the area from which the gas can escape. It is probable that different techniques would be required in different cases; hence, three methods, described below, are being tested.

1. Vertical wells, roughly 1500 ft (450 m) apart are drilled into the coal bed.
Hydraulic fracturing (i.e., injection of high-pressure water) may then be used to break up the coal mass and release the accumulated gas. Sand is often added to the water to prevent the fractures from closing when the water pressure is decreased.

2. Horizontal holes are drilled in several directions into the coal at the bottom of an existing shaft or heading (tunnel). In an unmined bed, a vertical shaft can be drilled for gas recovery and later used for mining. The horizontal holes may be up to 1000 ft (300 m) in length.

3. In coal beds containing identified natural major fractures or in which such fractures are made artificially (e.g., by high-pressure water or explosives), directional drilling from the surface (see **Petroleum Production**) can be used to intersect several fractures. Gas is then withdrawn from the directional well.

Tight Gas Formations. In parts of the western and midwestern United States, large quantities of natural gas are contained in extensive formations of too low permeability to permit recovery by conventional methods. The only practical way to release the gas from these "tight" formations, as they are called, is by fracturing the gas-bearing beds. Tests have been made of fracturing with nuclear explosives, but this approach has been abandoned. The most promising technique appears to be massive hydraulic fracturing involving the injection through drilled wells of large amounts of pressurized water and sand, with certain additives.

Devonian Shales. The Devonian shales, which occur in large areas of the eastern and midwestern United States, differ from the western **oil shales.** The eastern shales usually contain little or none of the oil-producing kerogen, but they do hold significant quantities of natural gas. Some of this gas is already being recovered from wells drilled into natural fractures which are enlarged with chemical explosives. However, massive hydraulic fracturing (see above) appears to be more effective in stimulating the flow of natural gas, and this technique is being pursued. Furthermore, large fractures, both natural and artificial, may be interconnected with advantage by directionally drilled wells.

Geopressured Aquifers. Numerous aquifers, containing hot saline water (i.e., brine) at a very high pressure, underlie the coasts of Texas and Louisiana at depths of roughly 2 to 5 miles (3.2 to 8 km) both on- and off-shore (see **Geothermal Energy**). These geopressured (i.e., earth-pressured) aquifers contain dissolved methane gas produced over millions of years by the decay of buried organic matter and trapped by impermeable mineral deposits above and below. The solubility of methane in water (or brine) at normal pressure is quite small, but it is increased at the high pressures—up to 1350 atm (135 MPa)—of the geopressured aquifers. If the water is brought to the surface and the pressure reduced, the methane gas is released from solution.

The methane content of the geopressured water is variable; it is generally 10 to 20 cu ft of gas (measured at normal temperature and pressure) per **barrel** of water (1.8 to 3.6 cu m gas/cu m water), but substantially higher values have been reported in short-duration tests. Rough estimates indicate that the total amount of natural gas contained in geopressured aquifers exceeds the more conventional natural gas resources of the United States. However, in view of the costs of drilling deep wells into highly pressurized reservoirs of hot brine, the amount of natural gas recoverable economically is uncertain. Long-term studies are required to assess the potential of geopressured aquifers as a source of natural gas.

See also **Substitute Natural Gas.**

NATURAL GAS LIQUIDS (NGL)

A mixture of light, normal paraffin **hydrocarbons,** mostly from propane (C_3H_8) through hexane (C_6H_{14}), obtained from wet **natural gas** or **casinghead gas.** It is commonly extracted either by compressing and cooling the gas or by absorption in a hydrocarbon oil from which the natural gas liquid is subsequently recovered by condensing the vapor obtained by heating the oil. Natural gas liquids are separated into two fractions: **natural gasoline,** which is liquid at normal temperature and pressure, and **liquefied petroleum gas,** which becomes a liquid under a moderate pressure at normal temperatures.

NATURAL GASOLINE

A liquid obtained from **natural gas liquids,** recovered from wet **natural gas** and **casinghead gas,** consisting mainly of the normal paraffin **hydrocarbons** pentane (C_5H_{12}) and hexane (C_6H_{14}). Natural gasoline has a relatively low **octane number** and is not used directly as an automobile fuel; however, its high volatility facilitates easy starting in cold weather and so it is a useful component of commercial **gasoline.**

NITROGEN OXIDES

Compounds of the elements nitrogen and oxygen, both of which are present in air. The combustion of **fossil fuels** in air is accompanied by the formation of nitric oxide (NO) which is subsequently partly oxidized to nitrogen dioxide (NO_2). The resulting mixture of variable composition is represented by the symbol NO_x, where x has a value between 1 and 2. Nitrogen oxides are present in stack gases from coal, oil, and gas furnaces and in the exhaust gases from **internal-combustion engines** (gasoline and diesel) and **gas turbines.**

Nitrogen oxides in the air have adverse respiratory and other effects, both directly and indirectly. Interaction of nitrogen oxides with **hydrocarbons** (mainly from automobile and truck exhausts) in the atmosphere in the presence of sunlight leads to the formation of eye-irritating photochemical smog. Furthermore, solution of the nitrogen oxides in water produces nitrous and nitric acids, thus contributing to acid rain which is destroying aquatic organisms. Because of these harmful effects, the U. S. Environmental Protection Agency has established standards for nitrogen oxide emissions from stationary and mobile sources.

Nitric oxide is formed in combustion processes in two ways: (1) combination of nitrogen and oxygen in the air and (2) reaction of atmospheric oxygen with chemically bound nitrogen in the fuel. Apart from the nitrogen content of the fuel, the extent of nitric oxide formation depends primarily on the combustion temperature. A useful approximate rule is that up to about 1830°F (1000°C) the formation of nitric oxide from air is negligible, but above 1830°F the amount increases with increasing temperature. As the combustion gases move away from the flame front and are cooled, the nitric oxide is slowly converted by oxygen from the air into nitrogen dioxide. This is how nitrogen oxides are formed in stack and other exhaust gases.

It is possible, in principle, to remove nitrogen oxides from combustion gases, but the process is difficult, unlike the removal of **sulfur oxides** (see **Desulfurization of Stack Gases**). Consequently, in order to meet the emission standards, the combustion conditions must be adjusted to minimize the formation of nitric oxide. The obvious requirements are low combustion temperatures and use of low-nitrogen fuels if possible. The temperatures in a continuous combustion engine, such as a gas turbine, are lower than those attained during the burning stage of intermittent combustion (e.g., gasoline or diesel) engines; hence, nitrogen oxide emissions are generally lower for the former.

Studies are being made of fuel-burner and furnace design and combustion conditions that will result in a decreased nitric oxide formation. The **fluidized-bed combustion** system under development for burning coal is of special interest in this respect.

NUCLEAR ENERGY

Energy released as a result of interactions involving atomic nuclei, that is, reactions in which there is a rearrangement of the constituents (i.e., protons and neutrons) of atomic nuclei; it is also called atomic energy (see **Atom**). Any nuclear process in which the total mass of the products is less than the mass before interaction is accompanied by the release of energy. The amount of energy released is related to the decrease in mass by the Einstein equation for the equivalence of mass and energy; this equation is

$$E = mc^2$$

where E is the energy released, m is the decrease in mass, and c is the speed of light.

If m is expressed in kilograms and c in meters/sec, the energy E is obtained in **joules.** As a general rule, however, nuclear masses are given in atomic mass units and E in electron volts. (The *electron volt*, equal to 1.602×10^{-13} joule, is the energy gained by a particle carrying a unit electric charge when it is accelerated by a potential of 1 volt.) In these units, the Einstein equation becomes

E (in MeV)

$\qquad = m$ (in atomic mass units) $\times$ 931

where MeV stands for million electron volts.

The masses of atoms (nuclei and electrons) and of nuclear particles are known with great accuracy from measurements and calculations. It is therefore a simple matter to determine if a given nuclear reaction will be accompanied by the release of energy or not. There are many nuclear reactions in which there is an overall decrease of mass and hence a release of energy. This is the case, for example, in all radioactive changes; the energy is carried by alpha and beta particles and gamma rays and eventually appears as heat (see

Radioactivity). The nuclear energy associated with radioactivity, however, is of limited application (see **Radioisotope Thermoelectric Generators**).

The nuclear reactions that are likely to prove useful as energy sources fall into two general categories: fusion and fission. In *fusion reactions*, the combination (or fusion) of two light nuclei leads to the formation of a heavier nucleus. In *fission reactions*, on the other hand, a heavy nucleus is split into two nuclei of intermediate mass accompanied by a number of neutrons.

Several different nuclear fusion and fission reactions are possible, but only a very few of each type are practical for use as energy sources. The fusion processes that may prove useful within the foreseeable future as practical energy sources involve nuclei of the two heavier isotopes of hydrogen, namely, deuterium and tritium (see **Atom**). Other fusion reactions occur in the sun and stars, but the conditions cannot be duplicated on earth. (For a review of the conditions required to utilize nuclear fusion as a terrestrial energy source, see **Fusion Energy**.)

The only useful fission reactions are those in which free neutrons cause fission of nuclei of certain isotopes of uranium and plutonium with odd mass numbers; they are, in particular, uranium-233, uranium-235, and plutonium-239. Of these three species, only uranium-235 occurs in nature; the other two are made artificially (see **Nuclear Fuel**). An important aspect of fission as an energy source is that it can be a self-sustaining process. A neutron can cause fission which is itself accompanied by the liberation of neutrons; these neutrons can cause further fissions, and so on. A fission chain reaction, with the continuous release of energy, is then possible (see **Fission Energy**).

NUCLEAR FUEL

In general, the material from which nuclear energy can be released by fission or fusion reactions (see **Nuclear Energy**).

More specifically, as used here, nuclear fuel is the material serving as the energy source in nuclear fission reactors (see **Nuclear Power Reactor**). Fusion fuels are described in the article on **Fusion Energy**. Nuclear reactor fuels usually consist of a mixture of a fissile species, namely, uranium-233, uranium-235, or plutonium-239 (and plutonium-241), and a fertile species, namely, uranium-238 or thorium-232 (see **Fission Energy**). These materials are all present in or derived from the two naturally occurring elements, uranium and thorium.

Uranium Fuels

Natural uranium consists of 99.29 weight percent of uranium-238 and 0.71 percent of the lighter isotope uranium-235 (see **Atom**). By far the most important reactor fuel material is uranium dioxide (UO_2) in which the average proportion of the fissile isotope uranium-235 has been increased to about 2.5 to 3 percent. The uranium as found in nature must therefore be enriched in the lighter isotope for use in this type of fuel. The enrichment processes utilize uranium in the form of the hexafluoride (UF_6) which is subsequently converted into enriched uranium dioxide. The fabrication of uranium fuels thus involves the following main stages: (1) production of uranium hexafluoride from uranium ores, (2) enrichment of the hexafluoride in uranium-235, and (3) conversion of the enriched uranium hexafluoride into the dioxide fuel.

Uranium Ores. A few ores are fairly rich (1 to 4 percent) in uranium, but the common (medium-grade) ores contain only 0.1 to 0.5 percent of uranium. The major available sources of these ores are the United States, Canada, and Australia. The principal uranium deposits in the United States are in the Colorado plateau (i.e., parts of Arizona, Colorado, New Mexico, and Utah), Wyoming, and Texas. The average uranium content of the medium-grade U. S. ores, expressed as the oxide U_3O_8, is 0.2 weight percent; that is, on the average,

2200 lb (1000 kg) of ore contains the uranium equivalent of 4.4 lb (2 kg) of U_3O_8. (It is the general practice to state the uranium content of raw materials in terms of U_3O_8 which contains 84.5 weight percent of the element uranium.)

As the medium-grade ores are depleted, low-grade ores may acquire economic significance in the next century. These low-grade sources include extensive phosphate deposits in Florida and Idaho, the Chattanooga shales in Tennessee, and the uraniferous lignites of Wyoming and the Dakotas. Some uranium is already being recovered as a by-product from phosphoric acid made from Florida phosphates. South African gold ores also contain small proportions of uranium which can be recovered economically because the costs of mining have been assigned to the gold.

Estimates of uranium resources are indefinite, partly because they are markedly dependent on the value of uranium as an energy source. In addition, exploration for uranium deposits has been relatively limited. To remedy this situation, the National Uranium Resource Evaluation (NURE) program was initiated in the United States in 1973. The objective of the program is to encourage private exploration for nuclear fuel resources based on a comprehensive and systematic evaluation of uranium distribution in all parts of the country. A report covering what are thought to be the most favorable areas is expected in the early 1980s.

Commercial **light-water reactors** (LWRs), such as are in general use in the United States for generating electricity (see **Nuclear Power Reactor**), each require roughly 200 short tons (180,000 kg) of U_3O_8 annually. Present indications are that sufficient uranium ore should be available at an economic price to provide the fuel requirements well into the next century of the 250 (or so) such reactors operating in the year 2000. If plutonium breeding becomes practical, the known uranium reserves should be sufficient for several hundred years (see **Breeder Reactor**).

Uranium Milling. In the United States, uranium ores are obtained in very roughly equal amounts from open-pit and underground mines. The ore is taken to a "mill" where the uranium is extracted into solution and concentrated by chemical procedures. (In a few cases, the uranium is extracted directly from the ore in the ground without mining.) Ores with a large proportion of calcium carbonate (limestone) are leached (i.e., extracted) with sodium carbonate solution to dissolve out the uranium, but most U. S. uranium ores have a low limestone content and are leached with sulfuric acid. The uranium is eventually precipitated as sodium diuranate ($Na_2U_2O_7$) from the carbonate leach solution or as ammonium diuranate [$(NH_4)_2U_2O_7$] from the acid solution. After filtering and drying, the solid product, called "yellowcake," contains the equivalent of 70 to 90 percent of U_3O_8.

Hexafluoride Production. Two different procedures, involving the same basic chemistry, are used to produce uranium hexafluoride from yellowcake. In each case, however, the essential steps are *(a)* reduction of the oxide in yellowcake (U_xO_y) to uranium dioxide (UO_2) by hydrogen (H_2) gas, *(b)* formation of uranium tetrafluoride (UF_4) by heating the dioxide in hydrogen fluoride (HF) gas, and *(c)* conversion of the tetrafluoride into uranium hexafluoride (UF_6) by fluorine gas (F_2), as outlined below.

$$U_xO_y \xrightarrow{H_2} UO_2 \xrightarrow{HF} UF_4 \xrightarrow{F_2} UF_6$$

Yellow cake　　Brown oxide　　Green salt

In the "wet" process, which must be used when the yellowcake contains sodium, the yellowcake is first dissolved in acid. The uranium is extracted from the solution and purified before conversion into the dioxide, so that no further purification is required in the later stages. In the alternative "dry" process, on the other hand, no purification is carried out until the final stage when the solid uranium hexafluoride is purified by a form of distillation. Upon heating the crude product, the hexafluoride vaporizes, and the vapor is condensed to the pure solid; the impurities are mostly nonvolatile and remain behind.

The next major stage in nuclear fuel production is to enrich (i.e., increase the uranium-235 content of) the uranium hexafluoride (see **Uranium Isotope Enrichment**). The enriched hexafluoride is then converted into enriched uranium dioxide, which is the actual fuel material.

Conversion to Dioxide. Uranium hexafluoride is first reacted with water (H_2O) to form uranyl fluoride (UO_2F_2), and then ammonium diuranate (ADU) is precipitated by addition of aqueous ammonia (NH_3). The solid is separated and heated to produce uranium trioxide (UO_3), which is finally reduced to uranium dioxide by hydrogen gas; thus,

$$UF_6 \xrightarrow{H_2O} UO_2F_2 \xrightarrow{NH_3} ADU$$

$$\xrightarrow{Heat} \underset{\substack{\text{Orange} \\ \text{oxide}}}{UO_3} \xrightarrow{H_2} UO_2$$

Since the isotopes of uranium have identical chemical properties, the uranium dioxide product has the same uranium-235 enrichment as the uranium hexafluoride starting material.

For the fabrication of reactor fuel, the solid uranium dioxide is powdered and cold-pressed into small cylindrical pellets, roughly 0.35 in. (0.9 cm) in diameter and 0.6 in. (1.5 cm) long. The pellets are then sintered (i.e., the particles are made to adhere) by heating at 3090°F (1700°C) to increase the overall density. Finally, they are ground to size and packed end-to-end in zirconium alloy (zircaloy) tubes to form long thin fuel rods.

Uranium Carbides. Normal (unenriched) uranium monocarbide (UC) has been proposed as the fertile component of **fast reactor** fuel. Furthermore, the dicarbide

(UC_2), with either uranium-235 (more than 90 percent) or uranium-233 (see later), may be the fissile component of **High-Temperature Gas-cooled Reactor** fuel. The carbides are made by heating the appropriate uranium dioxide with the required amount of graphite (carbon) in the absence of air. The density of the product can be increased by sintering or by hot or cold compaction.

Plutonium Fuels

Since plutonium does not exist in nature, except perhaps in minute traces, it is made artificially starting with uranium-238 (in normal uranium). The most important isotope is fissile plutonium-239, but, as explained below, it is always accompanied by heavier isotopes, some of which are also fissile. The major interest in plutonium fuel is for fast **breeder reactors** which are designed to generate more fissile material than they consume while producing useful power. The fuel elements in such reactors are commonly a mixture of 15 to 20 percent of fissile plutonium dioxide and 80 to 85 percent of fertile uranium-238 dioxide.

Plutonium Production. Fissile plutonium is made from uranium-238 by exposure to neutrons in a nuclear reactor. A uranium-238 (^{238}U) nucleus captures a neutron (n) to form uranium-239 (^{239}U); the latter is radioactive and decays, with a half-life of 23.5 min, by the emission of beta particles (see **Radioactivity**). The product, neptunium-239 (^{239}Np), undergoes beta-particle decay, in turn, with a half-life of 2.35 days to form plutonium-239 (^{239}Pu) as indicated below.

$$^{238}U + n \rightarrow {^{239}U} \underset{23.5 \text{ min}}{\xrightarrow{\beta}} {^{239}Np} \underset{2.35 \text{ d}}{\xrightarrow{\beta}} {^{239}Pu}$$

Plutonium-239 is an alpha-particle emitter with a half-life of 24,100 years; consequently, it decays very slowly and tends to accumulate in the reactor. However, the accumulation is limited by fission, since plutonium-239 is a fissile species, and nonfission reactions. In a succession of neutron captures, the plutonium-239 first forms nonfissile plutonium-240 and then fissile plutonium-241, and so on. Thus, plutonium as obtained from uranium-238 in a nuclear reactor consists mainly of plutonium-239, but it always contains the isotopes plutonium-240, -241, and -242 in proportions decreasing in this order.

The actual composition of the plutonium depends on the neutron exposure conditions. For example, **thermal reactor** fuels almost invariably include uranium-238; hence, some plutonium-239 (plus other isotopes) is formed during reactor operation. Typically, the plutonium in the spent fuel from a commercial light-water reactor would have the following approximate composition:

Isotope	Pu-239	Pu-240	Pu-241	Pu-242
Percentage	58	25	13	4

Both plutonium-239 and -241 are fissile and contribute to some extent to the energy produced in the reactor. The plutonium remaining in the spent fuel could be recovered (see below) and reused for reactor fuel.

In a fast breeder reactor, plutonium-239 is formed from fertile uranium-238 both in the core and in the blanket (internal and external). Since the objective of such a breeder reactor is to produce more plutonium-239 (and -241) than is consumed, the spent core and blanket materials must be treated for recovery of the plutonium.

Separation of plutonium from uranium (and fission products) is carried out by chemical methods. The most common procedure is by the Purex process (see **Nuclear Fuel Reprocessing**) in which the plutonium is obtained as a plutonium nitrate solution. The addition of oxalic acid leads to the precipitation of the oxalate; the solid is separated by filtration and heated to yield the common fuel material, plutonium dioxide (PuO_2).

As a reactor fuel, plutonium dioxide is used in a mixture with uranium dioxide. The mixed oxide is usually made by mechanical mixing of the powders before compression into pellets. Another method is to add an ammonia solution to an aqueous solution containing plutonium and uranium nitrates in the desired proportions. The resulting precipitate is separated and heated to form a mixture of the oxides.

Plutonium monocarbide (PuC) may eventually be used instead of the oxide in fast reactor fuel. If so, the carbide would be made by heating plutonium dioxide with carbon. It would then be mixed mechanically with uranium monocarbide made in a similar manner, as stated earlier.

Thorium and Related Fuels

Thorium as found in nature consists almost entirely of thorium-232; there is no fissile isotope as is the case with uranium. The importance of thorium as a nuclear fuel is that thorium-232 is a fertile species which can be converted into the fissile uranium-233. There are some drawbacks, as well as some advantages, to the use of the thorium-232/uranium-233 system in power reactors.

Hitherto this fuel combination has been used or tested in only a few reactors (see **High-Temperature Gas-Cooled Reactor; Light-Water Breeder Reactor; Molten Salt Breeder Reactor**), but thorium-based reactors are attracting interest because they have some proliferation-resistant potential.

Thorium Production. The major source of thorium is the mineral monazite, consisting mainly of a phosphate of several rare-earth elements. Nearly all monazite ores contain a few percent of thorium, as well as smaller amounts of uranium. Erosion of monazite rocks by water has resulted in the formation of sandy deposits, known as *monazite sands*, along beaches and dry river beds. Monazite sands are found in the United States in the Carolinas, Florida, Idaho, Montana, and South Dakota. Large quantities of monazite sands also occur in Brazil, India, South Africa, Canada, and elsewhere. The total resources of thorium may exceed those of uranium.

Thorium has been obtained as a by-product in the recovery of rare-earth elements from monazite sand. One procedure is to dissolve out the thorium (and rare earths) with sulfuric acid; partial neutralization results in the precipitation of impure thorium phosphate. The thorium is purified by dissolving the impure product in nitric acid and extracting with an organic solvent (see **Nuclear Fuel Reprocessing**); back extraction with dilute nitric acid yields a fairly pure solution of thorium nitrate. This may be further purified for removal of traces of rare-earth elements.

To obtain thorium dioxide (ThO_2) for use as the fertile component of nuclear fuel, the thorium nitrate solution is concentrated by evaporation, and thorium oxalate is precipitated by adding oxalic acid. Upon heating, the solid oxalate decomposes, leaving thorium dioxide. If thorium carbide (ThC or ThC_2) is required, as it may be for some reactors, the thorium dioxide is heated with carbon in the absence of air.

For use in high-temperature (gas cooled) reactors, the thorium is in the form of very small particles of the dioxide or carbide. The particles are coated with layers of pyrolytic carbon, by heating in a hydrocarbon vapor, to retain fission products of thorium-232 (by high-energy neutrons) and of uranium-233 generated from thorium-232. These fertile particles, together with similar fissile particles of uranium-233 or -235 oxide (or carbide), are incorporated in a graphite matrix and fabricated into cylindrical fuel elements.

Uranium-233. Fertile thorium-232 is converted into fissile uranium-233 by exposure to neutrons in a nuclear reactor. The process is similar to that described earlier for the conversion of uranium-238 to plutonium-239. Thus, thorium-232 (^{232}Th) captures a neutron, and this is followed by two stages of beta-particle emission:

$$^{232}Th + n \longrightarrow \,^{233}Th \xrightarrow[\text{22 min}]{\beta} \,^{233}Pa \xrightarrow[\text{27 days}]{\beta} \,^{233}U$$

where ^{233}Pa represents protactinium-233. Uranium-233 is an alpha-particle emitter with a half-life of 159,000 years and is therefore moderately stable.

Uranium-233 can be separated from thorium-232 (and any fission products that may be present) by the Thorex process which resembles the Purex process for separating plutonium from spent uranium fuels. Since the isotopes of uranium are essentially identical chemically, compounds of uranium-233, such as the dioxide and carbide for use as nuclear fuel, are prepared in the same manner as described earlier for uranium fuels.

From the neutronic standpoint, uranium-233 has some advantage over uranium-235 and plutonium-239 as the fissile material in thermal reactors. There is a major drawback, however, which probably accounts for the limited use of the thorium-232/uranium-233 system. In addition to conversion into uranium-233, part of the thorium in a reactor undergoes other neutron reactions and beta-particle decays that result in the formation of uranium-232. Since the isotopes of uranium are chemically identical, the uranium-233 separated from thorium invariably contains uranium-232.

Uranium-232 decays, with a half-life of 72 years, by emitting alpha particles; the product, thorium-228, with a half-life of 119 years, also emits alpha particles. Among the subsequent radioactive decay (or daughter) products are several short-lived species which are strong emitters of gamma rays. Consequently, uranium-233 as derived from thorium-232 develops a considerable gamma-ray activity; this increases with time as the uranium-232, thorium-228, and their daughters decay. Since gamma rays constitute a potential health hazard, all stages involved in the fabrication of fuel elements containing uranium-233 must be conducted by remote manipulation behind concrete shields.

Although the gamma-ray activity makes the operations more difficult and more costly, it might be advantageous in the

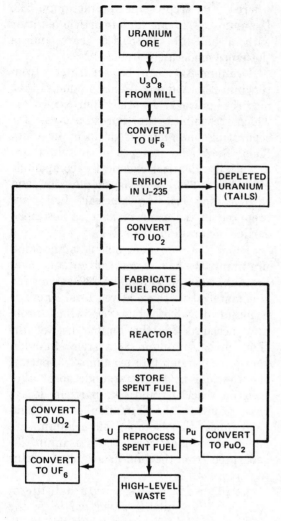

Fig. 93   Uranium fuel cycle.

respect that it greatly decreases the risks of unauthorized diversion of fissile uranium-233 for nuclear weapons.

NUCLEAR FUEL CYCLE

The path followed by **nuclear fuel** in its successive stages from mining of the uranium (or thorium) ore to the final disposal of the highly radioactive wastes from the reprocessing of spent reactor fuel (see **Nuclear Fuel Reprocessing**). The complete cycle for nuclear fuels derived from uranium is shown in Fig. 93. It is based on the supposition that the spent fuel removed from the reactor is reprocessed for

the recovery of the contained uranium and plutonium for subsequent recycling (i.e., reuse) as reactor fuel (see **Nuclear Power Reactor**).

The uranium recovered from **light-water reactor** (LWR) fuel contains about 0.8 percent of uranium-235, whereas an average enrichment of 2.5 to 3 percent is required for fresh fuel. For reuse as LWR fuel, the recovered uranium would be converted into uranium hexafluoride for enrichment in the fissile uranium-235 by the gaseous-diffusion or gas-centrifuge method (see **Uranium Isotope Enrichment**). Uranium recovered from spent **breeder reactor** fuel or blanket would not require enrichment; hence, it would be converted directly into the dioxide (UO_2) for refabrication. The plutonium, as the dioxide (PuO_2), would be mixed with depleted (or natural) uranium dioxide in the appropriate proportions for use as fuel for either LWRs (less than 5 percent of plutonium) or breeder reactors (15 to 20 percent).

The decision was made in the United States in 1977 to postpone the reprocessing of spent fuel from LWRs. Furthermore, there are no plutonium breeder reactors in the United States, other than test devices, and so fuel reprocessing, which is an essential aspect of breeding, has not been required to any extent. The current uranium fuel cycle for LWRs is therefore as shown within the broken lines in Fig. 93. The spent reactor fuel and the depleted uranium hexafluoride (tails) from the isotope enrichment operation are stored. This incomplete cycle is called a *once-through* cycle.

A thorium-based fuel cycle is outlined in Fig. 94; since there is no natural source of uranium-233, reprocessing of spent fuel for the recovery of this fissile species would be mandatory. In the early stages, at least before sufficient uranium-233 is available, some uranium-235 would be required to provide an adequate fissile content of the fuel. This would be supplied from an isotope-enrichment plant.

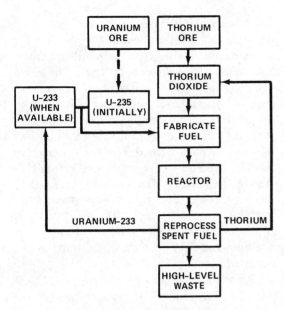

Fig. 94   Thorium fuel cycle.

Proliferation-Resistant Cycles

Material containing more than about 20 percent of fissile species could possibly be made into a nuclear explosive. The proliferation of nuclear weapons in the hands of national or subnational (terrorist or criminal) groups, by diverting fissile material, is thus an inherent potential of all nuclear fuel cycles. In some cycles, however, the diversion (or weapons proliferation) risk is reduced although it cannot be eliminated entirely; these are called *proliferation-resistant cycles*.

In the development of proliferation-resistant cycles for nuclear fuels, it is assumed that separation of the fissile and nonfissile isotopes of the same element (e.g., uranium) would be difficult, except perhaps for a nation with the necessary economic and technological capabilities. On the other hand, chemical separation of one element from another (e.g., plutonium from uranium or uranium from thorium) is considered to be feasible provided the material has a low **radioactivity** level. A high level of radioactivity, especially if accompanied by gamma rays, would add to the difficulty of weapon fabrication by requiring shielding with remotely controlled operations.

On the foregoing bases, the simplest proliferation-resistant fuel cycle is the once-through cycle enclosed by the broken lines in Fig. 93. This represents the current situation in the United States. Because of the radioactivity of the fission products in the spent fuel, extraction of the plutonium would be difficult, especially for a subnational group. However, after a few hundred years, the radioactivity would decay to the point where remote operations would not be required. Unless safeguarded, the spent fuel would then constitute a possible source of weapons material.

The drawback to the once-through cycle is that the fissile material remaining in the spent fuel (about 0.8 percent uranium-235 and 0.65 percent plutonium-239 and -241 in LWR fuel) is not utilized. In any event, breeder reactors would have no meaning without reprocessing for recovery of the bred fissile species. Hence, suggestions have been made for spent-fuel reprocessing procedures with reduced proliferation risks.

Civex Process. One proposal, intended primarily for fast plutonium breeder reactors, is called the *Civex process*. It is a modification of the Purex process in which the mixed uranium and plutonium are first separated from most (about 75 percent) of the fission products in the normal way (see **Nuclear Fuel Reprocessing**). In the next stage, instead of producing a solution containing mainly plutonium, the product stream would contain nearly all the plutonium as a mixture of less than 20 percent of this element with about 80 percent of uranium and a few percent of fission products. The by-product stream contains the remaining uranium and fission products, with a very small proportion of plutonium.

The 20 percent (or so) plutonium–80 percent uranium stream could be converted into mixed dioxides (PuO_2–UO_2) for use as fast reactor fuel. However, the radioactivity level of the associated fission products makes remote operation necessary. The moderate proportion of plutonium and the radioactivity level constitute the proliferation-resistant aspects of the Civex process. The uranium by-product could not be used in weapons, but it could be freed from fission products and recycled as breeder blanket material.

Denatured-Fuel Cycles. Another proliferation-resistant approach, which is intended for **thermal reactor** fuels, is the *denatured-fuel cycle*. A denatured fuel is one containing a few percent of a fissile isotope and a large proportion of a nonfissile isotope of the same element. A difficult isotope separation process would then be required to convert the fuel into a weapons material. In this sense, the common commercial reactor fuel, with roughly 2.5 to 3 percent of fissile uranium-235 and 97 to 97.5 percent of uranium-238, is a denatured fuel. However, reprocessing of the spent fuel is ruled out because it leads to the isolation of fissile plutonium (Fig. 93). In principle, this could be denatured by an excess of nonfissile (even mass number) plutonium isotopes, but the latter are not available.

An artificially denatured thermal reactor fuel would consist of a few percent of fissile uranium-233, denatured with uranium-238 (e.g., natural or depleted uranium), together with thorium-232 as the main fertile material (see **Nuclear Fuel**). Such a fuel could be used in a reactor moderated by light water, heavy water, or graphite (see **Nuclear Power Reactor**). The two latter are especially attractive because of the high degree of conversion of thorium-232 into uranium-233 that would be possible.

Since the fuel would contain both fertile species thorium-232 and uranium-238, the spent fuel after reactor operation would include uranium-233 and plutonium-239. The conditions would be such, however, as to produce much more uranium-233.

Reprocessing of the spent fuel would lead to removal of the fission products and extraction of a mixture of uranium-233 and -238 isotopes; this mixture could be used for fabricating fresh fuel, but the uranium-233 content would be too low for weapons. As is the case with all uranium-233 fuels, the presence of the radioactive decay products of uranium-232 makes remote operations necessary. The small proportion of plutonium in the spent fuel could be protected from diversion by allowing it to remain with the highly radioactive fission product wastes (see **Radioactive Waste Disposal**). Unused thorium could also be recovered in the reprocessing, but there would probably be sufficient thorium-228, from the decay of uranium-232, for remote handling to be required (see **Nuclear Fuel**).

As stated earlier, a thorium-based fuel cycle would require that uranium-235 be supplied, at least in the initial stages. It is doubtful, however, that a reactor using denatured fuel could produce sufficient uranium-233 to sustain operation with this fissile material alone. Instead of making up the deficiency with uranium-235 from an enrichment plant, a more economical procedure would probably be to produce uranium-233 in a special breeder reactor. Such a reactor, and an associated reprocessing plant, would produce essentially pure uranium-233 from thorium-232 in a high-security location. The uranium-233 would be denatured with uranium-238 prior to shipment as fuel for several power reactors in various parts of the country.

NUCLEAR FUEL REPROCESSING

The treatment of used (or spent) reactor fuel for the recovery of fissile and fertile materials free from radioactive fission products (see **Nuclear Fuel**). The spent fuel, representing one-fourth to one-third of the total fuel loading of a commercial **light-water reactor** (LWR), is discharged about once a year. This material still contains fissile species, but it is no longer suitable for use as fuel for the following reasons: (1) depletion in the fissile species (see **Fission Energy**); (2) accumulation of fission products and isotopes of heavy elements which act as neutron poisons, and (3) changes in dimensions and shape of the fuel rods.

The fresh fuel for an LWR consists of about 2.5 to 3 percent of fissile uranium-235 and 97 to 97.5 percent of fertile uranium-238. Roughly one-fourth of the original uranium-235 remains in the spent fuel, as well as fissile isotopes of plutonium which have been formed (and partly consumed) during reactor operation. The approximate composition of a typical LWR spent fuel is given in the table. Uranium-

Approximate Composition of LWR Spent Fuel

Substance	Weight percent
Uranium-238	95
Uranium-236	0.4
Uranium-235	0.8
Plutonium (fissile)	0.65
Plutonium (nonfissile)	0.25
Fission products	2.9

236 is formed by nonfission capture of a neutron by uranium-235. The fissile isotopes of plutonium are those with odd mass numbers, mainly 239 and 241, whereas the nonfissile isotopes have even mass numbers, chiefly 238, 240, and 242 (see **Atom**).

In the past, it had been generally assumed that the spent fuel from commercial nuclear power reactors would be reprocessed, mainly for the recovery of the plutonium for reuse (or recycling) as reactor fuel. Uranium would also be recovered for possible re-enrichment. However, reprocessing of spent commercial fuels in the United States was suspended in 1977, partly in order to avoid the production of plutonium, which could be diverted to unauthorized nuclear weapons (see **Nuclear Fuel Cycle**), and partly because of the uncertainty about the economics of reprocessing.

Spent fuel reprocessing may be undesirable or unnecessary at present, but it may have economic value some time in the future. Furthermore, reprocessing of fuel and blanket material is an essential aspect of **breeder reactors** which generate power and, at the same time, produce more fissile material than they consume.

Although commercial reactor fuels are not now being reprocessed, much is known about the procedures. A reprocessing plant for spent LWR fuel operated in the United States between 1966 and 1972. In addition, normal uranium reactor fuels are reprocessed for the recovery of plutonium for nuclear weapons. Highly enriched uranium fuels from special reactors are also reprocessed; these spent fuels contain only a very small proportion of plutonium and are treated for the recovery of the enriched uranium free from fission products.

Reprocessing Principles

Preliminary Stages. The first step in the treatment of spent reactor fuel is called the *cooling* phase. The fuel rod assemblies are removed from the reactor and transferred to a deep water-filled storage tank. The water serves to remove heat produced by radioactive decay of the fission products (see **Radioactivity**) and also provides radiation shielding. The spent fuel remains in the tank for at least 150 days to permit the decay of fission products (and some undesirable heavy isotopes) of short and moderately short half-lives. The decreased radiation emission simplifies handling of the spent fuel and reduces decomposition of the organic solvent used in the separation process.

While reprocessing is in abeyance, the spent fuel assemblies remain in water-tank storage. For reprocessing, however, the assemblies would be transported in strong, metal casks to a reprocessing plant. Here, the fuel rods are cut up, and the spent fuel material is dissolved out with concentrated nitric acid. The solution formed contains uranium, plutonium, and fission product nitrates.

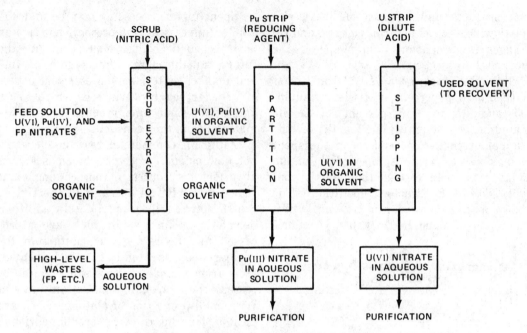

Fig. 95   Purex process for spent nuclear fuel.

Solvent Extraction. The solvent extraction process commonly used for recovery of uranium and plutonium from the nitrate solution is based on the following facts. Compounds of these elements can occur in different oxidation (or valence) states, the most important being the hexapositive (VI), tetrapositive (IV), and tripositive (III) states. In the (VI) and (IV) states, uranium and plutonium nitrates are soluble in certain organic solvents. On the other hand, neither the nitrates of the (III) state nor the fission product nitrates are soluble in these solvents.

Consequently, when an aqueous nitric acid solution containing the nitrates of uranium (VI), plutonium (IV), and fission products is brought into intimate contact with the organic solvent, the uranium and plutonium are extracted by the solvent while the fission products remain in the aqueous solution. The extraction process is favored by the presence in the original solution of excess nitric acid which acts as a so-called "salting" agent.

To separate (and remove) the uranium and plutonium now present in the organic solvent, the latter is brought into contact with an aqueous solution containing a substance that reduces the plutonium (IV) to the (III) state but leaves the uranium (VI) unchanged. Since the plutonium (III) nitrate is not soluble in the organic solvent, it is transferred to the aqueous medium. The uranium (VI) remaining in the organic solvent is finally extracted with water (or dilute acid); in the absence of the nitric acid salting agent, the uranium (VI) nitrate passes into the water (aqueous) solution. In this way, the uranium and plutonium are obtained in separate aqueous solutions essentially free from fission products.

Purex Process. In the Purex process, which has been used in the limited reprocessing of spent commercial reactor fuel and also for the production of plutonium for weapons, the organic solvent consists of the extractant tributyl phosphate (TBP) and a diluent, kerosine or similar hydrocarbon liquid. The main features of the Purex process, starting with a nitric acid solution of spent uranium fuel, are shown in the simplified flow diagram in Fig. 95. Because the organic solvent is less dense than water, an organic solution flows upward whereas an aqueous solution flows downward in each column.

The aqueous nitric acid feed solution of uranium (VI), plutonium (IV), and fission product (FP) nitrates enters at the middle of the first column, and the plutonium and uranium are extracted by the organic solvent in the lower part of the column. In the upper part, the upward-flowing organic solution is scrubbed with aqueous nitric acid (salting agent) to remove remaining fission products. The aqueous solution discharged from the bottom of the column, containing essentially all the fission products plus traces of uranium, plutonium, and other heavy elements as nitrates, forms the high-level radioactive waste of the reprocessing operation (see **Radioactive Waste Disposal**).

In the second (partition) column, the plutonium (IV) is reduced to the (III) state with a reducing agent in aqueous solution. The plutonium is thus extracted from the organic into the aqueous medium which is withdrawn from the bottom of the column. The uranium (VI) nitrate remaining in the organic solvent is finally back-extracted into water (or dilute nitric acid) in the third (stripping) column. The separate aqueous solutions of plutonium nitrate (from the second column) and uranium nitrate (from the third column) are subjected to further solvent extraction (or other treatment) to remove remaining impurities.

In the standard Purex process, as just described, a relatively pure plutonium nitrate solution is one of the products. In order to reduce the risks of diversion of this material to unauthorized use in nuclear weapons, a modification, called the Civex process, has been proposed especially for the reprocessing of **fast reactor** spent fuel and blanket materials. The plutonium would be recovered as a mixture with about 80 percent of uranium-238 and a few percent of fission products; the radioactivity of the latter would then make remote manipulation necessary (see **Nuclear Fuel Cycle**).

Thorex Process. The purpose of the Thorex process, which has been tested in a pilot plant but not used on a large scale, is to recover fissile uranium-233 from spent reactor fuels with thorium-232 as the fertile component. The Thorex process depends on the fact that thorium (IV) nitrate in an aqueous solution can be extracted into an organic solvent in the presence of aluminum nitrate as salting agent. However, unlike plutonium, thorium cannot be readily reduced into a (III) state; hence, the thorium and uranium are separated by adjustment of the salting agent.

The organic solvent in the Thorex process is a TBP-diluent mixture similar to that used in the Purex process. The feed solution, obtained by dissolving the spent fuel in nitric acid, is extracted with the organic solvent, which is scrubbed with an aluminum nitrate solution. The uranium and thorium pass into the organic solvent while the fission products remain in the aqueous waste solution. The thorium is extracted from the organic solution in the partition column by means of a moderately concentrated aqueous nitric acid solution. The uranium-233 (with other uranium isotopes) remaining in solution is finally extracted with water or dilute acid.

Provided fission products have been removed, the uranium and plutonium from the Purex process have little or no gamma-ray activity. Special shielding is then not required in subsequent fuel production operations with the recovered materials. This is not the case, however, for the products of the Thorex process.

Exposure of thorium-232 in a reactor leads to the formation of some uranium-232 which decays to thorium-228. Consequently, the recovered uranium-233 will include a proportion of uranium-232 and the thorium will contain thorium-228. Since the daughter products of uranium-232 and thorium-228 include intense gamma-ray emitters, both the uranium and thorium recovered in the Thorex process will develop gamma-ray activity in time. Special precautions would thus be necessary when the recovered products are used for fuel fabrication (see **Nuclear Fuel**).

NUCLEAR POWER REACTOR

An arrangement for the utilization of **nuclear energy** as a source of power, usually electric power. Although nuclear energy can be released in either fusion or fission reactions, the only nuclear power reactors in existence (or likely to exist before the end of the century) are based on fission (see **Fission Energy**). Consequently, the treatment here is concerned with nuclear reactors (in brief, reactors) of this type.

In a nuclear fission reactor, the conditions are such that fission energy is released at a controlled rate. The fission energy is converted into heat in the reactor, and this heat is utilized to raise steam, directly or indirectly. The steam then drives a turbine-generator to produce electricity in the conventional manner (see **Electric Power Generation**).

Fast and Thermal Reactors

Reactors fall into two broad classes, according to the energy (or speed) of the neutrons causing most of the fissions. (The energy of the neutron is kinetic energy, i.e., motion energy, which is related to its speed.) Nearly all the neutrons released in the fission process are high-energy (or fast) neutrons. Unless a material is present that removes much of their energy (see below), most of the neutrons in the reactor would be categorized as fast (or moderately fast) neutrons. The majority of fissions will then result from the absorption of such neutrons by fissile and, to some extent, by fissionable nuclei. In other words, fast neutrons are the main fission chain carriers. A nuclear reactor in which this situation exists is known as a **fast reactor**. The adjective "fast" refers only to the speed of the neutrons causing most of the fissions.

Neutrons of low energy (i.e., slow neutrons) have been found to be more effective than fast neutrons in causing fission in fissile nuclei; consequently, many reactors deliberately include a material, called a *moderator*, to slow down the fast neutrons. In colliding with the nuclei of the moderator, the fast (fission) neutrons transfer some of their kinetic energy to the nuclei. In a short time, the neutron energies are decreased to such an extent that the energy distribution is dependent (ideally) only on the temperature of the medium (i.e., the moderator). Neutrons whose energies are reduced to this extent by repeated collisions with nuclei are referred to as *thermal neutrons*.

Reactors which include a moderator to slow down the fission neutrons are known as **thermal reactors**. In such reactors most of the fissions are caused by low-energy neutrons. Rarely is sufficient moderator present to reduce the neutron energies to the level at which they are truly thermal. Nevertheless, the term thermal reactor is used to describe any reactor with a substantial proportion of moderator. Slow neutrons then cause the great majority of fissions.

Although numerous variations are possible in the composition and design of reactor systems, all reactors have a number of features in common. Some differences arise between fast and thermal reactors, and these are explained below.

Reactor Principles

Core: Fuel. All reactors have a central core in which fissions occur and most of the fission energy appears as heat (Fig. 96). The core contains the **nuclear fuel** consisting of a fissile species (i.e., uranium-233, uranium-235, or plutonium-239) and usually a fertile material (i.e., thorium-232 or uranium-238). In some thermal reactors, the fuel is natural uranium with roughly 0.7 weight percent of uranium-235 and 99.3 percent of uranium-238. Most commercial power reactors in the United States utilize a uranium fuel enriched to about 2.5 to 3 percent in uranium-235 (see **Uranium Isotope Enrichment**). For a fast reactor a

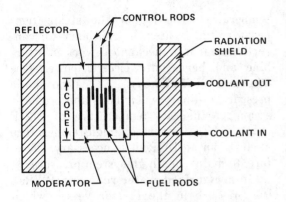

REFLECTOR

CONTROL RODS

RADIATION SHIELD

COOLANT OUT

CORE

COOLANT IN

MODERATOR

FUEL RODS

Fig. 96   Principle of nuclear reactor.

more highly enriched fuel is required. A typical fast reactor fuel might contain 15 to 20 percent of uranium-235 or plutonium-239; the remainder is uranium-238 to serve as fertile material.

Nuclear fuels for both thermal and fast reactors are generally fabricated from uranium dioxide (UO_2) and/or plutonium dioxide (PuO_2); these are ceramic materials capable of withstanding very high temperatures without melting. The powdered oxides are compressed into small cylindrical pellets which are packed end to end within a sealed cladding to form long, thin rods (or "pins"). The cladding retains the fission products and protects the fuel from chemical action of the coolant. The cladding material for thermal reactor fuel is commonly a zirconium alloy (zircaloy), whereas stainless steel is used in fast reactors. One of the factors determining the choice of cladding is the ability to withstand the action of water (in a thermal reactor) or liquid sodium (in a fast reactor) at high temperature.

Core: Moderator. In a thermal reactor, the core contains the moderator in addition to the fuel. Fission neutrons escaping from the fuel rods at high speed are slowed down as a result of collisions with moderator nuclei. The slow (approximately thermal) neutrons may then be absorbed, generally in other fuel rods, and there cause fissions. In this way, slow neutrons can maintain the fission chain. The actual energy (or

speed) distribution of the neutrons causing most of the fissions is determined by the nature and proportions of the fuel and moderator, as well as by the temperature.

The best moderators are materials consisting of elements of low mass number (see **Atom**), preferably with little or no tendency to capture neutrons. Ordinary water (H_2O), for example, is a common moderator; two of its three atoms are hydrogen with a mass number of unity. The great majority of power reactors in the United States (and elsewhere) are of the type known as **light-water reactors** because they use ordinary (or light) water as the moderator.

In some power reactors, especially in Canada, the moderator is deuterium oxide (D_2O), also called heavy water; this is a form of water containing deuterium, the heavier, naturally occurring isotope of hydrogen with a mass number of two. Such reactors are called **heavy-water reactors**. The only other moderator material used to any significant extent is graphite which is essentially pure carbon, with a mass number of 12 (see **Graphite-Moderated Reactor**).

Each of these three moderators has its advantages and disadvantages. Ordinary water is cheap, and light hydrogen nuclei are the most effective in reducing neutron energies. However, light hydrogen captures neutrons to such an extent that it is impractical to realize criticality (i.e., to maintain a fission chain) with natural uranium as fuel in a light-water moderator. Consequently, light-water reactors must use a fuel that is enriched in uranium-235 to attain criticality.

Deuterium (or heavy hydrogen) is the most effective moderator after light hydrogen; furthermore, it has very little tendency to capture neutrons. As a result, heavy-water reactors can, and generally do, utilize natural uranium (dioxide) as fuel. The main drawback to these reactors is the high cost of heavy water. Criticality is also possible with natural uranium in a graphite moderator, but the critical mass of fuel is very large. In some graphite reactors the critical mass is reduced by utilizing enriched uranium as the fuel. Graphite is not as good a moderator as heavy water, but it has the advantage of being a solid that can withstand high temperatures.

Reflector. The main purpose of the *reflector*, which surrounds the core, is to decrease the loss of neutrons. Neutrons escaping from the core enter the reflector where many collide with the reflector nuclei and are turned back into the core. The critical mass of fuel is then less than it would be without a reflector; consequently, the size and cost of the reactor are reduced. In thermal reactors, the same material usually serves as both moderator and reflector. Neutrons entering the reflector with high energies, as many do, are thus returned to the core as slow neutrons.

In a fast reactor, the reflector must be a material of high mass number to avoid slowing down the neutrons. The core is then surrounded by a layer (or *blanket*) consisting of uranium-238, either as natural uranium or uranium that has been depleted in uranium-235. (Depleted uranium is the residual material from the uranium-235 enrichment operation.) The uranium blanket acts as a reflector in returning some of the neutrons escaping from the core. However, an important purpose of the blanket is to serve as a fertile material; capture of neutrons by uranium-238, followed by two stages of beta-particle emission, results in the formation of fissile plutonium-239 (see **Nuclear Fuel**). In some fast reactor designs an additional stainless steel neutron reflector surrounds the fertile blanket.

Coolant: Thermal Reactors. The heat generated in the fuel by fissions is removed by circulation of a *coolant* through the reactor core. In the common thermal light-water reactors (LWRs), ordinary water is the coolant as well as the moderator and reflector. If the heat is to be used to produce steam for power generation, it is desirable to operate at the highest practical temperature; the higher the steam temperature the greater the efficiency for converting heat into mechanical work in a turbine and hence into electric power (see **Thermal Efficiency**). At atmospheric pressure, ordinary water boils and produces steam at the low temperature of 212°F (100°C). To generate high-temperature steam from an LWR, it is necessary, therefore, to maintain a high system pressure.

In **pressurized-water reactors** (PWRs), the pressure in the reactor vessel, which contains the core and coolant, is so high that the water does not boil. After passing through the core to remove fission heat, the high-temperature water is pumped through tubes in a **heat-exchanger** steam generator and returned to the reactor vessel. Heat is transferred from the reactor water to feedwater at a lower pressure surrounding the steam generator tubes; because of the lower pressure, the feedwater boils and produces steam to drive the turbines. The exhaust steam from the turbines is condensed and returned as feedwater to the steam generator.

Another type of LWR is the **boiling-water reactor** (BWR) in which the pressure is high but not high enough to prevent the water from boiling in the core. The steam is then produced by direct use of the fission heat, and there is no separate steam generator. The pressure in the BWR reactor vessel is roughly equal to that on the feedwater (and steam) side of a PWR steam generator. Hence, the temperature and pressure of the steam produced from a BWR and a PWR are about the same.

Heavy-water reactors (HWRs) resemble LWRs in the respect that the same substance, in this case heavy water, is the coolant as well as the moderator and reflector. There is, however, an important difference in the design of the two reactor types. The neutronic characteristics of heavy water are such that the fuel rods are more widely spaced in an HWR than in an LWR. Consequently, there is room for the fuel rods to be enclosed in wider tubes; the coolant then flows through the spaces around the rods but within the tubes. The heavy-water coolant, which is pressurized to prevent boiling, is thus kept separate from unpressurized heavy water outside the tubes which serves as the moderator (and reflector). The high-temperature and high-pressure coolant leaving the reactor is pumped to a heat-exchanger steam generator where steam is produced from ordinary water at a lower pressure just as in a PWR (see **CANDU Reactor**).

When graphite (solid) is the moderator (and reflector) in a power reactor, the coolant is usually a gas. (In a few cases ordinary water and liquid sodium have been used as coolants, but the sodium-cooled, graphite-moderated reactor concept has been abandoned in the United States.) In the United Kingdom, the common coolant is carbon dioxide gas, but in the United States helium, which is more expensive, is preferred for **graphite-moderated reactors.**

Helium is chemically inert and, unlike carbon dioxide, does not react with graphite even at very high temperatures. Furthermore helium has little or no tendency to capture neutrons. The gas is used as coolant under a moderate pressure, largely to increase its density and thus improve its heat-removal properties. The hot helium gas leaving the reactor core passes through the tubes in a steam generator and returns to the reactor vessel. Because of the high temperatures that are possible in a helium-cooled reactor, the steam temperature can be considerably higher than from LWRs or HWRs.

Coolant: Fast Reactors. Water cannot be used as coolant in fast reactors because of its moderating effect on neutron energy. Consequently, most current fast reactor designs are based on molten sodium as the coolant. Sodium has a mass number of 23 and so is not an effective moderator. The heat-transfer (or heat-removal) characteristics of sodium are superior to those of water, and its high boiling point under normal pressures means that liquid sodium

can be used at high temperatures without the need for pressurization to prevent boiling.

Sodium has, however, several drawbacks which must be allowed for in the design and operation of a fast-reactor system. For example, because of its chemical reactivity, exposure of sodium to oxygen (in air) or water must be prevented. Furthermore, the temperature everywhere must be maintained above 207.5°F (97.5°C), otherwise the liquid sodium will solidify and block pipes and pumps.

Another consideration is that as the sodium passes through the reactor core it captures neutrons to some extent and is converted into a radioactive isotope. The product, sodium-24, emits gamma rays as it decays and so constitutes a potential radiation hazard (see **Radioactivity**). Consequently, heat is transferred from the hot radioactive sodium leaving the core to non-radioactive sodium in a well shielded (intermediate) heat exchanger. Steam is then produced in a separate steam generator from the hot nonradioactive sodium leaving the heat exchanger (see **Liquid-Metal Fast Breeder Reactor**).

It is possible that helium gas may have some advantage over liquid sodium as a fast-reactor coolant (see **Gas-Cooled Fast Reactor**). Although helium has a mass number of only 4, its low density makes it ineffective as a neutron energy moderator. However, the future of the helium-cooled fast reactor is related to that of the corresponding graphite-moderated thermal reactor (see **High-Temperature Gas-Cooled Reactor**).

Reactor Control. Control of a reactor involves bringing the reactor up to a desired power level (i.e., rate of heat generation), maintaining that power level or varying it within certain limits, and finally decreasing the power for reactor shutdown. These conditions are achieved by adjusting the neutron density (i.e., number of neutrons per unit volume) in the reactor core (and sometimes in the reflector). For a given fuel, an increase in the neutron density means an increase in the fission rate and hence in the heat generation rate (or thermal power). Similarly, a decrease in the neutron density will result in a decrease in the thermal power.

The neutron density in a reactor core is generally varied by means of a *neutron poison* (i.e., a material which readily absorbs neutrons). Two such poisons, especially for slow neutrons, are the elements boron and cadmium. These substances have the advantage of not becoming radioactive as a result of neutron absorption. For thermal reactors, the control material is usually either boron (as the carbide B_4C) or cadmium (as an alloy). The preferred material for fast reactors is boron (as carbide); although this is a much less effective absorber of fast neutrons than of slow neutrons, it is adequate for fast reactor control.

When a reactor is shut down (or before it starts operating), the control rods are fully inserted into the core. The capture of neutrons by the control poison material then prevents the propagation of a self-sustaining fission chain. The reactor is now in a subcritical state (see **Fission Energy**).

In startup, the control rods are gradually withdrawn to permit the neutron density to increase. The reactor first becomes critical, when a chain reaction is just possible, and then supercritical when the number of fission chains (and neutron density) increases rapidly. When a desired thermal power is reached, the control rods are partially inserted so as to maintain a critical state. The number of fission chains, neutron density, and thermal power then remain essentially constant. To shut the reactor down, the control rods are reinserted and a subcritical condition is restored.

The disadvantage of reactor control by a poison is that valuable neutrons are lost by absorption in the poison. An alternative control procedure is being used in the experimental **Light-Water Breeder Reactor**. The neutron density is controlled by moving part of the reactor core. If the part is moved out of the core, neutrons escape and the density in the core is decreased, just as if a poison were inserted. However, the escaping neutrons are not lost, as they would be in a poison, but are captured in a surrounding blanket of fertile material which is thus converted into a useful fissile species.

Radiation Shielding. Although it has no effect on nuclear power generation, the radiation shield is an essential component of a reactor system. Its purpose is to protect those working in the vicinity of the reactor from the potentially harmful effects of radiation. Neutrons and gamma rays, in particular, can travel considerable distance in air and even in solids before they are attenuated to acceptably low levels. The reflector returns many neutrons to the core, but a substantial number escape from the reactor vessel. In addition, part of the gamma rays from the fission products formed in the core also escape. The most common method used to reduce the neutrons and gamma rays to an acceptable level is to surround the reactor vessel by a thickness of about 6 to 8 ft (1.8 to 2.4 m) of concrete.

Converters and Breeders

Another method of classifying reactors is as converters or breeders. Except for certain compact reactors, such as those on nuclear submarines, in which the fuel is highly enriched in uranium-235, all reactor fuels contain a substantial proportion of a fertile species. During reactor operation, some of the fission neutrons are taken up by the fertile nuclei and as a result a fissile species is formed (i.e., plutonium-239 from uranium-238 and uranium-233 from thorium-232). A reactor in which a significant amount of fertile species is changed into a fissile one is called a *converter reactor*. Thus, all commercial reactors are converters, at least to some extent.

A *conversion ratio* may be defined, for the present purpose, by

Conversion ratio =

$$\frac{\text{Fissile nuclei formed (from fertile nuclei)}}{\text{Fissile nuclei removed (in fission and nonfission reactions)}}$$

In most thermal reactors, the conversion ratio is less than unity; for example, it is about 0.55, on the average, in LWRs. This means that fewer fissile nuclei are formed by conversion than are lost in fission and nonfission reactions. Nevertheless, conversion of fertile into fissile material improves the efficiency of fuel utilization.

It is possible, however, under the proper conditions, for the conversion ratio to exceed unity; such a reactor can operate and generate power and, at the same time, produce more fissile material than is consumed. A reactor in which this occurs is called a *breeder* (see **Breeder Reactor**) and the conversion ratio is then referred to as the *breeding ratio*. The most highly developed breeder is a fast reactor with uranium-238 as the fertile species and plutonium-239 as the fissile species. Breeding of plutonium-239 is not practical in a thermal reactor. Breeding of uranium-233 from thorium-232 should be possible in both thermal and fast reactors, but it has not yet been demonstrated.

Safety Features

The radioactivity of the fission products which accumulate in the fuel during reactor operation has an important influence on the design of nuclear reactors of all types. Because radioactive material in the air or water constitutes a potential health hazard, special precautions are taken to ensure that any unavoidable releases to the environment during normal operation are at the lowest reasonably achievable levels. In addition, so-called "engineered safety features" are provided to minimize the escape of radioactivity in the event of a severe malfunction.

After a reactor is shut down, either deliberately or as the result of an emergency, heat continues to be generated by radioactive decay of the fission products present in the fuel. For a reactor which has been operating for some time, the rate of decay heat generation after shutdown is initially about 7 percent of the full reactor heat power. This decreases with time but is still significant after several days. Consequently, to avoid damage to the fuel by overheating, with the accompanying release of fission products, adequate cooling must be maintained for some time after the reactor is shut down. (For further descriptions of engineered safety features, see **Boiling-Water Reactor** and **Pressurized-Water Reactor**.)

O

OCEAN THERMAL ENERGY CONVERSION (OTEC)

The conversion of **solar energy** stored as heat in the ocean into electrical energy by making use of the temperature difference between the warm surface water and the colder deep water. The facilities proposed for achieving this conversion are commonly referred to as OTEC plants or sometimes as Solar Sea Power Plants (SSPP). Since the ocean waters are heated by the sun, they constitute a virtually inexhaustible source of energy. However, unlike direct solar energy, the ocean energy is available continuously rather than only in the daytime.

The operation of an OTEC plant is based on a well established physical (thermodynamic) principle. If a heat *source* is available at a higher temperature and a heat *sink* at a lower temperature, it is possible, in principle, to utilize the temperature difference in a machine or **prime mover** (e.g., a **turbine**) that can convert part of the heat taken up from the source into mechanical energy and hence into electrical energy (see **Electric Power Generation; Heat Engine**). The residual heat is discharged to the sink at the lower temperature. In the OTEC system, the warm ocean surface water is the heat source and the deep, colder water provides the sink.

The possibility of using the ocean water temperature difference to produce power was conceived in France in 1881 and verified in 1929 in an installation off Cuba. Subsequently, French scientists constructed an OTEC plant on a ship, which was stationed near Brazil, and a fixed-location plant off the West African coast. Although the principle of ocean thermal conversion was demonstrated, the projects were abandoned because of engineering problems, notably the difficulty in pumping large volumes of cold water from the ocean depths. Since the early 1970s, however, the demand for energy and the limited conventional resources, together with technological advances, have caused a revival of interest in the OTEC concept.

Open and Closed Cycles

Both open-cycle and closed-cycle turbine systems have been proposed in connection with OTEC plants. In the *closed-cycle* system (Fig. 97), a liquid working fluid, such as propane or ammonia, is vaporized in an evaporator (or boiler); the heat required for vaporization is transferred from the warm ocean surface water to the liquid by means of a **heat exchanger**. (The temperatures indicated in the figure are for ammonia as the working fluid; the values for propane are not greatly different.)

The high-pressure (HP) vapor leaving the evaporator drives an expansion turbine, similar to a steam turbine except that it is designed to operate at a lower inlet pressure (see **Vapor Turbine**). The turbine is connected to an electrical generator in the usual manner. The low-pressure (LP) vapor exhaust from the turbine is cooled and converted back into (LP) liquid in the condenser. The cooling is achieved by passing cold, deep ocean water, from a depth of 2300 to 3000 ft (700 to 900 m) or more, through a heat exchanger. The liquid work-

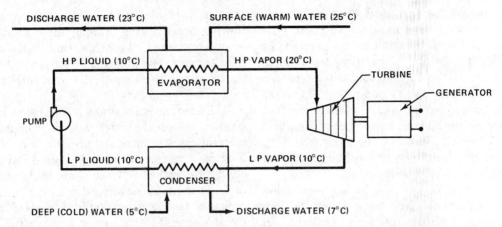

Fig. 97   Ocean thermal energy conversion.

ing fluid is then pumped back (as HP liquid) to the evaporator, thus closing the cycle.

In the *open-cycle* turbine system, water is the working fluid. The warm surface water is caused to boil by lowering the pressure, without supplying any additional heat. The low-pressure steam produced then drives a turbine, and the exhaust steam is condensed by the deep colder water and is discarded. A heat exchanger is not required in the evaporator (as it is in Fig. 97), and direct contact between the exhaust steam and a cold-water spray makes a heat exchanger unnecessary in the condenser. On the other hand, because of the low energy content of the low-pressure steam, very large turbines or several smaller units operating in parallel would be required to achieve a useful electric power output.

Modifications of the open-cycle systems have been proposed in which, instead of a steam turbine, electric power is generated by a **hydraulic turbine** operated in the usual manner by a gravity head of water (see **Hydroelectric Power**). The basic idea is to form a mixture of liquid water and vapor by vaporizing the warm, surface water at a reduced pressure. The mixture of liquid and vapor, being lighter than the bulk water, would tend to rise to a higher level, for example, in a pipe. At the top of the pipe the vapor would be removed, leav-

ing the water with a gravity head to operate a hydraulic turbine. The vapor would be condensed by cold, deep water, thus restoring the initial low pressure required for vaporizing the surface water.

Although both closed- and open-cycle turbine systems are being explored, it appears that closed-cycle systems offer the most promise for the near future. Each of the possible working fluids (i.e., ammonia and propane) has advantages and disadvantages. Ammonia has better operating characteristics than propane and it is much less flammable. On the other hand ammonia forms a noxious vapor and probably could not be used with copper heat exchangers (see below). Propane is compatible with most heat-exchanger materials, but it is highly flammable and forms an explosive mixture with air. Ammonia has been used as the working fluid in successful tests of the OTEC concept with closed-cycle systems made near the Hawaiian Islands.

Heat Exchangers

The maximum (or ideal) efficiency for the conversion of heat into mechanical work (or electricity) in a turbine depends on the drop in temperature of the working fluid in its passage through the turbine and the turbine inlet temperature (see **Thermal Efficiency**). In Fig. 97, the temperature drop in the turbine is 10°C (10K) and the inlet temperature is 20°C or 20 + 273 = 293K; hence, the maximum thermal efficiency is 10/293 = 0.034 or 3.4 percent. In an OTEC system, departure from ideal behavior in the turbine and allowance for the energy required to pump the cold water from great depths would reduce the net efficiency for electric power generation to 2 to 2.5 percent. This may be compared with the almost 40 percent efficiency of a modern, coal-fired power plant.

In the OTEC system, the low conversion efficiency would be compensated by the enormous amounts of heat available in ocean surface waters. But in order to utilize this heat to generate electrical energy in economically useful amounts, water must be pumped through the heat exchangers in both evaporator and condenser at very high rates. In a facility designed to produce 100 megawatts (see **Watt**) of electric power, for example, the total flow of water might be more than 500 million gallons (2.2 million cu m) per hour. The area of the heat exchange surfaces for both evaporator and condenser would be about 10 million sq ft (roughly 1 million sq m). (Note that 100 megawatts is only about one-tenth of the electric power generated by a single modern coal-fired or nuclear plant.)

The effectiveness and cost of the heat exchangers are regarded as critical for the OTEC concept. The electric power that can be generated depends, in the first place, on the rate of heat transfer from the warm ocean water to the working fluid in the evaporator. Furthermore, conversion of this heat into electrical energy with maximum efficiency requires that the temperature of the working fluid entering the turbine should be as high as possible and that of the fluid leaving the turbine as low as possible. All these requirements can be met only if there is effective heat transfer in the heat exchangers.

Special efforts are being made to improve the engineering design of heat exchangers suitable for OTEC use. In addition, the constructional materials must have good heat conductivity and be resistant to corrosion and erosion by rapidly flowing ocean water. Among the materials being considered, the prime candidates are (1) titanium, (2) aluminum (or an alloy), (3) an alloy of copper (90 percent) and nickel, and (4) plastic.

1. Titanium is resistant to corrosion and erosion by ocean water and it has good mechanical strength. At present, however, it is an expensive metal of limited availability. Nevertheless, since titanium is not a rare element in nature, the supply could probably be increased and the cost of the metal decreased if there were a sufficient demand.

2. Aluminum is considerably cheaper than titanium, but the common form of the metal is more susceptible to corrosion by ocean water. However, some alloys have been found to be suitable for use with salt water. Even if the aluminum alloy heat exchangers had to be replaced more often than those of titanium, they might be more cost-effective in the long run.

3. A 90/10 copper–nickel alloy has been used extensively in both land-based and shipboard power plant condensers with ocean water as coolant. It is less expensive than titanium although more than aluminum alloys. Although the copper–nickel alloy is resistant to corrosion by seawater, the copper is readily attacked if ammonia is present. Hence, copper heat exchangers could probably not be used with ammonia as the working fluid. In normal operation, the ocean water and ammonia would not be in contact, but small leaks in the heat-exchanger surface, which are almost impossible to avoid, would result in severe deterioration of the copper.

4. The possibility of fabricating heat exchangers from relatively inexpensive plastic material has attracted interest. The heat conductivity of plastics is normally too low for efficient heat exchange, but it could be increased by inclusion of graphite. But even the graphite-filled plastic is inferior to most metals in this respect and correspondingly larger heat-exchange areas would be necessary for a specified electrical

output. Plastics are expected to be resistant to corrosion by ocean water and ammonia, but some may be affected by propane. Materials with sufficient mechanical strength and resistance to erosion by high-velocity seawater remain to be developed.

Biofouling

The deposition and growth of microorganisms, called *biological fouling* (or *biofouling*), on the cooling-water side of the condenser heat exchanger, is a problem encountered in most power plants. It would also be expected to arise in both the evaporator and condenser heat exchangers of an OTEC plant. Biofouling is less with copper (or copper alloy) heat exchangers because traces of dissolved copper act as a biocide. Biofouling is important because it reduces the heat-transfer efficiency and is usually dealt with by chemical (chlorination) or mechanical (brushes or rubber balls) means. Increasing the flow rate of the water is advantageous because the organisms are less likely to become attached to the heat-exchanger surfaces. However, the flow rate must not be high enough to cause erosion.

Biofouling effects and ways of dealing with them are being studied in connection with the design and location of OTEC plants. Such effects are expected to be especially significant for the evaporator heat exchanger where the warmer water would be conducive to the growth of marine organisms.

Site Selection

In selecting a site for an OTEC facility, the primary consideration is, of course, a significant temperature difference—at least about 36°F (20°C)—between surface and deep ocean waters that will permit year-round operation. The greater the temperature difference, the lower will be the cost of generating electricity. The best sites are in the tropical belt between about 20°N and 20°S latitude. There are, however, several locations outside this area in U. S. territory that might be suitable for OTEC plants. Some of these are in the Gulf Stream off the Florida coast, in the Gulf of Mexico, and near the Hawaiian Islands. In choosing a site, consideration should be given to the potential for biofouling effects as noted earlier.

As a general rule, an OTEC plant would be located offshore in order to provide access to the deep colder water. However, an ideal situation might be one where the shoreline dropped steeply to a considerable depth. Most of the installation could then be more conveniently built on land.

Energy Utilization

If possible an OTEC plant should be less than about 20 miles (32 km) from shore. The electricity generated could then be transmitted inexpensively to land by submarine cable. At somewhat greater distances, the transmission costs would be increased but might be tolerable. If the plant is so far from shore that these costs become prohibitive, the electricity generated can be utilized at the plant site to produce energy-intensive materials.

One suggestion is to use direct electric current to decompose seawater by the process of electrolysis; the main products would be hydrogen and oxygen. The hydrogen could be liquefied and transported by tanker to a point where it could be used as fuel (see **Hydrogen Fuel**). Alternatively, the hydrogen could be combined with atmospheric nitrogen to form ammonia for use as fertilizer, thereby saving natural gas which is presently the main source of hydrogen for this purpose.

OCTANE NUMBER

A number indicating the antiknock quality (or knock resistance) of **gasoline** fuel in an internal-combustion, **spark-ignition engine.**

In the normal operation of such an engine, a flame (or combustion) front is initiated in a mixture of fuel vapor and air by a spark from the spark plug. The flame front then travels smoothly through the remainder of the fuel–air mixture causing progressive ignition to occur. Under some conditions, however, especially in engines operating at a high compression ratio (i.e., volume of the mixture of air and fuel vapor before compression divided by that after compression), the unburned air–fuel mixture ahead of the flame front is compressed and heated to the point at which a very fast explosion (or detonation) wave is propagated. This is accompanied by violent vibrations which cause a sharp metallic noise called *knock*. The knock condition leads to a loss of efficiency as well as the possibility of engine damage. Modern automobile engines operate at fairly high compression ratios for improved efficiency and hence fuels of good antiknock quality are required.

The octane number of a gasoline is based on an assigned value of 100 for the highly knock-resistant (antiknock) branched chain paraffin **hydrocarbon** 2,2,4-trimethyl pentane, commonly called isooctane, and zero for normal heptane, a straight-chain paraffin with poor antiknock quality. The octane number of a gasoline is defined as the volume percent of isooctane in a mixture with normal heptane that has the same knock (or antiknock) quality as the specified gasoline. Thus, a gasoline with an octane rating of 90 has the same knock characteristics as a mixture of 90 volume percent of isooctane and 10 percent of normal heptane.

The standard method for determining octane number utilizes a Cooperative Fuel Research (CFR) single-cylinder engine of specified dimensions in which the compression ratio can be varied. The engine is operated with a test gasoline and the compression ratio is increased until knocking occurs. If the engine has been calibrated with a number of mixtures of isooctane and normal heptane of known composition (or other fuels with previously determined octane numbers), the octane number of the test gasoline can be obtained directly

from the observed compression at which knocking occurs. There are variations in the techniques for determining octane numbers, but the general principles are the same.

The engine conditions used for determining octane numbers can affect the observed values. The research octane number (RON) is measured under mild operating conditions, with an engine speed of 600 rpm, whereas the motor octane number (MON) refers to more severe engine conditions and an engine speed of 900 rpm. The RON is usually larger than the MON. The (road index) octane value posted on gasoline pumps is commonly the average of the RON and MON. However, the MON is probably more representative of the behavior of the fuel in modern automobile engines which operate with lean (i.e., excess air) mixtures to reduce undesirable exhaust emissions.

The octane number of a gasoline can be increased by the addition of a small amount of an antiknock agent. The most common such agent is tetraethyl lead (TEL), an organic compound of lead, $(C_2H_5)_4Pb$. The related compound tetramethyl lead, $(CH_3)_4Pb$, has also been used as an antiknock agent. To prevent fouling of the spark plugs by lead oxides, a mixture of ethylene dichloride and dibromide is added to facilitate volatilization of the lead, so that it is discharged in the engine exhaust. The maximum addition before 1975 of 3 grams of lead (in TEL) per gallon of automobile gasoline increased the octane number from 75 to 90 for a poor antiknock gasoline and from 87 to 95 for one of better quality. However, efforts are being made to reduce substantially the lead content of gasoline (see below).

Because lead in gasoline can inactivate the catalyst in the converters used for emissions control, unleaded gasoline is mandatory for automobile engines with these converters. Furthermore, the toxic nature of lead exhausted to the atmosphere makes it desirable to reduce or eliminate the TEL in gasoline even when catalytic converters are not necessary. Numerous attempts to find an acceptable alternative to TEL that is an effective antiknock agent in small concentrations have not been successful.

As a general rule, the octane number of unleaded gasoline is maintained by increasing the proportion of isoparaffin and/or aromatic hydrocarbons. However, production of these fuels demands more crude oil and more refinery capacity than a leaded gasoline. An alternative possibility is to use up to 7 percent of the "octane booster" methyl tertiary butyl ether (MTBE) as a fuel component. Although MTBE is made from petroleum refinery products, the requirements are reported to be substantially less than for conventional high-octane, unleaded gasoline.

Some spark-ignition engines used in aircraft require gasolines with better antiknock properties than isooctane (octane number 100). This is achieved by adding sufficient TEL to high-quality gasoline. An octane number of more than 100 has no meaning in terms of the common definition, and so an effective octane number (or performance number) is used. It is defined as the maximum antiknock power (or pressure) for an engine using the given fuel compared with that from isooctane at the same compression ratio. Thus, a performance number of 130 means that the test fuel would produce 130 percent as much power (or pressure) before knocking as would isooctane. Aviation gasolines are rated by two performance (or effective) octane numbers; for example, a grade of 100/130 indicates a performance number of 100 for a lean mixture (low fuel-to-air ratio) and 130 for a rich mixture (high ratio).

OIL SHALE

A fine-grained, impervious sedimentary rock, consisting of compacted mineral matter (e.g., carbonates, sand, and clay) associated with an organic material called *kerogen*. Kerogen is a solid of widely vari-

able composition; it is a highly complex substance containing mainly carbon and hydrogen together with smaller amounts of nitrogen, oxygen, and sulfur. The kerogen from a typical Colorado shale contains a little more than 80 percent by weight of carbon, about 10 percent of hydrogen, and some 9 percent of the other elements.

Oil shale originated from sediments, which included substantial amounts of debris of living organisms, deposited in stagnant or stratified waters. The sediments were thus isolated from the atmosphere. The water basins were gradually filled with debris and remained quiescent for millions of years. During the course of time the organic material was converted into kerogen.

Upon heating oil shale in the absence of air, the kerogen undergoes **pyrolysis** (or destructive distillation); a liquid **hydrocarbon** mixture and a combustible gas are produced and a substantial residue of spent shale is left. The liquid product is an impure **synthetic crude oil** (or syncrude) that can be treated (upgraded) and then refined like petroleum into various fuel products (see **Petroleum Refining**).

Oil shales are classified according to the number of gallons (U. S.) of oil obtained from 1 short ton (2000 lb) of shale as determined by the **Fischer assay** (1 gal/ton = 4.2 liters/1000 kg). High-grade shales are considered to be those yielding at least 25 gal/ton; this implies that 1.7 tons of shale would yield one **barrel** (or more) of oil. Provided they form beds of sufficient thickness, these high-grade shales have the potential for development as economic sources of energy.

Oil shales are widely distributed throughout the world, with roughly two-thirds of the identified high-grade deposits being in the United States. Most of these deposits are in the Green River Formation where the states of Colorado, Utah, and Wyoming meet (Fig. 98). Most of the highest grade shales are found in a 600-sq mile (1550-sq km) area of the Piceance Creek Basin of Colorado, where the land is

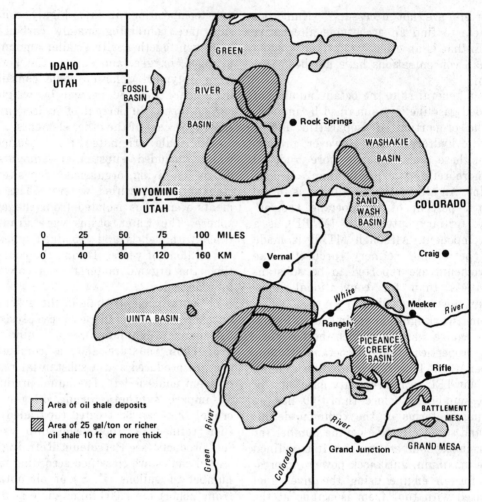

Fig. 98 High-grade oil shale areas.

largely under the control of the U. S. government. Substantial shale deposits, but of a lower grade, also occur in a roughly triangular region from Michigan to western Pennsylvania and Mississippi.

The use of oil shale as a fuel (or fuel source) was fairly common at one time in several countries, including the United States, before petroleum became readily available. However, with the increased cost of crude oil, there is a growing interest in the production of syncrude from shale. This is of special importance to the United States where the oil recoverable from high-grade shales is estimated to greatly exceed the demonstrated and inferred petroleum reserves. Unfortunately, there are many problems to be solved before oil shale can be utilized on a large scale.

Recovery of Oil from Shale

The pyrolysis of kerogen in oil shale is usually conducted at about 900 to 1000°F (480 to 540°C). The procedure for heating the shale is called *retorting*, after the name of the vessel (retort) in which substances are heated; the vapors produced are condensed and the liquid is collected. The potential processes may be considered in three categories: (1) surface retorting of mined shale, (2) underground (or in-situ) retorting without mining, and (3) modified in-situ retorting in which part of the shale is first removed by mining.

Surface Retorting. Several different surface methods have been proposed, but only a few have been demonstrated. Three of these are described below. In each case, the shale must be mined and transported to a crushing facility. Both surface (open-pit) and underground mining are possible, according to the depth of the overburden.

1. In the Union Oil Company process, the crushed shale is fed into the bottom of a vertical retort and is forced upward by an oscillating piston called a rock pump. After heating, the spent shale spills over from the top of the retort. The shale is heated from the top, and the vapors (and gases) travel downward and are partially condensed by the cool entering shale; the resulting oil and residual gases are drawn off from the bottom of the retort.

In the original (Retort A) process, heat was provided by introducing air at the top of the retort and burning some of the organic matter in the shale. This is referred to as heating by internal combustion. In the later (Retort B) form, there is no combustion in the retort; instead, indirect (or external) combustion is used to heat the shale. Part of the product gas from the bottom of the retort is passed through a separate heater and returned (recycled) to the top of the retort. Combustion in air of the **coke** (carbon) remaining in the spent shale provides the heat for the recycled gas. Indirect heating permits better control of the retorting temperature and the result is more oil of higher quality than from Retort A. In addition, the gas produced does not contain combustion products and nitrogen (from air) and has a much higher **heating value.**

2. The Paraho retort, which is a modification of the U. S. Bureau of Mines' Gas Combustion retort, differs from the Union Oil retorts in that the crushed shale enters at the top and descends under gravity. After heating, the spent shale is discharged from the bottom of the retort. The kerogen pyrolysis products travel upward and, after cooling by the entering shale, leave at the top as a mist of oil droplets suspended in gas. The liquid and gas are separated in an **electrostatic precipitator.**

In the earlier Paraho designs, the shale was heated by internal combustion (i.e., within the retort) of product (fuel) gas injected with air at three (or so) levels up the vertical retort. The temperature in the retort was controlled by adjusting the rate of fuel gas supply. In later modifications, indirect (external combustion) heating has been used; part of the product gas is heated and recycled to the bottom of the retort. Burning of some of the product gas provides the required heat.

3. In the TOSCO process, named for the Oil Shale Corporation (now Tosco Corporation), the retorting vessel (or pyrolyzer) is a horizontal rotating drum. Crushed shale is fed to one end of the drum together with hot ceramic balls which provide the heat required for retorting the kerogen. The balls are heated by combustion of the gas product in an external heating unit. The spent shale and ceramic balls leaving the other end of the pyrolyzer drum are separated by means of a screen; the spent shale passes through the screen for disposal whereas the balls are returned for reheating. The gases and vapors produced from the kerogen are cooled; the vapors are condensed to an oil product, and the residual gas is the fuel for the ball heater. A drawback to the TOSCO process is that the spent shale is a powder which requires considerable amounts of water for final disposal.

In-Situ Retorting. The objective of true in-situ retorting is to heat the shale in place underground. Not only are mining and crushing operations eliminated, but the process can be used for shales in locations where conventional mining techniques are not applicable. Furthermore, there is no problem of disposing of spent shale (see below). Heat for retorting is provided by combustion of part of the kerogen or product gas or in other ways (e.g., superheated steam or possibly microwave radiation). The vapors produced are condensed in the cooler regions of the shale bed and collected as a liquid which is pumped to the surface.

In-situ (and modified in-situ) processes are described as vertical or horizontal. In *vertical retorting*, the combustion front (or other heating zone) travels downward through the shale bed, as also do the product vapors and gases. In *horizontal retorting* the movement is in the horizontal direction; the general principles are then similar to those used in in-situ coal gasification (see **Coal Gasification: Underground**). As a rule, vertical retorting would be preferable in thick shale beds and horizontal retorting in thinner beds.

An essential requirement for in-situ retorting is adequate permeability of the shale between the heating zone and the oil collection region. This is necessary to permit movement of the heating vapors and gases as well as of the retorting products. One shale formation that has the required permeability is the "leach zone" in the Piceance Creek Basin, Colorado, where voids have been created by the dissolution in water of salts originally distributed throughout the shale bed. Tests have been conducted in the leach zone using superheated steam for heating the shale.

The majority of shale deposits, however, are not sufficiently permeable. Attempts are being made to increase the permeability by hydraulic (i.e., water pressure) and/or explosive fracturing of the shale from drilled vertical wells. Such fracturing must raise the ground surface to provide the required underground voids. Hence, as is to be expected, fracturing has proved more successful in beds at a moderate depth, with a less massive overburden, than in deep beds.

Modified In-Situ Retorting. In the modified in-situ processes, some 15 to 25 percent of the shale is removed by mining to provide spaces into which the remaining shale is broken up by blasting. One possibility is to drill a vertical shaft and excavate a number of horizontal galleries (or drifts) from it at different depths. Explosives are then detonated in the galleries to produce a large, fairly uniform volume of broken (or rubblized) shale with the required permeability. The formation of a suitable rubblized retorting zone, which is not a simple matter, is critical for the success of modified in-situ retorting.

In thick shale beds, up to several hundred feet in thickness, vertical retorting is used. After combustion has been initiated in the kerogen at the top of the broken shale by an external source, the latter is discontinued. A downward moving combustion front is then maintained by injecting air and possibly recycled product gas to serve as an alternative fuel. The hot combustion gases pyrolyze the kerogen and produce vapors which condense to oil on the cold shale below; the oil is pumped to the surface. The combustible gas produced at the same time may be recycled, as indicated above. If the shale bed is less than about 100 ft (30 m) thick, horizontal retorting would be preferred.

In some situations, it is economically advantageous to combine modified in-situ retorting with surface retorting of the mined shale. There is a substantial increase in total oil recovery, and the mixed product oil is of better quality than from surface retorting alone; in particular, the mixture has a lower **pour point** than surface-retort oil and flows more readily. Furthermore, if an external combustion process is used in the surface retort, the mixed gas product has a higher heating value than from modified in-situ retorting alone. The latter gas contains inert combustion products, nitrogen (from the air), and often carbon dioxide from decomposition of carbonates in the shale.

Retorting Products

The main product of oil shale retorting is an impure synthetic crude oil. The composition depends on the nature of the kerogen and on the retorting conditions (e.g., temperature, heating rate, and surface or in-situ). As a rule the oil is rich in paraffin hydrocarbons with smaller amounts of olefins and aromatics. However, it may contain larger proportions of undesirable sulfur and nitrogen compounds than natural crude oil. Furthermore, the oil from surface retorting has the drawback of a fairly high pour point [e.g., 70 to 90°F (21 to 33°C)].

To improve the quality of the crude shale oil, it is upgraded by **hydrotreating** (i.e., heating with hydrogen gas in the presence of a catalyst). In this process, the sulfur and nitrogen are removed as gaseous hydrogen sulfide (H_2S) and ammonia (NH_3), respectively, and the hydrocarbons undergo some chemical changes. The product can then be distilled and treated, either alone or mixed with natural crude oil, as in petroleum refining. The distillation fractions of higher boiling point can be subjected to catalytic **cracking** (i.e., heating to a high temperature with a catalyst) to yield additional amounts of lighter products (i.e., with lower boiling points), such as gasoline and other fuels.

Among the noncondensable (i.e., gaseous) products of kerogen pyrolysis are the combustible gases hydrogen, carbon monoxide, and simple paraffin hydrocarbons. In addition, there may be substantial amounts of inert carbon dioxide, as mentioned earlier, which reduces the heating value of the gas. When internal combustion provides heat for retorting, as it does in some surface processes and most true and modified in-situ processes, the product gas is further diluted by combustion products and atmospheric nitrogen. The gas then has a heating value of less than 100 Btu/cu ft (3.7 MJ/cu m). It is probable that most of the gas produced in shale retorting, regardless of its heating value, will be used locally to heat the shale and to generate the power required for the operation.

Waste Disposal

The spent shale remaining after retorting has 80 to 85 percent of the mass before retorting, but the volume, even after com-

paction, is at least 12 percent greater than the original shale in the ground. In surface retorting, all the spent shale remains and must be disposed of in an environmentally acceptable manner. One possibility is for disposal in underground mined-out areas (i.e., by backfilling). However, because of the larger volume of the spent shale, some will still remain for disposal elsewhere.

Several shale deposits are associated with minerals containing sodium carbonate and bicarbonate. In fact, large amounts of soda ash (sodium carbonate) for industrial use are obtained commercially from the Green River Basin (see Fig. 98). One proposal is to extract sodium carbonate (and alumina) from spent shale by leaching, thus recovering marketable products while reducing the volume of waste material. Complete underground disposal might then be feasible.

True in-situ retorting would eliminate the disposal problem, but it appears that this process will be of limited applicability. Modified in-situ retorting shows promise, but 15 to 25 percent of the shale, which is removed to permit the required permeability, must be retorted on the surface or disposed of without retorting. Some form of surface disposal, either of the spent or raw shale, is then required.

Current plans call for surface disposal of spent shale in a box canyon or gulch where the accumulated waste would regularly be compacted and leveled. The upper layers would then be leached with water to remove soluble salts that might interfere with plant growth. The area would be revegetated by seeding with local plants and grasses. Mulch and fertilizer would be added to provide a suitable seeding medium. In some cases, it may be desirable to cover the compacted spent shale with topsoil previously removed from the disposal site. Because of the scarce and uncertain rainfall in the areas where high-grade shales occur, irrigation may be required during the first year or so to ensure seed germination and establishment of root systems.

Water Requirements

It was thought at one time that water requirements would limit the production of oil from shale. Surface retorting was expected to require three or four barrels of water for every barrel of oil produced; almost half of the water would be used in surface disposal and revegetation of spent shale. True and modified in-situ retorting and disposal in mined-out areas would consequently greatly reduce the water requirements. Technological developments can also decrease the amount of water used in the retorting and related operations. It is now felt, therefore, that the availability of water should not seriously restrict the recovery of oil from shale.

Actually, a certain amount of water is produced in shale retorting, especially in true and modified in-situ processes; it is formed by decomposition of mineral and organic matter and by the combustion of kerogen and gaseous retorting products. However, this water is polluted with both inorganic and organic impurities. Methods are being studied for purifying the water to permit its use in the oil recovery process. The demand for water from external sources could then be decreased.

OTTO CYCLE

A repeated succession of operations (or cycle) representing an idealization of the processes in the **spark-ignition** (or gasoline) **engine** form of **heat engine** commonly used in automobiles. The stages of the Otto cycle are shown and described in Fig. 99. In the description each stage is assumed to have been completed before the next stage is initiated. However, in an actual engine there is a gradual, rather than a sharp transition from one stage to the next; hence, the sharp points in the figure would actually be rounded off.

In a spark-ignition (or Otto) engine, a mixture of fuel and vapor is formed in the cylinder and compressed adiabatically (see **Carnot cycle**) by inward motion of the piston (see Fig. 153). The partially heated

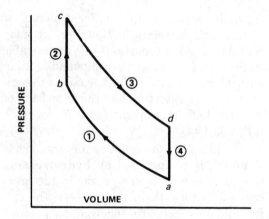

Fig. 99   Otto Cycle

1. **Adiabatic compression of the working fluid (gas) along** ab; the temperature and pressure are increased. Work is done on the gas to compress it.

2. **Heat addition along** bc **at constant volume;** the temperature of the gas increases. No work is done in this stage.

3. **Adiabatic expansion along** cd; the temperature and pressure decrease. Work is done by the expanding gas.

4. **Heat removal (rejection) along** da **at constant volume; the temperature and pressure decrease and the gas is restored to its initial condition.**

air-fuel mixture is ignited by a spark and burns at constant volume (stage 2). The resulting hot combustion gases expand adiabatically and push back the piston and thus do mechanical work (stage 3). In an actual engine, the final stage 4 is not completed because the exhaust gases are discharged to the atmosphere at d. A fresh charge of air (mixed with fuel vapor) is drawn into the cylinder to start the next cycle at a.

The net work done in the Otto cycle is the difference between the work done by the working fluid in stage 3 and that done on it in stage 1.   No work is involved (ideally) in stages 2 and 4. The **thermal efficiency** of the cycle (i.e., the fraction of the heat supplied in stage 2 that is converted into net mechanical work) is increased by increasing the temperature at c and/or decreasing that at d.   An equivalent statement is that an increase in the compression ratio (volume at a divided by volume at b) increases the thermal efficiency.

P

PEAT

A brown or dark brown soft, fibrous material formed by the partial decomposition of accumulated plant residues in moist areas. Peat is much younger geologically than coal and is commonly found in bogs at or near the ground surface. It is thought that most coals have been derived from peat by burial and the subsequent action of high temperature and pressure.

The **heating value** of an average air-dried peat, containing roughly 25 percent by weight of moisture, is about 6000 Btu/lb (14,000 kJ/kg). This is equal to the heating value of many varieties of lignite or soft brown coal (see **Coal**). Peat is used as a fuel in Ireland and especially in the U.S.S.R., where it is burned to produce steam for commercial **electric power generation**. Because of the large water content, and hence high transportation costs, the plants are located fairly close to the peat beds.

Instead of burning peat directly, it could be converted into gaseous fuel, especially **high-Btu fuel gas** (or **substitute natural gas**), in the same way as coal (see **Peat-Gas Process**). It could then be transported for considerable distances at a relatively low cost per unit of heat.

Peat resources in the United States are small compared with those of coal, and very little use has been made of peat as a fuel. However, there are certain parts of the country where large amounts of peat are available and could be used as a substitute for coal, particularly because of its low sulfur content. The largest peat-bed areas in the United States are in Minnesota, with smaller but significant areas in Wisconsin, Michigan, North Carolina, and other Atlantic states.

Both environmental and economic problems are associated with the use of peat as a fuel. The major environmental problems arise from the need to remove vegetation from and expose large areas of land to depths of 10 to 20 ft (3 to 6 m) for many years. Economic problems are mainly the costs of harvesting and drying the peat prior to direct use as a fuel or conversion into a fuel gas.

PEAT-GAS PROCESS

A process proposed by the Institute of Gas Technology for converting **peat** into a **high-Btu fuel gas** (or **substitute natural gas**). The Peat-gas process, which is a modification of the **HYGAS process** for **coal gasification**, would be conducted in two main stages. The dried peat is first reacted with hydrogen gas in an **entrained bed** at high temperature; the residual char from this stage is then gasified with steam and oxygen in a **fluidized-bed** reactor. The product is an **intermediate Btu fuel gas** which can be subjected to the **water-gas shift reaction**; if necessary, followed by **methanation** to yield a high-Btu gas.

PETROLEUM

Literally "rock oil", a naturally occurring flammable liquid consisting mainly—usually more than 90 percent—of a complex mixture of **hydrocarbons**; it is commonly called crude oil or, simply, crude. Several hundred different hydrocarbons, with up to at least 60 carbon atoms per molecule, are present in petroleum. Because hydrocarbons produce large amounts of heat upon combustion (i.e., some form of burning) in air, **petroleum refining** leads to the production of important fuels (see **Diesel Fuel; Fuel (Heating) Oil; Jet Fuel; Gasoline; Kerosine; Liquefied Petroleum Gas**). Petroleum is also the source of many chemical products called petrochemicals (see **Petroleum Products**).

Petroleum ranges in appearance from a thin (mobile), nearly colorless liquid to a thick (viscous), almost black oil. The **specific gravity** [at 60°F (15.6°C)] varies correspondingly from about 0.75 to 1.0, with most in the range from 0.80 to 0.95 (45 to 17°API). (See **API Gravity.**) The increase in density is generally associated with an increase in **viscosity** (i.e., resistance to flow).

Crude oil is found in many parts of the world, sometimes in "seeps" at or near the surface, but more commonly below the ground or the seabed. In 1859, Edwin L. Drake drilled the first commercially important oil well, near Titusville, Pennsylvania, for the specific purpose of producing petroleum as a source of kerosine fuel. Oil was encountered at a depth of just over 69 ft (21 m). As a general rule, however, petroleum occurs at much greater depths; in recent years, oil wells have been drilled down to about 30,000 ft (9000 m).

Although the types and relative amounts of the hydrocarbons and other chemical compounds (carbon compounds containing oxygen, nitrogen, or sulfur) present in petroleums from different sources vary widely, the elemental (or ultimate) compositions fall within fairly narrow limits, as shown in the table. Small proportions of many metallic elements have been found in petroleum. Of special interest are nickel and vanadium which are partly present as porphyrin complexes. The porphyrins were probably formed from

chlorophyll, the green coloring matter of plants. This suggests that petroleum is derived, at least to some extent, from plant organisms. There is also evidence of partial animal origin.

Sulfur compounds in petroleum often have unpleasant odors. Oils with less than 1 percent of sulfur usually have little or no sulfur odor and are commonly referred to as "sweet" crudes. "Sour" crudes are, in general, those with more than 1 percent of sulfur.

The hydrocarbons in petroleum fall into three main categories: (1) the paraffins (alkanes), (2) the cycloparaffins (cycloalkanes or naphthenes), and (3) the aromatics (benzene and derivatives). Nearly all crude oils contain hydrocarbons of all three types, but the proportions vary widely. Crude oils usually contain a substantial fraction of paraffins and/or cycloparaffins, whereas the aromatics are invariably present to a smaller extent than the other two hydrocarbon types.

Crude oils are sometimes classified as paraffin-base and asphalt- (or naphthene-) base crudes. The paraffin-base crudes are rich in paraffins, and they yield large proportions of high-quality lubricating oils and petroleum (paraffin) wax. The asphalt-base crudes, on the other hand, contain substantial amounts of cycloparaffins (and often aromatic hydrocarbons); the residue after distilling volatile components is a form of asphalt. Mixed-base oils are intermediate in character.

Approximate Elemental Composition Ranges of Most Petroleum Oils

Element	Weight percent
Carbon	82 to 87
Hydrogen	11 to 15
Nitrogen	up to about 1
Oxygen	up to about 2
Sulfur	up to about 6
Metals	up to about 0.05

PETROLEUM COKE

The solid, carbonaceous material remaining after severe thermal **cracking** (or coking) of **residual oils** from petroleum distillation. Petroleum coke can serve as a fuel, but it has several important industrial nonfuel uses (see **Petroleum Products**).

PETROLEUM: OCCURRENCE AND EXPLORATION

Occurrence

The formation and location of **petroleum** in amounts suitable for commercial exploitation. Petroleum generally occurs in a porous (or highly fractured) rock reservoir that is enclosed by relatively impermeable rocks so as to form a trap. Geological evidence indicates that petroleum was formed in another medium called the source rock. The oil was then forced under pressure into the reservoir rock where it accumulated because the trap prevented its escape. The word "pool", often used to refer to a petroleum reservoir (or a number of interconnected reservoirs), is misleading because it implies a body of liquid; actually, the oil is contained in crevices and other spaces between granules in the reservoir rock.

Petroleum reservoirs may also contain natural gas and water in various proportions. In some cases much of the gas is in a free state in the form of a separate layer (or phase) above the liquid; this is called a gas cap. Even if there is no free gas, there is always some gas dissolved under pressure in the oil; when the pressure is lowered, much of the gas is released. The water usually contains substantial amounts of dissolved salts, mostly common salt (sodium chloride); hence, waters associated with oil are often referred to as oilfield brines. Most of the water, which has a higher density than the oil, occurs in the reservoir in a layer below the oil. Some of the water may also be present together with oil (and gas) within the pores of the reservoir rock.

The most widely accepted theory is that petroleum was formed in ancient sea basins which gradually filled with a sediment of dead microscopic plant and animal organisms and small particles of clay, sand, and limestone brought down by rivers. As a result of complex physical, chemical, and possibly bacterial processes, the organic matter was converted almost completely into a mixture of **hydrocarbons**.

In the course of time, the sediments were covered by other rock layers, and the pressure of the overburden forced the oil to migrate to the reservoir rock. The sedimentary formations, which are the source rocks, are now largely shales and clay. Source rocks range in age up to 500 million years, but the actual formation of the petroleum probably occurred in not more than a few million years.

Reservoir rocks are commonly coarse-grained sedimentary rocks consisting of sandstones or carbonates (limestone and dolomite). In order to contain a substantial quantity of oil that can move fairly freely, a reservoir rock must be both porous and permeable. Porosity is a measure of the spaces between the rock grains in which the oil can collect; this is usually from 5 to 30 percent of the rock volume. Permeability in the present connection is the property of a porous (or fractured) solid medium that allows movement (or flow) of the contained oil; the essential requirement for permeability is that the pores (or fractures) should be interconnected.

The cap rock, that is, the top of a petroleum reservoir, is typically an impermeable clay or shale; however, limestone and marl (i.e., a mixture of clay and limestone) sometimes form cap rocks. The same material may also underlie the reservoir rock.

Traps, consisting of reservoir and cap rocks, in which the petroleum collects, fall into two general categories: structural traps and stratigraphic traps (or a combination). *Structural traps* result from some kind of distortion or disturbance of the underground rock strata. In *stratigraphic traps* the strata are often tilted but not deformed.

Structural traps occur chiefly in association with anticlines, salt domes, and faults. In an anticline, the upfolding of the rock layers causes the strata to slope in opposite directions from the crest. The manner in which the oil is trapped is shown in Fig. 100A. Petroleum occurs more frequently in anticlines than in any other type of reservoir trap. Salt domes are intrusions of rock salt, a few kilometers (or miles) in height and somewhat less in width, such as occur in the Gulf Coast region of Texas and Louisiana. The upward thrust of the salt produces distortion of the adjacent sedimentary strata so that oil traps can form (Fig. 100B). A structural trap may also result from displacement of the rock strata at a more-or-less vertical fault. The permeable reservoir rock may then be closed off by a relatively impermeable rock, as indicated in Fig. 100C.

Stratigraphic traps can arise in several different ways. A simple example is a permeable, porous reservoir rock layer, tapering at the edges, which has been deposited between impermeable layers in such a way as to form a lens-like trap (Fig. 100D). Ancient barrier reefs, similar to coral reefs, formed by living organisms, are often associated with stratigraphic traps.

Stratigraphic traps can also occur as a result of what is called a geological unconformity, that is, a lack of continuity between successive strata (Fig. 100E). An unconformity can arise when there is a substantial time delay between the deposition of the layers, so that a significant change has occurred in the lower layer. For example, an oil trap can be formed if the lower permeable layer (reservoir rock) has been tilted by folding and then smoothed off by erosion before deposition of an impermeable upper layer (cap rock).

Exploration

The first wells drilled for the specific purpose of producing petroleum were located where surface seeps indicated the presence of oil. When such evidence is not available, prospecting (or exploration) for oil is based on the identification of geologic structures, such as these described above, in which petroleum traps may exist. In some instances the identification can be made from surface observations that reveal the presence of anticlines, salt domes, or faults. As a general rule, however, geophys-

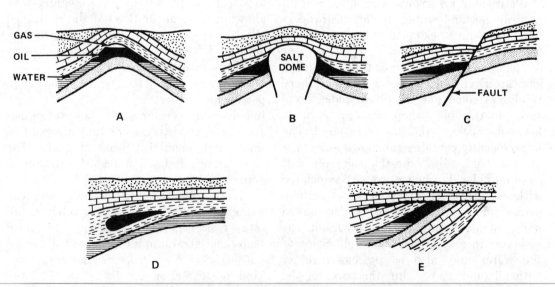

Fig. 100   Petroleum traps. A: Anticline; B: Salt dome; C: Structural trap; D: Lens; E: Unconformity.

ical methods are used in exploring for oil. When a suitable geologic structure is discovered, the presence (or absence) of oil can be determined only by drilling an exploratory well.

The chief geophysical methods used in prospecting for oil are based on (1) gravity, (2) magnetic, or (3) seismic measurements. Gravity and magnetic surveys of an area, either on land or under water, can usually be conducted at a relatively low cost. If the results of these surveys appear promising, a seismic survey will be undertaken to provide more detailed geological information.

The gravity method is especially useful for indicating the presence of salt domes or anticlines. The density (or **specific gravity**) of common (rock) salt is less than that of most other rocks; hence, the force of gravity is slightly less than average above salt domes. On the other hand, the gravitational force is generally more than average above an anticline because the uplift has brought a more dense rock layer nearer to the surface. The gravity method can also help in locating the thick sedimentary layers of coarse sandstone, limestone, or dolomite of the kind in which petroleum reservoirs can form.

A gravity survey over the ground is conducted by means of a sensitive gravity meter suspended from an aircraft flying in a definite pattern. Gravity surveys can be made under water by lowering the meter to the sea bottom or, less accurately, by a meter carried on a ship.

Sedimentary rocks have weak magnetic properties, but impermeable rocks, which might serve as cap rocks for a petroleum reservoir, are often more strongly magnetic. By observing the variations in the magnetic field strength over an area, it is often possible to locate sedimentary layers and to determine their approximate thickness. Unless the layer is thick enough, drilling for oil, even if present, would not be economically justified. Magnetic surveys over either land or sea are usually made with a magnetometer suspended from an aircraft.

Seismic exploration is more expensive than the other methods, but it provides by far the most useful information in the search for petroleum. In the most common seismic methods as used on land, a charge of explosive is detonated in a hole about 100 ft (30 m) deep. The explosion sets off seismic waves, similar to very weak earthquake waves, in the ground and their times of arrival at definite distances are determined by sensitive detectors (geophones).

In one (reflection) form of the seismic method, when the signals are recorded at moderately short distances, the seismic waves can be initiated by dropping a weight of a few tons or by striking the ground with a piston contained in a cylinder in which a gas mixture is exploded. These and similar means for producing seismic waves, as well as the detonation of explosive charges in the water, are used for offshore prospecting.

Because of the complexity of the geologic structures in which petroleum is found, the data obtained in the seismic exploration for oil are inevitably complicated. Their interpretation thus requires the efforts of experienced petroleum geophysicists. In the following descriptions, however, very simple situations will be considered in order to explain the basic principles of seismic prospecting. The techniques used are of two types: (1) the older refraction methods and (2) the more recent reflection methods.

In the seismic refraction procedure, part of the seismic wave is transmitted from the shot (explosion) point into the ground; when it strikes a layer of different material, the wave can travel along the interface before returning to the surface where its arrival is detected (Fig. 101A). From the arrival times at various distances, up to several miles from the shot point, the depth of the interface and the velocities of the seismic wave in the ground and along the interface can be determined. The refraction techniques can also be used

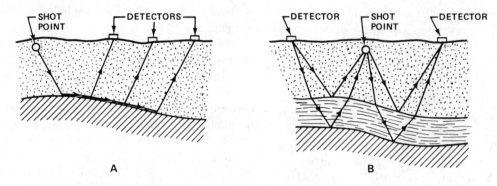

Fig. 101   Seismic exploration for petroleum. A: Refraction method; B: Reflection method.

to indicate the presence of sloping layers which might serve as oil reservoirs.

The reflection method is now the most widely used seismic technique in oil exploration. Measurements are made of the time taken for the seismic wave to travel from the shot point to an interface between two geologic layers and then directly back to detectors on the surface (Fig. 101B). Sloping layers may be indicated by having two detectors at the same distance from the shot point but in opposite directions; the times of arrival of the reflected waves will then be different at the two detectors. To determine the depths (and thicknesses) of the various geologic layers, the velocity of the seismic wave in the ground is required. This cannot be determined from the reflection data, but an experienced geophysicist can usually make a good estimate. Although the reflection method is simple in principle, interpretation of the results is often difficult because of the complexity of the geologic formations.

PETROLEUM PRODUCTION

The exploitation of petroleum reservoirs (see **Petroleum: Occurrence and Exploration**) to recover as much as possible of their crude oil content. In the United States, essentially all petroleum is recovered by drilling wells, which may be up to 30,000 ft (9000 m) deep, and pumping the oil to the surface.

Well Drilling

The most common oil-well drilling technique makes use of a rotary drill. The drilling bit is attached to the lower end of a length of drill pipe; rotation of the weighted pipe then causes the rock to be gradually worn away. Additional lengths of pipe, called *joints*, each roughly 30 ft (9 m) long, are attached at the top of the drill pipe (or *drill string*) as the hole is deepened (Fig. 102). Wear of the drilling bit makes replacement necessary from time to time. The drill string is then lifted up and *stands* of two or three joints are detached and stored vertically on the drilling rig until the bit is replaced.

The bit is lubricated and cooled, and the broken rock material is removed by circulating a specially formulated fluid referred to as drilling mud. The hydrostatic pressure due to the weight of this heavy fluid also serves to prevent escape of oil or gas that might be encountered in the drilling operation. A common drilling mud consists of a suspension in water of a clay material called bentonite plus various additives. Although the bentonite suspension is able to flow freely when pumped, it has the unusual property of setting to a semisolid gel (i.e., a jelly-like material) when stationary, for example, when the drill string is withdrawn.

Should the drill hole enter a reservoir where the pressure is greater than that exerted by the drilling fluid, there is

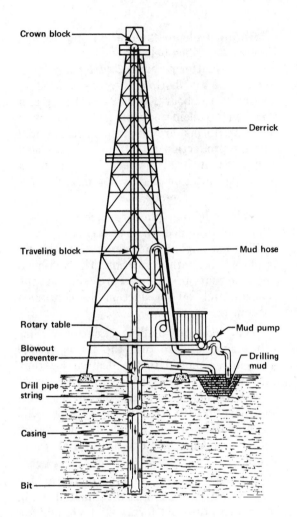

Crown block

Derrick

Traveling block

Mud hose

Rotary table

Mud pump

Blowout preventer

Drilling mud

Drill pipe string

Casing

Bit

Fig. 102   Petroleum drilling rig.

cedure called well logging. Several well-logging techniques have been used; they include measurements of spontaneous (or natural) electrical potentials, electrical resistance, sound wave transmission and reflection, gamma-ray emission from natural radioactive material, or gamma rays induced by neutron capture. An appropriate detector is lowered into the bore hole, and data are recorded at increasing depths. These measurements provide useful information about the local geology and indicate where subsequent drilling might be undertaken.

If there is evidence that the well is likely to be productive, it is cased completely with steel pipe. The casing is cemented in place by forcing a cement slurry into the space between the outside of the pipe and the wall of the drilled hole. When the hole reaches a substantial oil reservoir, the casing may not be extended to the bottom of the hole; oil can then enter the well through the bare (or uncased) portion. This scheme is known as "open hole" or "barefoot" completion. A string of pipes, called the *production string*, is extended down into the reservoir for removal of the oil.

Open-hole completion is used when the reservoir rock is fairly firm (e.g., limestone, dolomite, or hard sandstone). In a less rigid medium (e.g., loose sand) open-hole completion is not practical. The casing is then extended to the bottom of the reservoir and is perforated by means of special tools; alternatively, a slotted liner may be used as the extension. If well logging has indicated the presence of oil at two or more levels, the casing (or liner) is perforated at these levels. Up to three or four separate production strings are then lowered to the oil levels. Plugs are inserted in the casing at different locations so that production from individual reservoirs can be conducted independently.

When the well is completed, the top of the casing is closed with an arrangement of pipes and valves called a Christmas tree. The production pipe strings are attached to

danger of a blowout of oil or gas. In order to prevent this, blowout preventers are connected to the top of the casing inserted in the upper part of the drill hole at an early stage. The blowout preventers do not hinder the rotation of the drill pipe, but they close rapidly if an increase of pressure is indicated.

If the well being drilled is of an exploratory nature, undisturbed core samples may be taken from time to time by means of a special core barrel. These cores serve to identify the various formations encountered and also provide samples for porosity and permeability measurements.

Before the exploratory well is completed, however, it is examined by the pro-

this arrangement, and the oil is drawn off through the valves.

Directional Drilling. Oil wells on land are usually drilled vertically to facilitate casing. In some circumstances, however, directional (i.e., nonvertical) drilling is desirable. For example, if the reservoir is under water at a moderate distance from the shore, the simplest method of reaching the oil is to use directional drilling from a rig on land. Furthermore, in petroleum production at sea, several directional holes can be drilled from a single platform.

In directional drilling, the hole is first drilled vertically to a depth of about 1000 ft (300 m) in the normal manner. A wedge-shaped steel device, called a *whipstock,* is then inserted alongside the drill bit to change the drilling direction. The whipstock is then withdrawn and rein-serted from time to time while drilling is continued. The rate of directional change must be gradual enough to permit later insertion of the casing.

Another procedure for directional drilling makes use of a turbodrill in which the drill is driven by a downhole turbine operated by the drilling fluid. At a certain depth, the ordinary rotary drill bit is withdrawn and replaced by the turbodrill which is set to continue drilling at an angle to the vertical. When the desired direction is established, the turbodrill is removed and drilling continued with the rotary drill. In some countries all drilling, vertical and directional, is done with turbodrills.

Offshore Drilling. Large quantities of petroleum occur in offshore reservoirs that cannot be reached by directional drilling from land-based equipment. Drilling is then conducted from an offshore platform which may be supported from the sea bottom or moored in place. The drill rig, drilling procedures, and well completion operations are then similar to those employed on land. As many as 40 wells have been drilled from a single platform. Four general types of drilling platforms for offshore petroleum exploration and production are described below.

Fixed platforms supported by legs resting on the sea bottom have been used in waters up to about 800 ft (240 m) deep and are being considered for greater depths. Such platforms are used for drilling and also for production after the well has been completed. A fixed platform is built for a particular location and is not intended for later use elsewhere.

Mobile drilling platforms, on the other hand, are designed for repeated use. When drilling is completed at one location, the rig is moved to another location. The wellhead and blowout preventers are set on the seabed, and oil production is continued without the platform.

The jackup drilling platform has retractable legs, so that the rig is mobile and can be moved to the desired location. The legs are then extended until they rest on the sea bottom and the platform is at a safe level above the water surface. The jackup platform is designed for use in water less than about 330 ft (100 m) in depth.

Another type of mobile platform for offshore drilling is a semisubmersible rig in which the platform is mounted on large-diameter hollow columns that serve as floats. The rig is towed to the drilling location, and the floats are ballasted with water until the platform sinks to the desired level. It is then moored with cables attached to a number of heavy anchors.

For drilling in deeper water, a drill ship must be used. Such a ship has a central hole or pool through which the actual drilling is performed. In moderately deep water, the ship can be moored at the drill site; in deep water, computer-controlled propellers (or thrusters) maintain the ship's position.

Production

If the pressure in the petroleum reservoir is sufficient to force the oil to the surface, natural flow will occur as soon as the well is completed. The pressure may arise from gases collected above the oil layer (*gas-cap drive*), from dissolved gases released from the petroleum when the reservoir is opened up (*dissolved-gas drive*), or from the water beneath the oil layer (*water drive*). When this initial phase, called "flush" production, ceases, the oil must be pumped from the reservoir. In this stage, referred to as "settled" production, most of the oil is produced.

When the petroleum flow rate slows to about 10 barrels per day, "stripper" production is employed. Pumping may then be intermittent, with intervals to allow the oil to accumulate in the reservoir region near the well. Although the production from a single stripper well is small, the large number of such wells in operation produce a substantial quantity of oil.

If the production from a well is low, especially if it is the result of low permeability of the reservoir rock (see **Petroleum: Occurrence and Exploration**), the oil flow may be stimulated by opening up fissures and cracks around the bore hole. In carbonate (limestone or dolomite) formations, hydrochloric acid may be used to increase the permeability by dissolving the carbonate. Hydrofluoric acid can serve the same purpose in sandstone deposits.

Hydraulic (or massive) fracturing is also used to stimulate petroleum flow; fluid at a pressure up to 6000 to 8000 lb/sq in. (41 to 55 MPa) is pumped into the reservoir to enlarge the cracks. The fracturing fluid consists of water or oil with coarse sand or other granular material in suspension. The solid particles remain in the fractures and prevent sealing when the pressure on the fluid is reduced.

Oil–Gas–Water Separation

Gas and water (oilfield brine) present in the reservoir often accompany the oil reaching the wellhead. Furthermore, the decrease in pressure of the oil leaving the well results in the release of a large fraction of the gas dissolved in the oil. The gases are separated before the oil is pumped to a storage tank. Apart from small amounts of such extraneous constituents as nitrogen, carbon dioxide, and sometimes helium, petroleum gases consist mainly of the paraffin **hydrocarbons** methane (CH_4), ethane (C_2H_6), propane (C_3H_8), and butanes (C_4H_{10}) in various proportions. Of these, the propane and butanes can be readily liquefied under pressure to form **liquefied petroleum gas**. The methane and ethane, which remain in the gaseous form, constitute what is called associated **natural gas**, that is, natural gas associated with oil. The proportions of gas and oil produced vary widely from one location to another.

As a result of turbulence in pumping, the oil and water at the wellhead are frequently in the form of an emulsion of small oil droplets in suspension in the water. Sometimes the emulsion can be "broken" by heating; the water, which has a higher density, then separates by gravity. Various demulsifying chemicals may be added, with or without heat, to facilitate the breakup of the emulsion. Another method for separating the oil from the water is to pass the emulsion through a high-voltage electric field.

Because of the salt content of many oilfield waters, their disposal is a problem. The water is frequently returned to the ground by way of deep wells in the vicinity of the producing wells. Injection of water in this manner can contribute to the secondary recovery of oil (see below).

Secondary and Tertiary Recovery

When the rate of stripper production falls to the point where it is no longer economic, a substantial fraction of the original oil still remains in the reservoir rocks. The proportion of oil recovered varies with the nature of the geologic formation, but an average value in the United States is about 32 percent. Up to almost another 28 percent may then become available by secondary recovery which, however, increases the cost of the oil. In recent years, secondary recovery has contributed roughly 30 percent of the oil produced in the United States.

A common secondary recovery procedure is water-flooding: a number of injection wells are drilled in a definite pattern in relation to a producing well (or wells). Water is then pumped into the injection wells, and this forces the oil to flow toward the producing well. As much as possible of the water initially pumped with the petroleum is returned to the ground in the water-flooding operation. As an alternative to water injection, gas (air or **casinghead gas**) injection has been used to increase the reservoir pressure. In some cases secondary recovery has been achieved by returning the casinghead gas to the top of the reservoir, thereby producing gas-cap drive.

Heavy crude oil is so viscous that normal pumping results in the recovery of less than 10 percent of the oil present. Water (or gas) flooding is not effective; further recovery is achieved by heating the oil to decrease its viscosity and permit it to flow. Since the procedures are specialized, they are described separately in the section on **heavy crude oil**.

Tertiary recovery methods are being developed for use in lighter-oil reservoirs when secondary recovery has reached its economic limit. In the miscible drive process, carbon dioxide gas is injected into the oil reservoir; the carbon dioxide dissolves in the oil and decreases its viscosity and thus increases its mobility. Subsequent water-flooding results in additional recovery. In an alternative miscible drive process, a quantity of a light hydrocarbon, such as liquid propane or butane, is dissolved in the oil to increase its mobility.

A different approach to tertiary recovery is to introduce a "slug" of a surface-active material (or *surfactant*) in water. A surfactant is a compound similar to a detergent which tends to collect at the rock-oil interfaces and thus reduces the capillary forces that oppose the flow of oil through cracks or other narrow spaces. The surfactant addition is followed by a slug of water containing a *polymer*, and then by water-flooding. (A polymer is a very large molecule formed by the combination of many, often identical, small molecules.) The polymer increases the viscosity of (i.e., thickens) the water, so that the polymer slug acts as a sort of buffer between the surfactant and oil in front and the water-drive behind. (The surfactant-polymer technique is sometimes called *micellar-polymer flooding*; a micelle is an aggregate of surfactant molecules formed in water.)

Simpler tertiary recovery procedures, such as polymer-augmented water-flooding without a surfactant and use of a solution of sodium hydroxide (caustic soda) in water, are being investigated. These procedures lead to smaller recovery than by micellar-polymer flooding but the costs are much less.

PETROLEUM PRODUCTS

Products derived from **petroleum** (crude oil) and to some extent from **natural gas**. These products may be considered in three broad categories: (1) fuel materials, (2) finished nonfuel products, and (3) raw materials (or feedstocks) for the chemical industry (i.e., for petrochemicals).

Fuel Materials

From the energy standpoint, fuel materials are the petroleum products of major interest; they are treated in other sections under the following headings:
Diesel Fuel
Fuel (Heating) Oil
Gasoline
Jet Fuel
Liquefied Petroleum Gas (LPG)
See also **Gas Oil; Heavy (Fuel) Oil; Kerosine; Natural Gas; Petroleum Coke.**

Nonfuel Products

The chief nonfuel products from petroleum are solvents, lubricating oils and greases, petroleum (or paraffin) wax, petrolatum (or petroleum jelly), asphalt, and petroleum coke. There are also specialized products for the textile, metallurgical, electrical, and other industries.

Solvents. Many different **hydrocarbon** solvents are obtained as distillates in petroleum processing (see **Petroleum Refining**). The boiling ranges vary from about 110 to 310°F (45 to 155°C) for the most volatile solvents to about 320 to 570°F (160 to 300°C) for the least volatile. Some uses of petroleum (**naphtha**) solvents are for paints, lacquers, printing inks, rubber products, dry cleaning, and wax polishes, and for the extraction of oils from vegetable products.

Lubricating Oils and Greases. These are derived from the relatively viscous hydrocarbon oils of high boiling point obtained by the vacuum **distillation** of the residues remaining after removal of the more volatile constituents of crude oil. The vacuum distillate may include waxy material which is removed, as described below. The properties of lubricating oils vary according to the type of machinery for which they are to be used.

Oils for automobile engines are generally characterized by their SAE (Society of Automobile Engineers) **viscosity** numbers. Crankcase oils in common use are SAE 10W and 20W for winter driving and SAE 20 and 30 for summer. (The numbers 10, 20, 30 indicate increasing viscosities.) Multigrade lubricating oils, such as 10W–30 are intended for use over a wide range of ambient temperatures. They are made by adding various polymeric compounds to hydrocarbon oils. Essentially all automobile lubricating oils also contain other additives, including antioxidants (to reduce formation of acids and insoluble resins by atmospheric oxygen), antiwear products, and surface-active materials (detergents).

Semisolid greases consist of lubricating oils and a thickener, usually a soap (or soaplike) material made from an animal or vegetable oil and an alkali (e.g., calcium, lithium, or sodium hydroxide). Greases can be made by heating a mixture of a lubricating (petroleum) oil and animal or vegetable oil with the alkali. Alternatively, a prepared soap, such as aluminum or lithium stearate, is mixed with the petroleum oil to produce a grease. Various additives have been developed for extending the use of petroleum-based greases to high or low temperatures.

Petroleum Wax and Petrolatum. When the lubricating oil fraction obtained in the vacuum distillation of petroleum residues is cooled, wax that may be present tends to deposit in solid form. The wax is best separated from the oil by solvent extraction. The vacuum distillate is dissolved in a suitable solvent, such as liquid propane or a mixture of benzene, toluene, and methyl ethyl ketone. When the solution is chilled, the solid wax separates and can be removed by filtration. The solvent is recovered by distillation, leaving the dewaxed lubricating oil. The highest yields of wax and the better lubricating oils are obtained from paraffin-base crude oils (see **Petroleum**).

Petroleum (or paraffin) wax is used extensively in the paper and packaging industry and for making candles, matches, and polishes. A smooth blend of paraffin wax and a heavy petroleum oil, called petrolatum (or petroleum jelly), forms the basis for medical and toilet preparations and other products.

Asphalt. Asphalt occurs naturally, but a related complex hydrocarbon mixture, plus small amounts of sulfur, nitrogen, and oxygen compounds and other substances, remains after removal of volatile matter in the distillation of asphaltic (or naphthenic) crude oils. Asphalt has many uses, including paving roads and airfields, the manufacture of roofing materials and floor coverings, and in hydraulic engineering (e.g., to surface breakwaters, canals, reservoirs, etc.).

Petroleum Coke. This product remains after severe **cracking**, (called coking) of heavy residual oil from petroleum distillation. Petroleum coke is used in the manufacture of electrodes for dry cells and for the aluminum industry, and of "brushes" for electric motors. It is also used for the production of artificial graphite. If the sulfur content is small enough, petroleum coke serves as a relatively low-ash, solid fuel.

Petrochemical Feedstocks

The petrochemical industry produces a large number of substances by the chemical treatment of raw materials derived from petroleum (or natural gas). The major petrochemical feedstocks are the hydrocarbons ethane, ethylene, propylene, normal- and iso-butylenes, butadiene, and aromatics. Among the final products are plastics (including synthetic rubbers), synthetic fibers, drugs, and detergents.

Ethylene is made mainly by **steam cracking** of ethane that is separated from natural gas. In this process, the ethane is heated with steam at a temperature of roughly 1300 to 1500°F (705 to 815°C). The product is cooled and separated into liquid fractions by a form of distillation based on compression in stages.

Steam cracking of naphthas and heavier oils yields a wide variety of products, including ethylene, propylene, butylenes, butadiene, and aromatics (see below). Butylenes and butadienes are also made to a small extent by catalytic dehydrogenation (i.e., hydrogen removal) from normal butane. Propylene and butylene are by-products of catalytic cracking of petroleum distillates.

The aromatic hydrocarbons benzene, toluene, and xylene, commonly referred to as BTX, occur in some natural crude oils. They are also found in the **synthetic crude oil** (syncrude) obtained in **coal liquefaction** processes. In fact, coal tar was at one time the only source of aromatic hydrocarbons. At present, they are produced mainly by catalytic **reforming** or steam cracking (see above) of naphthas. The BTX mixture

is extracted by a suitable solvent from the fraction boiling below about 300°F (150°C). The solvent dissolves the aromatics but not the other hydrocarbons; the BTX is then recovered from the solution by distillation.

PETROLEUM REFINING

The treatment of **petroleum** (crude oil) to yield a variety of fuel and nonfuel products (see **Petroleum Products**). The refinery operations described here are mainly those concerned with the production of fuels. The crude oil is first fractionally distilled (see **Distillation**), and the direct (or straight-run) distillation products are usually subjected to further processing to improve their characteristics (e.g., increase the **octane number** of gasoline) or to change their composition (e.g., convert heavy oils of high boiling point to lighter fuels).

The proportions of straight-run distillation products are not necessarily consistent with the demand for these products. For example, straight-run gasoline constitutes an average of 20 to 25 percent by volume of crude oil, but the total amount of gasoline used in the United States represents almost half of the oil consumed. Furthermore, there are seasonal variations in the requirements for such products as gasoline and heating oils. Refinery operations are designed to provide the flexibility to meet varying demands by appropriate processing.

The raw products obtained from the various processing operations are separated by distillation in the usual manner into fractions in various boiling ranges. These fractions may include **refinery gases, gasoline, kerosine, diesel fuel,** and heavier **fuel (heating) oils.** They may be blended and marketed directly or used as feedstocks for other refinery processes.

Distillation

In the initial fractional distillation of petroleum, sometimes called "topping," the oil is passed through a series of tubes in a furnace where it is heated to a temperature of 600 to 750°F (315 to 400°C), depending on the nature of the crude oil and the desired products. The hot vapor mixed with some liquid enters the fractionating column where it meets the downflowing liquid reflux. Steam is often introduced at the bottom of the column, so that steam distillation and fractionation occur simultaneously.

As a rough general rule, the boiling temperature of a liquid **hydrocarbon** increases (i.e., the volatility decreases) as the number of carbon atoms in the molecule increases. Hence, when crude oil is fractionally distilled, the less volatile (or heavier) components, containing the more complex molecules with the largest number of carbon atoms, condense at the lower levels of the fractionating column (see Fig. 30). The simpler (lighter) components, with smaller numbers of carbon atoms per molecule, condense at successively higher levels. The various liquid fractions (or side streams) are drawn off at different levels for use or further treatment. Gases present in the petroleum are not condensed but are removed from the top of the column.

Some typical fractions, with the numbers of carbon atoms per molecule, approximate boiling ranges, and principal uses (after treatment), are tabulated below. The **specific gravities** and **viscosities** increase in the order given. Fractions called **naphthas** and **distillates** are sometimes collected; naphthas partially overlap gasoline and kerosine, and distillates overlap kerosine and fuel (or gas) oil.

To prevent excessive decomposition of the oil by heat, the distillation temperatures are usually maintained below about 650°F (345°C), where the lubricating oil fraction would begin to distill at atmospheric pressure. The liquid drawn from the bottom of the column, called "topped" or "reduced" crude, is then vacuum distilled at a lower temperature to obtain lubrication products (oil and grease). According to circumstances, the residue may be used as fuel oil or as asphalt; alternatively, it may be broken down (by cracking) into simpler products (see below).

Processing

Many different processes have been developed for changing the compositions and characteristics of petroleum distillation fractions. The more important of these processes are outlined here and are described more fully elsewhere under the indicated headings. The extent to which any process is used depends on the nature of the material subjected to treatment and the demand for the products.

Cracking. Thermal **cracking** (i.e., the breaking of large hydrocarbon molecules into simpler molecules by heat alone) was used extensively at one time in petroleum refining. However, it is now employed to a limited extent only (see **Flexicoking Process**). *Visbreaking* is a mild form of thermal

Typical Petroleum Distillate Fractions

Fraction	Carbon atoms per molecule	Approximate boiling range		Major uses
		°F	°C	
Gasoline (straight-run)	5–12	85–390	30–200	Gasoline engines
Kerosine	10–15	355–525	180–275	Diesel and jet engines
Fuel (or gas) oil	15–22	500–650	260–345	Heating
Lubricating oil	20–30	645–750	340–400	Lubrication
Residue	Large number	High	High	Heating, asphalt, etc.

cracking which causes a partial breakdown of the complex hydrocarbons in distillation residues and thus reduces the viscosity. Distillation of the resulting product yields some gasoline and kerosine, but the main fraction is a medium-heavy distillate fuel oil.

Coking, on the other hand, is an extreme form of thermal cracking used to convert heavy distillation residues into various lighter (lower molecular weight) oils, leaving a residue of **petroleum coke.** The lighter oils can serve as feedstocks for catalytic cracking.

Catalytic cracking is now in wide use for converting petroleum distillates with medium and high boiling ranges into more volatile products, especially gasoline. The catalyst is usually a special zeolite (aluminosilicate). The gasoline distilled from the catalytically cracked product has a high octane number because it contains a substantial proportion of isoparaffins and aromatic hydrocarbons. Some light olefins (propylene and butylenes) are also formed.

In *hydrocracking,* the petroleum product and hydrogen gas under pressure are passed over a dual catalyst, consisting of a cracking catalyst (e.g., silica-alumina or a zeolite) which serves as support for a hydrogenation catalyst (e.g., platinum or other metal). Nitrogen compounds in the hydrocarbon feed can poison (i.e., inactivate) the latter catalyst; consequently, in the two-stage hydrocracking process, the nitrogen is removed as ammonia gas in the first stage. The catalyst is then a cracking catalyst as support for a metallic sulfide as hydrogenation catalyst.

Hydrocracking has been applied to a wide variety of feedstocks, from light distillates (e.g., kerosine) to heavy residual oils. The products range from petroleum gases, through high-octane gasoline (rich in isoparaffins and aromatics), to middle distillate fuel oils.

Reforming. In catalytic **reforming,** the feedstock is a low-octane gasoline or light naphtha; it is passed with hydrogen gas over a heated catalyst consisting of platinum (with rhenium or iridium) distributed on an alumina base. Several different reactions take place; the most common are dehydrogenation (i.e., hydrogen removal) and isomerization (see below). The overall result is the conversion of normal (straight-chain) paraffins into isoparaffins (branched chains) and aromatic hydrocarbons and of cycloparaffins (naphthenes) into aromatics. In each case, the products have higher octane numbers than the feed materials; consequently, reforming is used mainly to increase the antiknock qualities of the gasoline fraction of crude oil.

Alkylation. In petroleum refining, **alkylation** is the reaction of the isoparaffin isobutane (i-C_4H_{10}) with the simple olefinic hydrocarbons propylene (C_3H_6), butylenes (C_4H_8), and amylenes (C_5H_{10}) in the presence of a catalyst (sulfuric or hydrofluoric acid). The product, called *alkylate,* is mainly a mixture of isoparaffins boiling in the gasoline range and having a high octane number. The isobutane feed is commonly obtained from the refinery gas butane by the isomerization process described below; the olefins are by-products of catalytic cracking and other refinery operations.

Isomerization. In the present context, **isomerization** refers specifically to the conversion of a normal (straight-chain) paraffin hydrocarbon into an isoparaffin (branched chain) without changing the numbers of carbon and hydrogen atoms in the molecule. The process involves passage of the normal paraffin vapor over a heated catalyst, which may be aluminum chloride (with hydrogen chloride) or platinum. Hydrogen gas is commonly included to prolong the life of the platinum catalyst.

The major applications of isomerization in petroleum refining are the conversion of normal butane (C_4H_{10}) into isobutane for use in alkylation and of normal pentane (C_5H_{12}) and hexane (C_6H_{14}) into branched-chain forms (isomers). Normal pentane and hexane, commonly found in **natural gasoline** and **straight-run gasoline,** have low octane numbers, but their branched-chain isomers have higher octane numbers. They also volatilize more readily and make engine starting easier in cold weather.

Polymerization. In petroleum processing, **polymerization** is the combination, in the presence of phosphoric acid as catalyst, of two simple olefin hydrocarbons, which may be the same or different, to form one or more complex olefins. For example, a mixture of propylene (C_3H_6) and butylene (C_4H_8) will polymerize to form olefins with six, seven, or eight carbon atoms per molecule. Polymerization was used at one time to make a gasoline component, but it has now been replaced by alkylation.

Hydrotreating. The general term **hydrotreating** covers a number of different processes involving the treatment of a petroleum product with hydrogen gas in the presence of a catalyst. The reaction conditions depend on the nature of the feed and the purpose of the treatment. The use of hydrotreating has increased in recent years because of the availability of by-product hydrogen from reforming processes.

One of the major objectives of hydrotreating is to remove combined sulfur by conversion into hydrogen sulfide gas (see **Desulfurization of Liquid Fuels**). Other applications are in the pretreatment of catalytic hydrocracking and reforming feeds to prolong the useful life of the catalyst by removing nitrogen and oxygen compounds, conversion of olefins into paraffins and naphthenes, and decolorization of liquid fuels.

Sweetening. The methods for removal of unpleasant odors from petroleum products are called **sweetening processes.** These odors are largely due to the presence of sulfur compounds called *mercaptans.* With the increasing use of hydrotreatment, which removes sulfur, sweetening has become less important. If the sulfur content of the raw product is low enough, hydrotreating may be eliminated; sweetening, if desirable, may

then be achieved by oxidizing the mercaptans (RSH) into essentially odorless disulfides (RS · SR).

PETROLEUM: STORAGE AND TRANSPORTATION

The storage and transportation of crude oil and refinery products. Petroleum and its liquid products are commonly stored in large, aboveground, cylindrical steel tanks at oil wells, trans-shipment points, and refineries. In some cases, excavated, concrete-lined reservoirs are used. Underground caverns, especially in salt formations, provide storage for large quantities of petroleum products. An emergency supply of crude oil (Strategic Petroleum Reserve) is being stored in a salt mine and caverns in salt domes in Texas and Louisiana. Oil pumped from offshore wells may be stored in various kinds of anchored-floating or bottom-supported tanks in the water adjacent to the wells.

Crude oil is transported from oilfields to refineries and from the refineries to consumers by water (tankers for overseas shipment and barges for inland waterways), by pipeline (overland or underwater), by railroad (tank cars), or by highway (tank trucks). The major modes of transportation are by pipeline and tanker ships.

Pipelines

Petroleum pipelines range in diameter from about 4 in. (10 cm) to 50 in. (1.27 m); the narrower lines are generally used for short-distance transportation of relatively small amounts of oil, whereas the wider lines carry large quantities, especially over long distances. To maintain the flow of the oil through a long pipeline, pumping stations are located roughly 50 to 150 miles (80 to 240 km) apart.

Oil pipelines fall into two general categories: crude-oil lines and product lines. There are more than 100,000 miles (160,000 km) of *crude-oil pipelines* in the United States. Of these, about 40 percent are gathering or feeder lines that carry oil from wells at several locations to a nearby refinery or to a trunk line transfer point. Trunk lines, which constitute the remainder of the crude-oil lines, then transport the oil to a more distant refinery. In the Persian Gulf oilfields, gathering and trunk lines carry the crude oil from inland wells to the coast for overseas shipment by tanker.

The Trans-Alaska Pipeline System, connecting the oil fields at Prudhoe Bay on the north coast of Alaska to the ice-free, deep-water port at Valdez in the south, is of special interest. The line, completed in 1977, is almost 800 miles (1280 km) long and 4 ft (1.22 m) in diameter. This pipeline rises from sea level to an elevation of 4800 ft (1460 m) in crossing the Brooks Range and later to 3500 ft (1070 m) in the Alaska Range before finally descending to sea level. It is designed to withstand the most severe earthquakes that may be experienced in Alaska. The initial capacity of the pipeline was about 1.2 million **barrels** per day, but it may be increased later to 2 million barrels.

In the United States, about 70,000 miles (112,000 km) of *product pipelines* carry finished petroleum products from refineries to various points for local distribution. A pipeline can transport several products simultaneously by introducing them in succession. The products are then arranged in order of increasing or decreasing density; for example, gasoline is followed by kerosine and then by diesel fuel, or the reverse. To minimize mixing at the interfaces, the rate of flow through the pipeline must exceed a minimum value, which varies with the pipe diameter and the nature of the transported products. The change from one product to another at the pipeline terminal is usually indicated by the change in density.

Tankers

A tanker is an ocean-going vessel used for the bulk transportation of petroleum and petroleum products. The capacity of a tanker is expressed in deadweight tons (DWT); this is the weight in *long tons* of the oil plus the stores (i.e., food, water, and fuel) required during the sea voyage. For larger tankers, the stores represent a small fraction of the total capacity, and the DWT is an approximate measure of the quantity of oil that can be carried. One long ton [2240 lb (1016 kg)] is equivalent to about 7.4 barrels, 310 U.S. gallons, or 1.2 cubic meters of crude oil.

Prior to 1950, the standard tanker had a capacity of 16,500 DWT; the maximum capacity was roughly 25,000 DWT. Subsequently, with the increased demand for oil, especially from the Middle East, larger and larger tankers were constructed. By the mid-1960s, tankers of 100,000 DWT were common. The closing of the Suez Canal in 1967 provided the impetus for the design of still larger tankers to transport oil economically from the Persian Gulf to Europe and America by the route around Africa. In the 1970s very large crude carriers (VLCC), with a capacity of 200,000 DWT or more, and ultra large crude carriers (ULCC) of 300,000 DWT came into use. A few still larger tankers with capacities up to more than 500,000 DWT have been built. However, with the reopening of the Suez Canal and the decreased worldwide consumption of petroleum products, interest in very large tankers has declined.

The increase in carrying capacity of tankers has been accompanied by an increase in the draft (i.e., the depth in the water when the tanker is fully laden). Deep-water ports are thus required to accommodate the larger (200,000 DWT or more) tankers. Several such ports have been developed, but there are none in the United States. Consequently, when very large tankers carry oil destined for the United States, the oil is transferred to smaller tankers at a suitable location in Canada or the Bahamas.

Two alternatives to deep-water harbors are called single-point mooring (SPM) systems and sea-island structures. In the SPM systems, the tankers are berthed at a mooring buoy anchored in deep water. A sea-island structure, on the other hand, is supported by legs resting on the bottom. In either case, a water depth of at least 80 ft (24 m) is required for tankers of 300,000 DWT or more. A submarine pipeline or small tanker is used to transfer the oil from the SPM or sea-island structure to storage facilities on shore. These methods can also be used for loading and unloading very large tankers when sufficiently deep ports are not available.

PIPELINE QUALITY

A fuel gas of pipeline quality must satisfy the following specifications of the U.S. Department of the Interior:

Minimum heating value	900 Btu/cu ft (33.5 MJ/cu m)
Maximum inert constituents	5 volume percent
Maximum carbon monoxide	0.1 volume percent
Maximum sulfur content	10 grains/cu ft (2.3 grams/cu m)

See **Heating Value.**

POLYMERIZATION

In general, the combination of two or more simple molecules to form a more complex molecule, called a *polymer*. Rubber (and other plastic materials) and artificial fibers are examples of polymers. In the petroleum industry, polymerization refers to the combination of two simple olefin molecules (see **Hydrocarbons**), especially propylene (C_3H_6) and butylenes (C_4H_8), to form olefin compounds containing six, seven, or eight carbon atoms per molecule. The olefin feedstock is a by-product of catalytic cracking operations (see **Petroleum Refining**).

Polymerization of gaseous olefins is usually conducted in the presence of phosphoric acid as catalyst. The catalyst may be absorbed in kieselguhr (a porous diatomaceous form of silica), or it may be used in the liquid form. With the kieselguhr-based catalyst, the operating temperature is roughly 390 to 480°F (200 to 250°C), and the pressure is in the range of 35 to 70 atm (3.5 to 7 MPa), depending on the design of the reactor.

Polymerization was used at one time in petroleum refineries to make a **gasoline** component, but this procedure has been abandoned because a superior product can be obtained by using the propylene and butylenes for the **alkylation** of isobutane. However, polymerization is still used to some extent for the production of petrochemicals.

POTASSIUM VAPOR CYCLE

A **heat engine** using potassium vapor as the working fluid in a condensing (**Rankine cycle**) turbine, similar in principle to a **steam turbine** except that the working vapor enters and leaves at much higher temperatures (see **Vapor Turbine**). The potassium vapor cycle was originally developed in the 1960s as a possible power source for space vehicles. At present, however, its main interest is as a potential **topping cycle** for a steam turbine in a **combined-cycle generation** system.

The steam (or heat source) temperature in a steam turbine is limited by the properties of materials to about 1000°F (540°C). Because liquid potassium has a higher boiling point even at atmospheric pressure [(1500°F (815°C)], the potassium vapor turbine can take better advantage than a steam turbine of the high-temperature heat produced by the combustion of coal or other fuel. By combining a potassium vapor topping cycle with a conventional steam turbine, the overall **thermal efficiency** (i.e., proportion of heat converted into useful mechanical work or electrical energy) could be increased by at least 10 percent. This would mean about 25 percent decrease in fuel consumption for the same electrical output.

In a potassium vapor cycle, the liquid potassium is vaporized at a moderate pressure in a boiler heated by the combustion in air of coal (possibly by **fluidized-bed combustion**) or of **low-Btu fuel gas** made from coal. The vapor at a temperature of roughly 1540°F (840°C) operates the turbine, and the exhaust is condensed to liquid in a **heat exchanger** using water under high pressure as the coolant. The heat removed from the condensing potassium vapor converts the water in the heat exchanger into superheated steam at a temperature near 1000°F (540°C), suitable for use in a steam turbine. The condensed liquid potassium is returned to the boiler for recirculation (Fig. 103).

cled as slurry oil. The remainder is a fuel oil. The heavy liquid residue after distillation is cooled and carbonized to yield a low-sulfur, low-ash coke. The more recent **Consol Synthetic Fuel** and **Solvent Refined Coal Processes** are improvements of the Pott–Broche process.

POUR POINT

The lowest temperature, expressed as a multiple of 5°F (or 3°C), at which an oil is observed to flow when it is cooled and examined under prescribed conditions. The pour point is determined by heating the oil

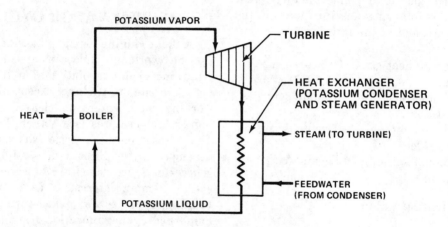

Fig. 103   Potassium-vapor topping cycle.

POTT–BROCHE PROCESS

A process invented by A. Pott and H. Broche in the late 1920s and developed by I. G. Farben for solvent refining of coal. It was used to some extent in Germany in World War II primarily to produce a raw material for fabricating carbon electrodes for aluminum plants.

Coal is formed into a slurry with a hydrogen-rich (hydrogen-donor) solvent oil and then heated at a temperature of about 800°F (425°C) and a pressure of 150 atm (15 MPa). Part of the coal is dissolved and hydrogenated, and the liquid product is separated from the residual solid. The liquid is distilled, and the major portion is hydrogenated with hydrogen gas and recy-

in a cylindrical test jar until it flows and then cooling it slowly. At intervals of 5°F (or 3°C), the jar is tilted; when the oil ceases to flow, it is held in a horizontal position for exactly 5 seconds. If there is no movement, the preceding temperature is taken as the pour point. The pour point of an oil depends on its composition; for example, oils containing significant amounts of wax have high pour points.

POWER

The rate with respect to time at which work is done or energy transformed (e.g., from mechanical work to electrical energy) or transferred (from one place to another);

that is, it is the amount of work done or energy transformed or transferred in unit time. The total amount of work done, etc., is the product of the power and the time during which the power is operating. (More exactly, the work done is the integral of the power over the operating time period.) The common unit of power is the **watt** which represents work done, etc., at a rate of 1 **joule** per second. The *horsepower*, often used to express the rate of work done by a machine, is equivalent to 746 watts.

An electric current of 1 ampere flowing under an electromotive force (or voltage) of 1 volt develops a power of 1 watt. Hence, in a direct-current circuit, the power in watts is equal to the product of the current in amperes and the electromotive force in volts. In an alternating-current circuit, the actual power transmitted is equal to the apparent power (i.e., the product of the effective current and effective power) multiplied by the **power factor** (see also **Electric Power Transmission**). Electric power is often expressed in kilowatts (1 kilowatt = 1000 watts) or megawatts (1 megawatt = 1 million watts). (Electrical energy is correspondingly stated in kilowatt-hours.)

POWER FACTOR

The ratio of the actual **power** expended or transferred in an alternating-current circuit to the apparent power. The actual power is stated in **watts** (or kilowatts), whereas the apparent power, equal to the effective (root-mean-square) electromotive force (or voltage) multiplied by the effective (root-mean-square) current, is in volt-amperes (or kilovolt-amperes). Hence, the power factor is expressed as watts/volt-amperes (or multiples thereof). The difference between the actual and apparent powers is due to the phase difference between the current and voltage of the alternating current, and this depends on the inductance and capacitance of the circuit. (For further discussion of the power factor, see **Electric Power Transmission.**)

PRESSURIZED-WATER REACTOR (PWR)

A thermal **nuclear power reactor** in which ordinary (or light) water is the moderator and coolant as well as the neutron reflector (see **Light-Water Reactor; Thermal Reactor**). The system is maintained under a high pressure to inhibit the water from boiling in the reactor core. The PWR concept was first tested in 1953 in a land-based prototype system for a nuclear submarine, and its success led to the development of commercial power reactors. The first such reactor commenced operation in the United States at the end of 1957. About two-thirds of the nuclear power reactors operating or under construction in this country in the early 1980s are PWRs.

Three organizations in the United States manufacture PWRs based on the same (or similar) principles, but with minor design differences. Numerical values given in the following description refer to a large PWR system with an electrical generating **capacity** of roughly 1000 megawatts (see **Watt**).

General Description

Core. The core of a PWR is approximately cylindrical, about 13 ft (4 m) high and 13 ft (4 m) across. It is made up of some 40,000 vertical fuel rods (or "pins"), combined in assemblies each containing 200 or more rods. The fuel rods contain small cylindrical pellets of uranium dioxide with an initial average uranium-235 enrichment of about 2.5 to 3 percent (see **Nuclear Fuel**). The pellets are stacked inside tubes of a corrosion-resistant zirconium alloy (zircaloy), called the *cladding*, to form 13-ft (4-m) long fuel rods close to 0.4 in. (1 cm) in external diameter. The total mass of fuel in a PWR core is about 120 short tons (110,000 kilograms). Approximately one-third is removed after each year's operation and replaced with fresh fuel.

Many of the fuel-rod assemblies have spaces for guide tubes into which control rods can be inserted. Up to 24 control rods are combined in a cluster, and 70 or more such clusters are distributed throughout the reactor core. Each control rod is a stainless steel (or other) tube containing a neutron poison (i.e., a strong neutron absorber), usually an alloy of cadmium with indium and silver. The control rod clusters are moved in or out of the core, as may be required, by a drive mechanism located above the top of the reactor vessel.

Reactor (Pressure) Vessel. The reactor core is surrounded by a steel *core barrel* (or shroud) and is supported in a large cylindrical vessel, called the reactor (or pressure) vessel (Fig. 104). This vessel is typically 45 ft (13.7 m) high and 15 ft (4.6 m) in internal diameter. It is made of carbon steel 8 to 9 in. (20 to 23 cm) thick and is lined in the interior with stainless steel. The vessel is designed to withstand high pressure, as well as corrosion by the high-temperature water moderator and coolant it contains. The water is maintained under a pressure of 153 atm (15.5 MPa) to inhibit boiling. [At this pressure water boils at 653°F (345°C).]

Coolant System. Water flowing upward through the reactor core is heated to about 625°F (330°C) and passes to a **heat-exchanger** steam generator. After giving up some of its heat to boil water and produce steam in the steam generator (see below), the high-pressure water is pumped back into the reactor vessel. It enters just above the core and flows down through the annular region, called the *downcomer*, between the core barrel and the pressure vessel wall. At the bottom of the core, the water reverses direction and flows upward through the core to remove the heat generated by fission.

The coolant system pressure is maintained within a limited range by means of a pressurizer connected between the reactor vessel and a steam generator. The pressurizer is a large cylindrical steel tank containing some 60 percent by volume of liquid water and 40 percent steam during normal operation. Electric immersion heaters in the lower section can heat the water to generate steam, whereas cold water sprayed into the upper (steam) section can condense steam. If the system pressure should drop, a low-pressure signal would

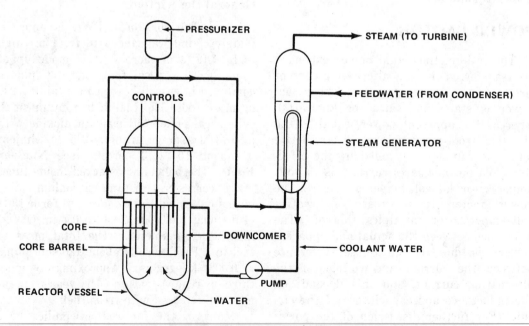

Fig. 104 Pressurized-water reactor.

actuate the heaters and the steam produced would increase the pressure. On the other hand, a high-pressure signal would operate the cooling water spray; condensation of steam would then result in a decrease of the system pressure. If this pressure should increase beyond the reduction capability of the spray, excess steam is vented through relief valves.

Steam System

A large PWR may have from two to four independent steam-generator loops in parallel. Most steam generators consist of a large number of inverted U-shaped tubes enclosed in a casing called the *shell.* The high-pressure, high-temperature water from the reactor flows through the inside of the tubes, and heat is transferred to water at a lower pressure [75 atm (7.6 MPa)] on the outside (shell side) of the tubes.

The water in the shell boils at the lower pressure and produces moist steam. Entrained moisture is separated in the upper part of the steam generator, and the steam at a temperature of about 555°F (290°C) proceeds to the turbine system. After passage through the steam generator tubes, the high-pressure water is pumped back to the reactor vessel at a temperature of 560°F (293°C).

The PWR **steam turbine,** which drives an electric generator, usually has high-pressure and low-pressure stages on the same shaft. Because of the relatively low steam temperature of 555°F (290°C), the gross **thermal efficiency** is not more than 33 percent, compared with 40 percent for a modern fossil-fuel steam–electric plant with a steam temperature of 1000°F (540°C). However, the net thermal efficiencies, after allowing for energy required to operate the plant, differ by a smaller percentage. (See **Electric Power Generation.**)

The exhaust steam from the low-pressure stage passes to the condensers, and the condensate provides the feedwater for the steam generators. Before the feed-water is pumped to the generators, its temperature is increased in a series of heaters by means of steam drawn from various intermediate sections of the turbine (see Fig. 160).

Control and Protection Systems

Coarse (or "shim") control of a PWR is achieved by the neutron poison (i.e., absorber) boron, as boric acid, dissolved in the reactor water. The boron compensates for the extra fuel present initially, and, as this is used up during reactor operation, the boric acid concentration is decreased. The control rods, referred to earlier, which can be moved in or out of the core, are used to start up the reactor and shut it down and for automatic fine adjustments during normal operation. Another use of the control rods is to make the heat (or power) distribution as uniform as possible throughout the core. Complete insertion of the rods will always cause the reactor to shut down.

The *reactor protection system* includes many instruments to monitor the conditions in the reactor and steam systems. If an abnormal condition that cannot be corrected automatically by the control rods is indicated, the reactor is immediately shut down (see **Boiling-Water Reactor**). In normal operation, the control rod clusters are held in place by electromagnetic clutches; but, if an emergency shutdown ("scram" or "trip") signal is received, the current to the magnets is cut off and all the rods drop into the core under the influence of gravity.

Engineered Safety Features

The purposes of the engineered safety features in a PWR are the same as described in the article on the boiling-water reactor. There are, however, some design differences.

The *emergency core-cooling system* of a PWR consists of three independent subsystems. In the event of a small break in the coolant system, the pressure would drop but still remain moderately high. The *high-pressure injection system* would then compensate for the coolant loss by pumping water containing boric acid into the reactor vessel. If the coolant loss should be considerable, the pressure would drop further and an *accumulator injection system*, operated by nitrogen gas pressure without a pump, would introduce more borated water. Finally, a *low-pressure injection system* would pump still more water into the reactor vessel.

The *containment structure* for a PWR contains the reactor vessel and its concrete radiation shield, the steam generators, pressurizer, coolant system pumps, and the accumulator injection system mentioned above. A common type of PWR containment structure has a cylindrical form with a domed top. The building is made of reinforced concrete, typically 42 in. (1.1 m) thick, with a 1.5-in. (3.8-cm) thick steel liner. The external height is approximately 210 ft (64 m), and the diameter is about 120 ft (37 m). In a modification used by one reactor manufacturer, the containment consists of a steel sphere enclosed in a concrete cylindrical building.

All of these structures contain water sprays to condense some of the steam produced by escaping coolant water, and thus reduce the pressure, and to remove part of the radioactivity in the interior atmosphere. In a few PWR containment structures, which are somewhat smaller than those referred to above, the upper part of the interior is lined with a lattice arrangement filled with pieces of ice. By condensing the steam and lowering the temperature, the ice prevents a potentially dangerous increase in pressure.

PRIME MOVER

Any device that can transform a natural form of **energy** into mechanical (or motion) energy. **Heat engines** are prime movers because they convert heat into mechanical work. Thus, the **steam engine, steam** and **gas turbines,** and **diesel** and **spark-ignition** (gasoline) **engines** are prime movers. So also are **hydraulic** and **wind-energy conversion turbines.**

PRODUCER GAS

A **low-Btu fuel gas** produced by the interaction of a mixture of air and steam with coal (or coke) at temperatures commonly in the range of 1500 to 2000°F (815 to 1090°C). Combustion of the carbon (in the coal or coke) with oxygen (in the air) generates the heat required to maintain the temperature for the carbon–steam reaction in which heat is absorbed (see **Coal Gasification**). The producer gas formed in this manner contains, on the average, roughly 25 volume percent of carbon monoxide, 15 percent of hydrogen, and 1 to 3 percent of methane (dry basis) as fuel constituents; the inert components are about 50 percent of nitrogen (from the air) and up to 10 percent of carbon dioxide. The **heating value** of producer gas is usually 130 to 180 Btu/cu ft (4.8 to 6.7 MJ/cu m). By increasing the pressure in the producer, the **methane** content and heating value of the gas are increased.

Producer gas was at one time utilized extensively in the United States as a cheap fuel for industrial heating, but as **natural gas** became increasingly available, the use of producer gas declined. With the current need for low-sulfur fuels, there has been a revival of interest in the efficient conversion of coal into a low-Btu fuel gas similar to producer gas.

The name producer gas is also given to the gas made by the limited combustion of coal in air. It may contain about 35 to 45 percent of carbon monoxide and a few percent of hydrogen and methane; the remainder is nitrogen and some carbon dioxide. The heating value is roughly 100 to 150 Btu/cu ft (3.7 to 5.6 MJ/cu m).

PUMPED STORAGE

An indirect method for temporarily storing substantial amounts of electrical energy by pumping water from a lower to a higher level. Pumped storage can be used in conjunction with electrical generating plants of all types, regardless of the energy source.

Large generating plants operate most economically at a constant output (or **load**). At night, between about 10:00 PM and 6:00 AM, and during weekends, however, the plant capacity can exceed the demand. By storing the excess energy for use when the demand is greater than the base load, load leveling is possible. The power plant can then be operated at an essentially constant load, that is, in the most economical manner (see **Energy Storage**).

In a pumped-storage facility, the power generated in excess of the demand is used to pump water from a lower reservoir (e.g., a lake, river, or underground cavern) to an upper reservoir. During periods of peak demand, when the power demand exceeds the normal generating plant **capacity,** water from the upper level is allowed to flow through a hydraulic turbine at the lower level (Fig. 105). The turbine then drives a generator to produce electricity in the usual way (see **Hydroelectric Power**).

In most pumped-storage plants in the United States, the turbine–generator system is reversible and can serve to pump water from the lower to the upper level, as well as to generate electricity. In the pumping mode, the generator becomes a motor, driven by electricity produced by a generator in the main plant, and the turbine then operates as a pump (see **Electric Motor**). Startup of the turbine/generator or reversal from motor/pump to turbine/generator requires only a few minutes, so that power could be restored after a short delay in the event of a failure in the main plant.

The altitude difference between upper and lower water levels (i.e., the waterhead) in pumped-storage facilities ranges from less than 100 ft (30 m) to somewhat more than 1000 ft (300 m). As a general rule, Francis-type reversible turbines are used, but for low heads propeller-type turbines are preferred (see **Hydraulic Turbines**).

The efficiency of a pumped-storage plant, that is, the percentage of electrical energy used to pump the water that can eventually be recovered, is commonly from 65 to 75 percent. In spite of the loss of 25 to 35 percent, pumped storage can be economical. The stored water generates power at times of peak demand when it would otherwise be supplied by a gas turbine or diesel engine that is expensive to operate. On the other hand, apart from capital and mainte-

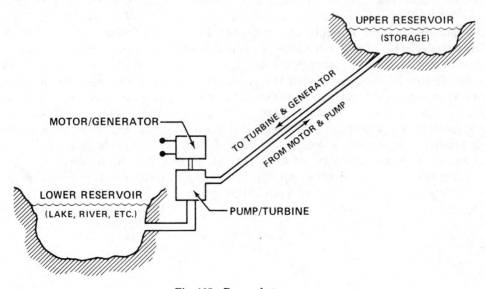

Fig. 105 Pumped storage.

nance charges, the costs of operating a pumped-storage plant are small.

The first pumped-storage facility in the United States was completed in Connecticut in 1928; it had a capacity of 50 megawatts (MW), where 1 MW = 1 million watts (see **Watt**). By 1960, the pumped-storage capacity had increased to only about 300 MW, but the development has been more rapid in recent years. In the early 1980s there were some 25 pumped-storage facilities (either completed or under construction) with an overall capacity of about 12,500 MW. This is roughly 2.5 percent of the total generating capacity of electrical power plants of all types in the United States.

Site Selection

Further use of pumped storage may be limited by the availability of sites, other than those where hydroelectric plants already exist. The topography must lend itself to the construction of a reservoir at a level of at least 100 ft (30 m), preferably more, above a river or lake with an ample water supply. Withdrawal of water by pumping must not jeopardize the normal downstream use.

The upper reservoir does not have to be very large, if the pumped storage is required to supply peak power for only a few hours daily. Nevertheless, some loss of land by inundation and possible adverse environmental effects may be inevitable.

Site selection for pumped-storage facilities is also limited by the requirement that, in order to avoid long transmission lines, the facility and its associated main power plant be close to a load center (i.e., where there is a demand for electric power).

The possibility is being studied of using underground caverns, at depths up to 4000 ft (1200 m) or so, as the lower reservoirs for pumped-storage systems. The upper reservoir would be at the surface, perhaps at some distance from the lower one. The siting of the facility would then not be dependent on the local topography, but rather on the possibility of excavating the underground reservoir. In some cases, natural caverns or abandoned mines could be used directly or as the starting point for further excavation.

PURISOL PROCESS

A process for removing **acid gases** (hydrogen sulfide and carbon dioxide) from fuel gases by physical absorption in the organic solvent N-methylpyrrolidone (NMP). The general principles of operation are similar to those for other processes based on physical solution in a liquid solvent (see **Desulfurization of Fuel Gases**). Passage of a stream of nitrogen through the spent solvent in the regenerator/ stripper facilitates removal of the dissolved hydrogen sulfide. This can be converted into elemental (solid) sulfur by the **Claus process.**

PYROLYSIS

The chemical decomposition (or breakdown) of a substance into simpler products by the action of heat alone. Thermal **cracking** of petroleum, whereby complex **hydrocarbons** are broken down into simpler ones by heat, is an example of pyrolysis. Destructive distillation, in which an organic material, such as coal, wood, or kerogen (in **oil shale**), is heated in the absence of air to produce gases and vapors, is a form of pyrolysis (see, for example, **Coal Pyrolysis**). Hydropyrolysis (e.g., of coal) is pyrolysis in the presence of hydrogen gas.

Q

Q

A unit used to express very large amounts of energy. One Q is one quintillion (i.e., 10^{18} or one billion billion) **British thermal units.** One Q is equal to 1.055×10^{21} joules or 2.93×10^{14} kilowatt-hours (see **Watt**).

See also **Quad.**

QUAD

A unit used to express very large amounts of energy. One quad is one quadrillion (i.e., 10^{15} or one million billion) **British thermal units.** One quad is equal to 1.055×10^{18} **joules** or 2.93×10^{11} kilowatt-hours (see **Watt**). The total annual energy consumption in the United States in the early 1980s is approximately 80 quads.

See also **Q.**

R

RADIOACTIVE WASTE DISPOSAL

The isolation of radioactive wastes in such a way as to prevent access by people and release to the environment (i.e., air, water, or food chain). The hazard from radioactive materials arises from the radiations (alpha or beta particles and gamma rays) they emit (see **Radioactivity**). All radioactive substances are a hazard inside the body (internal hazard), and gamma-ray emitters are also a potential hazard outside the body (external hazard) because of their long range in air.

Most radioactive wastes are categorized as either low level or high level. Low-level wastes, from normal **nuclear power reactor** operations and from medical treatment and research, are usually enclosed in steel drums and buried either at the reactor site where they are generated or at a federally licensed burial ground. A third category, the transuranium (TRU) wastes are those containing appreciable amounts of elements heavier than uranium, especially plutonium, americium, and curium. These wastes arise largely in the nuclear weapons program and in the fabrication of **fast reactor** fuels.

High-Level Wastes

High-level radioactive wastes are either the spent fuel that is removed periodically from a nuclear reactor or the wastes remaining from reprocessing of the spent fuel (see **Nuclear Fuel Reprocessing**). Spent fuel contains unused uranium as well as plutonium and other heavy elements formed by neutron capture and radioactive decay in the reactor during operation (see **Nuclear Fuel**), in addition to accumulated fission products (see **Fission Energy**). In reprocessing, about 99.5 percent of the uranium and plutonium is removed for further use, leaving a waste solution containing essentially all the fission products and about 0.5 percent of the uranium and plutonium in the spent fuel, plus small amounts of still heavier elements (americium and curium).

Large volumes of high-level liquid wastes from fuel reprocessing would be difficult to store. Consequently, the solutions would be solidified by heating to a high temperature, above 1000°F (540°C), to remove all the water and other volatile substances. The product (or calcine) is a mixture of oxides of fission products and heavy elements. The calcine is not suitable for ultimate disposal and is subjected to further treatment. The procedure favored in Europe is to convert the calcine into a glassy (borosilicate) material by heating it with a mixture of borax and silica. One of the advantages of the glass is that it is less leachable by water than the calcine; if water should accidentally reach the stored solidified waste, the radioactive material would then be extracted much more slowly from the glass than from the calcine.

A possible objection to the glassy material is that it is not fundamentally stable. At high temperature, such as might occur from the absorption by the radioactive material of its own radiations, the glass might change into a more readily leachable microcrystalline form. This drawback could perhaps be overcome by decreasing the fission product content of the glass or in other ways, thereby decreasing its heating rate. Nevertheless, efforts are being made in the United States and elsewhere to develop alternative, more stable forms for solidified high-level fuel reprocessing wastes. These include crystalline products similar to stable minerals and "cermets" consisting of ceramic (calcined) waste particles in a metal matrix.

Radioactivity of High-Level Wastes

From the radioactive standpoint, spent fuel and solidified wastes are similar; they both contain fission products, which decay

by the emission of beta particles often accompanied by gamma rays, and heavy elements, which mostly emit alpha particles. The main difference is that the spent fuel contains a much larger proportion of the heavy elements. For example, spent fuel contains initially about 200 times as much uranium and plutonium isotopes (see **Atom**) as the solid wastes that would result from reprocessing the same spent fuel.

Nearly all of the fission products have radioactive half-lives of not more than 30 years, whereas most of the heavy alpha-emitting elements have much longer half-lives. Consequently, during the first few hundred years of decay, both spent fuel and solid wastes are predominantly beta-particle and gamma-ray emitters. Because of the gamma rays, high-level wastes at this stage constitute a potential external as well as an internal hazard.

After about 600 years fission products will have decayed to insignificant levels, and the low residual radioactivity will be almost entirely due to alpha-particle emission by the heavy elements (Fig. 106). The

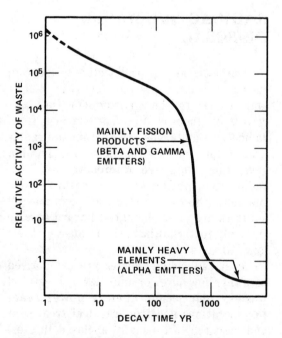

Fig. 106 Radioactive decay of high-level waste (or spent fuel).

potential external hazard is then small, but the internal hazard, especially from plutonium, is important. In this respect, spent fuel is a greater potential hazard, because of its higher plutonium content, than the equivalent solid reprocessing waste. Since plutonium-239 has a half-life of 24,100 years, it is necessary to prevent access of water, which might leach out some plutonium from the wastes, for thousands of years.

Storage of Wastes

Pending its final disposal, either in its existing form or by way of reprocessing wastes, spent fuel is stored in deep water-filled, steel-lined concrete tanks. The water, which is circulated through **heat exchangers,** serves as a means for removing heat generated by radioactive decay and also as a radiation shield. All nuclear power reactor installations have spent fuel storage tanks, but in due time, in the absence of reprocessing, these will be filled. It may then be necessary to construct large storage tanks, called away-from-reactor (AFR) pools, where spent fuel from several reactors can be stored. However, AFR storage can be only a temporary expedient.

It is generally accepted that more permanent storage of high-level radioactive wastes, either spent-fuel or solidified reprocessing wastes, could be found underground in a deep, dry geologic formation. Many formations have been dry and stable for millions of years, and it is reasonable to suppose that they will remain so for a long time into the future. Geologic formations that might be suitable for high-level waste storage include salt beds and domes, hard crystalline rocks (granite), carbonate rocks (limestone and dolomite), basalt, shales, and welded tuff. Salt formations have several advantages, but since salt is soluble in water, accidental entry of water into the waste repository could have serious consequences. It is expected that by 1985 exploratory shafts will have been completed in basalt, welded tuff, and salt

formations as possible locations for a high-level waste repository. One of these will be selected as the site for a Test and Evaluation Facility.

A waste storage facility would consist of a number of chambers or tunnels excavated in the selected geologic formation at a depth of 2000 ft (600 m) or more. Sealed, corrosion-resistant cylinders containing the waste would be inserted into vertical, metal-lined holes in the floor. The space between the waste cylinder and the metal liner would be filled with minerals that can retard the movement of substances dissolved in water. If water should ever reach the stored waste, the release of radioactive material to the surroundings would thus be inhibited. Furthermore, the distance of the waste repository from the ground surface would ensure a long transit time to the biosphere (i.e., where living organisms exist).

The dimensions of the repository and the spacing of the containment holes would depend on whether the stored high-level wastes are spent fuel or solid reprocessing wastes. Since the spent fuel contains some 97 percent of heavy elements, its volume is considerably greater than the equivalent reprocessing waste. The storage volume required for spent fuel is estimated to be at least ten times that for reprocessing wastes, but the radioactivity and associated heat-generation rate per unit volume would be correspondingly less.

If after a number of years of provisional storage no adverse effects were observed on the wastes themselves or on the geologic environment, the facility would be converted into permanent storage. The chambers or tunnels and entry shaft would be backfilled with mined-out material and sealed. However, the wastes should be retrievable by conventional mining techniques if more effective methods of disposal of long-lived heavy elements should be developed.

The radioactive character of the TRU wastes, referred to earlier, is similar to that of the high-level fuel processing wastes after the fission products have largely decayed. Hence, the TRU wastes might be converted into a form that would not be readily leached by water and stored in a similar manner in sealed containers.

RADIOACTIVITY

The spontaneous emission from unstable atomic nuclei (see **Atom**) of radiation, generally either an alpha particle or a beta particle often accompanied by gamma rays. As a result of the emission of a particle, the radioactive nucleus changes (or decays) into a nucleus of a different element. The latter, called the *daughter* (or *decay*) product, may be stable (i.e., not radioactive) or it may be radioactive and decay in turn. Eventually, a chain of two or more radioactive decays leads to a stable species called the *end product*.

The earliest studies of radioactivity indicated that the interior of the atom, now called the nucleus, might be an important source of energy. Actually, radioactive energy has made only a small contribution to the world's energy supply. Some 6 to 7 percent of the energy produced by nuclear fission is derived from the radioactive decay of the fission products (see **Fission Energy**). In addition, radioactive materials are used to a limited extent for the generation of electricity from heat (see **Isotope Power Systems; Radioisotope Thermal Generator**).

Elements with the highest atomic numbers (84 or more), including uranium (92) and plutonium (94), exist only as radioactive isotopes (see **Atom**); furthermore, the elements with atomic numbers of 81 (thallium), 82 (lead), and 83 (bismuth) occur naturally as both stable (nonradioactive) and radioactive isotopes *(radioisotopes)*. In addition to about 40 radioisotopes of 12 elements that are found in nature, some 1500 radioisotopes of all the elements have been produced by various nuclear reactions. These include more than 300 fission products.

Alpha-Particle Decay

Alpha particles are emitted in the decay of many isotopes of heavy elements (atomic number 83 or more), but only rarely in the decay of lighter radioisotopes. The alpha particle consists of two protons and two neutrons and is, in fact, identical with a helium nucleus. Because of its two protons, the alpha particle has a positive electric charge, but it soon picks up two electrons from the surroundings and becomes a neutral helium atom.

When it leaves a radioactive nucleus, an alpha particle has a considerable amount of kinetic (motion) energy and travels with a speed of about 1.4×10^7 to 2.0×10^7 m/sec (roughly 10,000 miles/sec). As the particle moves through matter, even air, it gradually loses energy (and speed) and is eventually converted into a helium atom. In the slowing down process, the kinetic energy of the particle is changed into heat.

The distance an alpha (or other charged) particle travels through a medium before losing its identity is called the *range* of the particle. The range depends on the nature of the medium and the energy of the particle. The ranges of alpha particles from most radioactive sources are from 1.2 to 3.2 in. (3 to 8 cm) in air. In body tissue (e.g., the skin) the range is about one-thousandth of the range in air. Thus alpha particles from a radioisotope on the exterior of the body cannot penetrate the outer layers of the skin.

When a radioactive nucleus expels an alpha particle, it loses two protons and two neutrons. It follows, therefore, that the product (or daughter) nucleus has an atomic number (i.e., number of protons) two units less and a mass number (i.e., number of protons and neutrons) four units less than the parent nucleus. Since the atomic number identifies the element, the daughter element is different from the parent. For example, when the common uranium isotope (atomic number 92, mass number 238) emits an alpha particle, it is converted into an isotope of thorium (atomic number 90, mass number 234).

Beta-Particle Decay

Beta particles are identical with electrons moving initially at very high speed, roughly ten or more times as fast as alpha particles. (Either positively or negatively charged beta particles may be emitted, but radioisotopes emitting negative beta particles are of major interest and only these will be considered here.) Like alpha particles, beta particles lose their energy, which is converted into heat, in their passage through matter. However, since beta particles have a much smaller mass (about $\frac{1}{7000}th$ that of alpha particles), they can travel much greater distances before being absorbed. The range of a beta particle depends on its initial speed, but it is usually a few meters (several feet) in air and a few millimeters (a fraction of an inch) in the body.

Although beta particles are electrons, they are not the electrons that normally surround the nucleus. Beta particles are expelled from radioactive nuclei where they are formed from neutrons. A neutron (neutral) can be converted into a proton (positive charge) and an electron (negative charge). The proton remains in the nucleus, but the electron is expelled immediately as a beta particle.

Since a neutron is changed into a proton in radioactive beta decay, the daughter nucleus contains one more proton than its parent but the same total number of protons and neutrons. Hence, the daughter has an atomic number one unit higher than the parent and is therefore a different element, but the mass number is unchanged. For example, thorium-234 (atomic number 90), resulting from the alpha-particle decay of uranium-238, decays by the emission of a beta particle. The product is thus an isotope of the element of atomic number 91 (protactinium) with the same mass number (234) as its thorium parent.

Gamma Rays

The expulsion of an alpha or beta particle from a radioactive nucleus is often, although not always, accompanied by emission of *gamma rays*. These rays are a form of **electromagnetic radiation** identical with high-energy x rays; the only difference between them is that gamma rays originate from the nucleus whereas x rays are produced outside the nucleus (e.g., by slowing down high-energy electrons).

Because they are not material particles and have no mass or electric charge, gamma rays can travel considerable distances in gaseous, liquid, or solid matter (e.g., the body). As they do so, the gamma rays gradually lose their energy, which is dissipated as heat.

As a rough general rule, the rate at which gamma rays lose their energy in passing through a material is roughly proportional to the density of the material. Hence, a thin layer of a high-density material (e.g., lead) is as effective as a thicker layer of a less dense material (e.g., water, wood, or concrete) in providing protection (or shielding) against gamma rays (see below).

Atomic nuclei can generally occur in several states of the same composition (i.e., unchanged numbers of protons and neutrons) but with different specific amounts of internal energy. (This is a form of potential energy, distinct from the kinetic or motion energy.) The lowest energy state is called the *ground state*, whereas the higher energy states are the *excited states*. An excited state will always have a tendency to emit part or all of its excess (or excitation) energy to form either a lower excited state or the ground state. This energy emission occurs as gamma rays.

In some radioactive decays, the expulsion of an alpha or a beta particle from the parent nucleus leaves the daughter nucleus in an excited state; the latter then emits one or more gamma rays as it converts into the ground state. Since there is no change in the numbers of protons and neutrons, the identity of the daughter nucleus is unaffected by the gamma-ray emission.

Radioactive Half-Life

Each radioactive species decays (i.e., emits alpha or beta particles) at a rate characteristic of that species. This rate depends only on the nature of the species and is independent of the external conditions, such as temperature, pressure, and physical and chemical state. Thus, a certain amount of uranium-238 decays at the same rate regardless of whether it exists as the metallic element (solid), as one or another of several solid oxides, as the vapor of uranium hexafluoride, or as a uranium salt in solution. (There are a few minor exceptions in special cases which are of no concern here.)

An important feature of radioactive decay is that, no matter how much of a given species is present in a sample, it always takes the same time for the activity (or rate of particle emission) to decrease to half of its initial value (Fig. 107). It is consequently possible to express the decay rate of a radioactive species in terms of its *half-life;* it is the time for any amount of that species to decay to half. Each radioactive species has its own characteristic and unchangeable half-life.   For different known species, the half-lives range from about a millionth part of a second to billions of years.

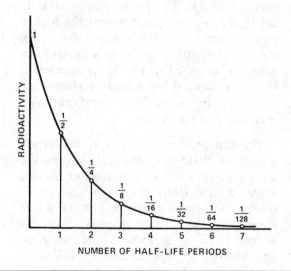

Fig. 107   **Radioactive decay and half-life.**

The fact that a given amount of radioactive material will decrease to half (50 percent) during one half-life period means that it will decay to one-fourth (25 percent) in two half-lives, to one-eighth (12.5 percent) in three half-lives, and so on. In 6.6 half-lives, only 1 percent of the initial amount will remain; the other 99 percent will have been converted into the daughter species.

Radioactivity and Nuclear Energy

The half-lives of some radioactive species that are useful for the release of nuclear energy by fission are given in the table. Of these, uranium-235, uranium-238,

Species	Half-life (years)
Uranium-233	1.59×10^5
Uranium-235	7.04×10^8
Uranium-238	4.49×10^9
Thorium-232	1.39×10^{10}
Plutonium-239	2.41×10^4

and thorium-232, which have long half-lives and decay very slowly, are found in nature. Uranium-233 and plutonium-239, however, not only have shorter half-lives but they do not have long-lived predecessors from which they can be regenerated. Hence, these two species no longer exist in nature, except perhaps in small traces, but they are made artificially from the long-lived thorium-232 and uranium-238, respectively (see **Nuclear Fuel**).

Apart from the stable end products, probably all the 300 or more different species that result from fission of heavy nuclei emit negative beta particles (often accompanied by gamma rays). Each of the species has a definite half-life, but the mixture of fission products does not. As a result of radioactive decay of the many species with different half-lives, ranging from a few seconds (or less) to 1.6×10^7 years (iodine-129), the composition of the mixture is changing continuously.   The overall rate of decay (i.e., the rate of particle emission) cannot then be expressed in a simple manner.

Radioactivity Hazard

Radioactive substances are a potential health hazard because the radiations they emit can damage or destroy body cells. Since alpha and beta particles are readily absorbed and do not penetrate very far into the skin, sources of these particles outside the body are not a hazard unless they are actually on the skin. Gamma rays, however, are highly penetrating and so can cause injury even if the source is outside the body.

If radioactive material should enter the body, by inhalation, ingestion, or possibly through a wound, it would constitute a hazard regardless of the type of radiation emitted. Alpha particles can then be especially effective because they have such a short range; they deposit all their energy within a very small region of the body where severe damage may occur.

The hazardous nature of radioactive materials requires that care be exercised in handling them. Precautions must be taken to prevent entry of such materials into the body, and some form of shielding is required when operating with them outside the body. A few inches of air provide adequate shielding from alpha particles, but the alpha-emitting materials are generally held in an enclosed space to minimize the inhalation risk.

Beta particles and especially gamma rays require more shielding. Shields are generally designed to provide protection against gamma rays; they are then more than adequate for beta particles. On the small (e.g., laboratory) scale, lead shields and containers are commonly used. On a large scale, or in various nuclear power operations involving substantial amounts of fission products, the shielding may be up to 8 ft (2.4 m) of concrete. Materials containing fission products (e.g., spent nuclear fuels) are usually stored under a depth of water to provide shielding from gamma rays (see **Nuclear Fuel Reprocessing**).

RADIOISOTOPE THERMOELECTRIC GENERATOR (RTG)

A device in which heat deposited by absorption of alpha or beta particles from a radioactive (or radioisotope) source is converted directly into electrical energy (see **Radioactivity**). The RTG was originally developed for use in space as a compact and reliable power source with no moving parts and requiring no maintenance. The first RTG, designated SNAP (Systems for Nuclear Auxiliary Power)-1A, was originally intended for a U.S. Air Force satellite; it was built and tested in the late 1950s but was not used in space.

Since 1961, however, RTGs have provided power for the U.S. Navy's Transit navigational satellite, for several NASA and Department of Defense spacecraft, and for instruments and radio transmitters on the surface of the moon and Mars. Other applications of RTGs have been to operate navigational beacons and remote weather stations on earth. Small RTGs, which have a longer life than conventional batteries, have been used to power cardiac pacemakers implanted in the human body.

The generation of electricity by an RTG depends on **thermoelectric conversion** using *thermocouple junctions* of two different materials (see below). Heat is supplied to one set of junctions while the other set remains at a lower temperature. The heat is then converted (partially) into electrical energy (see Fig. 176).

To provide a useful heat source for an RTG, a radioisotope should have a moderately long half-life. If the half-life is short, the radioactivity (and rate of heat generation) will decrease rapidly; the generator would then have a short useful life. On the other hand, if the half-life is long, the rate of heat generation will be small and large amounts of radioactive material would be needed to produce appreciable electric power. Another requirement is that the radioisotope emit alpha or beta particles of fairly high energy. However, there should be no emission of gamma rays (or only gamma rays of very low energy) by the radioisotope and its decay products. The alpha and beta particles are readily absorbed in the solid thermoelectric material, but the gamma rays are not, and their presence would make heavy shielding necessary.

The heat source for most terrestrial RTGs is strontium-90 (half-life 28 years) as strontium titanate ($SrTiO_3$), whereas plutonium-238 (half-life 88 years) as plutonium dioxide (PuO_2) is generally used in space. Strontium-90, extracted from fission products (see **Fission Energy**), is relatively inexpensive; it emits beta particles with an average energy of 0.55 MeV (see **Nuclear Energy** for definition) but no gamma rays. Although plutonium-238 is more costly to produce, it has some advantages; its high-energy (5.5 MeV) alpha particles are readily absorbed, and the longer half-life makes possible RTGs with longer lifetimes than those based on strontium-90.

In the early RTGs, with initial electrical outputs up to 60 watts (see **Watt**), the thermocouples have been n- and p-type **semiconductors** of lead telluride (see **Thermoelectric Conversion**). In the later Multi-Hundred Watt (MHW) generators, with initial power of 150 watts, the semiconductors are based on a silicon-germanium combination. Generators with initial electrical outputs close to 250 watts are being developed using special high-temperature semiconductors containing the element selenium.

Because of the decay of the radioactive material and the decrease in efficiency of the thermocouples with time, the power of an RTG decreases steadily. However, with plutonium-238 as the active species, there is a loss of less than 10 percent after 10 years of continuous operation.

Most RTGs have a cylindrical form with a central metal cylinder containing the radioactive heat source. Temperatures of 930 to 1110°F (500 to 600°C) have been attained. Short rods of thermocouple materials are arranged around the cylinder with their (hot) junction in contact with it. A typical RTG may have 100 or more semiconductor pairs in series (see **Fig. 177**). The outer ends, supported by the containing cylinder, are kept at a moderately low temperature by radiating heat to the surroundings. Large metal fins attached to the outside of the RTG facilitate heat removal. The output voltage may be increased to the desired value by an inverter-transformer combination; the inverter changes direct to alternating current and the **transformer** then provides the desired voltage.

See also **Isotope Power Systems.**

RANKINE (OR CLAUSIUS) CYCLE

A repeated succession of operations (or cycle) representing the idealization of the processes in certain **heat engines** in which the working fluid is a liquid and its vapor (e.g., water and steam). **Vapor** (or **condensing**) **turbines** operate on an approximate Rankine cycle. Vapor-compression refrigerators, air conditioners (see **Refrigeration**), and **heat pumps** use the Rankine cycle in reverse.

The stages in an ideal Rankine cycle are shown and described in Fig. 108. In the description each stage is assumed to have been completed before the next stage is initiated. However, in an actual engine there is a gradual rather than a sharp transition from one stage to the next; hence, the sharp points in the figure would actually be rounded off.

In a condensing turbine (e.g., a **steam turbine**) system, the working fluid in liquid form is pumped to a boiler or other vapor generator (stage 1). Heat is supplied at constant pressure (stage 2), and the liquid is converted into vapor and possibly superheated. (The source of the heat is immaterial; it may be a fossil or nuclear

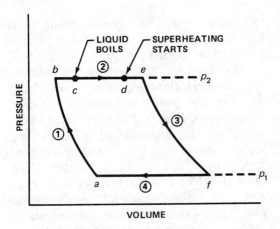

Fig. 108   Rankine cycle.

1. Adiabatic compression (pumping) of the liquid working fluid along *ab;* there is a substantial increase in pressure from p_1 to p_2, but only a small increase in temperature. Work is done on the liquid.

2. Heat addition to the liquid at constant pressure along *bcde;* the liquid boils at *c;* vapor and liquid coexist from *c* to *d;* at *d* the liquid has boiled off completely and the vapor may be superheated (i.e., the temperature raised above the normal boiling point at pressure p_2) from *d* to *e*. (Superheating is not essential to the Rankine cycle; if there is no superheating, stage 2 ends at *d*.)

3. Adiabatic expansion of the heated vapor along *ef;* work is done by the expanding gas, and the pressure decreases to the initial value p_1. The temperature decreases and part of the vapor is condensed to liquid.

4. Heat removal (rejection) along *fa* at the constant pressure p_1; the remainder of the vapor is condensed, and the initial (liquid) condition of the working fluid is restored.

fuel or solar energy.) The vapor is expanded adiabatically (see **Carnot Cycle**) in the turbine, and mechanical work is done (stage 3). Finally, the exhaust vapor passes to the condenser where heat is removed and the vapor is condensed to produce the liquid feed for the vapor generator (stage 4).

Nearly all the useful work in a condensing turbine is done by the expansion of the vapor in stage 3, and most of the work done on the working fluid is in pumping the liquid in stage 1. The net work is thus essentially the difference between stages 3 and 1. The **thermal efficiency** of the

Rankine cycle (i.e., the fraction of the heat added in stage 2 that is converted into net mechanical work) is increased by increasing the vapor temperature at *e* and decreasing it at *f*. This is one reason why superheated steam is commonly used to operate steam turbines; the temperature at *e* is higher than it would be without superheating (i.e., at *d*).

RECTISOL PROCESS

A process for removing **acid gases** (hydrogen sulfide, carbon dioxide, and carbon oxysulfide) from fuel gases by physical absorption in methanol (methyl alcohol) at a low temperature. The impure feed gas is pressurized to more than 20 atm (2 MPa) and brought into contact in an absorber column with methanol cooled to a temperature well below 0°F (−18°C). At the high pressure and low temperature, the acid gases are dissolved by the methanol. The clean gas leaves at the top of the absorber column. The operating principles are similar to those described for physical absorption processes in the **desulfurization of fuel gases** (see Fig. 25). By appropriate adjustment of the conditions, removal of hydrogen sulfide and carbon dioxide from the fuel gas may be carried out in separate stages. The Rectisol process can also be used for removing ammonia and other impurities from fuel gases.

REFINERY GASES

In general, any gas (or gas mixture) produced in **petroleum refining** operations; more specifically, however, refinery gases are the lightest paraffin **hydrocarbons, methane** (CH_4), ethane (C_2H_6), and propane (C_3H_8), which may be used as fuels for heating or for operating **gas turbines.** Other gases produced in a refinery, such as butanes, propylene, and butylene, are too valuable for use as fuel and are utilized in other refinery operations (see **Alkylation; Isomerization**).

Hydrogen gas is a by-product of **reforming** in petroleum refining; it is used to improve the quality of petroleum products by **hydrotreating**.

REFORMING

A **petroleum refining** process for converting liquid **hydrocarbons** with low **octane numbers** (e.g., normal paraffins and naphthenes) into others (e.g., isoparaffins and aromatics) having the same (or almost the same) numbers of carbon atoms but with higher octane numbers. Thus, reforming is a means for improving the antiknock quality of the **gasoline** (and light **naphtha**) fractions of petroleum distillation; the improved product is called *reformate*. The reforming catalyst is mainly platinum, and this has led to the trade names Platforming for the process and Platformate for the product. Reforming is also used in the petrochemical industry to convert paraffins into aromatic hydrocarbons, especially benzene, toluene, and xylenes (BTX).

Many variations of catalytic reforming have been patented. Basically, the process involves passage of the preheated feed (gasoline or light naphtha vapor) with hydrogen gas through a reactor containing the catalyst. The operating conditions vary, but the temperature is usually within the range of 850 to 1050°F (455 to 565°C) and pressure from 15 to 40 atm (1.5 to 4 MPa). The catalyst is platinum, with small amounts of additives (e.g., rhenium or iridium), dispersed on a predominantly alumina base. Deposition of carbon causes the catalyst to lose its activity, but it can be regenerated by burning off the carbon in air. In the older reforming processes, regeneration was carried out periodically, but continuous regeneration by circulating the catalyst through a separate unit is now in common use.

The catalyst is poisoned (i.e., inactivated) by compounds of nitrogen, oxygen, sulfur, and arsenic derived from the crude oil. These elements are removed from the reforming feed by pretreatment with hydrogen gas in the presence of a catalyst (see **Hydrotreating**).

The most common reforming reactions are **isomerization** (e.g., conversion of normal paraffins into isoparaffins) and dehydrogenation, that is, the removal of hydrogen from the hydrocarbon molecule (e.g., conversion of paraffins and naphthenes into aromatics). Consequently, large amounts of by-product hydrogen are available from reforming operations. The hydrogen gas required for hydrotreating petroleum products is usually obtained from this source.

See also **Steam Reforming**.

REFRIGERATION (AND AIR CONDITIONING)

A process for lowering temperatures below that of the environment. Mechanical refrigeration, which is widely utilized in the home, commerce, and industry, requires the use of energy as heat or electricity (or both). A refrigerator is essentially a machine operating in a manner opposite to that of a **heat engine**; it takes up heat at a lower temperature and discharges it at a higher temperature. In order to do this, energy must be supplied in some form. The medium (or material) from which the heat is taken up is thereby cooled, and the heat is discharged to the surroundings.

The most common refrigerators are of the vapor-compression or absorption types. In each case, the low temperature results from the evaporation of a liquid, called the *refrigerant*, which is accompanied by the absorption of the latent heat of vaporization. To permit the refrigerator to operate continuously, the refrigerant vapor is compressed and reconverted to liquid. The basic difference between the two refrigeration procedures lies in the manner in which the pressure of the vapor is increased. Another less common type of refrigeration system is based on an expansion turbine with a noncondensable gas as the working fluid.

Vapor-Compression Refrigeration

The *vapor-compression system*, outlined in Fig. 109, operates on a reversed **Rankine cycle.** It owes its name to the use of a mechanical (motor-driven) compressor to increase the pressure of the refrigerant vapor. In domestic refrigerators, freezers, and air conditioners, the refrigerant is usually a Freon or similar (chlorofluoromethane) compound, which is nonflammable and relatively innocuous. For industrial refrigeration, ammonia has been used extensively, largely because it is inexpensive. However, it has a penetrating odor and could be flammable and so is being replaced by Freon-type refrigerants.

The low-pressure (LP) refrigerant vapor leaving the evaporator (cooling) unit is compressed to a higher pressure (HP vapor). The compression increases the temperature of the vapor, and when the added heat is removed in the condenser unit, the HP vapor is condensed to HP liquid. In domestic refrigerators, freezers, and air conditioners, the condenser heat is removed by ambient air, but in large machines water cooling is required. The HP liquid passes through an expansion (or throttle) valve where its pressure is decreased; the resulting LP liquid is then vaporized in the evaporator where the latent heat is absorbed and cooling occurs.

Absorption Refrigeration

Absorption refrigeration is a modified reversed Rankine cycle system; it requires a liquid absorber, in addition to the volatile refrigerant. The vapor compressor is eliminated but it is replaced by a generator, a **heat exchanger,** an absorber unit, and a pump. The absorption refrigeration circuit indicated in Fig. 110 is seen to be more complex than the vapor-compression systems, but it uses less electrical energy for operation. However, some energy is supplied in the form of heat added to the generator (see later). Furthermore, an absorption refrigerator generally requires cooling water to remove heat from the condenser and the absorber vessel, since air cooling may not be sufficient. The water is circulated through a radiator or other cooling system (not shown in the figure) and is reused.

The refrigerant is usually either ammonia with water as the absorber, or water vapor is the refrigerant and a lithium bromide solution is the absorber. The lithium bromide–water vapor system has the advantage of requiring only moderate pressures and does not involve objectionable ammonia. The following description refers in particular to the

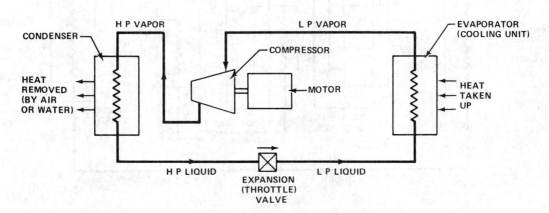

Fig. 109   Vapor-compression refrigeration system.

former system. The ammonia gas–water system operates in a somewhat similar manner, except that the dilute absorber (lithium bromide) solution in Fig. 110 has a high concentration of water and is thus equivalent to a concentrated ammonia solution. Similarly, a concentrated absorber solution in the figure is equivalent to a dilute ammonia solution.

In the absorption-refrigeration cycle, the cool, low-pressure (LP) water vapor leaving the evaporator, at the top right in Fig. 110, passes to the absorber vessel where it meets a cool concentrated lithium bromide (absorber) solution. The water vapor is absorbed and thus dilutes the solution; heat released in the dilution process is removed by cooling water. The cool dilute absorber is pumped to the generator by way of a heat exchanger where it takes up heat from the warm solution leaving the generator. Heat added to the generator causes the water to vaporize from the dilute solution, and the vapor at a relatively high pressure is condensed to form cool, pressurized liquid water in the condenser. Finally, the pressure of the liquid is decreased in passing through the expansion valve so that vaporization in the evaporator provides the desired cooling.

The major advantage of absorption refrigeration is that the heat can be added to the generator in several ways. For example, natural gas and low-pressure steam have been used as the heat source. In some cases, the cost is less than for the electricity required to operate a compressor. The utilization of water heated by solar energy as a heat source is of special interest (see **Solar Energy: Direct Thermal Applications**).

Intermittent Absorption Refrigeration

A modified method for absorption cooling, which operates intermittently rather than continuously, is based on a technique used at one time for food refrigeration in rural areas before electric power was readily available. In essence, the system consists of two vessels which function in two alternative modes. In one (regeneration) mode, one of the vessels is the generator and the other is the condenser of an absorption system (**Fig. 111A**). During this phase, heat is supplied to the generator by oil, gas, steam, or solar energy. In the

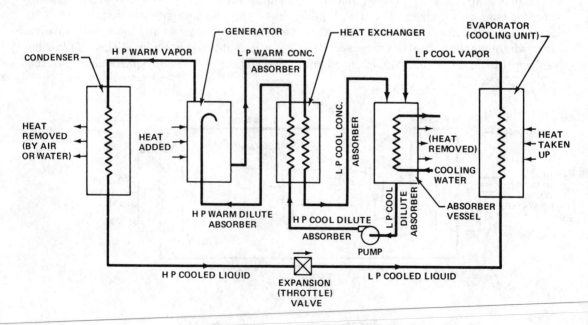

Fig. 110 Absorption-refrigeration system.

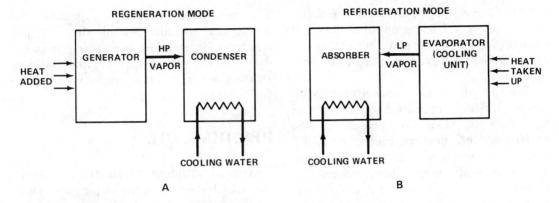

Fig. 111   Intermittent absorption refrigeration.

alternative (refrigeration) mode, the first vessel becomes the absorber and the other the evaporator (Fig. 111B). During this phase refrigeration occurs. The system operates in the regeneration mode for a few hours and is then changed (manually or automatically) to the refrigeration mode, and so on.

In the regeneration mode, heat is supplied to a dilute solution of lithium bromide in water contained in the generator unit. Water vapor at a moderately high pressure passes to the condenser unit and is condensed by the cooling water (Fig. 111A). When sufficient liquid water has collected in the condenser, the heat supply and cooling water are shut off and the refrigeration mode becomes operative. The lithium bromide solution in the absorber unit is cooled so that its vapor pressure is lowered. This causes the water in the evaporator to vaporize, and as a result cooling occurs. The relatively low-pressure water vapor is then absorbed by the solution in the absorber unit (Fig. 111B). After some time, the initial conditions are restored, and the system reverts to the regeneration mode.

Expansion Refrigeration

Cooling by expansion of a compressed, noncondensable gas was one of the earliest methods of mechanical refrigeration, but it is now used to a limited extent only. The system operates on a **Brayton cycle** which may be either open or closed. Since the working gas is often air, the operation has been referred to as an "air cycle." The working gas is compressed, and the heat generated by compression is removed in a heat exchanger cooled with atmospheric air or water. The compressed gas is then expanded in an **expansion turbine** (or turboexpander). The gas does work as it expands in the turbine, and as a result heat energy is taken up and the gas temperature falls.

Rating of Refrigeration Systems

Refrigeration systems (including home freezers, air conditioners, and heat pumps) have been rated in three ways: in (1) **British thermal units** (Btu) per hour, (2) tons of cooling capacity, and (3) horsepower. Room air conditioners are often rated in "Btu," where the quantity stated is actually the rate of heat removal in Btu per hour.

For large cooling systems (and heat pumps), the cooling capacity may be given in tons, where 1 ton of cooling represents the ability to take up heat at a rate that would melt 1 short ton (2000 pounds) of ice in 24 hours. In terms of heat units, 1 ton is equivalent to a heat removal rate of 12,000 Btu per hour, 3520 watts, or 3.52 kilowatts. (The **watt** is a unit of energy rate or power equivalent to 3.41 Btu per hour.) The horse-power of a refrigeration unit does not measure its cooling capacity but rather the rate at which energy is supplied to operate the system.

The Btu per hour, ton capacity, or horse-power does not indicate the efficiency of the machine. This is given by the **coefficient of performance** (or COP), defined as the rate of heat removal relative to the rate of energy input, where the energy-input rate is the total electric (or heat) power required to operate the refrig-eration machine. All rates are expressed in the same units (e.g., watts). If the sys-tem is used for space cooling, the energy-input rate would include the power required to operate air-circulation blowers. The cooling efficiency of home air condi-tioners (and sometimes **heat pumps**) is commonly expressed by the **energy effi-ciency ratio** (or EER), where the EER in Btu/watt-hour is 3.41 times the COP. The COP is a pure ratio and has no units.

In a vapor-compression refrigeration system, the horsepower of the motor (or motors) is a measure of the rate of energy input. Since 1 horsepower is equivalent to 746 watts, it follows that

$$COP = \frac{\text{Rate of heat removal in watts}}{\text{Horsepower} \times 746}$$

A system with a cooling capacity of 1 ton, for example, removes heat at a rate of 3520 watts. If a 2-horsepower motor is required to operate the machine, the COP is $3520/(2.0 \times 746) = 2.36$. The correspond-ing EER is $2.36 \times 3.41 = 8.0$ Btu/watt-hour.

REID VAPOR PRESSURE

A measure of the volatility of liquid fuels, especially **gasoline.** The Reid vapor pressure is defined as the vapor pressure of the liquid measured at 100°F (37.8°C) expressed in pounds per square inch (psi).

The Reid vapor pressure of gasoline for winter use is adjusted to about 12 psi, whereas for summer use it is somewhat over 9 psi. These values provide for easy starting while avoiding vapor lock in the fuel system.

RESIDUAL OIL

The oil withdrawn from the bottom of the **distillation** column after the more readily volatile constituents of petroleum (or petroleum products) have been removed (see **Petroleum Refining**). Residual oil is a mixture of viscous **hydrocarbons** of high molecular weight and high boiling point, generally above 650°F (345°C) at atmos-pheric pressure.

Residual oils are often classified as light, medium (or intermediate), and heavy; the **density, viscosity** (resistance to flow), and boiling range increase in this order: A major use of residual oil is in furnace fuels [see **Fuel (Heating) Oils**], but substantial quantities are processed in petroleum refineries (see **Cracking**) to obtain lighter products (i.e., lower density, molecular weight, viscosity, and boiling range).

Most of the nonhydrocarbon impurities present in the crude oil remain in the resid-ual oil. Of these, sulfur compounds must often be reduced before the oil can be used as a fuel (see **Desulfurization of Liquid Fuels**).

ROCKWELL COAL GASIFICATION (ROCKGAS) PROCESS

A Rockwell International Corporation process for **coal gasification** with oxygen gas (or air) in the presence of molten sodium carbonate. The process is similar to the **Kellogg Molten-Salt Gasification Pro-cess** except that, unlike this and nearly all other coal gasification processes, it uses lit-tle or no steam. The major fuel constituent of the product, called Rockgas, is carbon monoxide, formed by partial oxidation of carbon in the coal; there are smaller proportions of hydrogen and methane from **pyrolysis** of the coal. When oxygen gas is used in the process, the product is an **intermediate-Btu fuel gas;** with air, the product is a **low-Btu fuel gas.** The former can be converted into a **high-Btu fuel gas** (or **substitute natural gas**). Both caking and noncaking coals can be used without pretreatment.

Crushed coal and oxygen gas (or air) at a pressure up to 20 atm (2 MPa) are injected into a bath of molten sodium carbonate at a temperature of about 1800°F (980°C). Some of the coal undergoes pyrolysis, but most is oxidized to **carbon monoxide.** The latter reaction also generates the heat required to maintain the temperature in the gasifier. If the oxidizing (combustion) gas is air, no steam is required, but with oxygen a small proportion of steam may be added for temperature control.

Because the sodium carbonate removes most of the sulfur present in the coal, the product gas is essentially free of sulfur. A bleed stream of molten carbonate is withdrawn from the gasifier and regenerated as described for the Kellogg Molten-Salt process.

Most of the studies of the Rockwell International process have been made with air as the combustion gas. The product contains roughly 30 volume percent of carbon monoxide, 13 percent of hydrogen, and 1 to 2 percent of methane (dry basis) as the fuel constituents; about half of the gas consists of inert nitrogen from the air. The **heating value** is roughly 150 Btu/cu ft (5.6 MJ/cu m).

If oxygen gas is used instead of air, the product gas would contain little nitrogen and the proportions of the fuel components would be approximately double those given above. The heating value would be about 300 Btu/cu ft (11 MJ/cu m). This gas could be subjected to the **water-gas shift reaction** and **methanation** to yield a high-Btu fuel gas.

ROTARY (WANKEL) ENGINE

An internal-combustion, **spark-ignition engine** that generates rotary motion as distinct from the reciprocating motion of a conventional piston-type engine. Several different rotary engine designs have been proposed, but the Wankel engine has attracted the most attention. Because of its low weight and small size for a given power output, and the apparently simple design, the Wankel engine was given serious consideration for use in automobiles and other vehicles. However, interest has declined because of low fuel efficiency, high exhaust emissions, and mechanical problems.

The Wankel engine has a flat chamber, shaped somewhat like an ellipse with a slight waist; the chamber contains a three-sided rotor mounted on an eccentric shaft, as indicated in section in Fig. 112. Internal gears cause the rotor to rotate once for every three rotations of the crankshaft.

The rotor divides the chamber into three separate spaces which change in size as the shaft rotates; each space is equivalent to a cylinder of a piston engine with a rotor face serving as the piston. The chamber has an intake port and an exhaust port, shown at the left in Fig. 112; there are no valves, and the ports are opened and closed by passage of the rotor. Two spark plugs are located on the opposite side to the ports (i.e., at the right in Fig. 112).

Consider first the space *a*. In Fig. 112A, the gasoline–air mixture (or fuel charge) enters the space through the intake port; hence, this represents the intake stage of the operating cycle. As the rotor turns, the volume of space *a* decreases so that the fuel charge is compressed (Fig. 112B). When the minimum volume (or maximum compression) is approached, the spark plugs are fired; combustion of the fuel occurs, and the gases in space *a* expand (Fig. 112C). The work done on the rotor during the expansion stage causes the shaft to rotate. Finally, in Fig. 112D, the space is open to the exhaust port and the exhaust gases are forced out. (For the correspond-

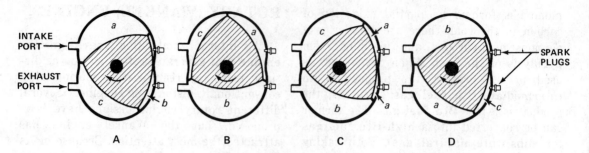

Fig. 112   Wankel engine.

ing stages in a spark-ignition piston engine, see Fig. 153.)

While space *a* is passing through the intake stage, space *b* has completed the compression stage. At the same time, space *c* is in the exhaust stage. Then, as the rotor turns, the other stages occur successively, as described above for space *a*. Thus, the Wankel engine has three power stages for each rotation of the rotor. But, since the crankshaft makes three rotations per rotation of the rotor, there is one power stroke for each rotation of the crankshaft. In the conventional four-stroke, spark-ignition engine there is one power stroke for two rotations of the crankshaft.

S

SCOT (SHELL CLAUS OFF-GAS TREATING) PROCESS

A process for removing elemental sulfur and sulfur compounds (including carbon oxysulfide, carbon disulfide, and hydrogen sulfide) from **Claus process** off-gas. The gas is passed together with hydrogen or a gas containing hydrogen and carbon monoxide (made by partial combustion of a fuel gas) over a cobalt molybdate (CoO–MoO_3) catalyst, on an alumina base, at about 570°F (300°C). In the catalytic reactor the elemental sulfur and sulfur compounds are converted into hydrogen sulfide. The cooled gas, containing up to 3 volume percent of hydrogen sulfide, is extracted with an aqueous solution of an alkanolamine (see **Desulfurization of Fuel Gases**). The off-gas from this process, which may still contain hydrogen sulfide, is recycled to the Claus unit.

SELEXOL PROCESS

A process for removing **acid gases** (hydrogen sulfide, carbon dioxide, and carbon oxysulfide) and other impurities from fuel gases by physical absorption in the organic solvent polyethylene glycol dimethyl ether, tradenamed Selexol. The general principles of operation are similar to those of other processes based on physical absorption of acid gases (see **Desulfurization of Fuel Gases**). The dissolved gases are removed and the solvent regenerated by lowering the pressure of the spent absorber and then stripping with a stream of nitrogen. The hydrogen sulfide in the off-gas can be converted to elemental (solid) sulfur by the **Claus process**.

SEMICONDUCTOR

A solid with an electrical conductivity between that of a good conductor (e.g., a metal) and a nonconductor or insulator (e.g., glass or plastic). Most semiconductors are based on the element silicon. An **atom** of this element has four outermost electrons; in a silicon crystal, the electrons from adjacent atoms are essentially all bound in such a manner that they are not free to conduct an electric current. (An electron carries a negative electric charge, and an electric current usually involves a movement of electrons in a particular direction.) Because there are few free electrons in pure silicon, the solid is a poor electrical conductor.

Silicon can, however, be converted into a semiconductor by introducing controlled amounts of certain impurity elements into the crystal. The process is commonly called *doping* and the impurity is referred to as the *dopant*. Suppose the dopant is an element containing five outer electrons per atom (e.g., phosphorus, arsenic, or antimony); four of these electrons will be bound to the surrounding silicon atoms, leaving one free electron per impurity atom. This electron can move through the solid, and hence the silicon becomes a semiconductor. Because the conductivity is due to the presence of mobile electrons, which have a negative charge, the material is called an *n*-type semiconductor.

Another type of semiconductor is produced by doping the silicon with an impurity element having only three outer electrons per atom (e.g., boron, aluminum, indium, or gallium). Since silicon atoms can bind four electrons, each impurity atom introduces a vacant site called a "hole" that can be filled by a bound electron. A bound electron from an adjacent atom can enter the hole, and, when it does so, it leaves a hole behind. The net effect is that the hole and the bound electron have exchanged positions; in other words, the holes and electrons can move in opposite directions.

After an electron has moved into a preexisting hole, it stays bound, but the hole remaining can again exchange places with a bound electron from an adjacent atom. Hence, a hole behaves as if it can migrate freely from one atom to another, although it is the bound electrons that actually move. The mobility of the electrons thus causes the silicon to become a semiconductor. Because a hole is associated with the absence of a negatively charged electron, it is equivalent to a positive charge; semiconductors with vacant electron sites (or holes) are thus said to be of the p-type.

In addition to the common silicon-based materials, similar n- and p-type semiconductors have been made by doping the element germanium; the atoms of this element, like those of silicon, have four outermost electrons. Other semiconductors for special purposes (e.g., gallium arsenide and lead telluride for high-temperature operations) have been made by adding suitable impurities to binary compounds (i.e., compounds of two elements) and in a few cases to ternary compounds (i.e., with three elements).

Various combinations of n- and p-type semiconductors are used extensively in so-called solid-state devices, such as calculators, computers, television and radio (transistor) receivers, and telephone switching systems. Applications of semiconductors in the energy field include solar cells (see **Solar Energy: Photovoltaic Conversion**), thermoelectric devices for converting heat directly into electricity (see **Thermoelectric Conversion**), rectifiers for changing alternating into direct current, and inverters for the reverse process.

SEPARATIVE WORK UNITS (SWU)

A measure of the effort expended in the separation (or enrichment) of an isotope of a given element (see **Atom**). The separative work concept was developed especially for the **uranium isotope enrichment** program. The amount of separative work (or number of SWU) required to obtain a certain mass of the product (enriched uranium) depends on the isotopic compositions in weight percent of the product, feed (natural uranium), and waste (tails).

The number of SWU required (or expended) is defined in such a way that it is equivalent to a mass. In the U. S. uranium enrichment program, masses are stated in kilograms (kg) and separative work is expressed in kg SWU. The capacity of an enrichment plant is given in kg SWU per annum.

Enrichment charges, which depend on the costs of building, operating, and maintaining the facility, are quoted in dollars per kg SWU. Hence, the enrichment cost of a certain mass of a specified enriched product is equal to the cost per kg SWU multiplied by the number of kg SWU required to make the product. The actual (or total) cost of the enriched material depends also on the cost of the feed and the value, if any, of the waste.

A **light-water reactor** power plant, operating at its full rated electrical generating **capacity** of 1000 megawatts (see **Watt**), will have an annual requirement of about 150,000 kg SWU for fuel enrichment, assuming natural uranium feed, a product of 3 percent enrichment, and tails containing 0.2 percent of uranium-235 (see **Nuclear Fuel**).

SHELL–KOPPERS PROCESS

A process under development by Shell Internationale in The Netherlands and

Krupp-Koppers GmbH in West Germany for **coal gasification** with steam and oxygen gas. The product is an **intermediate-Btu fuel gas** which can also serve as a **synthesis gas**. By subjecting the gas to the **water-gas shift reaction** followed by **methanation**, the product would be a **high-Btu fuel gas** (or **substitute natural gas**).

The Shell-Koppers process is essentially the same as the **Koppers–Totzek process**, except that the pressure in the gasifier is more than 30 atm (3 MPa) instead of atmospheric. The temperature in the combustion zone may also be higher. The product gas has approximately the same composition in both cases, but the carbon dioxide content may be somewhat smaller for the Shell-Koppers process. Because the operating pressure is higher than in the Koppers-Totzek process, the capacity of a gasifier of a given size is increased.

SNPA–DEA (AND DEA) PROCESSES

Similar processes for removing **acid gases** (hydrogen sulfide and carbon dioxide) from fuel gases by chemical absorption in a weakly alkaline solution. The absorber is an aqueous solution of diethanolamine (DEA) at a temperature of 90 to 120°F (32 to 49°C); the operating pressure may be up to 70 atm (7 MPa). The general principles are the same as in other processes based on chemical absorption (see **Desulfurization of Fuel Gases**). (SNPA in the title is an abbreviation for Société Nationale des Pétroles d'Aquitaine which developed one form of the process.)

SOLAR ENERGY

In general, the energy produced and radiated by the sun; more specifically, the term refers to the sun's energy that reaches the earth. Solar energy, received in the form of radiation, can be converted directly or indirectly into other forms of energy, such as heat and electricity, which can be utilized by man. Since the sun is expected to radiate at an essentially constant rate for a few billion years, it may be regarded as an inexhaustible source of useful energy. The major drawbacks to the extensive application of solar energy are (1) the intermittent and variable manner in which it arrives at the earth's surface and (2) the large area required to collect the energy at a useful rate.

Energy is radiated by the sun as electromagnetic waves (see **Electromagnetic Radiation**) of which 99 percent have wavelengths in the range of 0.2 to 4.0 micrometers (1 micrometer = 10^{-6} meter). Solar energy reaching the top of the earth's atmosphere consists of about 8 percent ultraviolet radiation (short wavelength, less than 0.39 micrometer), 46 percent visible light (0.39 to 0.78 micrometer), and 46 percent infrared radiation (long wavelength, more than 0.78 micrometer).

Solar Constant

The rate at which solar energy arrives at the top of the atmosphere is called the *solar constant*. This is the amount of energy received in unit time on a unit area perpendicular to the sun's direction at the mean distance of the earth from the sun. Because the sun's distance and activity vary throughout the year, the rate of arrival of solar radiation varies accordingly. The so-called solar constant is thus an average from which the actual values vary up to about 3 percent in either direction. This variation is not important, however, for most practical purposes. The National Aeronautics and Space Administration's standard value for the solar constant, expressed in three common units (see **Watt; British Thermal Unit**), is as follows:

1.353 kilowatts per square meter
116.4 langleys (calories per sq cm) per hour
429.2 Btu per sq ft per hour

Direct and Diffuse Radiation

The solar radiation that penetrates the earth's atmosphere and reaches the surface differs in both amount and character from the radiation at the top of the atmosphere. In the first place, part of the radiation is reflected back into space, especially by clouds. Furthermore, the radiation entering the atmosphere is partly absorbed by molecules in the air. Oxygen (O_2) and ozone (O_3), formed from oxygen, absorb nearly all the ultraviolet radiation, and water vapor and carbon dioxide absorb some of the energy in the infrared range. In addition, part of the solar radiation is scattered (i.e., its direction is changed) by water droplets in clouds, by atmospheric molecules, and by dust particles.

Solar radiation that has not been absorbed or scattered and reaches the ground directly from the sun is called "direct" radiation. It is the radiation which produces a shadow when interrupted by an opaque object. The radiation received after scattering, on the other hand, is generally referred to as "diffuse" radiation. Because the solar radiation is scattered in all directions in the atmosphere, diffuse radiation comes to earth from all parts of the sky.

The total solar radiation received at any point on the earth's surface is the sum of the direct and diffuse radiation. This is referred to in a general sense as the *insolation* at that point. More specifically, the insolation is defined as the total solar radiation energy received on a horizontal surface of unit area (e.g., 1 sq ft or 1 sq m) on the ground in unit time (e.g., 1 day).

The insolation at a given location on the earth's surface depends, among other factors, on the altitude of the sun in the sky. (The altitude is the angle between the sun's direction and the horizontal.) Since the sun's altitude changes with the date and time of day and with the geographic latitude at which the observations are made, the rate of arrival of solar radiation on the ground is a variable quantity even in the daytime.

There are, nevertheless, some general points that can be made. The smaller the sun's altitude, the greater the thickness of

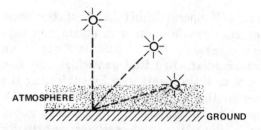

Fig. 113   Insolation and the sun's altitude.

atmosphere through which the solar radiation must pass to reach the ground (Fig. 113). As a result of absorption and scattering, the insolation is less when the sun is low in the sky than when it is higher. However, when scattering occurs, diffuse radiation constitutes a larger fraction of the total received.

On a clear, cloudless day, about 10 to 20 percent of the insolation is from diffuse radiation; the proportion increases up to 100 percent when the sun is completely obscured by clouds. When the humidity is high, the insolation on a cloudy day, consisting entirely of diffuse radiation, may be as high as 50 percent of the insolation on a clear day at the same time and place.

Surface Orientation Effect

The rate of receipt of solar energy on a given surface on the ground depends on the orientation of the surface with reference to the sun. A fully sun-tracking surface that always faces the sun (Fig. 114) receives the

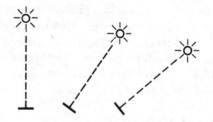

Fig. 114   Solar-tracking mirror.

maximum possible solar energy at the particular location. A surface of the same area oriented in any other direction will receive a smaller amount of solar radiation.

Because solar radiation is such a "dilute" form of energy, it is desirable to capture as much as possible on a given area. This can be done with fully tracking collectors that are adjusted continuously to be perpendicular to the sun's rays. Periodic adjustment is simpler to accomplish but is less effective. Both fully tracking and partially tracking systems have been developed for use in special circumstances when their cost may be justifiable.

The major economic application of solar energy at present and in the near future is for heating water for domestic use and for space heating of buildings. For these purposes, solar-energy collectors that require continuous or periodic adjustment would appear to be a drawback. It is the general practice, therefore, to use flat-plate collectors (see **Solar Energy Collectors**) with a fixed orientation that provides optimum insolation for the required purpose.

Where space heating (November through March) is the prime consideration, as it commonly is, the best efficiency is obtained with a fixed, flat-plate collector if it faces roughly south and slopes at an angle to the horizontal equal to the local latitude plus about 15°. For example, at a mid-latitude of 40°N, the optimum slope for space and water heating is roughly 55°. Although, in theory, a flat-plate collector facing south should be best, an orientation slightly west of south is preferable. This provides more exposure to the afternoon sun when the atmosphere is warmer than in the morning. If a substantial amount of solar energy is required in the summer (e.g., for air conditioning), a slope angle closer to the latitude value would be desirable.

Solar Energy Utilization: Historical Background

For many years, primitive use of solar energy has been made for heating dwellings, for drying agricultural products, for distilling liquids, and for other purposes. In the 1920s and later, solar water heaters were in common use in Florida and in the southwest United States, as well as in other countries, but their number decreased as other forms of energy became available at a moderate cost.

The application of scientific principles to the use of solar energy in space heating began to attract serious interest in the United States in the 1930s, and between 1939 and 1957 five solar-heated houses were constructed in Massachusetts. In addition, prior to the early 1970s, about a dozen more residences and other structures were built to demonstrate that solar energy could provide a large proportion of the space heating requirements in different parts of the country. The wider use of solar heating was inhibited, however, by the fact that components were not available commercially and so were expensive. This situation is now changing.

In another use of solar energy, a solar steam engine was demonstrated in France in 1878. The sun's radiation was focused on a water boiler, and the steam generated was used to operate a conventional **steam engine**. A few years later, John Ericsson, who designed the warship *Monitor*, built a hot-air engine driven by solar heat. During the next 50 years or so, many patents were issued for solar engines, some of which were built and served to operate irrigation pumps. Among the early concepts was the use of a volatile liquid, instead of water, as the working fluid in a heat engine, thus permitting operation at lower temperatures.

Although there is no evidence that it was implemented in practice, a patent was issued in 1893 for a combination of a solar engine with electric storage batteries. This idea was apparently novel in two respects: first, it implied the use of a solar engine to drive a generator for producing electric power, and second, storage of the electrical energy in batteries would provide a supply of electricity at all times and not only during periods of sunshine.

The conversion of solar energy into electrical energy without the use of machinery, by utilizing the thermoelectric effect, was proposed in 1888 and later years. If one junction between two different materials is heated (e.g., by the absorption of solar radiation) while the other junction is kept cool, an electrical voltage is produced (see **Thermoelectric Conversion**). Thermoelectric generators using solar radiation which have been built so far, however, have had a very low efficiency. The direct conversion of solar energy into electricity became practical in 1954 when the silicon solar cell was demonstrated in the United States. Although the cost of such cells was initially high, they proved to be invaluable as a source of electric power for spacecraft.

U. S. National Solar Energy Program

The recent interest in the development of solar energy in the United States can be attributed partly to the realization of the need for sources of energy to supplement the dwindling reserves of conventional fuels, notably gas and oil, and partly from the sharp increase in the price of imported oil at the end of 1973. However, in January 1972, before this increase had occurred, a Solar Energy Panel was organized by the National Science Foundation and the National Aeronautics and Space Administration. In its report at the end of 1972, the Panel identified a number of solar energy areas that showed promise from technical, economic, and energy potential standpoints.

A program for studying solar energy technology, initiated by the National Science Foundation, received impetus from several Congressional actions, including passage of the Solar Heating and Cooling Demonstration Act of 1974 and the Solar Energy Research, Development, and Demonstration Act of the same year. The latter authorized a comprehensive program, including the establishment of a Solar Energy Research Institute, aimed at providing the option of utilizing solar energy as a viable, future energy source. Upon its formation in early 1975, the Energy Research and Development Administration (ERDA) became responsible for the solar energy program and, in 1977, Golden, Colorado, was chosen as the site for the Solar Energy Research Institute. In October 1977, the activities of ERDA were transferred to the new U. S. Department of Energy.

The actual and proposed applications of solar energy may be considered in three general categories:

1. *Direct Thermal Applications* make direct use of heat, resulting from the absorption of solar radiation, for space heating (and cooling) of residences and other buildings, to provide hot-water service for such buildings, and to supply heat for agricultural, industrial, and other processes that require only moderate temperatures. These applications are described in the section on **Solar Energy: Direct Thermal Applications**.

2. *Solar Electric Applications* are those in which solar energy is converted directly or indirectly into electrical energy. Four general conversion methods are being investigated:

Solar thermal methods involve production of high temperatures, such as are required to boil water or other working fluid for operating turbines which drive electric generators. These are considered under **Solar Energy: Thermal Electric Conversion**.

Photovoltaic methods make use of devices (solar cells) to convert solar energy directly into electrical energy without machinery. See **Solar Energy: Photovoltaic Conversion**.

Wind energy is a form of solar energy that can be converted into mechanical (rotational) energy and hence into electrical energy by means of a generator. This indirect use of solar energy to generate electricity is described under the heading of **Wind Energy Conversion**.

Ocean thermal energy conversion depends on the difference in temperature between solar heated surface water and cold deep ocean water to operate a vapor-expansion turbine and electric generator. See **Ocean Thermal Energy Conversion**.

3. *Fuels from Biomass* refers to the conversion into clean fuels (or other energy-related products) of organic matter derived directly or indirectly from plants which use solar energy to grow. Biomass materials include agricultural, forest, and animal residues, as well as terrestrial and aquatic plants grown especially for the purpose. For a discussion, see **Biomass Fuels**.

Other Forms of Solar Energy. Water power, other than tidal power, also represents a utilization of solar energy (see **Hydroelectric Power**). Heat from the sun causes the evaporation of surface waters; the vapor rises, is condensed in the upper atmosphere, and falls as rain. The resulting water that collects at higher elevations has substantial (potential) energy than can be converted into electrical energy by means of a turbine-generator at a lower level.

It may be noted, too, that, since fossil fuels—coal, oil, and natural gas—originate from living matter, their energy is really solar energy that has been converted and stored for millions of years. In fact, only **nuclear energy, geothermal energy,** and, to a large extent, **tidal energy** do not originate in the sun.

SOLAR ENERGY COLLECTOR

A device for collecting solar radiation. The solar energy collector, with its associated absorber, is the essential component of any system for the conversion of solar radiation energy into a more usable form (e.g., heat or electricity). Solar collectors fall into two general categories: nonconcentrating and concentrating. In the nonconcentrating type, the collector area (i.e., the area that intercepts the solar radiation) is the same as the absorber area (i.e., the area absorbing the radiation). On the other hand, in concentrating collectors, the area intercepting the solar radiation is greater, sometimes hundreds of times greater, than the absorber area. By means of concentrating collectors, much higher temperatures

can be obtained than with the nonconcentrating type.

Flat-Plate Collectors

Where temperatures below about 200°F (90°C) are adequate, as they are for space and service-water heating, flat-plate collectors, which are of the nonconcentrating type, are particularly convenient. They are made in rectangular panels, from about 18 to 32 sq ft (1.7 to 2.9 sq m) in area, and are relatively simple to construct and erect. Flat plates can collect and absorb both direct and scattered solar radiation (see **Solar Energy**); they are consequently partially effective even on cloudy days when there is no direct radiation.

There are many flat-plate collector designs, but most are based on the principle shown in section in Fig. 115. The collec-

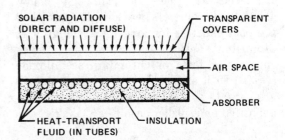

Fig. 115 Section through typical flat-plate collector.

tor consists of (1) a flat-plate absorber, which intercepts and absorbs the solar radiation energy, (2) a transparent cover (or covers) that permits solar radiation to pass through but reduces heat loss from the absorber, (3) a heat-transport fluid (air or water) flowing through tubes (or other channels) to remove heat from the absorber, and (4) a heat-insulating backing. The function of the insulation is to prevent loss of heat from the absorber and heat-transport fluid; standard insulating materials, such as fiberglass or styrofoam, are used for this purpose. The other three components of flat-plate collectors are described more fully on the following pages.

The Absorber. The absorbing plate is usually made of copper, aluminum, or steel coated with a heat-resistant black (carbon) paint; a thin layer of a black oxide (e.g., copper oxide on copper) is an alternative to paint. Solar radiation in the visible and infrared ranges, up to a wavelength of 2 micrometers (see **Solar Energy**), passes through the transparent cover (or covers) and is absorbed by the blackened surface. Here the radiation energy is converted into heat. As the temperature of the surface material increases, however, it tends to radiate energy in the infrared region, especially at wavelengths beyond 2 micrometers; this is called thermal infrared radiation and represents a possible source of energy loss.

The amount of energy emitted as thermal infrared radiation can be reduced by means of special "selective" coatings. These are good absorbers of radiation below about 2 micrometers wavelength, but they are poor emitters for longer wavelengths. A promising selective coating is "black chrome," a form of chromium sesquioxide (Cr_2O_3) with embedded chromium metal, in a layer 0.15 to 2 micrometers thick, electrodeposited on a nickel base.

Transparent Cover. The main purpose of the transparent cover of the flat-plate collector is to decrease heat loss without significantly reducing the incoming solar radiation. In the first place, the relatively still (or stagnant) air space between the cover and the absorber plate largely prevents loss of heat from the plate by convection. (Heat loss by convection refers to the removal of heat from a warm surface by a cooler moving fluid, such as air or water.)

Furthermore, if the cover is made of glass, it permits the passage of solar radiations with wavelengths less than 2 micrometers, but it is largely opaque to the longer wavelength thermal infrared. As a result, heat is trapped in the air space between the cover and the absorber plate in a manner similar to a greenhouse. The effect is to reduce the loss of heat from the absorber. However, since the enclosed air is inevitably warmer than the ambient air, there is some loss of heat to the surroundings from the top of the cover by convection, conduction, and radiation. The rate of heat loss increases as the temperature of the air space rises; as will be seen shortly, this affects the overall efficiency of the solar collector.

In many flat-plate collectors, the cover consists of two sheets of clear, tempered glass; the sheets are roughly ⅛ to ⅜ in. (3 to 4.5 mm) thick and are 0.4 to 0.8 in. (10 to 20 mm) apart. In warmer climates costs can be reduced, with only minor loss of efficiency, by using a single glass sheet as the cover. The disadvantage of glass is that it can be broken easily, although the breakage risk can be decreased, at some increase in cost, by utilizing specially tempered (strong) glass.

Transparent plastics have been used in place of glass, but they have some drawbacks. Most plastics are not as opaque as glass to the thermal infrared radiation and so permit greater loss of heat from the absorber. They also often suffer a decrease in transparency and sometimes break up in the course of time due to heating and the action of solar ultraviolet radiation. Efforts are being made to develop better plastic materials that might be used in solar collectors.

A certain proportion of the incident solar radiation is lost by absorption in the glass cover plates, but the loss can be kept small by using a clear ("water white") glass with a low iron content. A much larger loss occurs as a result of partial reflection. Two glass plates may reflect some 15 percent of the solar radiation coming from a perpendicular direction. The reflection loss increases as the direction of incidence departs from the perpendicular. The reflectivity of glass covers may be reduced by coating with thin films of certain substances (e.g., magnesium fluoride) or by gentle etching with a solution of hydrofluoric acid (Du Pont's Magicoat process). Such antireflective coatings add to the cost

ABSORBER ———→

Fig. 116   Heat-transport systems (in section) for flat-plate collector.

of the collectors but make them more efficient.

Heat-Transport System. The heat generated in the absorber is removed by continuous flow of a heat-transport (or heat-transfer) medium, either water or air. It is mainly in the design of the heat-transport system that flat-plate collectors differ. When water is used, it is most commonly passed through metal tubes with either circular or rectangular cross section; the tubes are welded to the absorber plate (or form an integral part of it) so as to assure effective transfer of heat to the fluid. Some examples are represented by the sections shown in Fig. 116. The tubes are connected to common headers at each end of the collector. In order to maximize the exposure to solar radiation, collectors are almost invariably sloped. Cooler water then enters at the bottom header, flows upward through the tubes where it is warmed by the absorber, and leaves by way of the top header (Fig. 117).

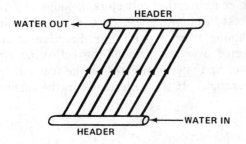

Fig. 117   Water flow in flat-plate collector.

In one simple type of flat-plate collector, the absorber is a blackened sheet with close corrugations running from top to bottom. The water flows downward through the grooves formed by the corrugations. A problem with this design is that, in cold weather, moisture may condense on the inside of the transparent cover plate and thus decrease the transmission of solar radiation.

Water is a very effective heat-transport medium, but it suffers from certain drawbacks. One is the possibility of freezing in the collector tubes during cold nights. Freezing is commonly prevented by using a solution of ethylene glycol (automobile antifreeze) in water, but this generally adds to the complexity of the heating system. Furthermore, the antifreeze solution is less effective than water for heat removal from the absorber. In some cases, the water is drained from the collector tubes if freezing is expected, but difficulties have been experienced in refilling all the tubes in the morning.

Another problem arises from corrosion of the metal tubes by the water; this is aggravated if the water is drained at night thus allowing air to enter. The oxygen in air increases the rate of corrosion of most metals. Corrosion can be minimized by using copper tubing. Aluminum is a less expensive alternative, although periodic chemical treatment of the water is desirable. Finally, leaks in a water (or antifreeze) circulation system require immediate attention.

Air has been used so far to a lesser extent as the heat-transport medium in solar collectors, but it may have some advantages over water. To decrease the power required to pump the necessary volume of air through tubes, wider flow channels are used. For example, the air may be passed through a space between the absorber plate and insulator with baffles arranged to provide a long (zigzag) flow path (Fig. 118).

The use of air as the heat-transport fluid eliminates both freezing and corrosion problems, and small air leaks are of less concern than water leaks. Moreover, the heated air can be used directly (or by way

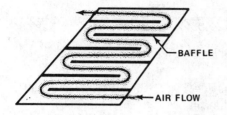

Fig. 118   Zig-zag airflow path in flat-plate collector.

of heat storage) for space heating. On the other hand, larger duct sizes and higher flow rates, with increased pumping power, are required for air than when water is the heat-transport medium. Another drawback is that transfer of heat from air to water in a hot-water supply system is inefficient.

The increase in temperature of the heat-transport fluid in its passage through a flat-plate collector has a bearing on the overall efficiency for conversion of the incident solar radiation into useful heat. The loss of heat to the surroundings from the top surface of the cover plates increases (and the net efficiency decreases) with increase in the temperature difference between the absorber and the environment. This difference is related to the increase in temperature of the heat-transport fluid. As a general rule, therefore, the greater the temperature increase of the fluid, the smaller is the overall efficiency of the collector.

The actual efficiency of a collector (i.e., the proportion of the solar energy absorbed as heat in the heat-transport fluid) also depends on the design details, but the following data provide a rough indication of what might be expected. For a temperature rise of 108°F (60°C), say from 77 to 185°F (25 to 85°C), the overall efficiency could be close to 30 percent. For an increase of only 54°F (30°C), however, from 77 to 131°F (25 to 55°C), the efficiency might increase to about 50 percent. On the average, the efficiencies of flat-plate, solar radiation collectors have been found in practice to be roughly 35 to 40 percent.

Concentrating Collectors: Focusing Type

The most efficient concentrating collectors of solar radiation are those of the focusing type. In these collectors, radiation falling on a relatively large area is focused onto a receiver (or absorber) of considerably smaller area. As a result of the energy concentration, fluids can be heated to temperatures of 930°F (500°C) or more.   An important difference between collectors of the nonfocusing and focusing types is that the latter concentrate only direct radiation coming from a specific direction. Since diffuse radiation arrives from all directions, only a very small proportion is from the direction for which focusing occurs.

Focusing collectors may be considered in two general categories: line focusing and point focusing types. In practice, the line is a collector pipe and the point is a small volume through which the heat-transport fluid flows. Because the sun has a finite size, focusing does in fact occur over a small area or volume, rather than a line or point.

Line-Focusing   Collectors: Parabolic Trough Reflector. The principle of the parabolic trough reflector, which is often used in concentrating collectors, is shown by the cross section in Fig. 119; solar radiation coming from a particular direction is collected over the area of the reflecting surface and is concentrated at the focus of the parabola.   If the reflector is in the form of

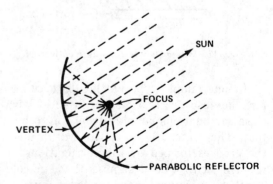

Fig. 119   Cross section of parabolic-trough collector.

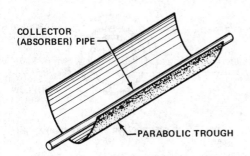

COLLECTOR
(ABSORBER) PIPE

PARABOLIC TROUGH

Fig. 120   Parabolic-trough collector.

a trough with a parabolic cross section, the solar radiation is focused along a line. The collector pipe, preferably with a selective absorber coating, is located on this line (Fig. 120).   The dimensions of parabolic trough collectors can vary over a wide range; the length of a reflector unit may be roughly 10 to 16 ft (3 to 5 m) and the width about 5 to 8 ft (1.5 to 2.4 m). Ten or more such units are often connected end-to-end in a row; several rows may also be connected in parallel. Parabolic trough reflectors have been made of highly polished aluminum, of silvered glass, or of a thin film of aluminized plastic on a firm base. Instead of having a continuous form, the reflector may be constructed from a number of long flat strips on a parabolic base.

For the solar radiation to be brought to a focus by a parabolic trough reflector, the sun must be in such a direction that it lies on the plane passing through the focal line and the vertex (i.e., the base) of the parabola (see Fig. 119). Since the elevation of the sun is always changing, either the reflector trough or the collector pipe must be turned continuously about its long axis to maintain the required orientation. Both schemes are used in different practical designs. Either the trough or the pipe is turned by partial rotation around a single axis parallel to the trough length.

Trough-type collectors are generally oriented in the east-west or north-south directions. For the east-west orientation, the collectors are laid flat on (or parallel to) the ground. For the north-south orientation, however, the north end of the

trough is raised so that the collectors are sloped facing south, just like flat-plate collectors. Ideally, the slope angle should be changed periodically; it is simpler, but less efficient, however, to use a fixed-angle design.

The north-south orientation permits more solar energy to be collected than the east-west arrangement, except around the winter equinox.   On the other hand, construction costs are higher for the north-south (sloping) type. Moreover, a system of such collectors requires a larger land area to allow for the shadowing effect of the sloping troughs.   The increased separation distance between rows of collectors also results in increased pipeline costs and greater pumping and thermal losses. Finally, the sunset position of an east-west reflector is essentially the same as the sunrise position, and little or no overnight adjustment is required.   For the north-south orientation, however, the trough (or receiver) must be turned through a large angle from sunset to sunrise. The choice of orientation in any particular instance depends on the foregoing and other considerations.

Mirror-Strip Reflector. In another kind of focusing collector, a number of plane or slightly curved (concave) mirror strips are mounted on a flat base. The angles of the individual mirrors are such that they reflect solar radiation from a specific direction onto the same focal line (Fig. 121). The angles of the mirrors must be adjusted

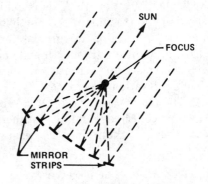

SUN

FOCUS

MIRROR
STRIPS

Fig. 121   Mirror-strip solar collector.

to allow for changes in the sun's elevation, while the focal line (or collector pipe) remains in a fixed position. Alternatively, as mentioned for parabolic trough collectors, the mirror strips may be fixed and the collector pipe moved continuously so as to remain on the focal line.

Fresnel Lens Collector. In addition to the reflecting collectors described above, a refraction-type of focusing collector has been developed. It utilizes the focusing effect of a Fresnel lens, as represented in cross section in Fig. 122. For a trough-type

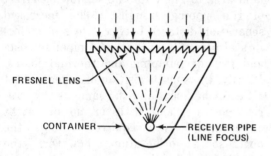

Fig. 122   Cross section of Fresnel lens trough collector.

collector, the lens is a rectangle, about 15.4 ft (4.7 m) in overall length and 3 ft (0.95 m) in width. It is made in sections from cast acrylic plastic and can probably be produced in quantity at low cost. The rounded triangular trough serves only as a container and plays no role in concentrating the solar energy.

To be fully effective, the Fresnel lens must be continuously aligned with the sun in two directions, namely, both along and perpendicular to its length. This is achieved by orienting the troughs in the north–south direction with rotation about the lengthwise axis; in addition, the north ends of the troughs are raised to increase the slope as the sun's elevation decreases (and vice versa). The total solar radiation energy that can be collected annually is about 30 percent greater than for an east–west orientation.

Receiver Pipe. The receiver pipe of a parabolic line-focusing collector, shown in

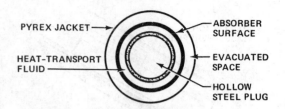

Fig. 123   Cross section of solar energy pipe receiver.

cross section in Fig. 123, has the same general characteristics as a flat-plate collector. The solar radiation absorber is a central steel pipe with a treated surface. A selective absorber surface, such as the black chrome referred to earlier, may be advantageous. A hollow steel plug within the absorber pipe restricts the flow of the heat-transport fluid to a narrow annular region. This results in a high flow velocity of the fluid and consequently a high rate of heat transfer from the absorber.

The absorber pipe is usually enclosed in a glass (Pyrex) jacket in order to decrease thermal losses by convection and radiation. The space between the pipe and the jacket is sometimes evacuated to reduce convection losses. The diameter of the glass jacket may be about 2 in. (5 cm) and that of the absorber pipe about 1.2 in. (3 cm). The annulus between this pipe and the plug may be as little as 0.1 in. (2.5 mm) wide.

In a Fresnel lens collector, the solar radiation is focused onto the absorber from the top, rather than from the bottom as in the parabolic (reflection) type. A modified absorber design is then possible (Fig. 124). Insulation at the bottom and sides of the absorber pipe and a flat glass plate over the top reduce thermal losses. A stainless

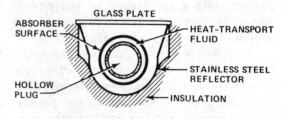

Fig. 124   Receiver for Fresnel lens collector.

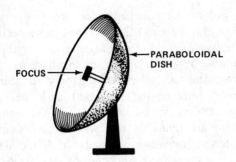

Fig. 125   Point-focus solar collector.

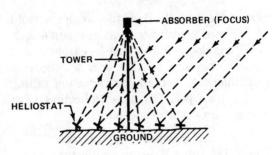

Fig. 126   Distributed heliostat point-focusing reflector.

steel reflector adjacent to the pipe reflects back emitted thermal radiation.

Point-Focusing Collectors. A paraboloidal dish reflector brings solar radiation to a focus at a point, actually a small central volume (Fig. 125). A dish 22 ft (6.6 m) in diameter has been made from about 200 curved mirror segments forming a paraboloidal surface. The absorber, located at the focus, is a cavity made of a zirconium-copper alloy with a black chrome selective coating. The heat-transport fluid flows into and out of the absorber cavity through pipes bonded to the interior. The dish can be turned automatically about two axes (up–down and left–right) so that the sun is always kept in a line with the focus and the base (vertex) of the dish. Thus, the sun can be fully tracked at essentially all times.

A system equivalent to a very large paraboloidal reflector consists of a considerable number of mirrors distributed over an area on the ground. Each mirror, called a *heliostat,* can be steered independently about two axes so that the reflected solar radiation is always directed toward a collector (absorber) mounted on a tower (Fig. 126). Since the mirrors have an appreciable size, the solar radiation from each mirror is not directed at a point but rather over a small area. Hence, the larger the size of the collector, the larger the practical dimensions of the mirror. A system of 1860 heliostats covering an area of about 74 acres (30 hectares) has been built near Barstow, California (see **Solar Energy: Thermal Electric Conversion**).

Concentrating Collectors: Nonfocusing Type

The simplest type of concentrating collector is the mirror-boosted, flat-plate collector. It consists of a flat plate facing south with mirrors attached to its north and south edges (Fig. 127). If the mirrors

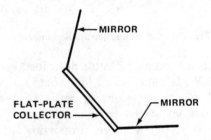

Fig. 127   Flat-plate collector augmented with mirrors.

are set at the proper angle, they reflect solar radiation onto the absorber plate. Thus, the latter receives reflected radiation in addition to that normally falling on it. The mirrors cut off part of the scattered radiation that would otherwise have reached the absorber plate, and only part of the scattered radiation falling on the mirrors will be reflected onto the absorber. Thus the concentration effect arises mainly from the increase in direct radiation reaching the absorber plate.

When a number of collectors are combined in two or more rows, as they often are, the rows must be set farther apart in the north-south direction to allow for the additional sunshading caused by the mirror extensions. Furthermore, in order for the mirrors to be effective, the angles should be adjusted continuously as the sun's altitude changes. For these reasons, and because they can provide only a relatively small increase in the solar radiation falling on the absorber, flat-plate collectors with mirrors are not widely used.

Compound Parabolic Concentrator (CPC). The CPC (or Winston collector) is a trough-like arrangement of two facing parabolic mirrors (Fig. 128). Unlike the single

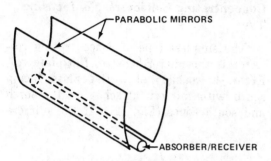

Fig. 128 Compound parabolic concentrator.

parabolic trough reflector described earlier, the CPC is nonfocusing, but solar radiation from many directions is reflected toward the bottom of the trough. Because of this characteristic, a large proportion of the solar radiation, including diffuse (scattered) radiation, entering the trough opening is collected (and concentrated) on a small area. In addition to collecting both direct and diffuse radiations, an advantage of the CPC is that it provides moderately good concentration, although less than a focusing collector, in an east–west orientation without (or only occasional) adjustment for sun tracking.

See also **Solar Furnace; Solar Pond**.

SOLAR ENERGY: DIRECT THERMAL APPLICATIONS

The direct use of the heat resulting from the absorption of solar radiation for space heating of buildings, for producing hot water, and for agricultural and industrial purposes. Solar heat can also be used indirectly for cooling.

Solar space-heating (or -cooling) systems may be classified as passive or active. *Passive heating systems* operate without pumps, blowers, or other mechanical devices; the air is circulated past a solar-heated surface (or surfaces) and through the building by convection (i.e., less dense, warm air tends to rise while more dense, cooler air moves downward). In *active heating systems*, fans and pumps are used to circulate the air and often a separate heat-absorbing fluid.

Passive solar thermal systems are more practical in locations where there is ample winter sunshine and an unobstructed southern exposure is possible. The building to be heated is an essential part of the system design. Active systems, on the other hand, can be adapted to almost any location and type of building; however, they are more expensive than passive systems to construct and operate. An advantage of active solar systems is that the building air temperature can be controlled in the same way as with conventional heating, but in most passive systems substantial temperature variations may occur in the course of the day.

In principle, it should be possible to provide all the heating (and cooling) needs of a building by solar energy. However, to do this, the heating system would have to be designed for minimum sunshine conditions and hence would be over-designed for the majority of situations. In most cases, solar-energy systems provide roughly 50 to 75 percent of the annual heating requirements. The remainder is supplied by an auxiliary-heating system using gas, oil, or electricity.

Active Space-Heating Systems

General Principles. Nearly all existing or proposed active solar space-heating and/or hot-water supply systems utilize three main components in addition to pumps and blowers: (1) a solar radiation collector with its associated heat-transport (or heat-transfer) fluid, (2) a heat-storage medium, and (3) a distribution system. The same arrangement of components can also be used to provide hot water for domestic and related use and, with the addition of other components, space cooling (air conditioning). In the collector, the solar radiation is collected and converted into heat. The heat-transport fluid removes the heat and carries it to the heat-storage system; the heat can then be withdrawn from storage and distributed throughout the building. A schematic simplified outline of a solar space-heating system is given in Fig. 129; the individual components are considered in the following sections.

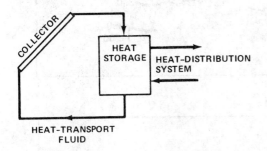

Fig. 129   Active solar-heating system.

Collectors and Heat Transport. Because of their simplicity, fixed-position, flat-plate collectors are almost invariably used in space-heating applications of solar energy (see **Solar Energy Collectors**). Since they can collect diffuse as well as direct solar radiation, flat-plate collectors are partially effective even when the sun is not shining. Either air or water can be used as the heat-transport fluid. The plumbing system is less likely to have problems in a system using air, but larger ducts and a larger heat-storage volume are required.

Many types of flat-plate collectors are available. As a general rule, these collectors are fabricated in panels, commonly about 6 to 8 ft (1.8 to 2.4 m) long and 3 to 4 ft (0.9 to 1.2 m) wide. The number of panels (or the total collector area) depends on the space to be heated (or cooled) and the local climate, as well as on economic factors, as will be seen shortly. In new home construction, the panels are mounted on a roof fac-

ing south (or south-southwest) with the optimum slope (e.g., latitude + 15° for heating). Otherwise, the panels with the required slope may be placed on the ground or on a flat roof, if one is available. The panels must then be spaced in such a manner that they do not shade one another significantly when the sun is low in the sky.

An essential piece of information for the design of a collector for space heating is the daily insolation, that is, the total amount of direct and diffuse solar radiation energy received per day on a horizontal surface of unit area on the ground (see **Solar Energy**). Weather stations in the United States report daily insolation expressed in langleys, where 1 langley represents 1 calorie per square centimeter (i.e., 4.18×10^4 joules/sq m, 0.0116 kilowatt-hours/sq m, or 3.69 Btu/sq ft). (See **British Thermal Unit, Watt,** and **Joule** for definitions of units.) From these data contour maps can be drawn showing lines of equal insolation on any given day. Two examples of such maps are given in Fig. 130; they show the daily insolations (in langleys per day) for January and July, respectively, averaged over several years. The heavier lines are for clear days and the broken lines for cloudy days; for the latter, most of the insolation arises from diffuse (scattered) solar radiation.

The data for January and July may be regarded as giving the annual minimum and maximum daily insolations, respectively. Hence, the average daily insolation in the United States ranges from about 100 to 750 langleys (1.16 to 8.7 kW-hours/sq m), depending on the location, the time of year, and the weather conditions. It should be noted that these data refer to a horizontal surface. For a surface inclined at an angle to the horizontal equal to the latitude + 15°, the solar radiation received is higher in January (winter) and lower in July (summer).

From a knowledge of the daily temperatures throughout the year, it is possible to determine the amount of thermal energy

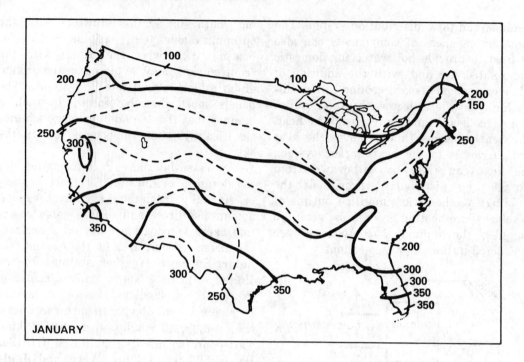

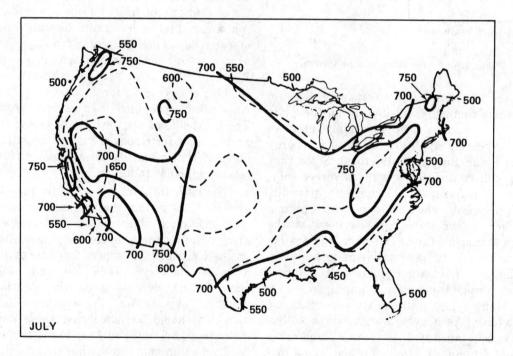

Fig. 130 Average daily solar insolation (langleys per day) contours for January and July (from *American Institute of Physics Handbook*, with permission of McGraw-Hill Book Co.).

(heat) required for space heating in the winter and cooling in the summer. By using the daily insolation at the given location, an estimate can then be made of the area of a flat-plate collector, sloping at a prescribed angle, that will supply a certain proportion of the heat required.

In a study performed by the Westinghouse Electric Corporation for the National Science Foundation, estimates were made of the collector areas for single-family homes with a floor area of 1600 sq ft (150 sq m) in five different climatic regions in the United States. Some of these estimates for two cases, in which solar energy provides 50 or 80 percent, respectively, of the requirements for heating only and for heating and cooling, are given in the table in square feet (ft^2) and square meters (m^2). The respective percentages are average values; there may be times in the winter when a system designed for 50 percent solar heating will provide all that is required. On the other hand, after several cloudy days, the system will supply a smaller proportion of the heating requirements.

The degree (or percentage) of dependence on solar energy for which a collector system is designed is determined by balancing the cost of the collectors against the estimated future costs of alternative energy sources. It is the future rather than the present costs that are important in this respect. The higher the expected cost of an alternative energy supply, the larger the economically acceptable collector area. Thus, in deciding on this area, the local climatic and economic factors must be considered. Furthermore, the collector area should be proportional to the area of the building to be heated (or heated and cooled). A very rough rule for moderately cool climates is a collector area about 50 percent of the floor area.

Heat Storage. A means for storing heat is necessary in active solar-energy systems to supply heating requirements during the evening and on cloudy days. The larger the heat-storage volume, the greater is the heat-storage capacity for a given material, and the longer the sunless period for which heat could be available. But increasing the volume means increasing the cost and

Collector Areas for Single-Family Homes
[Floor Area 1600 sq ft (or 150 sq m)]

| | Solar heating only | | | | Solar heating and cooling | | | |
| | 50 percent | | 80 percent | | 50 percent | | 80 percent | |
Location	(ft^2)	(m^2)	(ft^2)	(m^2)	(ft^2)	(m^2)	(ft^2)	(m^2)
Northeast (Wilmington, DE)	440	47	1120	121	680	73	1380	141
Southeast (Atlanta, GA)	240	26	820	88	560	60	1260	136
Gulf Coast (Mobile, AL)	120	13	240	26	580	62	1200	129
Great Lakes (Madison, WI)	1200	129	*	*	860	93	*	*
West (Santa Maria, CA)	160	17	220	24	100†	11†	160†	17†

*Exceeds available roof area.
†For domestic hot water only.

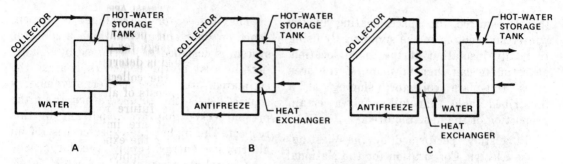

Fig. 131   Storage systems for solar heated water.

space requirements. As a compromise, heat storage tanks are generally designed to have a heat capacity equivalent to three days' normal demand. The common heat-storage materials are water and small pieces of rock.

When water (or an antifreeze solution) is used as the heat-transport fluid in the collector, a large, well-insulated tank of water provides the heat storage. The water may be pumped directly from the collector through the storage tank (Fig. 131A); it may pass through a **heat-exchanger** coil in the tank (Fig. 131B); or through a coil in a small tank connected to the main storage tank (Fig. 131C). When the heat-transport liquid contains an antifreeze compound (e.g., ethylene glycol), either B or C would have to be used.

If air is the heat-transport medium, the heat is usually stored in pieces of rock or pebbles, roughly 2 in. (5 cm) across, contained in an insulated tank through which the air circulates. The efficiency of transfer of heat from the hot air to the rock pieces is greater the smaller the size of the pieces. With decreasing size, however, the resistance to airflow increases and more pumping power is required to circulate the air. The selected size thus represents a compromise between two opposing factors.

A given volume of water can store substantially more heat than an equal volume of rock, assuming the same temperature increase in each case. Thus, water can store 62 Btu of heat per cubic foot per 1°F increase in temperature (4100 kilojoules per cubic meter per 1°C); on the other hand,

rock can store roughly 22 Btu per cu ft per °F (1460 kJ per cu m per °C), where the volume includes allowance for the spaces between the rock pieces. Hence, for the same heat storage, rock would occupy about three times the volume of water.

The following data provide an approximate guide to the relationship between the storage volume and the area of the flat-plate solar collector for water and rock storage:

Water: 0.17 to 0.25 ft^3 per ft^2 collector
(0.05 to 0.075 m^3 per m^2)

Rock: 0.56 to 0.84 ft^3 per ft^2 collector
(0.17 to 0.20 m^3 per m^2)

These numbers are average values for different climatic conditions; both smaller and larger volumes have been used in special circumstances.

If a home with a floor area of 1600 sq ft (150 sq m) has a solar-collector area of 915 sq ft (85 sq m), the storage volume of water would be in the range of 150 to 225 cu ft (4.2 to 6.3 cu m), whereas for rock, it would be 500 to 750 cu ft (14 to 21 cu m). The larger volume in each case would be desirable in a colder climate.

A considerable increase in heat-storage capacity, with a corresponding decrease in volume, could be achieved by using a suitable salt-hydrate, instead of water or air, as the storage medium. A salt-hydrate is a salt in which the solid crystals include a number of water molecules. Special interest has been focused on the relatively inexpensive salt-hydrate sodium sulfate decahydrate ($Na_2SO_4 \cdot 10H_2O$), commonly known as Glauber's salt.

When heat is added to the salt, the temperature increases to 90°F (32°C), called the transition point. At this point, the temperature remains constant while the salt changes to an aqueous solution plus solid Na_2SO_4 (i.e., without the water molecules); during this stage heat is stored, although the temperature does not rise. When all the $Na_2SO_4 \cdot 10H_2O$ has been changed to Na_2SO_4, the temperature can increase again as more heat is added. If heat is withdrawn from the system, the temperature falls to 90°F (32°C), where it remains while the stored heat is released and the $Na_2SO_4 \cdot 10H_2O$ is completely regenerated; thus,

$$Na_2SO_4 \cdot 10H_2O \underset{\substack{\text{Heat withdrawn} \\ \text{at 90°F (32°C)}}}{\overset{\text{Heat supplied}}{\rightleftarrows}} Na_2SO_4 \text{ (solid)}$$
$$+ \; H_2O \text{ (solution)}$$

In this way, the system offers the potential for storing heat at 90°F (32°C) in about one-third the volume of a water tank (or one-tenth the volume of rock) for the same heat capacity.

Salt-hydrate heat storage has been tested in solar-heating systems with air as the transport medium. The salt was contained in a number of stacked 5-gallon (19-liter) cans around which the air was circulated. The systems operated well for a time but gradually deteriorated. The reason was that as the solid Na_2SO_4 formed (from the $Na_2SO_4 \cdot 10H_2O$), it tended to deposit on the walls of the storage cans. As a result, the reverse process, in which the $Na_2SO_4 \cdot 10H_2O$ is regenerated from the Na_2SO_4, did not occur completely. Hence, the storage capacity decreased as the system was operated. Furthermore, the solid deposited on the walls decreased the heat-transfer efficiency. A possible means of overcoming both of these difficulties is by slow rotation of a drum containing the salt hydrate; the solid Na_2SO_4 formed would then remain in suspension in the solution.

Heat Distribution. The distribution system for solar space heating is much the same as for any other heat source; in fact, existing systems for gas or oil heat have been readily adapted to solar heating.

When water is the heat-transport fluid, it may be circulated through radiators in the building to be heated, and then back to the storage tank. Alternatively, a water-to-air heat exchanger can serve to heat air for distribution by way of conventional heating ducts. The air is blown (or drawn) across a pipe coil (heat exchanger) through which the hot water flows (Fig. 132). In a

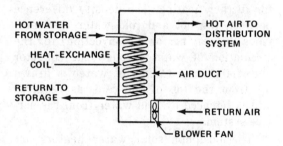

Fig. 132   Water-to-air heating system.

modification of this system, transfer of heat from hot water to air is achieved by passing the air through a container of rock pieces surrounding the storage tank.

If air is the heat-transport fluid (with rock or salt-hydrate storage), the heated air can be circulated through the building and back to the storage system. The building space is then part of a closed circuit.

Hot-Water Supply Systems

The simplest type of solar water heater is the *thermosiphon* system. A pipe from the upper header (see Fig. 117) of a sloping flat-plate collector leads to the top of the insulated storage tank. The bottom of the tank is at least 1 ft (0.3 m) above the top of the collector (Fig. 133). As the water is heated in its passage through the collector, its density decreases and hence it rises and flows into the top of the storage tank. Colder water from the bottom of the tank has a higher density and so tends to sink and enter the lower header of the collector

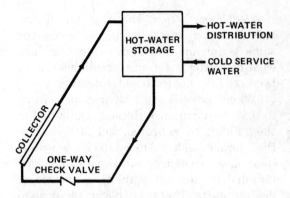

Fig. 133   Thermosiphon system for solar water heating.

for further heating. The density difference between the hot and cold water thus provides the driving force (convection) for the circulation of water through the collector and the storage tank. Hot water is drawn off from the top of the tank as required and is replaced by cold water from the service system.

Thermosiphon solar water heaters are passive systems and do not require a mechanical pump to circulate the water; however, a one-way check valve may be desirable to prevent reverse circulation and thus loss of heat at night. Such heaters are used extensively in rural areas of Australia, Israel, Japan, and other countries where electricity is expensive (or not available) and there is little danger of freezing.

By including an electric pump in the return circuit between the bottom of the storage tank and the lower header of the collector, the tank can be placed at a more

convenient level (e.g., in the house basement). This is now an active system. A control unit permits the pump to operate only when the temperature of the water at the bottom of the tank is below that of the water in the upper header.

When there is a danger of freezing, the water may be drained from the collector; alternatively, a slow reverse flow of the warmer water may be permitted through the collector on cold nights. The freezing danger can be overcome, although at some increase in cost, by using an antifreeze solution as the heat-transport medium, as described earlier. The heat is then transferred to water in the storage tank by way of a heat-exchanger coil (see Fig. 131B).

In colder climates, in particular, auxiliary heating will be required to supplement the solar water heater in the winter. An electrical resistance heater may be included in the storage tank, but it appears preferable to supply the auxiliary heat (by gas, oil, or electricity) to the water as it leaves the storage tank. The auxiliary system then operates only when the water from the storage tank falls below a certain temperature. A schematic representation of a solar hot-water system, with an antifreeze heat-transport fluid and a separate auxiliary heater, is shown in Fig. 134.

The area of a flat-plate collector for water heating depends on the climatic conditions and economic considerations. A rough rule is that a collector 10 sq ft (0.93 sq m) in area will provide about 10 U. S.

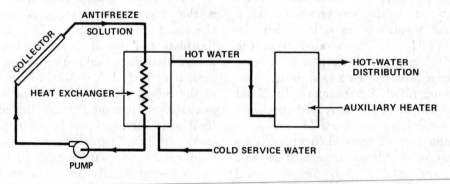

Fig. 134   Solar water-heating system with antifreeze.

gallons (38 liters) of hot water daily for domestic uses, with auxiliary heating when necessary. Thus, a collector area of 56 sq ft (5.2 sq m) is estimated to supply the requirements of a typical U. S. family of four persons, each using an average of 14 gallons (53 liters) of hot water daily. A well-insulated storage tank with a capacity of 40 gallons (150 liters) would probably be adequate for such a system.

Inexpensive types of solar water heaters (e.g., made from synthetic rubber tubing) have been designed for swimming pools. The pool itself then serves as the storage tank, and water is pumped from the pool through an adjacent sloping collector and back to the tank. The increase in temperature to be attained in the collector is relatively small, but considerable amounts of heat may have to be supplied to the water to compensate for the large losses from the surface. A transparent cover when the pool is not in use greatly reduces these losses.

Space-Heating and Hot-Water Systems

Many different concepts have been proposed (and tested) for using solar energy in space heating of buildings. An outline of an active heating system with a sloping flat-plate collector located on the roof of the building is given in Fig. 135. In this system, the heat-transport medium is an antifreeze solution in a closed circuit with the heat-

exchanger coil in the storage tank. The scheme shown in Fig. 131C could be used as an alternative.

Heat is transferred to the water in the storage tank, commonly located in the basement of the building. The solar-heated water from the tank passes through an auxiliary heater, which comes on automatically when the water temperature falls below a prescribed level. For space heating, the water may be pumped through radiators or it may be used to heat air in a water-to-air heat exchanger, as explained earlier. The hot-water supply system, with its own auxiliary heater, is independent of the space-heating system.

During normal operation, the three-way valves are set to permit solar-heated water to flow from the storage tank and auxiliary heater to the distribution system and back to the tank. If, after several cloudy days, the heat in storage is depleted, the valves will adjust automatically to bypass the storage tank. In this way, auxiliary heating of the large volume of water in the tank is prevented. Circulation of the antifreeze solution through the collector and associated heat exchanger occurs only when solar energy is available. If the temperature in the header at the top of the collector should fall below that at the bottom of the tank, the pump (at bottom left of Fig. 135) would be switched off automatically.

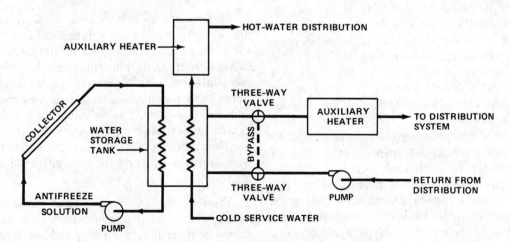

Fig. 135 Solar space-heating and hot-water system.

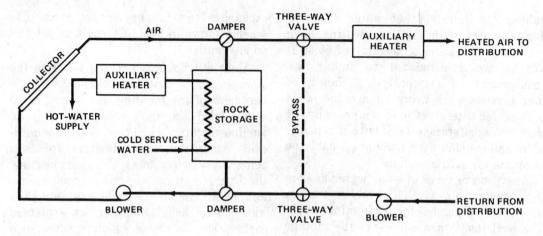

Fig. 136   Solar-heating system with heat-transport by air.

When water, rather than antifreeze, can serve as the heat-transport medium, the solar-heated water from the collector can go directly to storage. The situation is then similar to that described below for a system using air as the heat-transport medium. The heated water can be pumped directly to the distribution system, bypassing the storage tank, if necessary.

The use of air as the heat-transport and distribution medium in a space-heating system is outlined in Fig. 136. By adjusting the dampers, the heated air from the collector can be divided between rock storage and the distribution system, as might be required by the conditions. For example, when the sun shines after several cloudy days it would be desirable to utilize the available heat directly in the distribution system rather than placing it in storage. The three-way valves can be used to bypass the storage tank, as explained above. A hot-water supply may be provided by an air-to-water heat exchanger in the storage tank, as shown in Fig. 136.

In a hybrid system, advantage is taken of the good heat-transport properties of water (or an antifreeze solution) and the effectiveness of air for space heating. The antifreeze solution is circulated through a relatively small tank or a heat-exchanger coil surrounded by a large tank of rock pieces. The air for distribution in the liv-

ing space is then heated by passage through the large tank.

Passive Space-Heating Systems

The basic design principles of passive solar space-heating systems, that is, without mechanical components, fall into the following five general categories:

1. Direct gain
2. Thermal storage wall
3. Attached sunspace
4. Roof storage
5. Convective loop

There are modifications within each of these categories and two or more may be combined in a single building.

Direct Gain. In this system, the building has a south wall with a large number of windows; two layers of glass minimize heat losses. Solar radiation entering the windows falls on thick concrete, slate, or brick floors (and possibly walls) and is absorbed and stored as heat. The building air is then heated by radiation and convection from the floor and walls. Covering the windows with shutters or curtains at night reduces heat losses.

Thermal Storage Wall. A large wall-like mass, which absorbs solar radiation and stores heat, is placed directly behind large south-facing windows. The best known

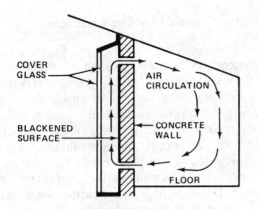

Fig. 137   Trombe wall for passive solar heating.

example of the storage wall design is the vertical *Trombe wall*, made of concrete, brick, or other masonry some 1 ft (0.3 m) or more thick. Alternatively, a storage wall may consist of a vertical stack of drums or other containers filled with water.

A Trombe wall arrangement is outlined in Fig. 137. Two layers of cover glass are sealed to the wall with an airspace between. The south face of the wall is blackened to absorb the solar radiation and the resulting heat is stored in the concrete (or other material). Heat radiated outward from the wall is trapped by the cover glass, so that the air in between is heated. The heated air rises and enters the adjacent room through vents at the top of the wall; the air circulates by natural convection and is returned by way of ducts at the bottom of the storage wall.

Attached Sunspace. A sunspace is any enclosed space, such as a greenhouse or sunporch, with a glass wall on the south side. A sunspace may be attached (or built on) to a thick south wall of the building to be heated by the sun. Vents near the top and bottom of the wall, as in Fig. 137, permit circulation through the main building of the air heated in the sunspace. Heat storage is provided by the thick wall, a concrete or masonry floor, water containers, and other materials in the sunspace. Thus, an attached sunspace system combines features of direct gain and storage wall concepts.

Roof Storage. A passive solar system, trade named Sky Therm, was designed for houses having a flat roof located in a mild climate (e.g., Arizona and Southern California). The heat is absorbed and stored in water about 10 in. (0.25 m) deep contained in plastic bags held in blackened steel boxes on the house roof (Fig. 138). In a later design, a layer of clear plastic sealed to the top of the bag provides a stagnant airspace to reduce heat losses to the atmosphere. Heat is transferred from the heated water to the rooms below by conduction through a metal ceiling.   Air circulation

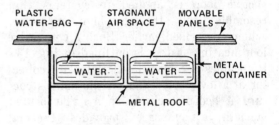

Fig. 138   Roof storage of solar heat.

may be aided by means of electric fans, but this is not essential. To prevent loss of heat during the night, thermal-insulator panels are moved, either manually or by a time-controlled electric motor, to cover the water bags. In the daytime, the panels, which are in sections, are removed and stacked one above the other.

In many parts of the southwest United States, the days are warm in summer but the nights are usually cool and clear. The Sky Therm system can then be used to provide summer cooling, as well as winter heating. In the summer, the water bags are covered in the daytime but are exposed at night; cooling then takes place by radiation to the environment. The cooling is enhanced if the water bags are immersed in a layer of water; vaporization into the dry atmosphere during the night results in cooling of the water by heat absorption (see later).

Convective Loop. In most passive solar space-heating systems, the heated air is circulated by convection, but the term con-

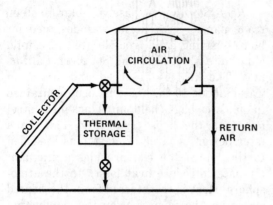

Fig. 139   Convective loop passive solar heating.

Space Cooling (Air Conditioning)

Evaporation Cooling. Evaporative coolers, colloquially called "swamp-box coolers," which utilize solar energy indirectly, have long been used in arid areas for space cooling in summer. The warm, dry ambient air is drawn (or blown) through a porous medium that is kept moist with water. The resulting evaporation is accompanied by the absorption of heat (latent heat of vaporization) from the air which is consequently cooled. The cooler moist air then enters the house.

Two drawbacks of evaporative coolers of this type are that (1) they are ineffective when the humidity of the ambient air is high since little evaporation then occurs, and (2) the cooled air, in any event, contains a substantial amount of water vapor and may be uncomfortable especially if the ambient air is moderately humid.

If the ambient air still has a useful capacity for evaporative cooling, the problem of uncomfortably high humidity of the cooled air can be solved by a two-stage system. Instead of drawing the moist cool air from the cooler directly into the house, it is used to cool ambient air in a heat exchanger (Fig. 140). The moist air is discharged to the surroundings and the cooled ambient air with no added moisture enters the house. Such a two-stage system would require two blowers (or fans) instead of the one used in the simple evaporative cooler.

vective loop is applied to systems that resemble the thermosiphon hot-water scheme described earlier. Such a convective loop heating system is outlined in Fig. 139. It includes a conventional flat-plate collector at a level below that of the main structure. A bed of rock, which may be located beneath a sunspace, provides thermal storage.

In normal operation, air passing upward through the collector is heated and enters the building through floor vents. The cool, denser air leaving the building returns to the bottom of the collector and is reheated. If more solar heat is available than is required for space heating, the floor vents may be partly (or wholly) closed. The heated air then flows through and deposits heat in the storage bed. Heat stored in this way may be used later, as needed, by transfer to the cooler air leaving the building.

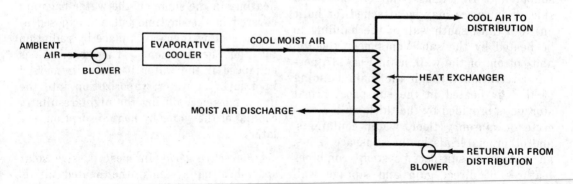

Fig. 140   Two-stage evaporative cooling system.

Mechanical Cooling. The major current interest is in mechanical cooling (or air-conditioning) systems that depend on solar heat for their operation and are unaffected by atmospheric humidity. The two most common **refrigeration** techniques are vapor compression and absorption and, in principle, both could be adapted for use with solar energy, although the temperatures required are higher than those adequate for space heating. In the former procedure, solar-heated water could vaporize propane or ammonia at a moderate pressure. The vapor could then drive a **turbine** which would, in turn, operate a vapor-compression cooling unit. Such a low-pressure vapor turbine, however, would inevitably have a very low efficiency. Furthermore, propane is highly flammable whereas ammonia forms a noxious gas.

Absorption cooling with solar energy, which is regarded as more practical, is possible with current technology although improvements in design would be desirable. The preferred system is based on the use of water as the vapor and lithium bromide as the absorber, as shown in Fig. 110. The heat required to vaporize the water in the vapor-generator unit is provided by solar-heated water passing through a heat-exchanger coil, as indicated in Fig. 141. Auxiliary heating may be necessary if the stored water temperature drops below a certain value (e.g., about 180°F, 82°C). The three-way valves permit the storage tank to be bypassed when necessary. Heat removed by the cooling water from the absorber and condenser units is discharged to the atmosphere by way of an external cooling tower, thus permitting reuse of the water.

The intermittent absorption - cooling method (see **Refrigeration**) could probably also be used with solar-heated water supplied to the generator (see Fig. 111). The disadvantage of intermittent operation could be overcome if water cooled in the evaporator were to be stored in a separate cold-water storage tank. The temperature fluctuations would then not be large. The water would be used to cool the air for distribution in the usual manner.

Combined Heating and Cooling

Systems have been designed and studied for using solar energy to provide space heating in the winter and continuous-type absorption cooling in the summer. However, the energy requirements in summer and winter are different in most regions of the United States. Hence, a solar collector–heat storage system designed to be optimum in one season will rarely be optimum in the other season. The design of an economically effective combined solar heating and cooling system is a complex problem that can best be solved by setting up a mathematical model. Such a model, programmed for a computer, can take into consideration economic factors, climatic

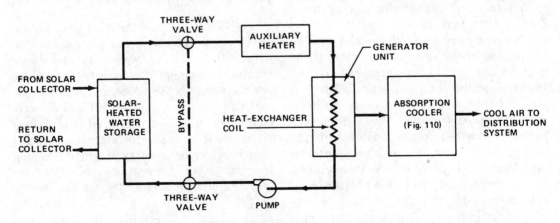

Fig. 141   Absorption refrigeration with solar heating.

conditions, and the characteristics and requirements of the building to be heated and cooled.

One general conclusion is that, since a continuous absorption cooling machine is expensive, it is not expected to be cost-effective in the northern United States where the cooling requirements are moderate. In these circumstances, it might be more economical, at least for the immediate future, to use an independent, electrically-operated conventional air conditioner of the vapor-compression type. However, this situation may change if efficient, low-cost cooling systems using solar energy are developed.

Solar-Assisted Heat Pump

In most parts of the United States, solar space-heating (and -cooling) systems require the use of an auxiliary energy source. If electricity is available, the solar-assisted **heat pump** appears to offer some promise. In outline, a heat pump consists of evaporator and condenser units connected by a compressor; heat is absorbed in the evaporator and given up in the condenser. In the winter, heat is taken up from an outside source by the evaporator and is released in the condenser to the air in the interior of the building. In the summer, the situation is reversed; heat is then taken up from the inside of a building and discharged to the outside (see Fig. 65).

A heat pump can provide any auxiliary heating that may be required by a solar space-heating system in the winter and all the cooling in the summer. If water is the heat-transport fluid, the heat required by the evaporator unit in the winter is obtained from the water in the solar heat-storage tank. By drawing heat from even partially warmed water instead of the colder outdoor air, as in a conventional heat pump, the performance is improved. In any event, apart from temperature considerations, the transfer of heat to the evaporator is more efficient when the heat is obtained from water rather than from air. The amount of auxiliary electrical energy used is thus kept to a minimum.

In the summer, the heat pump could operate in the cooling mode entirely on electric power with no assistance from solar energy. In the simplest case, heat removed by the heat pump from the building would be discharged to the outside air in the conventional manner. Alternatively, the heat could be discharged to the existing storage tank to be used, together with heat obtained from the solar collectors, to supply heat to the heat pump in the winter. In a concept being studied by the Brookhaven National Laboratory, the heat is stored in the ground by means of a system of buried pipes through which water is circulated.

Agriculture and Industrial Process Heat

By supplying heat, as hot water, hot air, or steam, for agricultural and industrial purposes, solar energy has the potential for replacing substantial amounts of fossil (nonrenewable) fuels. These applications of solar energy may be considered in three general categories, according to the temperature range within which the heat is supplied:

1. Low temperatures (below 212°F, 100°C)

2. Intermediate temperatures (212 to 350°F, 100 to 177°C)

3. High temperatures (above 350°F, 177°C)

Low-Temperature Applications. These are based on the use of flat-plate collectors, with either air or water as the heat-transport medium. The general principles are the same as for space and water heating. The hot water may be utilized directly or the heat may be transferred to air. Provision must be made for storage of excess heat in water or in rocks or gravel. Among the potential applications of low-temperature heat in agriculture are the following:

Heating and cooling of commercial greenhouses.

Space heating of livestock shelters, dairy facilities, and poultry houses.

Drying grain, soybeans, peanut pods, fruits, tobacco, onions, and forage.

Solar energy can also be used to convert salty (or other impure) water into potable water by distillation. Such solar stills have been operated for farm and community use in several countries. A simple, but widely used solar still consists of a shallow metal or concrete tray with a blackened interior surface. The tray has a roof-like, transparent cover of glass or plastic (Fig. 142).

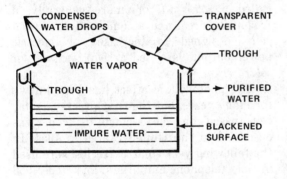

Fig. 142 Solar water still.

Solar radiation passes through the cover and is absorbed and converted into heat in the black surface. Impure water in the tray is heated and the vapor produced is condensed to purified water on the cooler interior of the roof. The transparent roof material transmits nearly all the radiation falling on it and absorbs very little; hence, it remains cool enough to condense the water vapor. The condensed water flows down the sloping roof and is collected in troughs at the bottom.

Intermediate-Temperature Applications. Food processing, textile, laundry, and other industries often require both hot water and low-pressure steam. For such applications, water can first be heated in flat-plate collectors followed by an array of parabolic-trough concentrating collectors. In this way, water, under moderate pressure, can be heated to temperatures above 212°F (100°C). Some of the processes which

fall in the category of intermediate-temperature applications are the following:
    Laundries
    Fabric drying
    Textile dyeing
    Food processing and can washing
    Alfalfa drying

High-Temperature Applications. Steam at temperatures above 350°F (177°C) is used extensively in industry, particularly in the generation of electric power. The same general methods for producing high-temperature steam with the aid of solar energy are applicable in all cases; they are described in the section on **Solar Energy: Thermal Electric Conversion.**

A different high-temperature application of solar energy is in pumping irrigation water. Most irrigated farms are located in the southwest United States where there is ample direct solar radiation. Hence, there is a considerable potential for the use of solar energy to operate irrigation pumps. Such pumps may be driven directly by a vapor-expansion **turbine,** similar to a **steam turbine** (see **Vapor Turbine**) or by an electric motor with power produced by a turbine–generator (see **Electric Power Generation**). The direct turbine-driven pumps, which are of moderate power and are suited to relatively shallow wells, are described below. For deeper wells, electric motor drive may be preferable (see **Solar Energy: Thermal Electric Conversion**).

A heat-transport fluid, such as a specially formulated hydrocarbon oil with a high boiling point and fairly good heat-transfer properties, is heated to a temperature below 600°F (316°C), by passage through an array of parabolic-trough (focusing) collectors. A north–south orientation of the troughs is preferred because considerably more solar radiation can be collected during the main irrigation season (spring and summer) than with an east–west orientation. Part of the heated hydrocarbon oil passes to a heat exchanger (or boiler) where the turbine working fluid (e.g., a Freon-type compound) is boiled under pressure. The vapor generated drives the turbine designed to operate at comparatively low pressures. The turbine is connected to the irrigation pump through a clutch and gear drive. Some of the pumped water is bypassed to cool the turbine-condenser. Although solar radiation provides the energy to operate the irrigation pump, an auxiliary (electrical) power supply is required for the pumps that circulate the heat-transport and working fluids.

A simplified outline of a turbine-driven pump system utilizing solar energy is shown in Fig. 143. In a particular system in New Mexico, the heat-transport fluid (Exxon Caloria HT 43) is heated to 420°F (216°C) in parabolic-trough collectors with a total aperture area of 6720 sq ft (624 sq m). Part of the heated liquid is stored for use when the sun is not shining. The turbine working fluid (Freon type, R-113) leaves the boiler and enters the turbine as vapor at a temperature of 320°F (160°C) and 15 atm (1.5 MPa) pressure. After expansion in the turbine, the vapor leaves at 200°F (93°C) and 0.7 atm (0.07 MPa); it is converted back to liquid in the condenser and returns to the boiler.

The irrigation pump operates at a rated power of 19 kW (25 horsepower) and delivers water at 500 to 600 gal/min (32 to 38 liters/sec) from a well roughly 100 ft (30 m) deep. The energy efficiency (i.e., percentage of solar energy collected that is converted into useful work) is 13 to 14 percent; this low value is largely a result of the relatively low temperature of the working fluid entering the turbine (see **Heat Engine**).

SOLAR ENERGY: PHOTOVOLTAIC CONVERSION

The direct conversion of solar energy into electrical energy by means of the photovoltaic effect, that is, the conversion of light (or other **electromagnetic radiation**) into electricity. A single converter cell is called a *solar cell* or, more generally, a *photovoltaic cell*; and a combination of such cells, designed to increase the electric power output, is called a *solar module* or *solar array*.

The photovoltaic effect has been known for many years, but the first practical solar cell, based on the element silicon, was described in the United States in 1954. In the following year solar batteries were used in field telephone amplifiers, and since 1958 they have provided electric power for many spacecraft. Solar cells have also been used to operate irrigation pumps, navigational signals, highway emergency call systems, railroad crossing warnings, automatic meteorological stations, etc., in locations where access to utility power lines is difficult. Hitherto, high cost has been the main hindrance to the more extensive use of

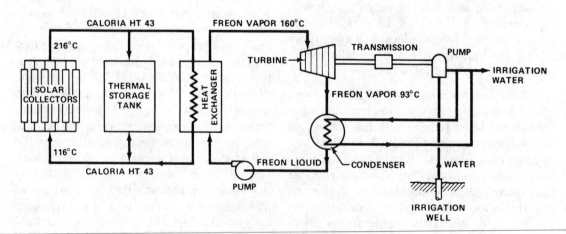

Fig. 143   Turbine-driven pump using solar energy.

solar cells, but research and development in progress are expected to lead to a substantial cost reduction during the 1980s.

Solar Cell Principles

Semiconductor Junctions. Modern solar cells make use of **semiconductor** materials, usually based on single-crystal silicon. When doped with phosphorus, arsenic, or antimony, the silicon becomes an *n*-type semiconductor; and when doped with boron, aluminum, indium, or gallium, it forms a *p*-type semiconductor. If a *p*-type semiconductor is brought into intimate contact with one of the *n*-type, they form a *p-n* (or *n-p*) junction. If the two semiconductor materials are derived from the same element (or compound), such as silicon, the system is referred to as a *homojunction*. It is also possible for a *p-n* (or *n-p*) junction to be formed from two different semiconductor materials, such as cadmium sulfide (CdS) and cuprous sulfide (Cu_2S); this is known as a *heterojunction*. The general behavior at the junction, as outlined below, is the same regardless of the type. The Schottky junction, consisting of a semiconductor and a metal, and other junctions are described later.

The free electrons (negative charges) of the *n*-type material and the holes (positive charges) in the *p*-type material can drift across the junction into the material of the opposite type. The electrons entering the *p*-type material will leave behind atoms with a net positive charge. At the same time, the holes entering the *n*-type material will leave atoms with a net negative charge (Fig. 144). The positive charges on the *n* side of the *p-n* junction and the equivalent negative charges on the *p* side constitute an electrical barrier. This barrier opposes the further drift of free electrons from the *n*-type to the *p*-type material and of holes in the opposite direction. However, the barrier would facilitate the passage of free electrons and holes in the respective reverse directions; that is, free electrons, if available, would readily cross the *p-n*

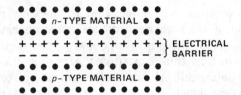

Fig. 144   Electrical barrier at an *n-p* junction.

junction into the *n*-type material whereas holes would cross into the *p*-type material.

Photovoltaic Conversion. Suppose the combination of *n*- and *p*-type materials is exposed to solar radiation; some of the radiation energy will be utilized to remove bound electrons from the semiconductor (e.g., silicon) atoms, thus producing free electrons. A hole (positive charge) is left at each location from which a bound electron is removed, and hence an equal number of free electrons and holes will be formed. The electrical barrier at the *p-n* junction then causes the free electrons in its vicinity to move into the *n*-type material and holes into the *p*-type material.

If electrical contacts are made with the two semiconductor materials and the contacts are connected through an external electrical conductor, the free electrons will flow from the *n*-type material through the conductor to the *p*-type material. Here the free electrons will enter the holes and become bound electrons; thus, both free electrons and holes will be removed. The flow of electrons through the external conductor constitutes an electric current which will continue as long as more free electrons and holes are being formed by the solar radiation (Fig. 145). This is the basis of

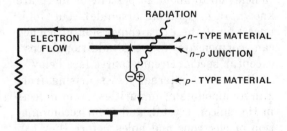

Fig. 145   Generation of electric current by a solar cell.

photovoltaic conversion, that is, the conversion of solar energy into electrical energy. The combination of *n*-type and *p*-type semiconductors thus constitutes a photovoltaic (or solar) cell. All such cells generate direct current which can be converted into alternating current if desired.

Conversion Efficiency and Power Output. For both theoretical and practical reasons, not all of the solar radiation energy falling on a solar cell can be converted into electrical energy. A specific amount of energy is required to produce a free electron and a hole in the semiconductor material. In silicon, for example, the energy minimum is 1.1 electron volts (see **Nuclear Energy** for definition of electron volt), and this is available only in radiation having a wavelength of 1.1 micrometers or less (see **Solar Energy**). Consequently, infrared radiation of longer wavelength has no photovoltaic effect in silicon but is largely absorbed as heat. Furthermore, even the energy of radiation with shorter wavelengths cannot be used completely; energy in excess of that needed to free a bound electron is simply converted into heat.

For the foregoing reasons, only about 45 percent of the energy in the solar radiation at sea level is capable of producing electrons and holes in silicon. However, because of the electrical resistance of the semiconductor material and other loss modes, the maximum practical efficiency for the conversion of solar energy into electrical energy in a silicon solar cell is estimated to be about 22 percent. Photovoltaic semiconductors with conversion efficiencies up to about 25 percent or more are known (e.g., gallium arsenide), but it is uncertain if the extra conversion efficiency can compensate for the additional cost, except in special circumstances (see below).

Because of internal losses arising from minute amounts of impurities, from defects in the silicon crystal, and from recombination of electrons and holes before they can be separated, and external losses from reflection, most commercial silicon solar cells have a conversion efficiency of roughly 10 to 14 percent (average about 12 percent). However, cells have been made with efficiencies of 18 percent, and this level will probably be attained in commercial cells.

The **power** output (in **watts**) of any generator of electricity, including a photovoltaic cell, is equal to the product of the voltage (in volts) and the current (in amperes). Theoretically, a silicon solar cell should have a voltage of 1.1 volts, from the 1.1 electron-volts energy of the free electrons produced. In practice, however, the maximum voltage is about 0.6 volt and this occurs on open circuit, when no power is produced. The maximum power of a silicon cell occurs at an output voltage of approximately 0.45 volt. In full sunlight, the current from a commercial cell is then roughly 270 amperes per sq m of exposed surface. The power is thus about $0.45 \times 270 = 120$ watts (or 0.12 kilowatt) per sq m (i.e., 13 watts/sq ft).

The rate at which solar energy reaches the top of the atmosphere (i.e., the solar constant) is 1.35 kilowatts/sq m (1.35 kW/sq m). Part of this energy is reflected back to space, and part is absorbed by the atmosphere. In full sunlight, the solar energy may reach the ground at a rate of roughly 1 kW/sq m. Since the electric power generated by a commercial silicon solar cell is 0.12 kW/sq m, the conversion efficiency of this particular cell is about 12 percent.

Solar Cell Production

Single-Crystal Cells. The very pure silicon required for semiconductors is obtained by refining elemental silicon which has itself been made from pure materials. In one process for making *n–p* junctions, the highly purified solid is melted [melting point, 2590°F (1420°C)], and a controlled amount of boron (as dopant) is added. From this melt, a *p*-type semiconductor is grown slowly as a single crystal in the form of a long cylinder about 3 to 4 in. (8 to 10 cm) in diameter. The cylinder is cut into disks, commonly 0.01 to 0.014 in. (0.25 to 0.35 mm) thick; this thickness is necessary in single-crystal silicon for effective absorption of the solar radiation. One face of each disk is then exposed to phosphorus vapor at a high temperature. The phosphorus (dopant) atoms diffuse into the silicon to form an n-type layer to a depth of about 1/25,000 in. (1 micrometer) or so over the p-layer. An n–p homojunction is thus produced.

The product, in which the n-side is exposed to sunlight, is called an n-on-p type solar cell. Solar cells of the p-on-n type have also been made, but they were found to be less resistant to space radiation. Since the main initial use of solar cells has been on spacecraft, commercial cells were (and generally continue to be) of the n-on-p type.

To complete the cell, electrical contacts, designed to cause minimum obstruction of sunlight (e.g., metal grids), are attached to the front and back of the semiconductor material. A transparent cover, with an antireflection coating to reduce loss of solar radiation by reflection, is placed in front of the cell, and the whole is encapsulated to provide protection from damage.

Cost Reduction. If solar cells are to be used more widely for generating electric power, their cost must be reduced substantially. Some reduction can be achieved by using automatic assembly procedures and improved fabrication techniques, such as ion implantation of dopants followed by laser annealing. However, more fundamental technical advances are still necessary. Considerable effort in this direction is being made by government and university laboratories, the semiconductor industry, and individuals. Because the situation is undergoing rapid change, only the general lines of investigation will be summarized here. They are conveniently considered in the following six categories.

1. Reduction in the cost of producing single-crystal silicon.
2. Development of less expensive forms of silicon.
3. Less expensive alternative semiconductor materials.
4. Utilization of a larger proportion of the solar energy.
5. Alternatives to conventional p–n junctions.
6. Concentrating collectors to increase the electric power output per unit area of semiconductor devices.

1. Among the methods being studied for decreasing the cost of single-crystal silicon is to grow the material as a continuous ribbon or possibly as a sheet of the required thickness. Small amounts of undesirable impurities are introduced, however, from the die through which the molten silicon is drawn. As a result, the conversion efficiencies of the solar cells are generally somewhat less than for cells made from single-crystal silicon. This loss of efficiency may be more than offset by the decreased cost of the cells.

2. Instead of expensive single-crystal silicon, it may be possible to make solar cells from cheaper polycrystalline silicon. (Polycrystalline material consists of an aggregate of small crystals or grains, rather than a single crystal.) To be effective in a cell, the individual silicon grains must exceed a certain minimum size for the solar radiation to be absorbed without crossing a grain boundary. Polycrystalline material of this type can be made by deposition from silicon vapor. Another possibility is to produce material with small grains, which is relatively simple, and then to increase the grain size using laser or electron-beam techniques.

A promising alternative to single-crystal silicon is the so-called amorphous (noncrystalline solid) silicon, which is actually an "alloy" of silicon and hydrogen. It is made most readily by passing an electric discharge through the gas silane, a compound of silicon and hydrogen (SiH_4), at low pressure; a thin layer of amorphous silicon–hydrogen alloy then deposits on a heated surface. The product can be doped with boron (to form a p-type semiconductor) or with phosphorus (for an n-type) by adding a controlled amount of a boron hydride (B_2H_6) or phosphorus hydride (PH_3), respectively, to the silane. The amorphous material is a much better absorber of solar radiation than crystalline silicon, and so a thickness of only 1 micrometer is sufficient for a solar cell. Homojunction p-n cells made from amorphous silicon are cheaper than single-crystal cells, but their photochemical conversion efficiencies have been low. Better results can be obtained with a Schottky junction as described in Item 5.

3. One of the best semiconductor materials for solar cells is gallium arsenide (GaAs) which can be doped like silicon; however, it is too expensive to use except in special circumstances. A less efficient, but much less expensive, solar cell utilizes a cadmium sulfide–cuprous sulfide (CdS–Cu_2S) heterojunction (maximum practical efficiency about 15 percent). This combination is a better absorber of solar radiation than crystalline silicon. Hence, photovoltaic cells can be made of thin microcrystalline films, obtained by spraying or vapor deposition of the sulfides on a conducting base. It should be possible to produce such solar cells in sheets that could be used as roof tiles.

Other semiconductor materials related to the cadmium sulfide–cuprous sulfide system are being investigated. For example, cadmium sulfide may be replaced by a combination of cadmium sulfide with the less expensive zinc sulfide. Other solar cell concepts are based on the use of cadmium sulfide with indium phosphide (InP) or copper indium diselenide ($CuInSe_2$) in place of copper sulfide. These cells are more expensive, but they may be capable of higher photovoltaic conversion efficiencies than the cadmium sulfide–cuprous sulfide cells.

4. A number of techniques are being studied for increasing the conversion efficiency of silicon (and other) solar cells beyond the apparently limiting practical value of 22 to 25 percent. One proposal is based on a combination of two or more cells using different semiconductor materials. In these materials different amounts of energy are required to produce free electrons and holes. The principle is that solar radiation that is not effective in one cell will be effective in a succeeding cell. One such combination of three cells, made from gallium arsenide, silicon, and germanium, for which the minimum energy requirements are 1.3, 1.1, and 0.8 electron volts, respectively, has a possible conversion efficiency of about 40 percent.

Another approach is to shift the solar radiation spectrum in a way that increases the proportion of energy in the effective range for producing free electrons and holes in the given semiconductor (e.g., 1.1 micrometer wavelength or 1.1 electron volts energy for silicon). In the *thermophotovoltaic cell*, this is achieved by absorbing the solar radiation in a refractory material which becomes hot [about 3270°F (1800°C)] and re-radiates energy in the infrared range (wavelength greater than 0.8 micrometer). Radiation that is not absorbed in the solar cell is reflected back to the radiator and is retained in the system. Overall conversion efficiencies of 30 to 50 percent are considered possible.

5. The cost of photovoltaic cells may perhaps be reduced by using alternatives to the conventional p-n junctions with two semiconductor materials. One promising possibility is the *Schottky junction* formed by depositing a thin layer of a metallic conductor (e.g., platinum) onto a p- or n-type semiconductor. A potential barrier is established just as in a conventional p-n junction. Free electrons and holes produced by solar radiation with the appropriate energy then travel in opposite directions under the influence of this potential, in the usual manner. Schottky junction photovoltaic cells made with the so-called amorphous silicon are more efficient than homojunction p-n cells of the same material.

The MIS (metal-insulator-semiconductor) solar cell is similar to the Schottky type except that a very thin layer (about 0.1 to 0.3 micrometer) of an insulator is deposited between the semiconductor and the metallic conductor. A conversion efficiency of more than 17 percent has been reported for an MIS solar cell made with single-crystal silicon.

A third type of solar cell under development makes use of what is essentially a Schottky junction with the metal replaced by a liquid "redox" electrolyte (see **Storage Battery**). A redox electrolyte contains an element (or group of elements) in two different oxidation states and behaves like a metal in being an electrical conductor and capable of acting as a source or remover of electrons. The common semiconductor materials (e.g., silicon and gallium arsenide) tend to disintegrate during cell operation, and several alternatives are being studied.

6. The electric power output of a photovoltaic cell is roughly proportional to the rate at which solar radiation falls on its surface. Hence, the output of a cell of a given area can be increased by combining it with a concentrating collector (see **Solar Energy Collectors**). Tracking collectors of the line-focus type can provide a concentration of a few hundredfold. With the compound parabolic (nonfocusing) concentrator (CPC), a concentration factor of about ten is possible without tracking. Since focusing collectors concentrate direct solar radiation only (but not diffuse radiation), they would be most useful in regions of high insolation (see Fig. 130). However, this limitation might not be applicable to the CPC, since it can collect diffuse (scattered) radiation.

Most of the solar energy that is not converted into electricity in a photovoltaic cell is absorbed as heat. In a commercial, single-crystal silicon cell, for example, with a conversion efficiency of about 12 percent, more than 80 percent of the incident solar energy appears as heat in the cell. In a flat-plate collector, the heat can be dissipated by radiation and convection to the ambient air. With concentrating collectors, however, special heat-removal systems, utilizing flowing air, water, or other fluid, may be necessary.

In spite of removal of heat in this manner, the temperature of the solar cells will inevitably rise to some extent. But as the temperature increases, the efficiency of an n-p homojunction silicon cell will decrease; for example, for a commercial cell it is about 8 percent at 212°F (100°C). Cells made of gallium arsenide are superior to silicon cells for operation at elevated temperatures; their higher cost may then be justified when used with concentrating collectors. The conversion efficiency of a practical gallium arsenide solar cell is roughly 18 percent at ordinary temperatures, 16 percent at 212°F (100°C), and 12 percent at 390°F (200°C). Gallium arsenide is a good absorber of solar radiation and n-p homojunctions can be made from thin films of polycrystalline material. High conversion efficiencies with concentrating collectors have been reported with cells made from a combination of gallium aluminum arsenide and gallium arsenide.

Solar Cell Modules

The optimum operating voltage of a photovoltaic cell is generally about 0.45 volt at normal temperatures, and the current in full sunlight may be taken to be 270 amperes/sq m, as already seen. If the exposed area of a cell is 40 sq cm (6.2 sq in.) or 40×10^{-4} sq m, the current would be 1.08 amperes, and the electric power output $0.45 \times 1.08 = 0.49$ watt in full sunlight. A decrease (or increase) in the solar radiation has little effect on the voltage, but the current and power are decreased (or increased) proportionately.

By combining a number of solar cells in series (i.e., in a string), the voltage is increased but the current is unchanged. For example, 110 volts, for operating commercial tools, motors, or domestic appliances, would require $110/0.45 = 244$ cells in series. To increase the current output at the same time, several strings of 244 cells would be connected in parallel, as depicted in Fig. 146. Suppose there were ten such strings in parallel; the current under optimum conditions would then be $10 \times 1.08 = 10.8$ amperes, and the power output would be $10.8 \times 110 = 1190$ watts or 1.19 kW. The so-called solar panels on spacecraft consist of modules (or arrays) of cells connected in series and parallel to provide the required voltage and power.

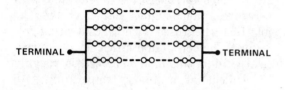

Fig. 146   Solar-cell arrangements in series and parallel.

If even a single cell in a string should fail, the whole string would become inoperative. The cells in the remaining strings would maintain the voltage, but the current (and power) output of the system would be decreased by the loss of one string of cells. A short circuit in a cell would not disable the string, although there would be a slight drop in voltage. There is a possibility that the other cells might cause current to flow in the wrong direction through any string having a reduced voltage. This danger is eliminated by including a diode, which permits current flow in one direction only, at the end of each string.

Instead of a number of strings of cells in parallel, the current (and power) could be increased by locating a single string of cells at the focal line of a sun-tracking, parabolic trough, concentrating collector. There would be some decrease in voltage because of the inevitable higher temperature of the solar cell material, but the current would increase approximately in proportion to the concentration factor of the collector. Thus, with a concentration factor of 100, the current from a single string would be increased to about 100 amperes. If the number of cells in the string is sufficient to produce 110 volts, the total electric power output would be approximately 12 kW.

Photovoltaic Applications

By the mid-1980s, the cost of solar cells should have decreased to the point where they are economically competitive with other sources of electricity in many cases. The potential applications of solar cells may be considered in two categories, depending on whether nonconcentrating (flat-plate) or concentrating collectors are used.

Flat-Plate Systems. The overall cost of a flat-plate system is roughly proportional to the electric power capacity; that is to say, the cost per kilowatt would be almost independent of the total power. Hence, there would be little or no advantage in size, as there probably would be in concentrating systems. Flat-plate collector arrays of solar cells could thus be used to provide from a few watts to several kilowatts of electric power. There are many possible applications of such power sources in homes, on farms, and for special purposes in isolated locations.

The drawback to photovoltaic conversion, as it is to other methods for the direct utilization of solar energy, is that it is effective only during daylight hours. In some situations, such as for operating irrigation pumps, this would be adequate, but in most circumstances power would also be required at other times. If an alternative power source is not available, some form of storage is necessary. The solar cell system must then be designed to generate more power than is used during the daytime; the excess is stored for use at night or on cloudy days.

The simplest means of storage on a small or moderate scale is in electric **storage batteries**, especially as solar cells produce the direct electric current required for battery charging. The stored energy can then be delivered as electricity upon discharge. The common lead–acid storage batteries, such as are used in automobiles, are not ideal for this purpose, but they are probably the best presently available. Extensive research in progress should lead to the development of more suitable storage batteries.

A possible alternative is to use the direct current from solar cells to decompose water (by electrolysis) into hydrogen and oxygen gases (see **Hydrogen Production**). These gases would be stored in a suitable form and utilized as needed to generate electricity in a fuel cell (see **Fuel Cell**).

Concentrating Systems. For the conversion of solar energy into electricity on a moderately large or large scale, concentrating collectors might prove economically advantageous; the costs of the solar cells and of the sun-tracking and heat-removal system per kilowatt of power generated could then be minimized. If all or part of the power were to be supplied to a utility, the direct current from the solar cells would be converted into alternating current by means of an inverter. The utility would then provide electric power at night and on cloudy days. If such a supplementary power source is not available, a storage system would be required. Batteries might be adequate for power storage on a moderate scale, but for large-scale storage other methods would have to be used (see **Energy Storage**).

Total Energy Systems. The overall efficiency of a solar photovoltaic conversion system could be greatly increased if the heat generated simultaneously by radiation absorption could be utilized. This would constitute a type of **total energy** (or **cogeneration**) **system**. In the Solar One house built at the University of Delaware in 1973, cadmium sulfide–cuprous sulfide solar cells in a flat-plate array on the roof provide warm air for space and water heating as well as electric power.

Cogeneration on a larger scale would be of special interest in industry where substantial amounts of process heat are used at temperatures below about 350°F (177°C). Such temperatures could be attained in the coolant for gallium arsenide cells with concentrating collectors.

Solar Power Satellite (SPS) Systems

A long-range concept is the generation of electric power by photovoltaic conversion of solar energy utilizing a satellite revolving in an orbit at an altitude of 22,300 miles (35,900 km) in the same direction as the earth rotates about its axis. If the orbital plane of the satellite coincides with earth's equatorial plane, the satellite will appear to remain stationary with respect to any location on earth. Such a satellite is said to be in a geostationary (or synchronous) orbit. From any given location on the earth, the geostationary satellite is always in the same direction.

With some minor exceptions, a satellite in geostationary orbit is always exposed to the sun; furthermore, radiation is received at the rate of the solar constant without any absorption loss. Hence, the total daily amount of solar radiation received per unit area is about six times the average at the most favorable locations on the ground in the United States. For a few days near the equinoxes, the satellite passes through the earth's shadow and does not receive sunlight for a maximum period of some 70 minutes around midnight. When averaged over the whole year, this represents a reduction of only 0.74 percent of the available solar energy.

A proposal has been made to construct an arrangement of two large solar-cell panels, each about 3.7 × 3.1 miles (6 × 5 km) in size, in low earth orbit and then to transfer it to the higher geostationary orbit. An attitude-control system would continuously adjust the orientation of the panels so that they always face the sun. The electricity generated would be converted into microwave radiation at a frequency of about 2.5×10^9 hertz (wavelength 0.12 meter) and then transmitted to a fixed station on the earth. Radiation of this frequency can be generated efficiently with existing equipment, and it suffers little absorption in passing through the ionosphere and atmosphere.

The microwave radiation would be collected by a large array of dish-type antennas, called *rectennas*, on the ground where it would be reconverted into direct current by the photovoltaic effect. A gallium arsenide semiconductor with a Schottky-type junction has been suggested for this purpose. The net electric power output on the earth from the two solar panels with the dimensions given above is estimated to be 5000 megawatts. This is equivalent to the output of five large fossil-fuel or nuclear power plants.

In the alternative SOLARES (an acronym for Space Orbit Light Augmentation Reflected Energy System) concept, the solar radiation would not be converted on the geostationary satellite, but would be reflected to earth by large, lightweight (aluminized polyamide) mirrors mounted on the spacecraft. The energy of the reflected solar radiation would then be converted into electrical energy by solar cells on the ground. This system would be simpler than the one described above, and the possible hazard associated with the transmission of microwave radiation would be avoided.

SOLAR ENERGY: THERMAL ELECTRIC CONVERSION

In general, the conversion of solar energy into electricity by way of thermal (or heat) energy. Heat can be converted directly into electrical energy by thermionic or thermoelectric methods (see **Thermionic Conversion; Thermoelectric Conversion**), but these techniques may not be suitable for use with sun-generated heat. The most practical thermal electric procedure for solar energy is to utilize the energy to heat a working fluid (e.g., a gas, water, or other volatile liquid). The heat energy is then converted into mechanical energy in a **turbine** and finally into electrical energy by means of a conventional generator coupled to the turbine (see **Electric Power Generation**).

For the efficient conversion of heat energy into mechanical energy and hence

into electricity, the working fluid should be supplied to the turbine at high temperature (see **Heat Engine**). To obtain such temperatures, above about 350°F (175°C), from solar energy requires the use of focusing concentrating collectors (see **Solar Energy Collectors**). Since these collectors concentrate direct solar radiation, but essentially no diffuse radiation, they would be most effective in locations where there is ample sunshine (e.g., in the southwest United States). Two basic arrangements have been proposed for converting solar radiation into electrical energy: (1) the central receiver system and (2) the distributed collector system.

In the *central receiver system*, known colloquially as the "power tower" design, an array of sun-tracking mirrors (heliostats) reflect solar radiation onto a receiver mounted on top of a central tower. Solar energy absorbed in the central receiver is removed as heat by means of a heat-transport fluid and converted into electrical energy in a turbine–generator.

The *distributed collector system* may consist of a number of parabolic trough-type (line-focusing) collectors or of paraboloid dish-type (point-focusing) collectors. The absorber pipes (or receivers) of the individual collectors are connected so that all the heated fluid is carried to a single location where the electricity is generated. The basic difference between the central receiver and distributed collector systems is that in the former the solar energy falling on a large area is transmitted to a central point as radiation, but in the latter, the energy is carried as heat in a fluid.

Analysis of the two systems indicates that they may have different preferred applications. The energy losses in transmission by radiation are less than in transport as a hot fluid. Hence, higher temperatures should be attainable at the turbine inlet with the central receiver design than if the same amount of fluid were transported from distributed collectors. Furthermore, in a large distributed collector system, the costs of the long pipelines and of the

SOLAR ENERGY: THERMAL ELECTRIC CONVERSION

energy required to pump the heat-transport fluid through them would be considerable. However, these costs are offset somewhat by the high cost of heliostats and the central receiver tower.

It appears, therefore, that the central receiver system is preferable for the large-scale generation of power for an electric utility. On the other hand, the distributed collector design may be more suitable for power plants of smaller electrical capacity, perhaps less than 2 megawatts (MW), where 1 MW is 1000 kilowatts (see **Watt**). The so-called **total energy** (or **cogeneration**) **systems**, intended to supply both electric power and heat or process steam to an institution, small community, or industry, may fall into the latter category.

Central Receiver Systems

The first stage in the U. S. Department of Energy's central receiver system for the conversion of solar heat into electricity is the 5-MW (thermal) Central Receiver Test Facility at the Sandia National Laboratories, Albuquerque, New Mexico. The purpose of this facility is to test components and to obtain experience useful for the 10-MW (electric) pilot plant at Barstow, California (see below). If the latter is successful, a larger demonstration plant should follow. The generation costs, per unit of electrical energy, decrease with increasing plant **capacity**, but with increasing size there is a decrease in the average effectiveness of the heliostats because of the greater distance from the central receiver. As far as can be estimated at present, the optimum electrical capacity of a central receiver solar thermal electric plant, from the economic standpoint, might be in the range of 50 to 300 MW. The height of the tower might be about 1000 ft (300 m) and the total land area of the plant roughly 800 acres (325 hectares).

Apart from the collector (i.e., heliostat) subsystem, other important subsystems of a solar central receiver system are (1) the receiver subsystem, (2) the heat-transport subsystem, and (3) the thermal storage subsystem. There is also an electric power generation subsystem (turbine and generator) which is integrated with the heat-transport and thermal storage subsystems.

Receiver Subsystem. The central receiver at the top of the tower has a heat-absorbing surface (e.g., panels coated with a heat-absorbing material) by which the heat-transport fluid is heated. Two basic receiver configurations have been proposed; they are the cavity and external receiver types. In the *cavity* type, pipes line the interior of a cavity; the solar radiation reflected by the heliostats enters through an aperture at the bottom of the cavity. On the other hand, in the *external receiver* type the absorber surfaces are on the exterior of a roughly cylindrical structure.

Heat losses by radiation and convection are generally less for the cavity than for the external receiver configuration, but focusing may be more critical because, in order to be absorbed, the radiation must enter through the cavity aperture. Furthermore, the external receiver has the advantages of simplicity, modular panel construction, easier access for maintenance, and a larger absorber area. The external receiver design has been selected for the 10-MW pilot plant.

Heat-Transport Subsystem. Water is a convenient heat-transport fluid, and in the form of steam it is the working fluid in the majority of existing turbines for generating electricity. Hence, water will be used, at least in the earliest solar thermal electric plants. Liquid water (i.e., turbine condensate) under pressure enters the receiver, absorbs heat energy, and leaves as superheated steam; typical steam conditions might be a temperature of 930°F (500°C) and a pressure of about 100 atm (10 MPa). The steam is piped to ground level where it drives a conventional turbine-generator system. As in all **steam-electric plants**, heat is rejected to the condenser cooling water.

In more advanced central receiver systems, the heat-transport fluid might be liquid sodium or a molten mixture of salts at about atmospheric pressure. At the bottom of the tower the high-temperature liquid is circulated through a **heat exchanger**, where the heat would be transferred to water to generate steam at a high temperature and pressure. Thus, a high efficiency of conversion into electricity could be achieved without the need for high pressure in the receiver. Another possibility is to use a gas as the heat-transport medium and also as working fluid in a gas turbine (see below).

Thermal Storage Subsystem. The purpose of the thermal storage subsystem is to store solar heat energy absorbed in the receiver for use at a later time. The stored energy can be utilized to produce steam when direct solar radiation is not adequate. Current designs are aimed at thermal storage sufficient to operate the turbine–generator at its rated load for at least 3 hours. To provide for heat storage during normal operation of the power plant, the capacity of the collector and receiver subsystems must exceed that of the turbine–generator.

Heat may be stored as sensible heat (e.g., in a liquid, rock, or ceramic material) or as latent heat (e.g., in a mixture of salts or of alkali metal hydroxides). Another concept for high-temperature storage is based on reversible chemical reactions. When heat is supplied, the chemical process occurs in one direction and the heat is absorbed; if the conditions are changed to permit the reverse reaction, the heat is released (see **Energy Storage**).

A simple example of the integration of thermal storage into a solar thermal power plant with steam–water as the working fluid is outlined in Fig. 147. High-temperature steam in excess of the immediate demand is diverted to a **heat exchanger** within the storage medium contained in a well-insulated tank. Heat is transferred to storage and the cooled fluid is returned to the central receiver. When heat is required, the stored heat is used to generate steam for the turbine.

The 10-MW Pilot Plant. The concepts of the McDonnell Douglas Corporation have been selected as the basis for the 10-MW pilot plant design. There are some 1800 heliostats, each consisting of six essentially rectangular, plane mirrors; the area of each heliostat is more than 400 sq ft (37 sq m). The heliostats are continuously steered automatically about two axes, so that the solar radiation is always reflected toward the central receiver regardless of the position of the sun.

The heliostats are distributed over a ground area of 74 acres (30 hectares) and the receiver is at the top of a 282-ft (86-m) tower located somewhat south of the center

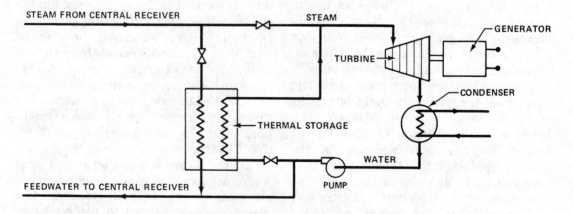

Fig. 147   Electric power generation using thermal storage.

of the area (Fig. 148). Most of the heliostats are thus north of the tower and face south. With this arrangement there is minimal mutual shading of the heliostats when the sun is low. An external-type receiver is used with high-pressure water flowing upward through the absorber pipes. The thermal storage medium consists of Exxon Caloria HT 43 oil with crushed granite and coarse sand; it has a capacity for generating 7 MW of electric power for 3 hours (or a lower power for longer periods).

Gas Turbine Systems. The U. S. Department of Energy and the Electric Power Research Institute are sponsoring conceptual, long-range studies of central receiver systems with a gas as the heat-transport and working fluid. In one such concept, the fluid, helium gas, would be heated to about 1110 to 1470°F (600 to 800°C) in its passage through the absorber tubes. The high-temperature gas would drive a closed-cycle turbine to generate electricity (see **Gas Turbine**). The associated thermal storage system would store energy as latent heat of a mixture of sodium and zinc fluorides melting at 1185°F (640°C). The proposed system would have a high efficiency for converting solar heat into electricity, and materials capable of withstanding the very high temperatures are being tested.

In another concept, air serves as the heat-transport and working fluid in an open-cycle gas turbine. In addition to the cost advantage over helium, the open-cycle turbine does not require the gas leaving the turbine to be cooled prior to compression. However, the turbine inlet temperature and hence the efficiency with air would probably be lower than with helium gas. A special aspect of the air-turbine system is that thermal storage could be eliminated. Instead, a combustor unit, using oil or gas as fuel to heat the air, would be integrated with the air turbine. The combustor could be started up rapidly when solar-heated air was not available.

Distributed Collector Systems

As noted earlier, distributed solar energy collectors may be preferred for thermal power plants of moderately small capacity (about 2 MW or less). Among the potential applications of such small plants, two under study are (1) total energy systems and (2) pumping water for irrigated farms.

Total Energy (or Cogeneration) Systems. The total energy concept is based on the sequential use (or cascading) of solar energy at two different temperature levels. First, a working fluid at a higher temperature can be used to drive a turbine and generate electricity in the usual manner. Then the heat discharged in the condenser cooling water at a lower temperature can be utilized for space heating and cooling. Alternatively, in a cogeneration system, the

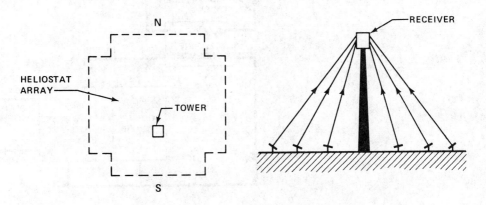

Fig. 148   Central receiver solar power plant.

turbine discharge may be used to provide process steam for industry. For the discharged heat to be useful, the temperature should be higher than in a conventional steam-electric power plant. As a result, the efficiency for electricity generation is decreased, but this can be more than offset by the economic value of the discharged heat.

Possible uses of total energy (including cogeneration) systems are to provide electric power and space heating and cooling for large buildings, for a group of family residences, or for a mixed community of residences and buildings of various sizes. They might also supply power, hot water, and process steam for industry. There may be capacity limits, however, within which total energy, distributed collector systems could prove economic. If the design capacity is too large, the costs of piping and land for the collectors become prohibitive. At the other extreme, too small a capacity may result in excessive unit costs of electricity generation and thermal storage.

The Solar Total Energy Test Facility at the Sandia National Laboratories, Albu-

querque, New Mexico, is being used to test various focusing collectors and other components of total energy systems to permit design and operation of demonstration systems. In a proposed scheme, shown in simplified form in Fig. 149, a high-boiling-point hydrocarbon (Monsanto Therminol 66) is the heat-transport fluid in a distributed array of focusing, trough-type solar collectors. The fluid, heated to about 600°F (316°C) by passage through the collectors, is circulated through a high-temperature thermal storage tank. Hot fluid is passed through the heat exchanger (or boiler) where toluene, the turbine working fluid, is converted into vapor at about 580°F (305°C). The toluene vapor is expanded in the turbine which drives the generator and produces electricity.

Cooling water leaves the turbine condenser at roughly 190°F (88°C), compared with 105°F (40°C) or less for a conventional steam-electric power plant. This warm water passes to a storage tank for use as required in space heating and to supply hot water (see **Solar Energy: Direct Thermal Applications**). The temperature of 190°F

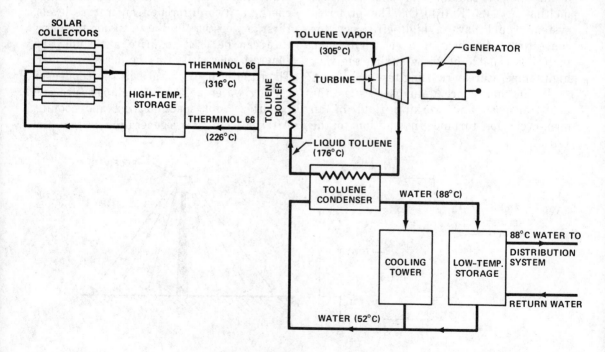

Fig. 149 Solar total-energy test system.

(88°C) should also be adequate for operating an absorption-type air conditioner (see **Refrigeration**).

If the solar total energy system is to be completely independent of electric power from a utility, auxiliary heaters burning oil or gas would be required to provide operational continuity. Such heaters could be started up quickly when the solar radiation is not adequate and the storage facilities are exhausted. One auxiliary heater would maintain the temperature in the heat exchanger (i.e., the toluene boiler) and another that of the water in the distribution system for space heating and cooling (low-temperature storage). If the temperature in either case should fall below the design value, the appropriate auxiliary would be turned on automatically.

Irrigation Water Pumping. Irrigation pumps of relatively low power can be driven directly by a vapor turbine using solar energy, as described under direct thermal applications. For deep wells, pumps of higher power are required, and these are preferably operated by electric motors. In a deep-well irrigation project in Arizona, solar energy is used to heat a hydrocarbon oil to about 550°F (290°C) by passage through an array of parabolic trough-type concentrating collectors. The heated fluid passes, by way of a thermal storage tank, to a heat exchanger where toluene is boiled to provide vapor for operating a turbine–generator combination. Pumped water is used as coolant in the turbine condenser, as in Fig. 143, and the temperature is more like that in a conventional power plant. The electric power generated (150 kW) drives electric motors to operate the deep-well pumps.

SOLAR FURNACE

A device for obtaining very high temperatures [up to about 6870°F (3800°C)] by means of a system of reflectors that collect solar radiation energy from a large area and focus it into a small volume. The prin-

ciple of the solar furnace is outlined in Fig. 150. A number of heliostats (see **Solar Energy Collectors**) are arranged in terraces on a sloping surface (e.g., on a hillside) so that, regardless of the sun's position, they always reflect solar radiation in the same direction onto a large paraboloid (or spherical) reflecting collector made up of many fixed mirrors attached to the face of a structure. The collector then brings the radiation to a focus within a small volume (receiver).

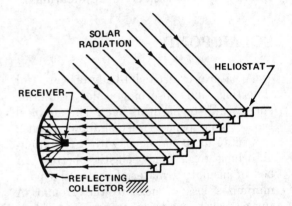

Fig. 150   Principle of solar furnace.

The first large solar furnace, with a thermal power of 45 kilowatts (kW), was completed in France in 1952. A similar furnace, with a power of about 35 kW, was constructed for the U. S. Army at Natick, Massachusetts, in 1958; this was moved to the White Sands Missile Range, New Mexico, in 1976 where it is used to study thermal radiation effects. Solar furnaces have also been built in Japan and the U.S.S.R.

The world's largest solar furnace, with a design thermal power of 1000 kW, commenced operation at Odeillo in the French Pyrenees in 1970. It consists of 63 heliostats, each having an area of 484 sq ft (45 sq m), arranged in eight tiers. The heliostats reflect the solar radiation in a parallel beam toward a paraboloidal mirror 131 ft (40 m) high and 177 ft (54 m) maximum width. This mirror is made up of 9500 pieces which are curved to produce a minimum image of the sun at the focus.

The maximum temperature, attained at the focal point of the mirror, is estimated to be over 6870°F (3800°C); the maximum heat flux is 16,000 kW/sq m (1500 kW/sq ft).

Solar furnaces are used in high-temperature studies of refractory (i.e., very high melting point) oxides and metals in a clean environment. Components for solar thermal electric systems, in which solar energy is concentrated onto absorber pipes containing a heat-transport fluid, can also be tested in solar furnaces (see **Solar Energy: Thermal Electric Applications**).

SOLAR POND

A natural or artificial body of water for collecting and absorbing solar radiation energy and storing it as heat. The simplest type of solar pond is very shallow, about 2 to 4 in. (5 to 10 cm) deep, with a radiation-absorbing (e.g., black plastic) bottom. A bed of insulating material under the pond minimizes loss of heat to the ground. A curved cover, made of transparent fiberglass, over the pond permits entry of solar radiation but reduces losses by radiation and convection (i.e., air movement). In a suitable climate, all the pond water can become hot enough for use in space heating and in agricultural and other processes (see **Solar Energy: Direct Thermal Applications**).

In a shallow pond, as described above, the water soon acquires a fairly uniform temperature. In a deeper pond, however, temperature variations generally exist. The warmer water tends to rise, while cooler water sinks. Loss of heat from the surface, especially at night, then results in circulation of the water by convection. The situation is changed if the pond contains salt water at the bottom with a layer of fresh water above it. Because of its salt content, the solar-heated bottom water is more dense than the cooler fresh water at the top, and hence it does not tend to rise. A relatively stable layer of heated salt water is thus produced at the bottom of the pond with a lighter layer of cooler fresh water, which acts as a heat insulator, above it.

Such a two-layer configuration has been established in uncovered solar salt ponds about 8 to 10 ft (2.4 to 3 m) deep. One proposed use for these ponds is space heating and cooling for a housing project. The heat stored in the bottom (salt) water in the summer is used for heating in the winter, whereas the colder (fresh) surface water provides cooling in the summer.

Another possibility is to utilize the temperature difference between upper and lower layers in a solar salt pond to operate a **heat engine**, in particular a low-temperature, closed-cycle condensing turbine (see **Vapor Turbine**). An attached electrical generator can then produce electric power in the usual manner (see **Electric Power Generation**). The arrangement is similar to the **ocean thermal energy conversion** (OTEC) system, except that heat is taken up from the warmer bottom of the solar salt pond and discharged to the cooler surface layer.

SOLVENT REFINED COAL (SRC) PROCESS

A process studied initially by the Spencer Chemical Company and later developed by the Pittsburg and Midway Coal Mining Company, both subsidiaries of the Gulf Oil Corporation. The original purpose of the process, like that of the related **Pott–Broche Process,** was to produce a clean, low-ash and low-sulfur solid boiler fuel from coal high in both ash and sulfur. This is now known as the SRC-I process. In a later development, called SRC-II, the main product is a clean liquid fuel. Both processes can use caking or noncaking coals without pretreatment.

SRC-I

Pulverized coal is slurried with solvent oil, made in the process, and injected with hydrogen-rich gas into a preheater at

800°F (425°C). It then passes to a reactor/dissolver where the temperature is about 850°F (455°C) and the pressure 100 atm (10 MPa) or more. Several decomposition and hydrogenation reactions occur, and as a result some 90 percent of the coal goes into solution, leaving solid unreacted coal together with mineral matter (ash).

Gases formed in the reactor/dissolver are separated from the liquid and solid and treated for the removal of sulfur, mostly present as hydrogen sulfide (see **Desulfurization of Fuel Gases**). Fuel gases, consisting of light **hydrocarbons,** are recovered from the clean gas, and hydrogen which remains is recycled to the coal slurry feed (Fig. 151).

The residual solid in the reactor/dissolver is separated from the liquid by filtration. It is sent to a gasifier where it reacts with oxygen and steam to produce hydrogen-rich gas for use in the process (see **Hydrogen Production**). The liquid is distilled to yield **naphtha, fuel (heating) oil,** and recycle oil for slurrying the feed coal. The heavy residue from the distillation operation is cooled to 350 to 400°F (175 to 205°C) to form the major process product, clean solvent refined coal.

The **heating value** of this solid is about 16,000 Btu/lb (37 MJ/kg) compared with some 12,000 Btu/lb (28 MJ/kg) for the feed coal. A typical bituminous coal, containing 7 percent ash and 3.4 percent sulfur, gave a refined product with 0.2 percent ash and 0.8 percent sulfur.

SRC-II

A slurry of pulverized coal is fed with hydrogen-rich gas to a preheater and then to a dissolver/reactor, just as in the SRC-I process. The temperature and pressure are much the same in both cases. The difference, however, is that the slurrying liquid in SRC-II is not a distillate oil, as it is in SRC-I, but part of the liquid–solid product from the dissolver/reactor (Fig. 152). This means that part of the dissolved and unreacted coal is returned to the dissolver/reactor with the feed coal for further decomposition and hydrogenation. The net result is a substantial increase in the proportion of liquid in the reaction products.

The gases leaving the dissolver/reactor are treated the same way as in SRC-I; sulfur is removed, light hydrocarbon fuel

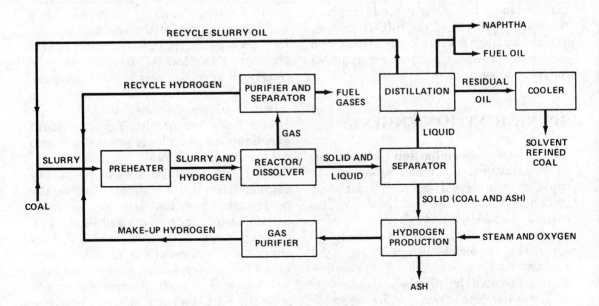

Fig. 151 SRC-I process.

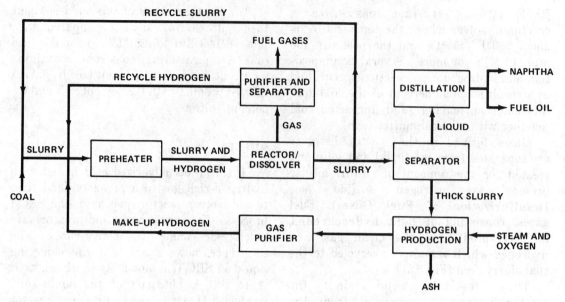

Fig. 152   SRC-II process.

gases are recovered, and the residual hydrogen is recycled. However, the liquid and solid are not separated as they are in SRC-I. Part of the mixture is used to form the coal slurry, as described above, and the remainder is separated into a light oil and a thick residual slurry of unreacted coal and mineral matter in a heavy oil. The light oil is fractionally distilled to yield naphtha and a low-sulfur fuel oil; these are the major products of the SRC-II process. The thick slurry is reacted with steam and oxygen to provide the hydrogen required for the process.

SPARK-IGNITION ENGINE

An **internal-combustion engine** in which a mixture of air and a fuel gas or vapor is compressed in a cylinder and ignited by means of an electric spark. The modern spark-ignition engine, with **gasoline** as fuel, is used extensively in automobiles, motor bicycles and boats, light aircraft, lawn mowers, etc. The gasoline engine, invented by Gottlieb Daimler in 1885, was developed from the first successful internal-combustion engine patented by Nikolaus Otto in 1876 in which the fuel was coal gas. The gasoline engine is thus sometimes called the Otto engine.

The spark-ignition engine is a **heat engine** (i.e., one that converts heat partially into mechanical work) operating on an approximation to the idealized **Otto cycle.** The distinguishing characteristic of this cycle is that combustion of the fuel, resulting in addition of heat to the working fluid (air), occurs at essentially constant volume. In the somewhat similar **Diesel cycle,** heat addition takes place at constant pressure. Nearly all spark-ignition engines use gasoline as the fuel, but it is possible to use mixtures of gasoline and an alcohol, an alcohol alone (see **Alcohol Fuels**), **natural gas, liquefied petroleum gas,** or hydrogen gas (see **Hydrogen Fuel**).

With few exceptions, such as the **rotary engine,** spark-ignition engines are of the reciprocating type; that is to say, a tightly fitting piston moves in and out of a cylinder in which the fuel burns. The in-and-out (or reciprocating) motion is converted into rotary motion by connecting the piston to a crankshaft. Automobile engines generally have 4, 6, or 8 cylinders attached to a single crankshaft. The cylinders may be arranged in-line or in V form; opposing arrangements, with equal numbers of cylinders in one plane, either horizontal or vertical, on each side of the crankshaft are commonly used in propeller-driven (nonjet) aircraft. A **flywheel** on the crankshaft of a spark-ignition engine smooths out irregularities in the rotation arising from intermittent power strokes (see below).

Four-Stroke and Two-Stroke Engines

Four-Stroke Cycle. A motion of the piston in or out of a cylinder is called a stroke. Spark-ignition engines can operate on either two or four strokes per power cycle, the four-stroke cycle being by far the more common. In a four-stroke engine there are two inward and two outward strokes per cycle, but only one is a power stroke in which the piston does useful work. Hence, only one power stroke occurs for two rotations of the crankshaft. (An in-and-out motion of a piston represents one rotation of the shaft.)

The cylinder of a four-stroke engine has intake and exhaust ports (openings) for the admission of the air–fuel mixture and expulsion of the burned gases, respectively. The ports are opened and closed by valves operated at the appropriate times by cams on a camshaft geared to the crankshaft. Proper timing of the valves is essential to the operation of a four-stroke engine.

The stages in the power cycle of a four-stroke, spark-ignition engine are shown and described in Fig. 153; each stage of the cycle corresponds to one piston stroke, down (outward) in stages 1 and 3, and up (inward) in stages 2 and 4. Once the engine has been started by rotating the crankshaft, usually by a battery-operated electric motor (or self-starter), the energy of the power stroke carries the piston motion through the other strokes of the cycle.

Two-Stroke Cycle. In the two-stroke engine either the exhaust valve alone or both inlet and exhaust valves are replaced

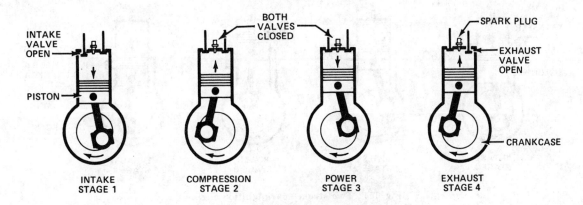

Fig. 153 Four-stroke spark ignition engine.

1. *Intake.* The piston moving downward (i.e., out of the cylinder) draws a mixture of air and fuel into the cylinder by way of the open intake valve. The exhaust valve is closed.
2. *Compression.* The intake valve is closed, and the piston moving upward (i.e., into the cylinder) compresses the air–fuel mixture.
3. *Power.* Just before the top of the compression stroke, with both valves closed, the air–fuel mixture is ignited by an electric spark from a spark plug. The fuel burns rapidly, producing hot gases that expand and push the piston downward. This is the power stroke in which mechanical work is done; not all of this work is available, however, since part is expended in the other strokes, especially the compression stroke.
4. *Exhaust.* The piston moving upward pushes the burned gases out through the open exhaust valve.

by ports in the cylinder wall which are opened (uncovered) or closed (covered) by motion of the piston. In the completely valveless two-stroke engine there are three ports: (1) the intake port through which the air–fuel mixture is admitted to the sealed crankshaft chamber (or crankcase); (2) the transfer port through which the mixture passes from the crankcase to the cylinder; and (3) the exhaust port through which the burned gases are discharged.

The operation of a valveless type of two-stroke engine is explained in Fig. 154; stages 1 and 2 occur during the up (inward) stroke and 3 and 4 during the down (outward) stroke. There are thus two strokes (one of which is a power stroke) for each rotation of the crankshaft.

Since a four-stroke engine has only one power stroke for two crankshaft rotations, whereas a two-stroke engine has a power stroke for each rotation, the latter might be expected to develop twice the power of the former at a given speed in engines of the same dimensions. In practice, however, this is not the case because two-stroke, spark-ignition engines are less efficient than the four-stroke type.

The following are some of the factors that lead to power losses in the two-stroke design. (1) The intake port is uncovered for such a short time that only a small charge of air–fuel mixture enters the crankcase in each cycle. (2) The exhaust port is also uncovered for a short time and substantial amounts of burned gases remain in the cylinder; the fresh charge is thus preheated with a consequent decrease in the compression ratio and the thermal efficiency. (3) Since the exhaust port is uncovered while the air–fuel mixture enters the cylinder through the transfer port, part of the fresh charge is lost with the exhaust. In some engines this loss is minimized by means of a small ridge (or baffle) on the top of the piston (see Fig. 154) which tends to direct the entering charge upward.

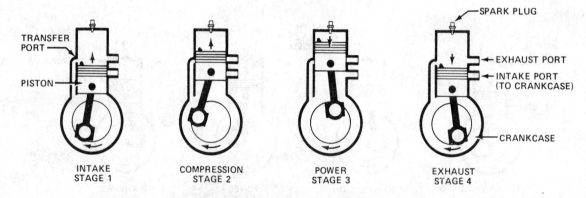

Fig. 154 Two-stroke spark ignition engine.

1. *Intake.* The air–fuel mixture, drawn into the crankcase in the preceding down (outward) stroke (see stages 3 and 4) of the piston, enters the cylinder through the transfer port which is uncovered. Note that the exhaust port is also open but the intake port is closed.

2. *Compression.* The piston moves up (inward), covering all the ports, and the air–fuel mixture is compressed.

3. *Power.* Just before the top of the compression stroke, the air–fuel mixture is ignited by an electric spark from a spark plug. The fuel burns rapidly, producing hot gases that expand and push the piston downward. This is the power stroke in which mechanical work is done, part of which is expended in the compression stroke. In the early stages of the power stroke, the intake port is uncovered and a fresh air–fuel charge enters the crankcase.

4. *Exhaust.* Just before the bottom of the power (down) stroke, the exhaust port and transfer port are uncovered; fresh air–fuel mixture enters the cylinder through the transfer port and pushes out the burned gases through the exhaust port.

The main advantages of the two-stroke engine are its mechanical simplicity, since it can be built without valves, and light weight. In small sizes, the two-stroke engine may deliver more power than a four-stroke engine of the same weight although with a lower fuel efficiency. Two-stroke engines have thus been used for low-power requirements (e.g., some lawn mowers and small chain saws), where light weight is more important than fuel consumption, since this is low in any event.

Air–Fuel Charge

The quantity of air required for the complete combustion of gasoline (to carbon dioxide and water) depends to some extent on the composition of the fuel. Based on an average composition, the calculated weight (mass) of air is about 14 times the weight of the gasoline; this weight ratio required for complete combustion is called the air-to-fuel *stoichiometric ratio*. (The ratio is sometimes expressed in the reverse manner as the weight of fuel relative to that of air; it is then 1/14 or about 0.071 for gasoline.) A rich mixture is one containing more fuel (less air) than corresponds to the stoichiometric ratio, whereas a lean mixture contains less fuel (more air) than the calculated ratio.

The maximum efficiency for utilizing the heat energy supplied by the fuel (i.e., the maximum fuel economy) should be attained in stoichiometric or in lean mixtures, since complete combustion is possible. However, a rich mixture provides the maximum power from a given engine at the expense of fuel economy. Since there is insufficient air for complete combustion in a rich mixture, the exhaust contains substantial amounts of hydrocarbons and carbon monoxide which are atmospheric pollutants.

Automobile engines built before 1973 generally operated with rich mixtures to maximize the power output with little regard for fuel economy or atmospheric pollution. In recent years, however, both of these considerations have become important and carburetors are adjusted to produce lean mixtures, with an air-to-fuel weight ratio somewhat larger than the stoichiometric value. Catalytic converters, which oxidize hydrocarbons and carbon monoxide in the exhaust gases, make possible the use of richer mixtures than would otherwise be permissible.

In most spark-ignition engines the composition and supply of the air–fuel mixture are controlled by the *carburetor* and its associated *choke* and *throttle valves* (Fig. 155). Modern automobile carburetors

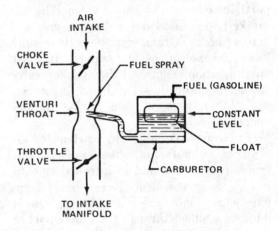

Fig. 155 Air–fuel supply to spark-ignition engine.

are complex, but basically they are designed to introduce a spray of fine droplets of fuel into an air stream in which the fuel is vaporized. The mixture of air and fuel is distributed among the cylinders by way of the intake manifold. In order to facilitate vaporization of the fuel, the manifold may be warmed by means of part of the engine exhaust.

The choke valve is located in the air intake ahead of the carburetor whereas the throttle valve is between the carburetor and the intake manifold. During normal engine operation, the choke valve is fully open and the throttle valve adjusts the amount of air–fuel mixture reaching the manifold. The throttle valve, operated in an automobile by the accelerator pedal, thus controls the engine speed. The carburetor is designed to provide a roughly constant air-to-fuel ratio regardless of the setting of the throttle valve, that is, at most engine speeds. The actual value of this ratio depends on the dimension of the air intake and the size of the fuel metering orifice in the carburetor.

When the engine is cold, only the most readily volatile components of the fuel are vaporized in the manifold; the less readily volatile components thus remain in the liquid state in which they do not burn well. To compensate for this, the proportion of fuel in the air-fuel mixture is increased by partially closing the choke valve. The air intake is decreased but a drop in pressure across the carburetor spray nozzle results in an increase in the fuel supply. In most spark-ignition engines, the choke valve opens automatically after the initial warm-up; it then remains open during normal operation.

In some engines, the carburetor is replaced by a *fuel-injection system*. The fuel is pressurized, and a controlled amount is sprayed in the form of very small droplets either into the air just before it enters a cylinder through the intake port or directly into the intake port when the valve is open to admit air. The low pressure in the cylinder during the intake stroke then causes vaporization of the fine droplets of fuel. An advantage claimed for fuel injection is a more uniform distribution of the air-fuel mixture to the cylinders than is possible with a carburetor and intake manifold.

The power output of a spark-ignition engine may be increased by supercharging, that is, by supplying the air at a pressure above atmospheric (see **Supercharger**). The air pressure may be increased by means of a pump or blower driven from the engine crankshaft or by a compressor operated by a turbine (turbocharger) using the engine exhaust as the working gas. As a result of the increased pressure (and density), a larger mass of air enters the engine intake; there is a proportional increase in

the quantity of fuel drawn in from the carburetor so that the air-to-fuel ratio is largely unaffected. The increase in mass of the air-fuel charge is then equivalent to an increase in cylinder size and power output.

A drawback to supercharging in a spark-ignition engine is that the increase in pressure (and temperature) of the air causes the engine to knock at a lower compression ratio than in the absence of supercharging (see below). Because of the need to operate at a lower compression, there is some loss of power (and efficiency). However, superchargers have been used for many years in high-compression engines for racing cars with special high-octane fuels to inhibit knocking. Supercharging has now been adapted to smaller automobile engines with moderate compressions to increase the power output without a substantial increase in weight.

A major use of supercharging has been in spark-ignition (piston type) aircraft engines. At increasing altitudes the ambient air density decreases, and without a supercharger there would be a loss of power because of the decreased mass of air-fuel mixture in the cylinder. By increasing the intake air pressure to normal (or above) atmospheric, the supercharger permits the aircraft engine to maintain its ground-level power at higher altitudes.

Compression and Engine Knock

The compression ratio is the ratio of the volume of the fuel-air charge at the beginning of the compression stage (stage 2 in Fig. 153) to that at the end of this stage. In theory, and to a great extent in practice, the **thermal efficiency** (i.e., the fraction of the heat supplied by combustion of the fuel that is converted into useful work) of an internal-combustion engine increases with the compression ratio. Automobile engines usually operate at a compression ratio in the vicinity of from 10 to 12; at higher compressions knocking (i.e., explosive combustion) occurs and the efficiency falls off (see **Octane Number**). An increase in the octane number of the fuel permits operation at a higher compression without knocking and hence at a higher efficiency.

In the past, an increase in octane number has been achieved by adding a small amount of tetraethyl lead to the gasoline, but the lead poisons (i.e., inactivates) the catalytic converters now used to reduce the pollutants in the engine exhaust. Consequently, lead-free (or no-lead) gasoline must be used in most modern automobile engines (see **Gasoline**). This fuel has a lower octane number than high-octane gasoline containing lead, and the engine compression ratio must be reduced accordingly.

As explained later, knocking can be controlled to some extent by adjusting the time at which the air-fuel mixture is ignited by a spark.

Valve Timing

Appreciable times are required for the air-fuel charge to be drawn into the cylinder through the intake valve and for the burned gases to be expelled through the exhaust valve. Consequently, in a four-stroke engine the valves are opened and closed at times that do not coincide precisely with the beginning and end of the piston strokes. The timing is determined by the locations on the rotating camshaft of the cams that control the opening and closing of the valves. In designing the engine, these locations are chosen to permit optimum power development under common operating conditions.

As a general rule, the intake valve starts to open just before the end of the exhaust stroke; it remains open throughout the intake stroke and does not close until some time into the compression stroke. Since the intake valve is open before exhaust is complete and after compression has started, there is some loss of power. But with proper timing, this is more than offset by the additional air-fuel charge drawn into the cylinder.

Similarly, the exhaust valve is timed to open before the end of the power stroke and to close after the end of the exhaust stroke. Here also there is some loss during the power stroke, but this is more than made up by the decreased work done on the piston in expelling the exhaust gases.

In a two-stroke engine without valves, the times of opening and closing of the ports are determined by their locations in the cylinder wall. There is then less scope for variation than in a four-stroke engine.

Ignition

Just before the end of the compression stroke (stage 2 in Figs. 153 and 154), an electric spark from a spark plug ignites the air-fuel charge in the cylinder. The spark initiates a combustion wave (or flame front) that travels rapidly through the combustible mixture. Expansion of the hot gases produced by the burning fuel then causes the power stroke. The high electrical voltage applied to the spark plug to generate a suitable spark is provided by the ignition system, which may be of either the battery-operated or magneto type.

Since a **storage battery** is required for starting and lighting, battery ignition is invariably used on automobiles. (The abbreviation SLI, for starting-lighting-ignition, is often applied to these batteries.) Direct-current pulses, produced by repeated closing and opening of breaker points, are passed through the primary winding of the spark (or induction) coil. As a result, a succession of high-voltage pulses is induced in the secondary (or ignition) winding. The secondary voltage pulses are applied in the required firing order to the spark plugs by means of the distributor. The breaker points are located in the distributor where the timing is controlled.

In the magneto ignition system, the primary current is generated by rotating a permanent magnet adjacent to the primary winding of the spark coil. The changing magnetic field causes a current to flow in this winding. Making and breaking of the current induces high-voltage pulses in the secondary winding in the same general manner as described above. Since a battery is not required, magneto ignition is used in lightweight equipment, such as lawn mowers, chain saws, outboard motors, and sometimes in small aircraft, which must be hand-cranked for startup.

The timing of the ignition spark has an important bearing on engine operation. Because of the appreciable time it takes for the fuel to burn, a high efficiency requires that ignition be initiated before the top of the compression stroke. This early timing is called *spark* (or *ignition*) *advance*. The optimum timing varies to some extent with the operating conditions (e.g., speed) and design of the engine. In automobiles, the timing is adjusted automatically to permit efficient engine operation over a limited range of conditions.

By decreasing the spark advance from the value for optimum engine power, the tendency for knocking to occur at a given compression ratio is reduced. Consequently, the octane number requirement of the fuel is decreased. The power output of the engine is somewhat less than with fuel of higher octane number, but this may be offset by the lower cost.

See also **Rotary Engine; Stratified Charge Engine.**

SPECIFIC GRAVITY (SP GR)

The mass (or weight) of a given volume of a substance at a specified temperature compared to the mass of an equal volume of pure water at a standard temperature. In the petroleum industry, the standard temperature is usually 60°F (15.6°C); in some cases, the temperature at which water has its maximum density [i.e., 39.2°F (4°C)] is taken as the standard. An alternative definition of specific gravity is the **density** of a substance at a specified temperature relative to that of water at the same temperature. The advantage of expressing density in terms of specific gravity (or relative density) is that it is

dimensionless (i.e., independent of units), whereas density depends on the units of mass (e.g., pounds, kilograms, etc.) and volume (cu ft, cu cm, cu m, etc.), which must be stated in each case.

See also **API Gravity.**

STEAM CRACKING (OR STEAM PYROLYSIS)

A process for the production of unsaturated (olefin) **hydrocarbons** by heating a light saturated (paraffin) hydrocarbon gas or light naphtha, obtained in **petroleum refining,** with steam to a temperature of about 1200 to 1500°F (650 to 815°C). See **Petroleum Products.**

STEAM-ELECTRIC PLANT

A facility in which electric power is generated by using a **steam turbine** to drive an **electric generator.** Apart from **hydroelectric power** plants, essentially all base-load **electric power generation** installations are steam-electric plants. Steam is usually produced by burning a **fossil fuel** or by **nuclear energy** (see **Steam Generation**). Solar energy (see **Solar Energy: Thermal Electric Applications**) or **geothermal energy** can also be used to produce steam. In addition to the steam generator, turbine, and electric generator, a steam-electric plant has a means for condensing the turbine exhaust steam and for heating the condensate to provide feedwater for the steam generator.

STEAM ENGINE

In general, a **heat engine** that converts heat energy from steam into useful mechanical work. The **steam turbine** is a type of steam engine, but the term is usually restricted to reciprocating engines in which steam causes a piston to move back and forth in a cylinder. The reciprocating motion can be converted into rotary motion by connecting the piston to a crankshaft.

Steam engines in a wide range of power ratings were used extensively at one time for a great variety of purposes, including ship propulsion, railroad locomotives, industrial operations, and electric power generation. However, in recent years, the steam engine has been largely replaced by steam and gas turbines, gasoline and diesel engines, and electric motors.

The steam engine operates in an approximation to the **Rankine cycle.** Pressurized (and possibly superheated) steam generated in a boiler is admitted to the cylinder through an inlet valve when the piston is near the end of its inward stroke (Fig. 156A). At a cutoff point, the valve is closed and the confined compressed steam expands; the piston is pushed out and mechanical work is done (Fig. 156B). During the expansion, the temperature and pressure of the steam fall. Near the end of the outward (or power) stroke, an exhaust valve is opened and the steam is discharged (Fig. 156C) to the atmosphere (noncondensing engine) or to a condenser (condensing engine). In either case, the exhaust valve is closed during the return (inward) stroke (Fig. 156D), and the residual steam is compressed. (Partial retention of the steam is not essential, but its compression has a cushioning effect that favors smooth operation.) Near the end of the inward stroke, steam is admitted to the cylinder from the boiler, and the cycle is repeated.

In condensing engines, the exhaust steam is condensed, and the resulting liquid is returned as feedwater to the boiler. Like all heat engines, the steam engine must reject (or discharge) some heat, and the lower the discharge temperature, the higher the **thermal efficiency** (i.e., the fraction of the heat taken up from the steam at a higher temperature that is converted into useful work). Condensing engines permit heat to be discharged at a lower temperature than noncondensing engines and consequently have a higher thermal efficiency.

Condensing engines require a supply of cooling water or other means of removing heat from the steam and have generally been used for ship propulsion and electric power generation. Noncondensing engines, on the other hand, were preferred for locomotives; they carried a supply of boiler feedwater which was replenished at intervals.

Most steam engines are of the double-acting type; the cylinder, which is closed at both ends, serves effectively as two cylinders with a single piston. While steam

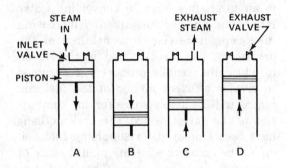

Fig. 156   Principle of simple steam engine.

is being admitted to the cylinder on one side of the piston, it is being exhausted from the other side, and vice versa. Each stroke of the piston, in one direction or the other, is thus a power stroke; there are then two power strokes for each revolution of the crankshaft.

The arrangement and timing of inlet and exhaust valves are an important aspect of steam-engine design. Several different systems, with either sliding or lifting valves, have been developed. One of the earliest, which was used extensively in locomotive engines, is the D-shaped slide valve. The valve slides back and forth along the length of the cylinder, alternately covering and uncovering ports through which steam is admitted or exhausted.

STEAM GENERATION

Conversion of water into steam at high temperature, primarily for use to operate turbines in electric utility power stations (see **Electric Power Generation**). The heat required to produce steam is usually obtained by combustion of a **fossil fuel** (i.e., coal, oil, or gas) or by using **nuclear energy; solar energy** and **geothermal energy** may also be used. At present, coal is by far the most common fuel in **steam-electric plants**; this situation is expected to persist for many years to come.

General Principles: Fossil-Fuel Boilers

In a fossil-fuel plant, the fuel is burned in air to produce heat for converting water into steam and for subsequently increasing the temperature (i.e., superheating) of the dry steam. The furnace, that is, the region in which the heat is produced, is designed to provide efficient mixing of the fuel and air, as well as ample space for the combustion of the gases formed. Several methods have been employed for supplying fuel and air to the furnaces of steam plants; some of these are described later. The heat generated is transferred to water (or steam) partly by radiation from the burning fuel and combustion gases in the furnace and partly by convection from the hot gases leaving the furnace flowing past tubes containing water (or steam).

The efficiency for the conversion into electrical energy of heat supplied as steam to a turbine-generator increases with the steam temperature (see **Steam Turbine; Thermal Efficiency**). Hence, steam should be produced at the highest practical temperature. The boiling point of water, and hence of the steam generated, can be increased by raising the pressure in the boiler. Steam in direct contact with boiling water is called *saturated steam*, and the maximum (or critical) temperature of saturated steam is 705°F (374°C). However, by separating the steam from the boiling water and supplying additional heat, it is possible to obtain *superheated steam* at a substantially higher temperature. In practice, the maximum temperature of superheated steam, as permitted by the corrosion and mechanical properties of materials, is about 1000 to 1050°F (538 to 566°C). Operation at higher temperatures would be possible with special alloys, but this is not economical.

The steam generating and superheating system consists of a number of interconnected components (Fig. 157). The feedwater, previously heated in the economizer (see below), first flows through a number of vertical pipes (or tubes) lining the interior of the large furnace volume. In these water-tube walls ("waterwalls"), the water is heated mainly by radiation and convection from the burning coal and hot gases. In most modern steam generators all the steam is produced in the walls; in some systems, however, an additional bank of boiler tubes is heated by gases leaving the furnace at a later stage.

In a saturated steam (subcritical) boiler, the steam-water mixture formed in the boiling region enters a cylindrical steam drum where steam is separated and the water is recirculated through the waterwall tubes. The steam then flows to the primary and secondary superheater tube banks where it is heated by convection from increasingly hot furnace gases. Between stages, the temperature of the superheated steam is often controlled by an *attemperator*, in which part (or all) of the steam is passed through tubes immersed in the boiler water.

The highest (or critical) pressure at which water boils in the normal manner to produce saturated steam is 218 atm (22 MPa). Some steam plants, however, operate at higher (supercritical) pressures, up to about 340 atm (34 MPa) or more. The water is then converted into a vapor (steam) with the same density as the liquid. Under these conditions, there is no distinction between liquid water and vapor; consequently no steam drum is required to separate them. All the feedwater is converted into high-temperature and high-pressure steam in a continuous system of pipes in which the steam is first formed and then superheated by hot furnace gases in the usual manner.

After passage through the steam-generating and heating units, the combus-

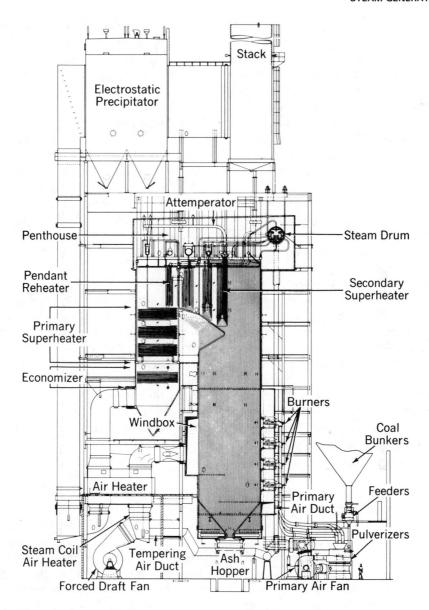

Fig. 157 Pulverized-coal fired boiler for superheated steam generation. (Courtesy Babcock & Wilcox.)

tion gases are still at a fairly high temperature, usually in the range of 600 to 800°F (315 to 425°C). The hot gases are passed through a bank of *economizer* tubes where incoming feedwater is heated, and then through a heater to increase the temperature of the combustion air. The temperature of the residual gases has now fallen to about 250 or 300°F (120 to 150°C).

Before discharge through a stack, these gases may be treated for removal of particulate matter (see below) and, if necessary, of **sulfur oxides** (see **Desulfurization of Stack Gases**).

Fly Ash Removal

When coal is burned, an ash remains (see **Coal: Mineral Matter**). Part of the ash falls into an ash pit beneath the combustion region; this is referred to as *bottom ash*. The remainder, depending in amount on the nature of the coal burned and the type of burner, is carried over with the flue gas in the form of extremely small particles called *fly ash*. The particles range in size from about 300 micrometers (μm) to less than 1 μm (1.2×10^{-2} to 4×10^{-5} in. or less). In order to comply with emission standards for particulate matter, up to 99 percent (or more) of the fly ash may have to be removed from the stack (or flue) gas before discharge to the atmosphere.

When wet scrubbers are used to absorb sulfur dioxide from the stack gases, fly ash is generally removed in the scrubbers. For the removal of very small fly ash particles, less than about 2 μm (8×10^{-5} in.), the particles must impinge on the water with a high velocity. This is achieved by admitting the gas to the scrubber through a venturi "throat". In scrubbers with throwaway wastes, the ash is disposed of with the waste material. Otherwise, the ash is separated by filtration or centrifugation from the scrubbing solution before treatment of the latter for recovery of the sulfur dioxide absorber.

If the sulfur dioxide content of the stack gas is low enough to meet existing standards without scrubbing, other means of fly ash removal are required. **Electrostatic precipitators** are most commonly used for coal-burning plants; **cyclone collectors** and **bag filters** (or baghouses) are used to a lesser extent in special circumstances. These methods for removal of fly ash (or other particulate matter) are described under the indicated headings.

Coal Burners

All coal-fired, electric utility steam generators operate along the general lines stated above. There are differences, however, in the way in which the coal is admitted to and burned in the furnace. The most important techniques are: pulverized-coal

firing and cyclone furnaces (or burners) for large installations, and mechanical stokers for smaller ones. In addition, coal-fired fluidized-bed steam generators are under development.

Pulverized-Coal Firing. The coal is fed into a hopper from which it enters a pulverizer where the coal is ground so that roughly 70 percent can pass through a 200-mesh screen (i.e., with a spacing of $1/200$ in. or 74 μm). The finely powdered coal is carried continuously by a forced draft of primary air to burners where the coal is ignited.

Secondary air (about 80 percent of the total) is supplied through ducts to mix with the primary air and coal a short distance beyond the point at which ignition occurs. In some furnaces, several burners are mounted around the walls with the flames directed horizontally toward the central axis. In another design, the flames are directed tangentially to the circumference of an imaginary circle in the interior; the air and powdered coal form a vortex in which the coal burns efficiently.

A possible drawback to pulverized-coal burners is the large proportion of fly ash produced. However, if the coal forms ash with a relatively low fusion temperature (softening point below about 2200°F or 1205°C), "slag-tap" or "wet-bottom" furnaces can be used. Some 50 percent of the ash is then removed as a molten slag, leaving 50 percent as fly ash. The slag is quenched with water to form a granular solid. Coals with less fusible ash are burned in "dry-bottom" furnaces; the fly ash may then constitute 80 percent of the total, with the remainder falling into the ash pit.

Cyclone Furnace. The cyclone furnace is intended largely for burning coals of lower rank (see **Coal**) with a high content of readily fusible ash. The coal, crushed to a size of about ¼ in. (6.4 mm) or less, is introduced into a burner directed toward the central axis of the furnace. Primary air, constituting about 20 percent of the total air, enters the burner tangentially and causes a whirling ("cyclone") motion of the incoming coal in the cylindrical combustion chamber [diameter 6 to 10 ft (1.8 to 3 m)]. Secondary air, admitted in the same tangential direction from the top of the chamber, sustains the motion. A small proportion (up to about 5 percent) of combustion air, called tertiary air, enters at the center of the burner. The number of burners used depends on the steam capacity of the installation.

The smaller coal particles (or *fines*) are ignited in the turbulent, rotating air, whereas the larger pieces are thrown to the walls of the combustion chamber by centrifugal force. The coal adheres to the slag-wetted surface and burns in the secondary air. The hot combustion gases pass from the burners into the furnace where they heat the water. Because of the relatively low melting point of the ash and the high temperature in the combustion chamber, most of the coal ash (85 percent or more) melts and is drawn off as slag. The fly ash thus constitutes less than 15 percent of the total. It should be recalled, however, that cyclone furnaces generally burn high-ash coals.

Mechanical Stokers. In a mechanical stoker, the coal is burned on a grate, similar in principle to the old-fashioned, hand-fed furnace, with the coal supplied and the ash removed continuously. The most widely used type of mechanical stoker is the spreader stoker, described below. Other types are the underfeed stoker, in which fresh coal is forced under the burning coal bed; the chain-grate stoker, in which the grate consists of a number of links joined together in a moving endless belt; and the water-cooled, vibrating-grate stoker with a grate inclined to the horizontal down which the burning coal travels. Advantages of mechanical stokers are that they can operate with coal of almost any type and that special sizing of the coal feed, apart from crushing, is not necessary.

In the *spreader stoker*, crushed coal, fed at a controlled rate from a hopper, falls onto a set of rapidly revolving paddles which throw the coal onto the perforated furnace grate. Air is introduced both below (primary air) and above the grate (secondary air).

The coal fines burn while still suspended in air in the hot combustion zone, whereas the coarser material falls and burns on the grate. About 25 percent of the ash leaves the furnace as fly ash; the remainder is removed continuously from the fuel bed by rocking or moving the grate. The combustion gases with fly ash in suspension are passed through a cyclone-type dust collector where the finer particles are removed for discharge. The larger particles, which contain unburned carbon from the coal, are returned to the furnace for more complete combustion.

Fluidized-Bed Combustion. The **fluidized-bed** system for coal combustion is under development. A bed of crushed coal and residual ash is fluidized by air blowing up from nozzles at the bottom of the bed. Limestone (or dolomite) is included in the bed to remove sulfur oxides and thus reduce the stack gas emission. Steam produced in the usual manner is superheated in tubes located within the combustion bed. For a fuller description see **Fluidized-Bed Combustion.**

Gas and Oil Burners

Burners have been designed for use with gas or oil fuels in conventional pulverized-coal or cyclone furnaces. In fact, the burners are often constructed so that they can burn either gas, oil, or coal. Gas and oil can also be used simultaneously. In oil-burning furnaces of the pulverized-coal type, the oil is atomized (i.e., dispersed as fine droplets) by a stream of air or steam. In cyclone-type burners, the oil is atomized by the combustion airflow. A major difference between coal-fired furnaces and those using gas or oil is that the latter produce little or no ash.

Nuclear Steam Generation

The heat produced by nuclear fission in **nuclear power reactors** is being used to

intake steam pressure may be as high at 235 atm (24 MPa), there may be more than 50 stages to provide a gradual decrease in pressure and thus make efficient use of the energy in the steam. High-power turbines are commonly divided into separate high-pressure (HP) and low-pressure (LP) sections on the same shaft; in some cases there is also an intermediate-pressure (IP) section.

Both impulse and reaction principles are utilized in steam turbines. Initial stages of an HP turbine are often of the impulse type with reaction (or hybrid) types in the later stages. In the hybrid type, the rotor blades are so shaped that impulse predominates at the inner (shaft) end and reaction at the outer end.

Steam Superheating and Reheating

Liquid water which may form in a turbine by condensation of steam as the temperature falls has deleterious effects; wet steam can reduce the mechanical efficiency of the turbine and also cause erosion of the blades. However, condensation can be reduced (or eliminated) by heating the turbine intake steam to a temperature above that at which condensation would occur. This can be done by using superheated steam to supply the initial (HP) turbine section and by reheating the steam leaving this section before it enters the next (IP or LP) section. (Superheated steam is steam at a temperature above that at which condensation can occur at the existing pressure.)

In a fossil-fuel steam plant, superheated steam is produced by using furnace gases to heat the saturated steam from the boiler (see **Steam Generation**). Exhaust steam from the HP turbine section is reheated in a similar manner, as indicated in Fig. 159. But in nuclear power plants based on **light-water reactors**, such as are commonly used in the United States, very little (if any) superheating is possible. The moisture that inevitably forms in the HP turbine is then removed from the exhaust steam in a moisture separator. The residual steam is reheated usually in two stages: first, by steam bled (i.e., extracted) from an intermediate stage of the HP turbine and then by steam from the steam generator (Fig. 160).

Regenerative Feedwater Heating

In principle, water formed by condensation of the steam in the condenser could be supplied directly as feed to the steam generator, as indicated in Fig. 158. It is more effective, however, to use steam bled off at various stages of the turbine to heat the feedwater. This process, represented schematically in Fig. 161, is called *regenerative feedwater heating*. The feedwater flows in one direction, from the condenser to the steam generator, while the

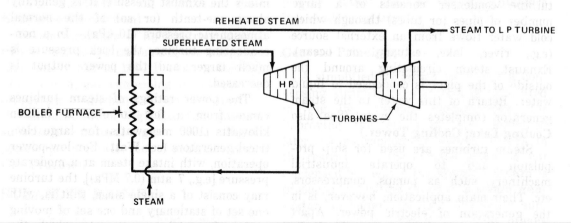

REHEATED STEAM

SUPERHEATED STEAM

STEAM TO L P TURBINE

H P

I P

BOILER FURNACE →

TURBINES

STEAM

Fig. 159 Turbine steam supply from fossil-fuel plant.

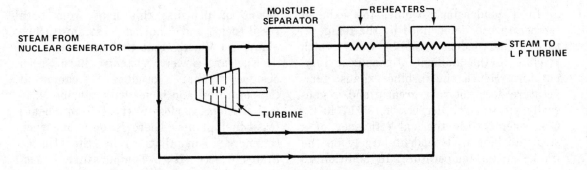

Fig. 160 Turbine steam supply from nuclear plant.

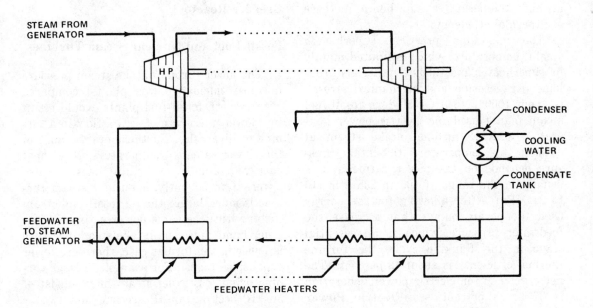

Fig. 161 Steam generator feedwater reheating.

steam used for heating flows in the opposite direction. The residual steam and accompanying condensed water eventually return to the condenser and form part of the feedwater system.

Although regenerative heating draws steam from the turbines and thus utilizes heat that could be converted into mechanical work, it nevertheless improves the overall cycle efficiency since less heat is required in the steam generator. Another advantage of regenerative feedwater heating is that the reduced steam flow through the later stages of the turbine permits the use of smaller units and decreases the condenser cooling requirements.

Turbine Thermal Efficiency

As with all heat engines, the overall **thermal efficiency** of a steam turbine (i.e., the proportion of the heat energy supplied to the steam generator that is converted into useful mechanical work) depends on the upper (intake) and lower (exhaust) steam temperatures. A high turbine intake steam temperature and a low exhaust temperature favor high thermal efficiency. Since the exhaust steam temperature is higher in a noncondensing (or back-pressure) turbine than in a condensing turbine, the efficiency is lower, assuming the same intake temperature.

In a condensing turbine, the exhaust temperature is determined by the temperature of the condenser cooling water which is close to the ambient atmospheric temperature. Hence, the turbine exhaust temperature does not vary greatly and is generally within a few degrees of 105°F (40°C). Consequently, the thermal efficiency of a steam turbine is dependent largely on the intake steam temperature. In addition to decreasing condensation in the turbine, superheating and reheating provide steam at high temperatures and hence increase the thermal efficiency.

The maximum practical superheated steam temperature, which is limited mainly by superheater and turbine materials problems (e.g., erosion and mechanical stress), is about 1000°F (540°C). Modern fossil-fuel steam plants used for electric power generation provide turbine intake steam at close to this temperature; the intake pressure depends on the power rating and is usually in the range of 100 to 235 atm (10 to 24 MPa). After allowing for the energy used to operate pumps that increase the feedwater pressure, which is an essential stage in the Rankine cycle, the turbine thermal efficiency is about 42 percent. (The net efficiency for electric power generation is 37 to 38 percent; see **Electric Power Generation**). The remaining 58 percent of the heat produced by combustion of the fuel is dissipated in the stack gases (about 10 percent) and in the condenser cooling water (about 48 percent).

As already stated, nuclear power plants commonly used in the United States cannot generate appreciable superheated steam. The turbine intake steam temperature is then in the vicinity of 555°F (290°C); the pressure is roughly 75 atm (7.6 MPa). Because of the lower steam temperature than in modern fossil-fuel plants, the turbine efficiency is lower at 34 percent. (The net electrical generation efficiency is about 32 percent.) There are no stack losses in a nuclear power system, and some 66 percent of the heat generated is discharged to the condenser cooling water. (For possible means of utilizing this heat from both fossil-fuel and nuclear steam-turbine plants, see **Waste Heat Utilization**.)

Nuclear power systems have been designed that produce steam at temperatures, and hence with turbine thermal efficiencies, close to those from modern fossil-fuel plants. There is only one such commercial installation in the United States (see **High-Temperature Gas-Cooled Reactor**), but others are under development (see **Liquid-Metal Fast Breeder Reactor**).

Fossil-Fuel and Nuclear Steam Turbines

The lower turbine intake steam pressure in most nuclear power plants, compared with modern fossil-fuel plants, would result in a lower mass rate of steam flow in a turbine of given size. Furthermore, because of the lower steam temperature, the heat energy available per unit mass is also lower. Consequently, a nuclear steam turbine is much larger than a fossil-fuel steam turbine with the same power rating. On the other hand, since the intake steam pressure is lower in the former case, there are fewer expansion stages. For example, a large turbine using fossil-fuel steam may consist of an HP section, an IP section, and two or more LP sections. However, in an equivalent nuclear plant there is usually no IP section.

Some other differences between a modern fossil-fuel plant and the common nuclear steam-turbine plants are the following. (1) The different methods of reheating the exhaust steam from the HP section have already been described. (2) Since steam temperatures are lower in a nuclear-power facility, there is less regenerative heating of the feedwater. There are commonly six stages of water heating in nuclear plants and eight or more in fossil-fuel plants. (3) Because there is little or no steam superheating in a nuclear steam plant, liquid water forms in the turbine at all stages of steam expansion (and cooling); provision must therefore be made for removal of the water.   Hence, the exhaust from the HP section is passed through a moisture separator before reheating (see Fig. 160), and the rotor blades have grooves for drawing off water during turbine operation.

STIRLING CYCLE

A repeated succession of operations (or cycle) representing the idealized behavior of the working fluid in the **Stirling engine**. The Stirling cycle is illustrated and described in Fig. 162. In the description, each stage is assumed to have been completed before the next stage is initiated. However, in an actual engine there is a gradual rather than a sharp transition from one stage to the next; hence, the

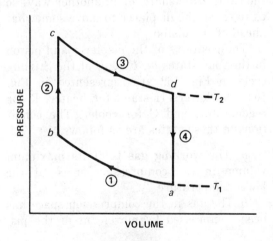

Fig. 162   Stirling cycle.

1. Isothermal compression of the working fluid (gas) along *ab* at the lower temperature T_1; work is done *on* the gas, and heat must be removed to maintain a constant temperature.
2. Heat addition to the gas at constant volume along *bc*; the pressure rises, and the temperature increases to the upper temperature T_2.
3. Isothermal expansion of the gas along *cd* at the temperature T_2; mechanical work is done *by* the gas, and heat must be supplied to maintain the constant temperature.
4. Heat removal (rejection) from the gas at constant volume along *da*; the temperature falls to T_1, and the gas returns to its initial state at *a*.

sharp points in the figure would actually be rounded off.

Work is done by the gas when it expands in stage 3 and on the gas when it is compressed in stage 1; no work is done in stages 2 and 4. Hence, the net work done in the cycle is the difference between stages 3 and 1. The **thermal efficiency** (i.e., the fraction of the heat supplied in stage 3 that is converted into net mechanical work) is increased by increasing the upper working temperature T_2 and/or by decreasing the lower temperature T_1 (see **Heat Engine**).

Figure 162 for the ideal Stirling cycle is superficially similar to Fig. 99 for the ideal **Otto cycle**; however, stages 1 and 3 in the Stirling cycle are isothermal, whereas in the Otto cycle they are adiabatic. As a consequence, the thermal efficiency of the Stirling cycle is greater than that of the Otto cycle for the same operating temperatures.

STIRLING ENGINE

A **heat engine** of the reciprocating (piston) type that differs from others of this type (e.g., **internal-combustion engines**) in having a working gas and a heat source that are independent. Hence, in principle, the Stirling engine can be designed to utilize any heat source (e.g., fossil or nuclear fuel or solar energy) and any convenient working gas (e.g., air).

In internal-combustion (gasoline or diesel) engines, compression and expansion of the working fluid occur in the same chamber (i.e., the cylinder). But in the Stirling engine the working gas is compressed in one region of the engine and is transferred to another region where it is expanded. The expanded gas is then returned to the first region for recompression. The working gas thus moves back and forth in a closed cycle (see **Stirling Cycle**).

The basic principle of the Stirling engine was patented by Robert Stirling, a Scottish minister, in 1816. The air engine, as the Stirling engine was called because air was the working gas, was improved and used, especially for pumping water, in the first half of the 18th century. Heat was supplied by burning wood or coal. With the development of steam and particularly internal-combustion engines, interest in the Stirling engine waned, but it was revived around 1940 at the Philips Research Laboratories in The Netherlands. The original purpose was to develop a reliable engine for generating electricity in remote locations. Subsequently, it was realized that the Stirling engine had the potential for use in road vehicles as well as in many other ways.

General Principles

Several configurations of the Stirling engine have been proposed, but they all have the same basic components: a compression space and an expansion space with a heater, regenerator, and cooler in between. Heat is supplied to the working gas at a higher temperature by the heater and is rejected at a lower temperature in the cooler. The regenerator provides a means for storing heat deposited by the hot gas during one stage of the cycle and releasing it to heat the cool gas in a subsequent stage.

As in all heat engines, the **thermal efficiency** (i.e., the proportion of the heat supplied that is converted into useful mechanical work) is increased by supplying heat at the highest possible temperature and rejecting it at the lowest possible temperature.

Stirling engine configurations fall into two general categories: single acting and double acting. In *single-acting* systems there are two piston-like moving components which may be in one cylinder or in two connected cylinders. In the simplest single-acting configuration, one of the moving components, called the *displacer*, serves to transfer the working gas from the expansion space at one end of the cylinder to the compression space at the other end and back again. Only a small amount of work is required to operate the displacer. The other moving component is a conventional working piston. The piston does work on the gas to compress it, and work is done on the piston by the expanding gas. Both displacer and piston may be connected to the same crankshaft (or in another way) so as to cause the displacer to move somewhat ahead of the piston.

The positions of the displacer and piston in the four states (*a, b, c, d*) of the Stirling cycle in Fig. 162 are represented in Fig. 163. The letters H stand for heater, R for regenerator, and C for cooler. The conditions in these states are as follows:

a. The working gas is at its maximum volume in the compression space at the lower temperature T_1.

b. The gas in the compression space has been compressed by movement of the pis-

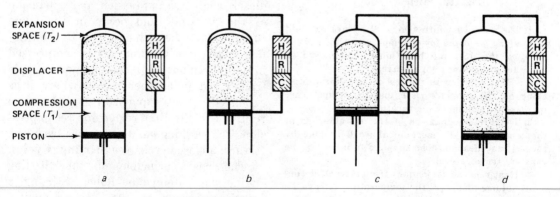

Fig. 163   Single-acting Stirling engine with displacer and piston.

ton; work has been done on the gas. The heat generated by compression is removed by the cooler so that the gas temperature remains essentially constant at T_1.

c. The compressed gas has been transferred to the expansion space, through the regenerator and heater, by movement of the displacer. Heat is taken up from both regenerator (see below) and heater, and the gas temperature increases to T_2.

d. The hot gas in the expansion space is expanded at temperature T_2, and work is done by pushing back the piston. Heat supplied by the heater maintains the temperature at T_2.

Subsequently, movement of the displacer transfers the gas back to the compression space, through the regenerator and cooler, to restore the initial state a and thus complete the cycle. Heat from the working gas is deposited in the regenerator and is later recovered when the gas passes from state b to state c.

Double-acting systems have two or more cylinders, each containing a piston which acts as both displacer (at one end) and working piston (at the other end). In a configuration proposed by H. Rinia in The Netherlands in 1946, the compression space in each cylinder is connected, through a cooler, regenerator, and heater, to the expansion space in an adjacent cylinder.

A simple form of the Rinia arrangement with two cylinders is shown in Fig. 164; the in and out motions of the two pistons are half a stroke apart. (A stroke is a complete in or out motion of a piston in a cylinder.) There are two separate quantities of working gas; one is contained in the upper part of cylinder A and the lower part of B, whereas the other is in the lower part of A and the upper part of B. With this system, two compression and two expansion (or working) strokes occur in each cycle.

Automobile Engines

The Stirling engine is attractive for automobiles for the following reasons: (1) low air pollution levels that are possible, (2) low noise levels because there are no explosions as in internal-combustion engines, and (3) ability to use a variety of fuels (i.e., heat sources), such as natural or synthetic gaseous or liquid **hydrocarbons,** stored solar energy, or even possibly powdered coal. Some of the special features of automotive Stirling engines are outlined below.

Heater. Stirling engines of the double-acting type are preferred because of their high specific power (i.e., power per unit mass). Four cylinders can be grouped together in a closely spaced arrangement that permits use of a single fuel burner to heat the working gas for all the cylinders. In the heater the gas passes through a number of tubes surrounded by the hot combustion gases produced by the burning fuel. Because the heat source is independent of the working gas, the burner and air supply are designed to ensure essentially complete combustion of the fuel. The residual hydrocarbons, carbon monoxide, and particulates in the exhaust gases can thus be kept at lower levels than with internal-combustion engines. Moreover, the flame temperatures are lower in the Stirling engine; hence, the **nitrogen oxides** content of the exhaust is also lower.

The heater tubes, which are in direct contact with the burning fuel, may determine the life of a Stirling engine. In the past, these tubes have been made from an

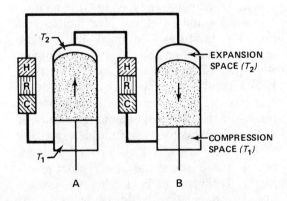

Fig. 164   Simple Rinia arrangement of Stirling engine.

expensive cobalt alloy, but a cheaper material would be necessary for a practical automobile engine. Efforts are under way to develop less costly alloys and refractory materials (e.g., silicon carbide or nitride) for heater tubes as well as for the hot ends of the cylinders.

Cooler. In the cooler, heat is removed by cold water flowing around a bundle of parallel tubes through which the working gas passes. The heat taken up by the water is dissipated to the atmosphere by circulation through a radiator. In order to maintain the thermal efficiency, the water must return to the cooler at the lowest possible temperature.

The radiator in a Stirling engine would be similar in design to a conventional automobile radiator, but it must be much more efficient in transferring heat to the air. In an internal-combustion engine, the cooling water plays no part in the heat engine cycle; heat is rejected with the exhaust gases, and the purpose of the cooling water is to prevent overheating and weakening of the cylinder walls. But in the Stirling engine, heat rejection to the cooling water and its dissipation in the radiator are essential to the operation of the cycle.

Regenerator. The regenerator is usually smaller than either the heating or cooling system. It may consist of a stack of fine-mesh wire screens or a network of metallic strips or wires.

Working Gas. Although the nature of the working gas is not material to the Stirling cycle, certain physical properties (i.e., heat capacity, thermal conductivity, density, and viscosity) affect the specific power. Hydrogen gas can best meet the requirements for an automobile engine, but it has some drawbacks. Hydrogen can diffuse through hot metals and can cause embrittlement; furthermore, the gas is flammable and can form an explosive mixture with air (see **Hydrogen Fuel**). It is thought, however, that these difficulties can probably be overcome. If this is not possible, helium will be the preferred working gas (see later).

Speed Control. In an internal combustion engine, adjustment of the quantity of fuel admitted to the cylinder results in an immediate response in the operating speed. Control by fuel adjustment is not practical in a Stirling engine for an automobile, especially because the response would be slow. The method adopted in most experimental engines is based on changes in the pressure of the working gas or in the "dead" volume (i.e., the volume of gas outside the expansion and compression spaces) or in a combination of both. An increase in pressure or a decrease in the dead volume produces an increase in power output. Another approach to speed control is based on a design which permits variation in the length of the piston stroke.

Startup. Cold startup of a Stirling engine is slower than a gasoline engine. The fuel is ignited by means of an electric spark or a hot wire, and the burner is allowed to operate for several seconds. When the working gas has become hot enough to produce power, the engine is started with a battery-operated electric motor in the same way as in a conventional automobile.

Heat Storage. The use of stored heat energy, rather than energy generated by burning fuel, may have advantages for operating Stirling engines in automobiles. The heat would be stored in a molten salt (or salt mixture) contained in a well-insulated vessel on the vehicle (see **Energy Storage**). Lithium fluoride, which melts at 1560°F (849°C) and has a high heat storage capacity per unit mass, has been proposed for use with a Stirling engine. A mixture of sodium and magnesium fluorides, melting at 1530°F (832°C), is a cheaper alternative, but it has a somewhat lower storage capacity. Transport of heat from the storage unit to the engine could be achieved by using the **heat pipe** principle with liquid sodium as the transfer medium.

The heat storage unit in a Stirling engine would have to be replaced periodically, possibly at intervals of about 150 miles (240 km). The heat could be provided by off-peak electric power, by burning coal, or possibly by solar energy.

Stationary Engines

The requirement of a relatively high power output per unit mass for an automobile engine does not apply to a stationary engine; hence, there are fewer constraints on the design of a stationary Stirling engine. For example, such an engine can use helium rather than hydrogen as the working gas; problems of safety and of permeability and embrittlement of metals at high temperature are then avoided. The pressure of the working gas can be lower in a stationary engine, thus permitting a greater choice of materials for the cylinders. There is also more flexibility in the use of fuels and heating methods, especially in conjunction with the heat-pipe principle for heat transport. Finally, several mechanical features are simplified in a stationary engine because a compact configuration is not essential, as it is in an automobile.

There are many potential applications for a Stirling engine capable of making efficient use of nonscarce fuels (e.g., coal) or possibly solar energy as the heat source. Stirling engines have been proposed for generating electric power in remote locations (see **Electric Power Generation**); the fuel would then presumably be a hydrocarbon oil. The Stirling engine is also being considered for driving the compressor of a **heat pump** for space heating and cooling. The overall efficiency is expected to be better than when an electric motor is used, as is generally the case.

The Stirling engine can be incorporated into a **total energy system** to meet both the electrical and thermal demands (i.e., for hot water and space heating and cooling) of a building or community. The engine would drive a generator to produce electric power

in the usual manner, and the heat normally dissipated to the atmosphere would satisfy the thermal demands.

The major waste heat sources from a Stirling engine are (1) the heat discharged to the water in the cooler and (2) the exhaust gases from combustion of a hydrocarbon fuel in the heater. For simplicity, these waste heat sources will be described separately, although various combinations are possible to meet local requirements.

In normal operation, the cooling water temperature of a Stirling engine is maintained as low as possible [e.g., 90°F (32°C)] to improve the thermal efficiency of the heat-engine cycle. By allowing the temperature to increase to about 140°F (60°C), there would be some loss of efficiency, but the water leaving the cooler would be suitable for domestic (and similar) purposes.

Even with an air preheater, the temperature of the combustion exhaust gas leaving a Stirling engine heater would be at least 500°F (260°C). By means of a **heat exchanger,** these gases could be used to heat air or water sufficiently for direct space heating. Water near its normal boiling point or low-temperature steam could provide cooling (air conditioning) by an absorption system (see **Refrigeration**). Process steam might also be generated for industrial use.

Reversed Stirling Cycle

When the Stirling engine is operated in reverse, that is, instead of producing power, the engine is driven by an external power source, heat is taken up at the lower temperature and is discharged at the upper temperature. The uptake of heat from a material (or space) at the lower temperature results in cooling whereas the heat discharged at the upper temperature can provide space heating. The reversed Stirling cycle can thus be used as an alternative to the reversed **Rankine cycle** for refrigeration or in a heat pump.

Commercial refrigeration systems based on the reversed Stirling cycle have been used successfully for cooling to very low temperatures. But other refrigeration applications, which do not require such low temperatures, are possible. Electric motors are generally used to drive the compressors for vapor-compression refrigerators and heat pumps, but consideration is being given to use of a combination of direct and reverse Stirling engines. The direct engine would supply the power, and the reverse engine would provide cooling or heating, as required.

STONE & WEBSTER/ IONICS DESULFURIZATION PROCESS

A wet, regenerative process of Stone & Webster Engineering, Inc., and Ionics, Inc., for the **desulfurization of stack gases.** The absorber is a solution of sodium hydroxide (NaOH) which is converted first into sodium sulfite (Na_2SO_3) and subsequently into sodium bisulfite ($NaHSO_3$) by the chemical absorption in a scrubber of sulfur dioxide (SO_2) from stack (flue) gases. Sulfur trioxide is also removed. The spent absorber solution is treated with sulfuric acid (H_2SO_4) or sodium bisulfate ($NaHSO_4$), which may be thought of as a mixture of sulfuric acid and sodium sulfate. Both sulfite and bisulfite are decomposed by the acid; sulfur dioxide gas is evolved (with water vapor) and the residual solution contains essentially only sodium sulfate (Na_2SO_4). The sulfur dioxide can be recovered and used in industry or converted into elemental (solid) sulfur or sulfuric acid (see **Desulfurization: Waste Products**).

The next (regeneration) stage represents the novel feature of the process. The sodium sulfate solution is electrolyzed; that is to say, a direct electric current is passed through it. The products are sodium hydroxide at one electrode (cathode) and sulfuric acid (or sodium bisulfate) at the other (anode). The sodium hydroxide is recycled to the stack gas scrubber, and the acid is used to treat the spent absorber (sulfite and bisulfite) solution. The merit of this method is that there is no problem of sodium sulfate disposal.

STORAGE BATTERY

A combination of electric storage (or secondary) cells; a storage cell is one that can be recharged after discharge by passing a direct current through the cell in the opposite direction to the discharge current. The chemical reaction taking place in the cell when it is charged is reversed when the cell is discharged. Thus, in the charged cell, electrical energy is stored as chemical energy which can be recovered as electrical energy when the cell is discharged.

In a storage battery, individual cells are connected in various ways. Like all electric cells, a storage cell has positive and negative electrodes to which the respective terminals are attached. Individual cells may be combined in series with the positive electrode of each cell connected to the negative electrode of the adjacent cell (Fig. 165). The total electromotive force (emf) or voltage of the battery is then the sum of the separate voltages. In most automobile batteries, for example, six cells each with

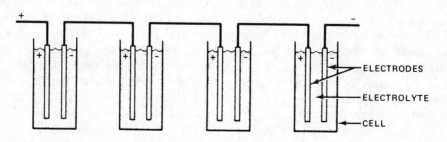

Fig. 165   Electric cells connected in series.

an emf of close to 2 volts are connected in series to provide a 12-volt output. The total current (in amperes) drawn from the series of cells is, however, the same as that drawn from each cell.

Storage cells in a battery can also be combined in parallel; all the positive electrodes of the individual cells are connected together and so also are all the negative electrodes (Fig. 166). The battery voltage is now the same as that of a single cell, but the current is the sum of the currents supplied by the individual cells. By combining appropriate numbers of cells in both series and parallel (see Fig. 146), the battery can deliver the desired voltage and current. The **power** of a battery (i.e., the rate at which stored energy is withdrawn) in **watts** is equal to the product of the emf in volts and the current in amperes; thus,

$$\text{Power (watts)} = \text{EMF (volts)} \times \text{Current (amperes)}$$

Hence, an appropriate combination of cells can provide the desired power output.

The energy storage capacity of a battery is often expressed in ampere-hours; this is the product of the current (in amperes) the battery can deliver multiplied by the delivery time (in hours). The usable energy (in watt-hours) stored is the ampere-hour capacity multiplied by the emf of the battery in volts. An automobile battery, for example, with an emf of 12 volts and 50 ampere-hour capacity, delivers $12 \times 50 = 600$ watt-hours (or 0.6 kilowatt-hours) of energy. (For an explanation of power in watts and energy in watt-hours, see **Watt**.)

The practical storage capacity of a battery depends on the discharge rate (or discharge time). Increase in the discharge rate (or decrease in discharge time) for a given battery results in a decrease in the amount of electrical energy that can be delivered. This effect is very marked in the lead–acid battery commonly used in automobiles. For example, a battery with an energy capacity of 0.6 kilowatt-hour (kW-hr) at 2.5-amperes discharge rate might have a capacity of only 0.24 kW-hr at 150 amperes. However, upon standing after a rapid discharge, the battery may experience a partial recovery and deliver more of its stored energy. In a lead–acid cell, the recovery arises mainly from local readjustments in the acid concentration which decreases when the cell is discharged (see below).

Since storage batteries are generally portable, they are common mobile sources of energy. A major use, for example, is the SLI (starting-lighting-ignition) system of automobiles. Among the many other applications are those in the operation of mine locomotives, forklift trucks, golf carts, road vehicles, and submarines and other underwater craft. If maintained in a charged state, storage batteries are a dependable source of electrical energy; they are consequently used in stationary emergency power systems. Furthermore, instruments and other components in the safety systems of many industrial plants are operated by batteries.

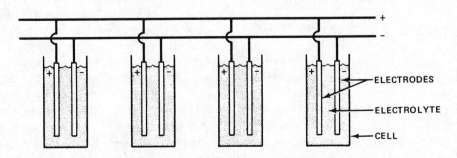

Fig. 166   Electric cells connected in parallel.

Storage batteries are expected to play an important role in energy resource conservation. Among the potential uses are (1) storage of electrical energy produced from solar energy (see **Solar Energy: Photovoltaic Conversion**) and wind energy (see **Wind Energy Conversion**), (2) propulsion of road vehicles (see **Electric Vehicles**), and (3) load leveling and peak shaving in electric utilities (see **Electric Power Generation; Energy Storage**).

Lead–Acid Battery

The lead-acid battery, invented by Gaston Planté in France in 1859, was the first practical storage battery; it is still, in spite of its defects, the one in widest use, especially for the SLI systems of road and farm vehicles. The name arises from the chemical nature of the electrodes (lead) and the electrolyte (acid) in which the electrodes are immersed. In the charged battery, the active material of the positive electrode is lead dioxide (PbO_2) whereas that of the negative electrode is metallic lead (Pb). The electrolyte is an aqueous solution of sulfuric acid.

When the cell is discharged, the active materials of both electrodes are converted into lead sulfate ($PbSO_4$). The process is reversed when the cell is charged; lead dioxide is regenerated at the positive electrode and lead at the negative electrode. The net chemical reactions taking place upon discharging and charging the battery are as follows:

$$PbO_2 + Pb + 2H_2SO_4 \underset{\text{charge}}{\overset{\text{discharge}}{\rightleftharpoons}}$$
$$\underset{(+)}{} \quad \underset{(-)}{}$$

$$\underset{(+)}{PbSO_4} + \underset{(-)}{PbSO_4} + 2H_2O$$

The (+) and (−) signs below the symbols refer to positive and negative electrodes, respectively.

The emf of a lead-acid cell is independent of the amounts of active material present on the electrodes, but it does depend on the concentration of the sulfuric acid. In practice, the concentration is indicated by the **specific gravity** of the electrolyte. In a fully charged cell, the specific gravity is usually about 1.26 to 1.28 at 77°F (25°C); the emf of a lead-acid cell is then close to 2.1 volts (at a low discharge rate). An increase or decrease in the specific gravity (i.e., in the acid concentration) is accompanied by an increase or decrease, respectively, in the emf.

When a lead-acid cell is discharged, sulfuric acid is consumed at both electrodes as the lead dioxide (positive) and lead (negative) are converted into lead sulfate. Furthermore, as seen in the equation given above, water (H_2O), which dilutes the acid, is produced in the discharge; this occurs at the positive electrode. As a result of the decrease in the acid concentration (and specific gravity), the emf of a lead-acid cell decreases steadily as it is discharged. The cell is generally regarded as being fully discharged when the specific gravity is about 1.08 (at 77°F) and the emf is roughly 1.7 volts. Both acid concentration and emf are restored when the cell (or battery) is charged.

The electrodes in a lead-acid cell are commonly in the form of plates upon (or within) which the active materials are deposited. Although the amounts of these materials do not affect the cell emf, they determine the energy storage capacity of the cell (or battery). In order to increase the storage capacity, it is necessary, therefore, to increase the accessible amount of active material. However, for efficient utilization of this material, the plates must not be too thick. Hence, the storage capacity is increased by increasing the area of active material exposed to the electrolyte. This is achieved by connecting in parallel a substantial number (ten or more) of positive plates alternating with a set of connected negative plates (Fig. 167). A large area (and mass) of accessible active material can thus be obtained within a cell of reasonable size.

For automobile batteries, the plates are commonly made from a cast grid (or lattice) of lead, alloyed with small amounts of

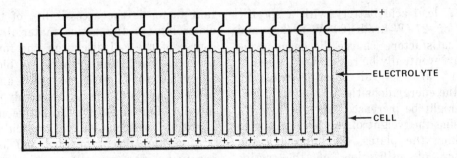

Fig. 167 Single cell with positive and negative plates in parallel.

other elements (e.g., antimony, calcium, etc.) to improve corrosion resistance and mechanical properties. The spaces in the grid are then filled with a paste made from lead monoxide (PbO), some finely divided lead, and aqueous sulfuric acid. These are called *pasted plates*.

The plates are now "formed" by treating them as if they are being charged at a low rate in dilute acid. The material on the positive plates is oxidized to active lead dioxide and that on the negative plates is reduced to spongy (active) lead. The plates are assembled, with porous separators between them, to form a unit cell of the desired storage capacity. Sulfuric acid electrolyte of the proper concentration (specific gravity) is then added. Finally, the unit cells, each with an emf of about 2 volts, are connected to form a battery of the required voltage (e.g., six cells in a standard 12-volt automobile battery).

Except when starting the engine, the current drain from an automobile SLI battery is relatively small; moreover, the battery is nearly always maintained in a charged state by the generator which operates when the engine is running. Batteries for emergency or portable power systems, vehicle propulsion, and electric utility applications, however, may be subjected to almost complete (or deep) discharge before being recharged. Special heavy-duty batteries are needed to withstand a series of such deep charge–discharge cycles.

The negative plates in heavy-duty batteries are usually of the pasted type, but the positive plates are also made in other ways. In the Planté type plates, for example, the lead grid has many grooves to provide a large exposed area; the lead is then oxidized to active lead dioxide electrolytically (i.e., by means of an electric current) in sulfuric acid. Electrolytic oxidation of lead is also used in Manchester-type positive plates; the plates have many circular holes filled with crimped lead spirals to provide a large area.

Tubular positive plates are often utilized in heavy-duty batteries. The "plates" are actually a set of connected tubes, made from an inert porous material (e.g., woven glass or plastic fibers), containing chemically prepared active lead dioxide. A central spine of lead (alloyed with antimony) provides electrical connection. Batteries with tube-type positive electrodes have been more expensive than those of other types, but potential cost reductions make them possible candidates for large-scale energy storage.

A disadvantage of the lead–acid battery, especially for vehicle propulsion, is its low *specific energy*, frequently called the energy density; this is the energy stored (and recovered) per unit mass. It is commonly expressed in watt-hours of stored energy per pound (W-hr/lb) or per kilogram (W-hr/kg). The best lead–acid batteries manufactured commercially in the United States have specific energies in the range of 16 to 18 W-hr/lb (35 to 40 W-hr/kg). Even at 18 W-hr/lb, a vehicle weighing 2200 lb (1000 kg), for example, with a moderate range and speed for urban driving, might require 827 lb (375 kg) of lead–acid batteries. A lead-acid battery with a specific energy of 27 W-hr/lb (60 W-hr/kg) and with a satisfactory charge-discharge cycle life may eventually be realized.

There are several possible ways in which the energy densities of lead-acid batteries might be increased. They include (1) decreasing the weight of the lead grid used in making the plates, (2) improving the efficiency of utilization of the active materials, and (3) increasing the concentration of the acid electrolyte.

1. Grids, required to support the active material, must be strong enough to maintain their integrity throughout the service life of the battery, in spite of inevitable corrosion by the concentrated sulfuric acid (about 38 weight percent in a charged cell). The grids must also provide electrical conduction throughout the mass of active material. So far grids have been made of lead (or lead alloys) because it is the only metal of reasonable cost that is not severely attacked by the acid. Even lead is corroded to some extent, so that the grids must be fairly massive.

Attempts are being made to design a lightweight grid for lead-acid batteries. The most promising approach appears to be the use of lead-coated aluminum. An undercoat of a resistant material would retard corrosion of the aluminum even if the outer lead coating is not perfect. A grid of this type would make possible a marked decrease in battery weight for the same storage capacity. There would also be a decrease in the electrical resistance, thereby reducing energy lost as heat.

2. When a lead-acid cell is completely discharged for all practical purposes, a substantial proportion, often about half, of the active material remains on the plates. The fraction of the active material actually consumed in the production of energy, called the *utilization coefficient*, depends on the design of the cell, the rate of discharge, and the temperature.

Since sulfuric acid is required for the electrochemical discharge reactions at both positive and negative electrodes, an impor-

tant factor is the accessibility of the acid electrolyte to the active material. In the course of the cell discharge, the formation of solid lead sulfate tends to block the pores in the active material and thus prevent access of the acid electrolyte to the interior. Consequently, the active material is not utilized completely. Low acid concentration, high discharge rate, and low temperature enhance sulfate blockage and reduce the utilization coefficient. In addition, repeated deep charge-discharge cycles and standing in the discharged state cause physical changes in the electrode materials that decrease the utilization.

An improvement in the utilization coefficient and an equivalent increase in the specific energy should be attainable. This would probably require new concepts in plate design and a better understanding of the details of the electrode processes. Although the overall electrochemical reactions are well known, there are many complex physical changes in the materials that occur when a cell is charged and discharged. Research is in progress aimed at resolving these complexities and thereby finding methods for increasing the utilization coefficient.

3. The emf of a lead-acid battery can be increased by increasing the acid concentration (or specific gravity); there is consequently an increase in the specific energy for a given discharge (in ampere-hours). Because the corrosion of lead increases with the acid concentration, the maximum specific gravity in a fully charged cell has been taken to be 1.30 at 77°F (25°C). However, it is reported that acid with a specific gravity of 1.4 has been used in Japan, with an accompanying increase in specific energy. The specific energy of a lead-acid cell can be increased by operating at higher temperatures, up to about 104°F (40°C), but this would also result in greater corrosion.

Nickel (Oxide)–Alkali Batteries

A number of storage cells are based on the same nickel oxide positive electrode and an alkaline electrolyte, usually an aqueous solution of potassium hydroxide. The active material in the charged state is a hydrated form (i.e., with combined and/or attached water molecules) of a higher oxide (or hydroxide) of nickel, represented approximately by $NiOOH \cdot H_2O$. The active negative electrode material may be iron, cadmium, zinc, or hydrogen gas. The cell emf is about 1.2 to 1.3 volts, in each case, at moderate discharge rates.

The specific energy attainable, with an iron, cadmium, or zinc negative electrode, is at least 23 W-hr/lb (50 W-hr/kg), which is better than for most lead-acid batteries. The main advantage of some nickel-alkali batteries would be a superior charge-discharge cycle life. The overall specific energy of a cell with a hydrogen electrode is uncertain because it depends on how the hydrogen is stored.

Nickel-Iron Cell. The nickel-iron alkaline cell was invented by Thomas A. Edison in the early 1900s. The electrochemical reactions occurring when discharged and charged are represented (approximately) by:

$$2NiOOH \cdot H_2O + Fe \underset{charge}{\overset{discharge}{\rightleftharpoons}}$$
$$(+) \qquad\qquad (-)$$

$$2Ni(OH)_2 + Fe(OH)_2$$
$$(+) \qquad\quad (-)$$

where $Ni(OH)_2$ is a hydrated form of a lower nickel oxide (NiO) and $Fe(OH)_2$ is hydrated ferrous oxide (FeO) formed when the cell is discharged. As in all electrical cells when producing current, reduction (in this case from a higher to a lower oxide) takes place at the positive electrode and oxidation (from metal to the oxide) at the negative electrode.

Unlike the lead-acid cell, the electrolyte, potassium hydroxide solution, is not involved in the electrochemical reactions. Hence, the concentration does not change appreciably when the cell is discharged (or charged). The specific gravity of the electrolyte cannot then be used to indicate the extent of discharge, as it can in the lead-acid battery. (The role of the electrolyte in an alkaline cell is to carry electrical charges from one electrode to the other through the cell.)

The positive plates of the Edison cell are made of a nickel-plated steel grid holding perforated tubes of the same material; the tubes are packed with alternating layers of nickel oxide and metallic nickel flakes to improve the electrical conductivity. The negative plates are also nickel-plated steel grids with long, rectangular perforated pockets containing iron oxide powder; this is reduced electrolytically to finely divided iron metal. Alternating positive and negative plates are connected in parallel (as in Fig. 167) to form a single cell.

The Edison battery has been used for railroad car and mine lighting, for mine locomotives, and for industrial trucks, largely because of its rugged construction and moderately high specific energy [about 24 W-hr/lb (53 W-hr/kg)]. The weight of the nickel-iron batteries is less, although the volume is greater, than for lead-acid batteries of equal energy storage capacity. Another advantage of the Edison batteries is that they can be stored for long periods in the uncharged state without suffering damage; because of physical changes in the lead sulfate, this is not the case with lead-acid batteries.

Major drawbacks of the nickel-iron battery are that, upon standing, it loses its charge (i.e., moderately rapid self-discharge) and requires addition of water from time to time. Both effects are due to the chemical reaction of the active iron with the electrolyte solution; the water is decomposed with the liberation of hydrogen gas and conversion of the iron into inactive ferrous hydroxide. Liberation of hydrogen gas may also occur when the cell is being charged. Another drawback of the Edison cell is that its performance falls rapidly at temperatures below about 50°F (10°C); some form of heating would thus be required in cold weather.

Some uses of the nickel–iron cell have declined in favor of the nickel–cadmium cell (see below) which is less subject to self-discharge. The latter cell is, however, too expensive for use in road vehicles. Since the Edison cell is cheaper, it is being reconsidered as a possible candidate, especially for buses and trucks. Efforts are being directed at decreasing self-discharge and the development of less expensive high-efficiency nickel oxide electrodes (see below).

Nickel–Cadmium Cell. The nickel–cadmium alkaline cell is similar to the Edison cell and was developed at about the same time by W. Jungner in Sweden. The main difference is that the active negative electrode material is cadmium metal rather than iron. The electrochemical reactions are approximately

$$2NiOOH \cdot H_2O + Cd \underset{\text{charge}}{\overset{\text{discharge}}{\rightleftharpoons}}$$
$$\underset{(+)}{2Ni(OH)_2} + \underset{(-)}{Cd(OH)_2}$$

where $Cd(OH)_2$ is a hydrated cadmium oxide (CdO). The potassium hydroxide electrolyte is again not involved directly in the cell reactions.

In the early nickel–cadmium cells, both positive and negative plates were of the same type as the negative plates of the Edison cell, that is, with relatively long, rectangular pockets containing the active materials. These are called "pocket-type" cells.

A later development is the "sintered plate" electrode. Fine metallic nickel powder is molded around a nickel (or nickel-plated) screen and sintered (i.e., heated to cause particle agglomeration) to form a porous nickel matrix. Because a large part of the active material in an electrode does not participate in the charge–discharge processes, the nickel powder may be mixed with graphite before sintering. The porous matrix is then impregnated with a solution of a nickel salt (for the positive electrode) or of a cadmium salt (for the negative electrode) and treated to form the respective hydroxides.

Sintered plate cells are more expensive than the pocket type, but they have a smaller internal resistance, a higher specific energy, and better utilization of the active electrode materials. Sintered plates are being proposed for use with other cells having nickel (hydroxide) electrodes.

Advantages of the nickel–cadmium cell over the Edison cell are a much lower rate of self-discharge and longer life. Unlike iron, cadmium does not react with the potassium hydroxide electrolyte. There is little loss of water and no hydrogen evolution either when the cell is standing in the charged state or when it is being charged. Consequently, nickel–cadmium cells are often sealed.

Because of their high initial cost, nickel–cadmium batteries are used where the long life and reliability compensate for the high cost. Small sealed cells are used for hearing aids, power tools, pocket calculators, etc., whereas larger batteries are employed in spacecraft, aircraft, and in landbased auxiliary power systems. There appears to be little prospect for the use of nickel–cadmium cells for road vehicle propulsion or electric utility applications.

Nickel–Zinc Cell. The electrochemical characteristics of zinc and cadmium in an alkaline electrolyte are similar. Since zinc is much cheaper than cadmium, efforts are being made to develop a nickel–zinc alkaline cell. The major problem encountered, which appears to be common to all cells with a zinc negative electrode in an alkaline electrolyte, is a detrimental shape change of the electrode and the formation of dendritic (i.e., treelike) growths of metallic zinc upon repeated charge–discharge cycling. These effects tend to reduce the cell life.

Various procedures are being investigated for overcoming the zinc electrode problem. General Motors Corporation, which is studying an electric car with a nickel–zinc battery, claims to have achieved some success by the use of special porous electrode separators and of additives to the zinc electrodes. The nickel electrode is of the sintered plate type similar to that described for the nickel-cadmium cell. The specific energy of the nickel-zinc cell is at least 30 W-hr/lb (65 W-hr/kg).

Nickel-Hydrogen Cell. An alkaline storage cell with a hydrogen gas negative electrode (and a nickel oxide positive) is an attractive prospect. It would not suffer from the drawbacks of the nickel-iron and nickel-zinc cells and might be less expensive than the nickel-cadmium cell. The future of the nickel-hydrogen battery depends on the production of an efficient hydrogen electrode of reasonable cost. The problem is the same as encountered in the hydrogen-oxygen **fuel cell.** Because of the need for hydrogen gas storage, the nickel-hydrogen cell might be more useful for stationary (utility) applications than for vehicle propulsion.

Silver (Oxide)–Alkali Cells

Storage cells, similar to those described above with nickel (oxide) positive electrodes and an alkaline electrolyte, are possible with the silver oxide AgO as the active positive electrode material. Silver oxide storage cells have a higher emf (1.4 to 1.5 volts) and a higher specific energy (about 45 W-hr/lb or 100 W-hr/kg) than the corresponding nickel oxide cells. The cost of the silver makes the cells too expensive for general use and hence they are restricted to certain (usually military) applications where their relatively light weight is a major factor. It is improbable that silver oxide cells would be used for electric vehicles or by utilities.

Of the possible silver oxide-alkali cells, the one with a zinc negative electrode is the most common. The discharge and charge reactions are as follows:

$$\underset{(+)}{AgO} + \underset{(-)}{Zn} + H_2O \underset{charge}{\overset{discharge}{\rightleftharpoons}}$$

$$\underset{(+)}{Ag} + \underset{(-)}{Zn(OH)_2}$$

The zinc hydroxide, $Zn(OH)_2$, dissolves to some extent in the alkaline electrolyte solution to form zincate ions, possible ZnO_2^{2-}, but this is not of immediate significance. As is the case with the nickel-zinc storage cell, changes in the zinc electrode limit the cycle life of the silver-zinc cell. These cells are consequently used for special purposes where a long life is not essential.

A corresponding silver oxide-cadmium cell, that is, with cadmium replacing zinc as the negative electrode, has also been developed. It has a longer cycle life than the silver oxide-zinc cell but is even more expensive. Although cells with silver oxide positive electrodes and either iron or hydrogen as the negative electrode are possible, they have no particular merit in situations where their high cost can be justified.

Oxygen (Air) Electrode Cells

In both the nickel oxide and silver oxide storage cells, the oxide is effectively a source of oxygen when the cell is discharged. It is feasible, however, to use oxygen gas itself, rather than the oxide, as the active positive electrode material. An important aspect of the oxygen electrode is that it does not require a gas storage system; oxygen is obtained from the ambient air when the cell is discharged (i.e., producing current) and is expelled to the air when being charged. Consequently, cells of high specific energy are possible.

To provide electrical contact with the oxygen gas and the electrolyte, usually an alkaline solution, the oxygen (air) electrode frequently consists of porous carbon through which air can pass. A catalyst to facilitate the electrochemical reactions at the electrode is deposited on the surface (see **Fuel Cell**). Platinum (or related) catalysts have been used, but a less expensive material (e.g., silver) may be preferred for practical cells. Carbon dioxide, which could combine with the alkaline electrolyte and increase its resistance, should be removed from the air supplied to the electrode.

The oxygen (air) electrode can serve as the positive in storage cells with iron, cadmium, zinc, or hydrogen negative electrodes in an alkaline electrolyte. Of these, cadmium is too expensive and is not being considered. In principle, the oxygen (air)-hydrogen cell could serve as a secondary, rechargeable cell; but storage of the hydrogen produced when the cell is charged, for use in the discharge phase, presents a problem. It is preferable, therefore, to treat this as a fuel cell with hydrogen supplied as required to generate electric power. The oxygen (air)-aluminum combination is also a type of fuel cell.

Oxygen (Air)-Zinc Cell. The cell with a positive oxygen (air) electrode and a zinc negative in an alkaline electrolyte was initially developed as a primary cell. It is rechargeable, however, and the air–zinc storage battery is being considered for vehicle propulsion because of its low cost and potentially high specific energy.

When the cell is discharged, the oxygen at the positive electrode is reduced to water and the zinc on the negative electrode is oxidized to zinc hydroxide. The reverse reactions occur when the cell is charged; thus,

$$\underset{(+)}{\tfrac{1}{2}O_2(\text{air})} + \underset{(-)}{Zn} + 2H_2O \underset{\text{charge}}{\overset{\text{discharge}}{\rightleftarrows}}$$

$$\underset{(+)}{H_2O} + \underset{(-)}{Zn(OH)_2}$$

Part of the zinc hydroxide dissolves in the alkaline electrolyte to form zincate ions, as noted earlier. The cell emf for low discharge currents is close to 1 volt, but it decreases as the discharge rate is increased.

The chief drawback to the zinc–air battery has been its short life, due mainly to deleterious changes in the zinc electrode. Solutions to this problem have been proposed based on a unique feature of storage cells with air electrodes. When other cells are charged, both positive and negative electrodes must be restored; in air-electrode cells, however, the positive material (i.e., oxygen) is always available from the ambient air and only the negative electrode needs restoring.

One possibility is simply to replace a failed zinc electrode with a fresh electrode. This mechanical replacement has been tried and found not to be economically practical. An alternative approach, reported to be successful in France and Japan, is to use a circulating slurry of zinc powder in potassium hydroxide solution, rather than a solid zinc plate, as the negative electrode material. Reduction of the zinc hydroxide back to zinc metal when the cell is charged can then take place outside the cell.

In essence, a single cell consists of a tube of porous carbon (with catalyst) that serves as the positive (air) electrode. A fan (or blower) provides an adequate supply of oxygen from the air. A copper-plated steel collector, along the tube axis, is the negative electrode. Zinc slurry is circulated through the tube while current is being drawn from the cell. Several cells in series, with the slurry flowing from one cell to the next, are combined to form a battery.

During the discharge process, the zinc powder in the slurry is converted into solid zinc hydroxide and zincate ions in solution. To recharge the battery, the slurry is circulated through another vessel called the electrolyzer. Here the charging current reduces the zinc hydroxide and zincate to zinc powder for reuse in the cells.

Oxygen (Air)-Iron Cell. This cell, being developed by the Westinghouse Electric Corporation for use in **electric vehicles,** uses low-cost materials and has the potential for a high specific energy, at least 50 W-hr/lb (110 W-hr/kg) and possibly 64 W-hr/lb (140 W-hr/kg). The electrode reactions are similar to those in the oxygen–zinc cell; thus,

$$\underset{(+)}{\tfrac{1}{2}O_2(\text{air})} + \underset{(-)}{Fe} + 2H_2O \underset{\text{charge}}{\overset{\text{discharge}}{\rightleftarrows}}$$

$$\underset{(+)}{H_2O} + \underset{(-)}{Fe(OH)_2}$$

Unlike zinc hydroxide, the ferrous hydroxide, $Fe(OH)_2$, formed at the negative electrode does not dissolve in the alkaline electrolyte. The problems of shape changes and dendritic growths that occur at the zinc electrode do not arise at the iron electrode. The discharge emf of the air–iron cell is approximately 0.9 volt.

The iron electrodes are made from sponge iron powder produced by reducing ferric oxide (Fe_2O_3) with hydrogen gas. The powder is consolidated into sheets by sintering at a temperature above 1200°F (650°C). Optimum cell conditions were observed when the density of the resulting porous material was about 23 percent of the normal density of solid iron.

The oxygen (air) electrode consists of carbon plates coated with a silver electrochemical catalyst. This electrode is designed to form an integral part of the cell casing, so that it also serves as the container for the electrolyte. The latter is a 25 weight percent aqueous solution of potassium hydroxide, with the possible addition of lithium hydroxide. The operating temperature is around 105°F (40°C).

Like the Edison cell, the air–iron cell is subject to some self-discharge upon standing because of reaction of the iron with the electrolyte solution. However, evolution of hydrogen gas when the cell is being charged can be avoided by control of the charging current.

Chlorine Electrode Cells

The chlorine (gas) positive electrode offers the prospect of storage cells with high emf and specific energy using relatively low-cost materials. A battery with a chlorine positive and a zinc negative electrode in an aqueous zinc chloride solution as electrolyte has an emf of about 2 volts. A specific energy of 36 to 45 W-hr/lb (80 to 100 W-hr/kg) has been predicted, but this apparently does not allow for the weight of auxiliary equipment for refrigeration and pumping (see below). A zinc-chlorine battery has been used successfully to propel a demonstration road vehicle.

The chlorine electrode consists of chlorine gas dissolved in the electrolyte, with a suitable conductor, such as graphite (carbon), to provide electrical contact. The negative electrode is also of graphite on which metallic zinc is deposited. The charge and discharge reactions are then

$$\underset{(+)}{Cl_2} + \underset{(-)}{Zn} \underset{\text{charge}}{\overset{\text{discharge}}{\rightleftarrows}}$$

$$\underset{(+)}{2Cl^-} + \underset{(-)}{Zn^{2+}} \rightleftarrows ZnCl_2$$

where the zinc (Zn^{2+}) and chlorine (Cl^-) ions form zinc chloride ($ZnCl_2$) in solution. Formation of dendritic zinc growths is much less in zinc chloride solution than in the alkaline cells described earlier.

A problem with the zinc-chlorine cell is the storage and supply of chlorine gas. One proposed solution is based on the use of two compartments; one compartment contains the electrodes and the zinc chloride electrolyte solution, whereas the other, which can be refrigerated, provides storage of the chlorine as the solid hydrate, usually $Cl_2 \cdot 6H_2O$, at a temperature below 48°F (9°C). When the cell is charged, metallic zinc is deposited on the negative electrode and chlorine gas is released at the positive electrode. The chlorine passes to the cooled storage compartment where it comes in contact with a small quantity of zinc chloride solution, and the solid hydrate is formed and retained.

To discharge the cell and draw current from it, the positive electrode must be kept supplied with chlorine. This is done by pumping warm electrolyte solution from the electrode compartment to the storage compartment. As the temperature in the latter increases, chlorine is released from the hydrate and is carried by the electrolyte back to the electrode compartment.

The highly corrosive nature of wet chlorine (e.g., in solution) is a serious drawback to the zinc-chlorine cell. Titanium is one of the few metals that is resistant to attack by wet chlorine, but it is expensive. Possibly plastic materials may be found that can withstand the action of chlorine.

Redox Cells

A simple type of oxidation–reduction (or reduction–oxidation) system, referred to in brief as a *redox system*, consists of a solution containing compounds of the same element in two different oxidation states. An inert current collecting material, such as carbon, inserted in such a solution constitutes a redox electrode. An example of a redox electrode would be a carbon rod or plate in an aqueous solution of ferric and ferrous chlorides. A combination of two different redox electrodes forms a redox cell with the more powerful oxidizing system as the positive electrode. Many such combinations are possible but only a few are suitable for practical application.

In a redox battery being investigated by the National Aeronautics and Space Administration, the two redox systems are aqueous solutions of ferric and ferrous chlorides ($FeCl_3$ and $FeCl_2$) and chromous and chromic chlorides ($CrCl_2$ and $CrCl_3$). Of these, the ferric–ferrous solution is the stronger oxidizing system and forms the positive electrode in the cell, as indicated below. The broken vertical line represents a membrane that separates the two solutions. The discharge emf of this cell is 0.8 to 1.0 volt.

$$\text{carbon} \left| \begin{array}{c} FeCl_3, FeCl_2 \\ \text{solution} \end{array} \right| \begin{array}{c} CrCl_2, CrCl_3 \\ \text{solution} \end{array} \left| \text{carbon} \right.$$
$$(+) \qquad\qquad\qquad\qquad\qquad\quad (-)$$

When the cell is discharged, the ferric chloride is reduced to ferrous chloride at the positive electrode and chromous chloride is oxidized to chromic chloride at the negative electrode. The reverse processes take place when the cell is recharged. The overall cell reactions are thus:

$$\underset{(+)}{FeCl_3} + \underset{(-)}{CrCl_2} \underset{\text{charge}}{\overset{\text{discharge}}{\rightleftharpoons}} \underset{(+)}{FeCl_2} + \underset{(-)}{CrCl_3}$$

Since the electrochemical processes actually involve charged entities (ions), a more exact representation is:

$$\underset{(+)}{Fe^{3+}} + \underset{(-)}{Cr^{2+}} \underset{\text{charge}}{\overset{\text{discharge}}{\rightleftharpoons}} \underset{(+)}{Fe^{2+}} + \underset{(-)}{Cr^{3+}}$$

The membrane separating the two solutions in the cell is of a special type, called "ion selective." Its purpose is to prevent mixing of the solutions while permitting selective passage of chlorine ions (Cl^-). In order to maintain electroneutrality (i.e., to balance the charges), these ions must be transferred from the positive to the negative compartment when the cell is discharged and in the opposite direction when it is charged. (In acidic solutions, the same result could be achieved with positive hydrogen ions moving in the respective reverse directions.)

When the cell is fully charged, the positive-electrode solution contains mainly ferric chloride (with a little ferrous chloride) and the negative-electrode solution contains chromous chloride (with a little chromic chloride). As the cell is discharged, the ferrous chloride solution formed by reduction of the ferric chloride is circulated to a storage tank from which it is replaced by fresh ferric chloride solution (Fig. 168). At the same time, the chromous chloride is oxidized to chromic chloride; this is circulated to a separate storage tank and replaced by fresh chromous chloride. The cell discharge can continue until essentially all the ferric chloride is reduced and the chromous chloride oxidized.

The cell is recharged in the usual manner by supplying electrical energy from an external source. By passing current between the electrodes in the opposite direction to the discharge current, the ferrous chloride is oxidized back to ferric chloride and the chromic chloride is reduced to chromous chloride. The circulation of the two solutions between the electrode compartments and the respective

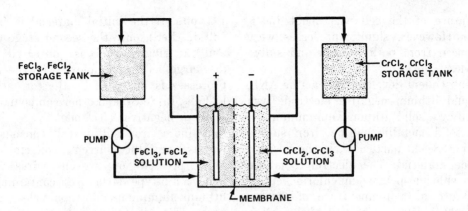

Fig. 168   Principle of a redox storage cell.

storage tanks is the same for the discharge and charge processes.

The amount of energy (in kW-hr) that can be stored in a redox cell (or battery) is proportional to the quantities of the active materials (i.e., equivalent amounts of ferric chloride and chromous chloride) that can be stored as solutions in the respective tanks. Because of the need for tanks and pumps, it appears that redox batteries would be better suited to stationary (utility) applications than to vehicle propulsion.

The redox battery is appealing because of its essential simplicity and for other reasons. All the active materials are in solution and hence there are no solids to undergo physical and chemical changes when the cell is charged, discharged, or allowed to stand. The active materials are inexpensive and are continuously renewable without appreciable loss. There are thus no apparent limitations to the charge-discharge cycle life, although the carbon current collectors and membranes may require periodic replacement.

A major problem of the $FeCl_3$, $FeCl_2$ ¦ $CrCl_2$, $CrCl_3$ redox battery is the production of a suitable ion-selective membrane. Ideally, such a membrane should have a good selectivity for chlorine (or possibly hydrogen) ions, have a low electrical resistance in the cell, not be degraded by the solutions, and be inexpensive to produce in large sheets. Furthermore, the membrane must prevent intermixing (or diffu-

sion) of the two electrode solutions. Another problem is the development of an inexpensive process for the manufacture of chromous chloride.

High-Temperature Batteries

Storage cells (and batteries) operating at high temperatures, with nonaqueous electrolytes, are being studied in the United States and several other countries for possible use in electric vehicles and for utility energy storage. For a variety of reasons, including the absence of water from the electrolyte, the low densities of the active materials, and the fairly high emfs, some of the high-temperature cells have specific energies as high as 68 to 91 kW-hr/lb (150 to 200 kW-hr/kg). The most promising cells of this type are the lithium alloy-metal sulfide and sodium-sulfur systems. Other high-temperature cells have been proposed but they have received less attention.

Lithium (Aluminum) Alloy-Metal Sulfide Cells. A study of the lithium sulfur system for a high-temperature cell was initiated at the Argonne National Laboratory (ANL) in 1968. In the earliest cells, the negative electrode was liquid lithium retained in a porous metallic support; the positive electrode was molten sulfur. The electrolyte was a liquid eutectic (i.e., minimum melting point) mixture of lithium chloride (about 60 molecular percent) and potassium chloride melting at 666°F (352°C). The operating temperature of the cell was about 750°F (400°C). However, significant losses were experienced from both lithium and sulfur electrodes.

In subsequent developments at the ANL, the liquid lithium negative electrode was replaced by a solid lithium-aluminum alloy, and a solid metallic sulfide, iron sulfide (FeS or FeS_2) in particular, is the positive electrode material. The electrolyte is the lithium chloride-potassium chloride eutectic mixture at a temperature of 750 to 840°F (400 to 450°C). The emf of the cell is less than that of the original lithium-sulfur cell, but the advantages of the lithium (aluminum)-iron sulfide cell are sufficient to compensate for the lower voltage (and specific energy).

The electrochemical reactions in the cell are very complex and are given here in simplified terms. When the cell is discharged, the lithium at the negative electrode forms lithium ions (Li^+). These ions migrate through the molten electrolyte to the positive electrode where they react with the iron sulfide to form lithium sulfide (Li_2S) and iron. The reverse processes take place when the cell is charged. (Note that, with the high-temperature cells described here, it is more convenient to consider the negative electrode material before the positive material; this is the reverse of the procedure used earlier.)

If the active positive electrode material is iron disulfide (FeS_2), the electrochemical reactions occur in two stages: one is

$$2Li \underset{(-)}{} + \underset{(+)}{FeS_2} \underset{\text{charge}}{\overset{\text{discharge}}{\rightleftharpoons}} \underset{(-)}{Li_2S} + \underset{(+)}{FeS}$$

for which the cell emf is close to 1.6 volts; the other, involving the ferrous sulfide (FeS), is

$$2Li \underset{(-)}{} + \underset{(+)}{FeS} \underset{\text{charge}}{\overset{\text{discharge}}{\rightleftharpoons}} \underset{(-)}{Li_2S} + \underset{(+)}{Fe}$$

at an emf of 1.3 volts. Thus, a cell with iron disulfide as the positive electrode material has discharge levels (or plateaus) at 1.6 and

1.3 volts. If the initial material is ferrous sulfide (FeS), only the second stage occurs and a single emf is observed upon discharge. (The complexity of the actual processes is indicated by the formation of Li_2FeS_2, $Li_4Fe_2S_5$, and other compounds.)

The negative electrode in a lithium (aluminum)-iron sulfide cell consists of a porous, metallic current collector plate (e.g., pressed or woven wires or a honeycomb-type structure) containing the lithium-aluminum alloy particles in the pores. The alloy composition ranges from about 20 weight percent of lithium in the fully charged state to 2.5 percent when discharged. From the standpoint of cost, availability, and ease of fabrication, the most suitable materials for the negative electrode current collector plates are low-carbon and stainless steels; nickel is a more expensive alternative.

The iron sulfides are poor electrical conductors, and hence efficient current collector plates are required to achieve satisfactory utilization of the positive electrode material. The electrodes can be made by pressing the sulfide into a porous (e.g., honeycomb or similar) metal matrix.

Iron disulfide (FeS_2) is very corrosive at high temperature and molybdenum is one of the few metals not attacked by it. Consequently, molybdenum has been used for the current collector of positive electrodes; but since this metal is expensive and difficult to fabricate, alternative materials are being sought. One possibility is a porous structure consisting of an intimate mixture of iron disulfide with graphite, the latter being a relatively good conductor. With ferrous sulfide (FeS) as the positive electrode material, corrosion is less severe, and it is expected that steel or nickel will be satisfactory for the current collector plates.

A key component of lithium (aluminum)-iron sulfide cells is the material placed between the positive and negative electrodes to prevent them from touching. The separator material must be a good electrical insulator and porous enough to contain the molten salt electrolyte and permit the passage of lithium ions. In addition it must be resistant to attack by lithium, the electrolyte, and iron sulfide at high temperatures. A woven boron nitride (BN) cloth has been used so far in most experimental cells, but this is expensive and does not completely prevent the loss of particles from the electrodes. The development of a cheap and efficient separator material is one of the key aspects of the lithium (aluminum)-iron sulfide storage battery.

In constructing a battery, the positive and negative electrode plates, with separators between them and electrolyte filling all the pores, are packed closely in a steel container. The plates may be arranged vertically, as proposed for a vehicle propulsion battery, or horizontally, for stationary energy storage. The container is surrounded by a thermal insulating layer, so that the outside is cool while the interior is at the operating temperature of 750 to 840°F (400 to 450°C).

The high operating temperature of the battery is attained initially or after standing idle by means of an external electrical resistance heater. During operation, either when discharging or charging, heat generated within the cell as a result of its internal resistance serves to maintain the temperature. A large battery operating at high power may then require an air blower to prevent overheating.

Because the materials problems are less severe, the first demonstration lithium (aluminum)-iron sulfide batteries will use FeS as the positive material. Such a battery, with an energy capacity of 65 kW-hr and an overall specific energy of 27 to 34 W-hr/lb (60 to 75 W-hr/kg), is proposed for propulsion of a van. The peak design power is 40 kW and the sustained power is 16 kW, with a discharge time of 4 hours.

Although iron sulfides have received major attention, studies are being made of other metal sulfides for positive electrode material to be combined with lithium (aluminum) negative electrodes. Nickel disulfide (NiS_2) and cobalt disulfide (CoS_2) either alone or as additives to iron disulfide show promise.

Lithium (Silicon) Alloy-Metal Sulfide Cells. The high-temperature cell being developed by Rockwell International Corporation is similar to the cells described above except that a lithium-silicon alloy replaces lithium-aluminum as the negative electrode material. The major advantage of this modification appears to be the potential for a 25 percent increase in specific energy, mainly because of the higher weight percent of lithium in the silicon alloy in the charged cell (about 50 weight percent compared with 20 percent in the aluminum alloy). The emf of the cell is also somewhat higher with the lithium-silicon negative electrode.

On the other hand, lithium-silicon has a smaller electrical conductivity than lithium-aluminum, and hence a more extensive current collector is necessary. A stainless steel honeycomb structure has been tested, but diffusion of silicon into the steel at high temperature results in embrittlement. There is evidence that certain additives to the silicon may be beneficial; but, if this approach is not successful, structures of titanium or molybdenum, which are known to be resistant to silicon, may be necessary. Another possibility may be to use carbon steel coated with one of these metals.

Sodium-Sulfur Cells. Work on the development of a high-temperature storage cell, with sodium as the negative electrode material and elemental sulfur as the positive material, was initiated by the Ford Motor Company around 1963, but the results were not published until 1967. Because of its inherent simplicity, the sodium-sulfur cell has attracted wide interest, especially for vehicle propulsion. Demonstration road vehicles have been reported from the United Kingdom and Japan, and the cell is being considered for railroad transportation. The sodium-sulfur battery could also be utilized for stationary energy storage systems.

The operating temperature of the cell is between 570 and 750°F (300 and 400°C), which is well above the melting points of sodium and sulfur; hence, both active electrode materials are in the liquid state. The sodium–sulfur cell is unusual in this respect as well as in having a solid electrolyte that also serves to separate the liquid electrode materials.

In most designs, the electrolyte is a ceramic called beta-alumina, consisting mainly of sodium oxide (Na_2O) and alumina (Al_2O_3). The general formula of beta-alumina is $Na_2O \cdot xAl_2O_3$, where x is in the range of about 5 to 11. Sometimes small amounts of lithium oxide (Li_2O) or other oxide may be added to improve the strength and the electrical conductivity. At the cell operating temperature, sodium ions can move through the beta-alumina in the same direction as the current. In the Dow Chemical Company's sodium–sulfur cell (see below), the electrolyte is a borate glass that is also a sodium ion conductor.

When a sodium–sulfur cell is discharged, the sodium forms sodium ions (Na^+) at the negative electrode; these ions travel through the solid electrolyte to the positive electrode where they react with the sulfur. The process is reversed when the cell is charged. In the simplest terms, the electrochemical reactions may be represented by:

$$2\underset{(-)}{Na} + \underset{(+)}{S} \underset{\text{charge}}{\overset{\text{discharge}}{\rightleftharpoons}} Na_2S$$

In practice, however, the sodium sulfide (Na_2S) interacts with excess sulfur at the positive electrode to form polysulfides having the general formula Na_2S_x, where x usually ranges from 5 (fully charged) to 3 (discharged). Within this range, the sodium polysulfide is soluble in the liquid sulfur. When x is less than 3, a solid polysulfide separates and interferes with the cell operation.

The cell emf is about 2 volts for very small discharge currents, but it is less for larger currents. Overall specific energies of

more than 45 W-hr/lb (100 W-hr/kg) should be possible for sodium–sulfur batteries.

In a common cell design, a tube made of beta-alumina is sealed to the open bottom of a stainless steel reservoir; both tube and reservoir contain molten sodium. Since sodium is a good electrical conductor, it serves as its own current collector. An aluminum wire inserted into the sodium, which does not attack aluminum, provides the negative electrode connection.

The beta-alumina tube is surrounded by molten sulfur in a suitable metal, possibly stainless steel, container. Sulfur is a very poor electrical conductor, and hence an effective current collector is required. Carbon (graphite), in the form of felt, foam, or specially shaped structures, immersed in the sulfur has been generally used for this purpose. The metal container then provides the positive electrode connection (Fig. 169).

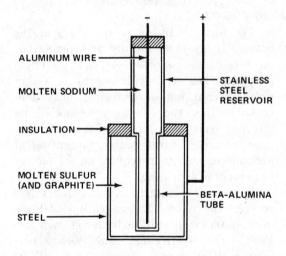

Fig. 169   Sodium–sulfur cell with beta-alumina electrolyte/separator.

The high operating temperatures are attained in the same way as described earlier for lithium (aluminum)–iron sulfide cells.

One of the problems of the sodium–sulfur cell is corrosion of the metal container by the hot, molten polysulfide. Proposed solutions to this problem are based on the use of refractory liners with good electrical conductivity (e.g., graphite or chromium oxide, Cr_2O_3). Another idea, being developed in the United Kingdom, is to reverse the electrode positions; the molten sulfur (polysulfide) is held in the interior of the beta-alumina electrolyte tube with the sodium on the outside in a stainless steel container.

The future of the sodium–sulfur battery depends mainly on materials development. In particular, commercial production at a reasonable cost of beta-alumina with adequate electrical conductivity, mechanical strength, and lifetime is essential. In addition, a low-cost, corrosion-resistant material is required for sealing the stainless steel sodium reservoir to the beta-alumina tube.

The Dow Chemical Company's sodium–sulfur cell is similar in principle to that described above, but the separator-electrolyte material consists of thousands of vertical, hollow fibers (i.e., very thin tubes) of a borosilicate glass. The upper ends of the hollow fibers are open to the sodium reservoir whereas the lower ends are sealed. The sodium-filled glass fibers are surrounded by molten sulfur, with aluminum foil distributed among them to serve as current collectors (Fig. 170). The foils are attached to the metal

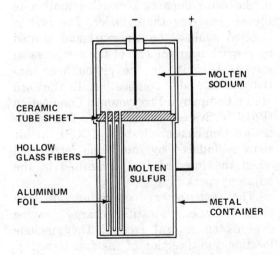

Fig. 170 Sodium–sulfur cell with glass electrolyte/separator.

container which provides the positive electrode connection. Specific energies as high as 68 W-hr/lb (150 W-hr/kg) should be attainable.

STRAIGHT-RUN GASOLINE

The **gasoline** fraction, boiling in the approximate range of 85 to 390°F (30 to 200°C), obtained by the distillation ("topping") of petroleum (see **Petroleum Refining**). The composition of straight-run gasoline depends on the nature of the crude oil, but the components are **hydrocarbons** with mostly five to nine carbon atoms per molecule. As a general rule, straight-run gasoline contains at least 50 percent of paraffin hydrocarbons; the remainder consists mostly of naphthenes (cycloparaffins) and possibly some aromatic hydrocarbons. Straight-run gasoline usually has poor antiknock quality (low **octane number**), but this is commonly increased by the **reforming** process in petroleum refining.

STRATIFIED-CHARGE ENGINE

A type of **spark-ignition engine** in which the fuel–air mixture (or charge) is not uniform, as it is in a conventional spark-ignition engine. In a stratified-charge engine the charge may be considered as consisting of two more or less separate parts; a small proportion of the charge has a high fuel-to-air ratio (i.e., a rich mixture) whereas the main charge has a low fuel-to-air ratio (i.e., a lean mixture). The rich mixture is ignited by a spark plug in the usual manner, and the resulting hot combustion gases then initiate burning in the main charge (lean mixture).

Potential advantages of the stratified-charge engine are better fuel economy [i.e., more miles per gallon (or kilometers per liter)] and lower air pollution levels than a conventional spark-ignition automobile engine. Since the main fuel charge in a stratified-charge engine is very lean (i.e., contains excess air), fuel combustion is more complete than in a conventional engine. As a result, the fuel efficiency is increased, and the carbon monoxide and unburned hydrocarbon pollutants in the exhaust gas are decreased. Furthermore, because of the lower temperature of the burning fuel in excess air, the **nitrogen oxides** in the exhaust are also reduced.

The tendency for a spark-ignition engine to knock with increase in the compression ratio (i.e., volume of the charge at the beginning relative to that at the end of the compression stroke) can be reduced by using a lean fuel-air mixture. Normally, lean mixtures are difficult to ignite, but this difficulty does not arise in a stratified-charge engine. Hence, the improvement in the **thermal efficiency** (i.e., fraction of the heat of combustion of the fuel that could be converted into useful mechanical work) associated with a high compression ratio should be realized in a stratified-charge engine with a less expensive, low octane gasoline (see **Octane Number**).

Engine Development

Research on the stratified-charge engine was conducted by H. Ricardo in England in the 1920s, but the automotive industry made little effort to develop the concept. However, in recent years the situation has changed, largely because the stratified-charge engine has the potential for good fuel economy and for meeting emission standards without special means (e.g., catalytic converters) for removing pollutants from the exhaust gases. Two general ways are being considered for separating the fuel-air charge into rich and lean parts: the dual (or divided) chamber and the single (or open) chamber designs.

Dual-Chamber Engine. The CVCC engine of the Honda Company of Japan has a small auxiliary chamber (or *prechamber*) at the top and to one side of the cylinder that is the main combustion chamber. The volume of the main chamber is at least ten times the volume of the prechamber. The two chambers are connected by a narrow throat called the *torch opening*. Each chamber has its own carburetor and intake valve, but there is only one exhaust valve to discharge burned gases from the main chamber.

A rich fuel mixture is admitted to the prechamber and a lean mixture to the main chamber. A spark plug inserted in the prechamber readily ignites the rich mixture near the end of the compression stroke. The flame produced then passes through the torch opening and ignites the fuel charge in the main chamber. The combustion gases (and residual air) expand and operate the engine in the usual manner.

Single-Chamber Engine. In engines of this type, there is no physical separation between the rich and lean parts of the fuel charge. In principle, separation is achieved by injecting the fuel, which may be gasoline or diesel fuel, into the top of the cylinder close to the spark plug. A rich mixture is consequently formed locally and is ignited by a spark. The hot gases produced then initiate combustion of the remaining unburned fuel in a large volume of air admitted through the intake valve.

In the MCP engine of the Mitsubishi Heavy Industries, Ltd. (Japan), a special intake valve imparts a swirling motion to the air entering the cylinder. The fuel is injected against the air swirl and ignited by a spark near the end of the compression stroke. A cavity in the piston head contributes to stable combustion. In the Ford Motor Company's Programmed Combustion (PROCO) System and in Texaco, Inc.'s Controlled Combustion System (TCCS), the air swirl is induced by the flame jet formed when the injected fuel is ignited by the adjacent spark plug.

The satisfactory operation of the single-chamber, stratified-charge engine depends on several factors. They include location and direction of the fuel injection, location of the spark plug, and shape of the piston head.

STRETFORD PROCESS

A process for removing hydrogen sulfide (and partial removal of organic sulfur compounds but not carbon dioxide) from fuel gases by oxidation to elemental sulfur (see **Desulfurization of Fuel Gases**). The impure feed gas at a pressure up to 20 atm (2 MPa) entering the absorber column from below meets a downflow of a weakly alkaline solution of sodium carbonate containing sodium vanadate and the sodium salt of anthraquinone disulfonic acid (ADA) at a temperature of 70 to 110°F (21 to 43°C). The hydrogen sulfide dissolves in the alkaline solution and is then oxidized to elemental solid sulfur by the vanadate. The clean gas leaves at the top of the absorber column. The spent solution passes from the bottom of the column to a tank where the sulfur is removed by froth flotation. The reduced vanadium compound is reconverted to the vanadate form by the ADA, which is finally regenerated by passing air through the solution.

SUBSTITUTE (OR SYNTHETIC) NATURAL GAS (SNG)

A fuel gas, similar in composition and **heating value** to **natural gas**, usually made from coal (see **Coal Gasification**) or a petroleum product (see **High-Btu Fuel Gas**). A similar gas is obtained by the bacterial decomposition of vegetable matter (see **Biomass Fuels**). Like natural gas, SNG contains at least 85 volume percent of **methane**, plus small amounts of other combustible constituents such as light **hydrocarbons** and possibly hydrogen. The heating value is close to 1000 Btu/cu ft (37 MJ/cu m). For distribution by pipeline, SNG must meet the specifications for **pipeline quality** gas.

A flow diagram for the production of SNG from coal is shown in Fig. 171. The coal is first converted into a **synthesis gas**, consisting mainly of carbon monoxide and hydrogen, by the action of steam and oxygen (or sometimes air). The molecular ratio of carbon monoxide to hydrogen is then adjusted to 1 to 3 by the **water-gas shift reaction** with steam. The resulting gas is treated for removal of carbon dioxide and especially sulfur compounds to prevent poisoning (i.e., inactivation) of the nickel catalyst used in the next (**methanation**) stage. In this stage, methane and water vapor are formed; the water is removed by condensation, leaving a substitute natural gas.

SULFIBAN PROCESS

A process for removing **acid gases** (hydrogen sulfide and carbon dioxide) from

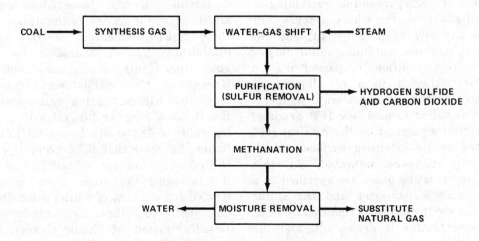

Fig. 171   **Production of substitute natural gas from coal.**

fuel gases by chemical absorption in a weakly alkaline solution. The absorber is an aqueous solution of monoethanolamine (MEA) at a temperature of 80 to 110°F (27 to 43°C) and a gas pressure up to 70 atm (7 MPa). The operational principles are similar to those described for chemical absorption in the **desulfurization of fuel gases**.

SULFINOL (SHELL) PROCESS

A process for removing **acid gases** (hydrogen sulfide, carbon dioxide, and carbon oxysulfide) and organic sulfides from fuel gases by combined physical and chemical absorption in a weakly alkaline solution (see **Desulfurization of Fuel Gases**). The absorber is a mixture of a physical solvent, tetrahydrothiophene dioxide (trade named Sulfolane), and an aqueous solution of the alkanolamine diisopropanolamine (DIPA), a weak alkali, as chemical absorber. The pressure in the absorber column may range up to 70 atm (7 MPa), and the temperature is about 110°F (43°C). The acid gases are released from the spent absorber solution partly by lowering the pressure and partly by heating in the regenerator column (see Fig. 26). The regenerated solution is cooled and recycled to the absorber column.

SULFREEN PROCESS

A process for removing the small amounts of hydrogen sulfide remaining in the **off-gas** from the **Claus process**. The gas is partially oxidized by air, and the resulting mixture, containing sulfur dioxide and hydrogen sulfide, is passed over a catalyst, either a form of alumina or activated carbon, at a moderate temperature. The sulfur formed (see **IFP process**) is adsorbed as a solid on the catalyst (i.e., retained on its extensive surface), which eventually becomes saturated. At this stage, the reacting gases are switched to a fresh catalytic converter and the sulfur from the spent catalyst is removed by heat. The liquid sulfur is drawn off, and the regenerated catalyst is reused.

SULFUR OXIDES

Compounds of the elements sulfur and oxygen, in particular, sulfur dioxide (SO_2) and sulfur trioxide (SO_3). When a **fossil fuel**, especially coal or fuel oil, is burned in air (or oxygen gas), the sulfur present is first converted into sulfur dioxide; from 1 to 3 percent of the dioxide is further oxidized to the trioxide (see **Coal: Sulfur Content**). These oxides, which are commonly present in the stack gases from coal- and oil-fired furnaces and in engine exhaust gases, have deleterious effects on humans (respiratory diseases), animals, and plants. Sulfur oxides in water form sulfurous and sulfuric acids which can corrode many structural materials. These acids in acid rain are also causing destruction of aquatic life.

The most serious health hazard associated with sulfur oxides is thought to be due to the formation of sulfate (including sulfuric acid) aerosols on suspended particulate matter in the atmosphere. (Sulfates with the general formula MSO_4 or M_2SO_4, where M is hydrogen or a metallic element, are formed by oxidation of sulfur dioxide to trioxide and subsequent interaction with moisture and dust particles.) Sulfate aerosols are especially hazardous because they penetrate deep into the lungs whereas sulfur dioxide gas tends to be absorbed in the upper respiratory tract.

Because of the potentially harmful effects of sulfur oxides in the atmosphere, the U. S. Environmental Protection Agency establishes emission standards for stack gases from plants burning fossil fuels. As a rough rule, these standards can be met by using coal with a **heating value** of 12,000 Btu/lb (28 MJ/kg) or fuel oil with a heating value of 18,000 Btu/lb (42 MJ/kg) containing not more than 0.7 percent of sulfur. If the sulfur content of the fuel exceeds this amount, the stack gases must be treated for removal of sulfur oxides before discharge to the environment (see **Desulfurization of Stack Gases**). However, even for fuel of 0.7 percent sulfur content, a power station rated at 1000 electric megawatts, operating at a **capacity** of 70 percent, would discharge about 33,000 short tons (3×10^7 kg) of sulfur oxides annually.

SUPERCHARGER

A device for increasing the mass of the air charge supplied to the cylinders of an **internal-combustion engine** by increasing the intake pressure (and density).

The amount of fuel that can be burned in the power stroke, and hence the power output, is limited by the mass of air present in the cylinders. If this mass is increased by supercharging, the fuel charge can be increased correspondingly, resulting in an increased power from an engine of a given size. Supercharging can be used with both gasoline (**spark-ignition**) and **diesel engines**. An important application of supercharging is to maintain the cylinder air pressure (and power output) in piston-type aircraft engines when flying at altitudes at which the outside air pressure is substantially less than at ground level.

Several different types of air blowers, pumps, and compressors have been designed for supercharging. They may be driven directly through a gearbox by the engine shaft or by an independent **gas turbine** operated by the hot exhaust gases. In the latter case, the supercharger is called a *turbocharger*. An advantage of turbocharging is that it utilizes much of the energy contained in the engine exhaust which would otherwise be wasted.

SUPERCONDUCTIVITY

The ability of certain metallic elements, alloys, or compounds at extremely low temperatures to carry an electric current with essentially no resistance. Once an electric current is established in a superconducting loop, it will continue almost indefinitely.

As the superconducting material is cooled, its resistance falls to zero at a cer-

tain *transition* (or *critical*) *temperature*. For nearly all superconducting metallic elements, the transition temperature is below 10 K, that is, within 18°F (10°C) of the absolute zero, the lowest conceivable temperature; some metallic alloys and compounds, however, have transition temperatures up to about 22 K. [A temperature of T kelvins (K) is equivalent to $-273.15 + T$ in °C or $-459.67 + 1.8T$ in °F.]

The recorded transition temperatures for superconductors are those measured at low (essentially zero) current and magnetic field. An increase in strength of the current or of the surface magnetic field results in a decrease in the transition temperature. Above a certain *critical current* or *critical magnetic field* strength, the normal conducting state is restored and persists no matter how low the temperature. Hence, superconductivity can occur only if the current and surface magnetic field are both less than the respective critical values.

Superconducting materials are classified as Type I and Type II. Those of Type I are mostly pure metals; they have relatively low critical currents and magnetic fields and are of little practical value. Type II superconductors, on the other hand, which are either compounds or solid-solution alloys of two or more metals, have certain characteristics that make them potentially useful materials. As a general rule, Type II superconductors retain their superconductivity at high magnetic fields and high currents. However, the critical temperature above which superconductivity ceases is decreased as the magnetic field and the current are increased.

Of the high-current Type II superconductors known in the early 1980s, two, which are available commercially, are of special interest. One is a solid-solution alloy of 30 to 40 atom percent of niobium (Nb) with titanium (Ti), represented by Nb-Ti; its transition temperature is about 10 K. The other is a compound of niobium and tin (Sn), formula Nb_3Sn, with a transition temperature of 18 K. Not only does the transition to the superconducting state occur at a higher temperature, which is advantageous from the cooling standpoint, but the critical current and magnetic field are higher for Nb_3Sn than for Nb–Ti. On the other hand, Nb_3Sn is a brittle material that has proved difficult to fabricate (see later).

Applications

Several energy-related applications of high-current Type II superconductors are being studied. In an electromagnet, a magnetic field is produced by passing a current through a conductor, usually a copper coil wound around an iron core. One factor determining the field strength is the current strength, and this is limited by heating of the coils. The heating rate is equal to I^2R, where I is the current strength and R is the resistance of the coil. In a superconducting coil, R is essentially zero and large currents can flow with little or no heating.

A thin wire (or film) of superconductor can thus carry more current than a much thicker normal conductor without overheating. Hence, if a sufficiently low temperature can be maintained, electromagnets with compact superconducting coils can be designed to produce strong magnetic fields. Such electromagnets have potential applications in **electric generators, magnetohydrodynamic conversion,** and **magnetic-confinement fusion**.

The amount of electric power that can be transmitted by an underground line of normal conductor is limited by resistance heating. The virtual absence of resistance heating in superconductors thus makes these materials of interest for underground **electric power transmission**. Another possible use of superconductors in the power industry is for electrical **energy storage**. The stored energy is equal to VI, where V is the voltage and I is the current strength. Since large values of I are possible, substantial amounts of energy can be stored.

Fabrication

Production of superconductors for practical use is complicated by the significant changes in their properties that accompany apparently minor variations in fabricating conditions. Nevertheless, both Nb–Ti and Nb_3Sn have been produced in useful forms as thin filaments or layers in a matrix of a normal conductor (e.g., copper). The latter serves as a heat sink to prevent undesirable local temperature increases that would cause the superconductor to revert to the normal state and to provide an alternative temporary path for the current in the event of a loss of superconductivity.

Niobium–titanium superconductor composites are fabricated by embedding thin rods of this material in copper and drawing out into wires containing fine superconducting filaments. This is possible because Nb–Ti is ductile. Several wires may be twisted or braided together to form a multifilament superconductor. Flat tapes or ribbons can be produced in a similar manner.

Since Nb_3Sn is brittle, direct wire drawing is impractical. The problem has been solved by producing filaments of niobium, which is a ductile metal, and then converting them into Nb_3Sn. For example, rods of niobium are embedded in a copper–tin alloy and drawn into wires; upon heating, tin diffuses into the niobium filaments to form thin surface layers of Nb_3Sn. Alternatively, filaments of niobium in copper are coated with tin and heated; the tin diffuses through the copper and converts the niobium into Nb_3Sn. The outer layer of tin, which may be undesirable, is then removed by acid. Superconducting tapes (or ribbons) required for certain applications are fabricated in a somewhat analogous manner by rolling sheets rather than drawing wires.

Low Temperatures

For cooling to the superconducting state, both Nb–Ti and Nb_3Sn require the use of liquid helium as refrigerant; at normal atmospheric pressure the temperature is 4.2 K. This temperature may be necessary for Nb–Ti when carrying a substantial current, but it is probably lower than required by Nb_3Sn. Since helium is an expensive and relatively rare material, a more readily available alternative refrigerant would be desirable. The only possibility is liquid hydrogen, but since its temperature is 20.3 K at atmospheric pressure, it cannot be used with either Nb–Ti or Nb_3Sn.

Consequently, a search is being made for superconductors with transition points above 20 K. The compound of niobium and germanium, Nb_3Ge, has a zero-current transition temperature of 22 to 23 K. However, this compound is not formed as readily as the analogous Nb_3Sn, and, in any event, a temperature below 20 K would be required to maintain superconductivity with large currents. This might be possible by using a refrigerant consisting of a slurry of solid and liquid hydrogen, which forms at 14 K.

SWEETENING PROCESSES

Methods for removing objectionable odors from "sour" petroleum products, especially gasoline and kerosine. These odors are largely due to volatile sulfur compounds, mainly hydrogen sulfide (H_2S) and mercaptans (or thiols). The latter have the general formula RSH, where R is a **hydrocarbon** radical.

If, in the course of **petroleum refining**, the product has been subjected to **hydrotreating**, most of the sulfur will have been removed and sweetening would not be required. However, if the low sulfur content or other circumstances make hydrotreating unnecessary, unpleasant odors may be removed (or greatly reduced) by oxidizing the mercaptans to the much less objectionable disulfides (RS·SR). Among the sweetening oxidants used are sodium plumbite, that is, lead oxide dissolved in sodium hydroxide solution ("doctor" treatment), cupric chloride, and air in the presence of a catalyst (see **Merox Process**).

SYNTHANE PROCESS

A Pittsburgh Energy Technology Center (U. S. Department of Energy) process for **coal gasification** in steam and oxygen gas (or air). When oxygen gas is used, the primary product is an **intermediate-Btu fuel gas**; the major purpose of the Synthane process is to produce a **high-Btu fuel gas** (or **substitute natural gas**). If air is used instead of oxygen gas, the product is a **low-Btu fuel gas**.

If the coal to be gasified is of a caking variety, it is crushed and heated to 800°F (430°C) with oxygen gas (or air) and steam in a pretreater. The pressure in the pretreater and in the subsequent gasifier stage is 40 to 70 atm (4 to 7 MPa). The coal undergoes partial devolatilization (i.e., removal of volatile matter) and partial oxidation, thereby destroying its caking properties. The resulting coal, together with released gases and excess steam, overflows to the **fluidized-bed** gasifier. Additional steam and oxygen gas are introduced on the next page (Fig. 172). Noncaking coals can be fed directly to the gasifier without pretreatment.

The fluidized coal mixes with the hot gases in the upper part of the gasifier and is further devolatilized. The carbon in the residue (**char**) then undergoes gasification reactions with steam and oxygen in the lower part of the gasifier at a temperature of 1600 to 1800°F (870 to 980°C). Reaction of carbon with steam produces mainly carbon monoxide and hydrogen; the reaction products with oxygen are carbon monoxide and some dioxide.

The heat released in the carbon–oxygen reactions provides that required for the carbon–steam reaction. The unreacted char, representing about 30 percent of the original coal, is removed from the bottom of the gasifier; it can be burned in the usual manner in a boiler to generate steam.

The product gas from the gasifier is scrubbed to remove tar and dust. The composition of the clean gas depends on the nature of the feed; typically, it contains roughly 32 volume percent of hydrogen, 15 percent of methane, and 13 percent of carbon monoxide (dry basis) as fuel components; most of the remainder is inert car-

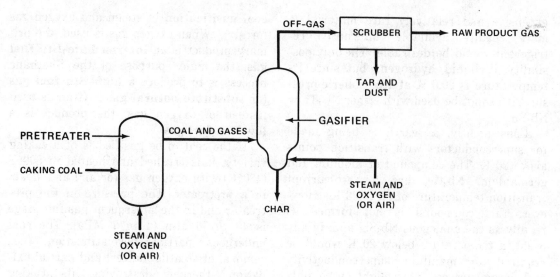

Fig. 172 Synthane process.

bon dioxide. The **heating value** is about 300 Btu/cu ft (11 MJ/cu m). This intermediate-Btu fuel gas can be converted into a high-Btu gas by the **water-gas shift reaction** followed by **methanation**.

If air is used instead of oxygen gas for combustion, the product gas contains 40 to 50 percent of inert nitrogen (from the air). The heating value is roughly 160 Btu/cu ft (6.0 MJ/cu m).

SYNTHESIS GAS

A gas consisting mostly of a mixture of carbon monoxide and hydrogen in various proportions, depending on the details of the production process. The name originates from use of the gas to synthesize various fuel products, including liquid **hydrocarbon** fuels (see **Fischer–Tropsch Process**), methanol and other **alcohol fuels**, and methane for **substitute natural gas**. Synthesis gas is also used as a source of hydrogen gas to serve as a fuel (see **Hydrogen Fuel**), for the synthesis of ammonia fertilizer, or in hydrogenation and **hydrotreating** processes (see **Coal Liquefaction; Petroleum Refining**).

Nearly all **intermediate-Btu fuel gases** are forms of synthesis gas. Hence,

methods for manufacturing those fuel gases are also applicable to synthesis gas. In essence, these methods involve reaction of the carbon in coal with steam; the heat required for this process is provided by the combustion of carbon in oxygen gas or, in a few cases, in air (see **Coal Gasification**). Other methods for making synthesis gas are catalytic **steam reforming** of light hydrocarbons and partial oxidation of a wide variety of hydrocarbons (see **Hydrogen Production**).

SYNTHETIC CRUDE OIL (SYNCRUDE)

A complex mixture of **hydrocarbons**, somewhat similar to natural crude oil (**petroleum**), obtained by **coal liquefaction**, from **synthesis gas** (a mixture of carbon monoxide and hydrogen), or from **oil shale** and **tar sands**. Syncrudes generally differ in composition from natural crude oil and may require pretreatment with hydrogen (see **Hydrotreating**) before they can be distilled and refined like natural crude (see **Petroleum Refining**). Syncrude from coal usually contains more aromatic hydrocarbons than natural crude oil.

SYNTHETIC FUEL (SYNFUEL)

A gaseous, liquid, or solid fuel that does not occur naturally. Synfuels can be made from coal (see **Coal Gasification; Coal Liquefaction**), petroleum products (see **High-Btu Fuel Gas**), **oil shale, tar sands**, or plant products (see **Biomass Fuels**). Among the synfuels are various fuel gases, including but not restricted to **substitute natural gas**, liquid fuels for engines (e.g., **gasoline, diesel fuel**, and **alcohol fuels**), and burner fuels [e.g., **fuel (heating) oils**].

SYNTHOIL PROCESS

A Pittsburgh Energy Technology Center (U. S. Department of Energy) process for **coal liquefaction** by direct hydrogenation with hydrogen gas. The process was designed primarily for the conversion of coal, caking or noncaking and regardless of its sulfur and ash content, into a low-sulfur, low-ash boiler fuel. It can, however, be adapted to the production of a **synthetic crude oil** (syncrude) that can be refined to yield various lighter distillate fuels (see **Petroleum Refining**).

The crushed coal, formed into a slurry with oil generated in the process, is pumped into the bottom of the hydrogenation reactor with a high-speed flow of a hydrogen-rich gas (Fig. 173). The reactor contains a catalyst consisting of pellets of cobalt molybdate (CoO–MoO_3) on a predominantly alumina support. The temperature is about 850°F (450°C), and the pressure is in the range from 140 to 270 atm (14 to 27 MPa). Turbulent flow of the gas propels the slurry upward through the catalyst and provides favorable conditions for hydrogenation (and liquefaction) of the coal.

The gases and hydrocarbon vapors leaving the reactor are cooled, and the condensed oil (with carried over solids) is separated from the gases. After the solids are removed, some of the oil is recycled to the coal slurrying operation, while the remainder is part of the low-sulfur fuel oil product. The solid, which contains unconsumed coal and **char**, may be pyrolyzed (see **Pyrolysis**) to yield more product oil; the remaining char and additional coal are

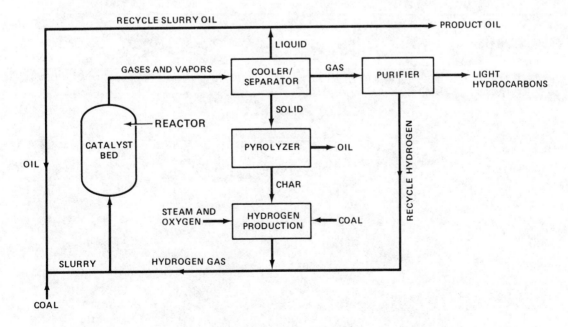

Fig. 173   Synthoil process.

used to generate the hydrogen-rich gas required for the process (see **Hydrogen Production**). The **heating value** of the oil is roughly 17,000 Btu/lb (39 MJ/kg); this approaches the value for a heavy petroleum **fuel (heating) oil**.

The off-gas from the cooler/separator consists mainly of light **hydrocarbons**, together with hydrogen sulfide, from most of the sulfur in the coal, and unreacted hydrogen. The hydrogen sulfide is removed in the gas purifier (see **Desulfurization of Fuel Gases**), and the hydrocarbon gases are liquefied by cooling; the residual hydrogen is recycled to the reactor. The hydrocarbon gases can be used as a fuel (see **Liquefied Petroleum Gas**) or subjected to **steam reforming** to generate hydrogen.

T

TAKAHAX PROCESS

A process for removing hydrogen sulfide from fuel gases by indirect oxidation to solid elemental sulfur (see **Desulfurization of Fuel Gases**). The oxidizing (absorbing) agent is the sodium salt of 1,4-naphthoquinone-2-sulfonic acid in a weakly alkaline aqueous solution of sodium carbonate. The impure fuel gas at about atmospheric pressure is brought into contact with the solution in an absorber column at ambient temperature. The hydrogen sulfide is removed by reaction with the sodium carbonate to form sodium sulfide (and bicarbonate). Clean gas is withdrawn from the top of the column. The sodium sulfide in the solution is oxidized by the naphthoquinone compound, with the precipitation of finely divided (solid) sulfur, while the naphthoquinone is reduced to the corresponding naphthohydroquinone. The spent absorber solution is transferred to a regeneration column where air is bubbled through it (see Fig. 26). The oxygen in the air converts the hydroquinone back to the original quinone, and oxidation of the sulfide is completed. The solid sulfur is removed by filtration, and the regenerated oxidizing and absorbing solution is recycled to the absorber column.

TAR SAND

Also called oil sand or bituminous sand; a sand or sandstone impregnated with an asphalt-like, highly viscous (i.e., very low mobility), complex **hydrocarbon** mixture (bitumen). The bitumen can be extracted from tar sand and treated to yield a synthetic crude oil (or syncrude) that can be refined, after preliminary treatment, like natural crude oil to yield a variety of fuel products (see **Petroleum Refining**). Tar sands containing more than 6 percent by weight of bitumen are thought to have commercial potential, but most of the materials exploited so far have contained 10 to 12 percent of bitumen. One ton of this tar sand should yield roughly 20 to 25 gallons (76 to 95 liters) of syncrude.

Very large deposits of tar sand are found in Canada, the U.S.S.R., and Venezuela. Deposits in the United States are much smaller, but they are nevertheless significant. Most of the U. S. tar sands are in Utah, with lesser amounts in California, Texas, and elsewhere. Early attempts to recover oil from tar sands in the United States were not very successful and were abandoned. However, the need to supplement petroleum supplies has resulted in a revived interest in tar sands as a source of syncrude.

For practical purposes tar sand beds may be considered in two categories: shallow deposits which are accessible by surface (or open-pit) mining and those at greater depths for which this type of mining is not economical. In both Canada and the United States, the deeper deposits constitute roughly 85 percent of the total. The process used for extracting the bitumen from tar sands depends upon whether open-pit mining is possible or not.

Shallow Deposits

In Alberta, Canada, where commercial production of syncrude from tar sands is under way, the open-pit-mined material is mixed with hot water containing a small amount of caustic soda (sodium hydroxide). The bitumen is released from the sand and is carried to the water surface as a froth with air. The residual sand falls to the bottom of the vessel and is removed. The bitumen is separated from the water and is subjected to **pyrolysis** (or destructive distillation). The vapors are condensed to produce a synthetic crude oil. Prior to refining, the oil is subjected to **hydrotreating** (i.e., action of hydrogen gas in the presence of a catalyst) to remove sulfur and upgrade the quality.

It is doubtful that the method just described for separating bitumen from tar sands can be applied to most U. S. deposits. The hot-water process depends on the fact that Alberta tar sands are "water wet," that is to say, each sand grain is surrounded by a thin film of water and the bitumen is not in actual contact with the grains. In Utah and elsewhere, the sands are usually "oil wet," with the bitumen (oil) in direct contact with the individual sand grains. Separation of the bitumen by hot water (and caustic) does then not appear to be practical.

A possible method for treating Utah tar sands is by extraction with a solvent. The solvent mixes with and thins the bitumen so that it can flow out of the sand. The solvent can be recovered for reuse by distillation, leaving the bitumen for further treatment. Another possibility is to pyrolyze the bitumen by heating the whole of the sand, just as is done in the surface retorting of **oil shale**. Condensation of the vapors would yield a syncrude.

Deep Deposits

In tar sand deposits that are too deep for surface (open-pit) mining to be practical, in-situ (i.e., in-place) extraction, without mining, appears to be necessary to recover the bitumen. As a general rule, the treatment involves heating the sand until the bitumen becomes mobile enough to flow; the oil can then be pumped to the surface. In this respect, in-situ recovery of oil from tar sands is similar to the recovery of **heavy (crude) oils**.

Several different in-situ tar sand processes have been (or are being) tested in the United States and Canada. The methods for heating the sand fall into three general categories; they are (1) steam, (2) combustion, and (3) others. It is probable that no single process will be found to be suitable for all deep tar sand deposits.

1. Steam heating may be continuous or discontinuous. In continuous steam heating (or steam flooding), superheated steam is injected into the tar sands from a control well; the heated (mobile) bitumen is collected from surrounding wells and is pumped to the surface. This scheme requires a permeable region between the injection and collection wells. Such a region may exist naturally or it may be created by hydraulic (high-pressure water) or explosive fracturing.

In the discontinuous (or cyclic) steam process, superheated steam is injected through a well into the tar sands for a period of 4 to 6 weeks or more. The steam flow is then stopped, and the molten bitumen is pumped from the injection well for several months. The injection and pumping phases can be repeated as long as it is economic to do so. If a breakthrough should occur from one well to another, the continuous steam flow process may be used between them.

2. For in-situ combustion heating of tar sand, the bitumen is ignited at the bottom of a well, and burning is sustained by introduction of air. Water or steam is injected at the burning front, and this forces the liquid bitumen to flow toward another (production) well. A permeable region between the wells may be created by fracturing or by reverse combustion. In the latter case, the process is similar to that described in the section on **Coal Gasification: Underground**. Reverse combustion for producing a permeable path is followed

by forward combustion to heat the bitumen and permit it to flow.

3. Other proposed methods of heating tar sand, which might be advantageous, are by electric current or microwaves. An entirely different in-situ approach is to introduce a solvent into the tar sand deposit to thin the bitumen and make it mobile. This is similar to a proposed method for tertiary recovery of petroleum (see **Petroleum Production**).

TEXACO COAL GASIFICATION PROCESS

A Texaco Development Corporation process for **coal gasification** with steam and oxygen. The product is an **intermediate-Btu fuel gas** which can be converted into a **high-Btu fuel gas** (or **substitute natural gas**) or used for hydrogen production. The process is applicable to almost any type of coal, caking or noncaking, and also to heavy **residual oils** (or "bottoms") from **petroleum refining** or **coal liquefaction**.

A thick slurry of pulverized coal in water is fed with pressurized oxygen gas into the top of a downflow, **entrained-bed** gasifier. Partial oxidation of the coal leads to temperatures of about 2450 to 2700°F (1340 to 1480°C) at which the carbon (in coal) reacts very rapidly with steam (from the water) to produce mainly hydrogen and carbon monoxide gases. Operating pressures have been from 20 to 80 atm (2 to 8 MPa), but higher pressures are being considered. At the high existing temperatures, the mineral matter in the coal melts to form a slag which, after quenching with water, is removed from the bottom of the gasifier. The water is recycled to the coal slurrying operation (Fig. 174).

In the *gas-cooler mode* of operation, the hot gas from the gasifier is used to generate steam in a waste-heat boiler. The cooler gas leaving the boiler is then treated for removal of sulfur (see **Desulfurization of Fuel Gases**) and particulate matter. The product is a clean fuel gas containing 45 to 50 volume percent of carbon monoxide and about 35 percent of hydrogen (dry basis) as the fuel components; the remainder is mostly inert carbon dioxide. The **heating value** is 250 to 270 Btu/cu ft (9.3 to 10 MJ/cu m).

In the alternative, *direct-quench mode*, the gas is cooled with water before it leaves the gasifier. The product is then saturated with steam and is suitable for the **water-gas shift reaction**. Interaction of carbon monoxide and steam results in the

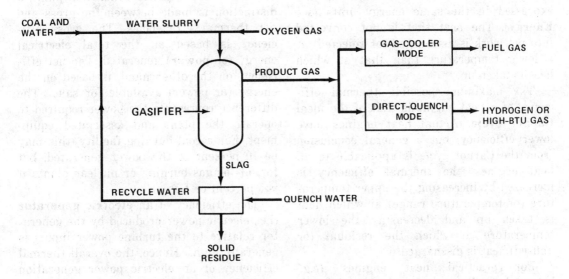

Fig. 174 Texaco coal gasification process.

formation of a mixture of hydrogen and carbon dioxide. Removal of the latter, by absorption in alkali or by cooling, leaves a gas consisting mainly of hydrogen. Several coal liquefaction processes require hydrogen gas which can be obtained in this manner, utilizing the high-boiling point liquid residues rather than coal as the feed material.

Although the two operational modes have been described separately, they can be conducted simultaneously, in variable proportions, with a single gasifier. The gaseous product, after being subjected to the water-gas shift reaction, would be suitable for **methanation**, thereby yielding a high-Btu fuel gas.

THERMAL EFFICIENCY

As applied to a **heat engine**, the proportion (fraction or percentage) of the heat taken up (or supplied) that is converted into useful (or net) work; that is,

Thermal efficiency

$$= \frac{\text{Heat converted into net work}}{\text{Heat taken up}}$$

An equivalent definition is the ratio of the net work done to the heat taken up, where the work done and heat taken up are expressed in the same energy units (see **Energy**). The heat that is not converted into net work is discharged (or rejected) at a lower temperature than that at which heat is taken up.

The maximum possible thermal efficiency of a heat engine is that of the ideal **Carnot cycle**. Actual heat engines have lower efficiency, but a general conclusion from the Carnot cycle is applicable to all heat engines: the thermal efficiency is increased by increasing the upper temperature (or temperature range) at which heat is taken up and decreasing the lower temperature at which the residual (or unused) heat is discharged.

For practical heat engines (e.g., **internal-combustion engines, gas tur-**bines**, and **steam turbines**), the definition of thermal efficiency is based on the heat produced by complete combustion of a fossil fuel (see **Heating Value**) or by nuclear fuel undergoing fission (see **Fission Energy**), rather than on the heat actually taken up by the engine. In a coal-burning, steam-turbine plant, there is a significant difference between these heat quantities because about 10 percent of the heat of combustion is discharged with the stack gases. In a nuclear steam plant, however, nearly all the fission heat is transferred to the steam.

In **electric power generation**, the useful work done (or heat converted into useful work) is taken to be the electrical energy generated in a certain time; the thermal efficiency is then defined by

Thermal efficiency

$$= \frac{\text{Electrical energy generated}}{\text{Heat produced by fuel consumed}}$$

where electrical energy and heat produced are expressed in the same energy units (e.g., kilowatt-hours, joules, etc.). The heat produced is based on the fuel, fossil or nuclear, consumed for the period over which the energy is generated.

In an electric power generation plant, a distinction is made between the gross and net thermal efficiencies. The gross efficiency is based on the total electrical energy (or **power**) generated. The net efficiency, on the other hand, is based on the energy (or power) available for sale. The difference represents the power required to operate the plant and associated equipment. For a coal-burning facility, this may be 10 percent of the power generated, but for oil- or gas-burning or nuclear plants it is 5 percent or less.

The efficiency of an **electric generator** (i.e., electric power produced by the generator relative to the turbine power input) is generally high. Hence, the overall thermal efficiency of an electric power generation plant is not very different from the efficiency of the turbine itself. The thermal efficiency of a **steam-electric plant** (i.e., a combination of steam turbine and electric generator) is thus determined by the temperature of the turbine steam intake (upper temperature) and of the condenser discharge (lower temperature).

See also **Heat Rate**.

THERMAL REACTOR

A nuclear fission reactor in which most of the fissions are caused by the absorption of thermal (i.e., relatively slow or low-energy) neutrons (see **Nuclear Power Reactor**). The neutrons released in the fission process are high-energy (i.e., fast) neutrons and in a thermal reactor they are slowed down by collision with the atomic nuclei of a moderator, that is, a material containing an element (or elements) of low mass number (see **Atom**). The most common moderators are ordinary water (H_2O), heavy water (or deuterium oxide, D_2O), and graphite, in which the elements of low mass number are ordinary hydrogen, deuterium (heavy hydrogen), and carbon, respectively.

The adjective "thermal" arises from the fact that, if the neutrons were allowed to come to equilibrium (i.e., reach a steady state) with their surroundings, the neutron energy distribution (or spectrum) would depend only on the temperature of the moderator. In fact, equilibrium is rarely attained, and the energy spectrum is not strictly thermal. Nevertheless, any reactor with a substantial proportion of moderator in the core, so that the neutron energy distribution approaches a thermal spectrum, is called a thermal reactor.

The majority of commercial power reactors are thermal reactors in which the moderator is ordinary (or light) water; the water also serves as the coolant for removing the heat generated by fission in the fuel. These are called **light-water reactors** (or LWRs); the two types of LWRs are **pressurized-water reactors** and **boiling-water reactors**. The fuel for these reactors is uranium dioxide enriched in the fissile species uranium-235 to an average of about 2.5 to 3 percent (see **Nuclear Fuel; Uranium Isotope Enrichment**).

In the **CANDU reactor**, used in Canada and some other countries, the moderator-coolant is heavy water. Fewer neutrons are captured by the deuterium in heavy water than by the hydrogen in light water; consequently, **heavy-water reactors** can operate with normal (i.e., unenriched) uranium dioxide as the fuel.

Graphite-moderated reactors, like the Magnox, carbon dioxide gas-cooled reactors in the United Kingdom, can use natural uranium metal as the fuel. In the Advanced Gas-Cooled Reactor, however, which operates at a higher temperature, the fuel is somewhat enriched uranium dioxide.

In all the reactors described above, the fuel consists of fissile uranium-235 and fertile uranium-238 (see **Fission Energy**). Thermal reactors of the **high-temperature gas-cooled reactor** (HTGR) type differ in the respect that the fertile species is thorium-232. The moderator is graphite, and the coolant is helium gas; this combination permits operation at higher temperatures than in other thermal reactors. In addition to high-temperature operation, an objective of the HTGR is to convert fertile thorium-232 into fissile uranium-233 which would eventually replace uranium-235 in the fuel.

Breeding of plutonium-239 from uranium-238 (i.e., the production of more fissile material than is consumed during reactor operation) is not practical in thermal reactors (see **Breeder Reactor**). Consequently, since nearly all existing thermal reactors utilize uranium-238 as the fertile material, they are not breeders but converters (i.e., they produce less fissile material than they consume).

On the other hand, breeding of uranium-233 from thorium-232 should be possible in a thermal neutron spectrum. A modified LWR, known as the **light-water breeder reactor**, is being tested as a possible uranium-233 breeder, and the CANDU (heavy water) and HTGR (graphite) reactors could probably also be designed as similar breeders. The graphite-moderated **molten-salt breeder reactor** is unusual in the respect that it utilizes a molten salt as fuel and coolant.

THERMIONIC CONVERSION

The **direct conversion** of heat energy into electrical energy (i.e., without a conventional **electric generator**) utilizing the *thermionic emission* effect, that is, the emission of electrons (i.e., negatively charged subatomic particles) from heated metal (and some oxide) surfaces. (The Edison Effect, discovered by Thomas A. Edison in 1883, namely, the ability of an electric current to pass between a hot and a cold electrode in an evacuated space, is now known to depend on thermionic emission by the hot electrode. The emitted electrons carry the current between the electrodes.)

The energy required to remove an electron completely from a metal is expressed by its *thermionic work function* (or simply, *work function*), which varies with the nature of the metal and its surface condition. In principle, a thermionic converter consists of two metals (or electrodes) with different work functions sealed into an evacuated vessel. The electrode with the larger work function is maintained at a higher temperature than the one with the smaller work function.

The hotter electrode (or emitter) emits electrons (i.e., negative charges) and so acquires a positive charge, whereas the colder electrode (or collector) collects electrons and becomes negatively charged. A voltage (or electromotive force) thus develops between the two electrodes and a direct electric current will flow in an external circuit (or load) connecting them (Fig. 175). The voltage, which may be 1 volt (or so), is determined primarily by the difference in the work functions of the electrode materials.

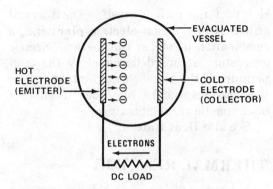

Fig. 175 Principle of thermionic conversion.

The emission of electrons from the hotter electrode is inhibited by a *space charge* resulting from the accumulation of electrons in its vicinity. To reduce the space charge, a small quantity of cesium metal is introduced into the evacuated vessel containing the electrodes. At the existing high temperature, the cesium vaporizes and ionizes; that is to say, it forms a mixture of positive ions (i.e., positively charged atomic particles) and electrons. By combining with the electrons in the vicinity of the emitter electrode, the positive ions help to decrease the space charge.

Deposition of solid cesium on the colder (collector) electrode is also beneficial in decreasing the work function and thus increasing the output voltage. A common thermionic electrode combination is a tungsten emitter and a cesium-coated tungsten collector.

A thermionic converter is a form of **heat engine**; in principle, heat is taken in at the upper (emitter) temperature, part is converted into electrical energy, and the remainder is discharged (or removed) at the lower (collector) temperature. As with other heat engines, the **thermal efficiency** (i.e., the fraction of the heat taken up that is converted into electrical energy) is favored by a high upper temperature and a low lower temperature. The thermionic converter will continue to generate electric power as long as heat is supplied to the emitter and a temperature difference is maintained between it and the collector.

To achieve a substantial electron emission rate (per unit area of emitter), and hence a significant current output as well as a high efficiency, the emitter temperature in a thermionic converter containing cesium should be at least 1830°F (1000°C). The efficiency is then about 10 percent. Higher efficiencies, possibly up to 40 percent, can be obtained by operating at still higher temperatures. Although temperature has little effect on the voltage generated, the increase in current (per unit emitter area) associated with a temperature increase results in an increase in power. (Electric **power** is the product of voltage and current.)

In principle, any heat source (e.g., a fossil or nuclear fuel, a radioactive material, or solar energy) can be used in a thermionic generator, provided a sufficiently high emitter temperature can be attained. Thus, thermionic conversion can be utilized in many different situations; several applications have been suggested for remote locations on the earth and in space.

During the 1960s, the U. S. National Aeronautics and Space Administration sponsored the development of thermionic generators using nuclear fission heat (see **Fission Energy**) to produce electricity for extended space missions. When interest in this project declined in the early 1970s, attention turned to the possibility of using **solar energy** and high-temperature **fossil fuel** combustion gases as heat sources.

In the latter case, thermionic conversion would be used in a **topping cycle** as part of a **combined-cycle generation** system. The hot combustion gases would supply heat to the emitter at about 3000°F (1650°C), and the heat discharged from the collector at around 1000°F (540°C) would produce superheated steam to operate a turbine-generator in the conventional manner (see **Electric Power Generation**). The overall efficiency for conversion of heat into electricity might then be about 50 percent. A major problem is to design a thermionic converter that would withstand the hot, corrosive combustion gases.

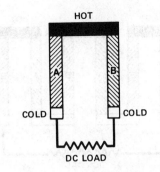

Fig. 176 Simple thermocouple arrangement.

THERMOELECTRIC CONVERSION

The **direct conversion** of heat energy into electrical energy (i.e., without a conventional **electric generator**) based on the Seebeck *thermoelectric effect*. Suppose two different electrical conductors (or **semiconductors**) are joined at the ends to form a loop; if one junction is kept hot and the other cold, an electric current will flow in the loop. Such a combination of two materials capable of producing electricity directly as a result of a temperature difference is called a *thermocouple*.

A simple arrangement for utilizing the Seebeck effect is shown in Fig. 176. The thermocouple materials A and B are joined at the hot end, but the other ends are kept cold; an electrical voltage (or electromotive force) is then generated between the cold ends. A direct current will flow in a circuit or load (e.g., a motor, resistance, etc.) connected between these ends. The current will continue to flow as long as heat is supplied at the hot junction and removed from the cold ends. For a given thermocouple, the voltage and electric power output are increased by increasing the temperature difference between hot and cold ends. In a practical thermoelectric converter, several couples are connected in series to increase both voltage and power (Fig. 177).

If the output voltage is not sufficient to operate a particular device or equipment, it can be increased, with little loss of power, by an inverter-transformer combination.

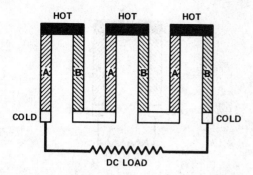

Fig. 177   Thermocouples in series to increase voltage.

The direct current generated by the thermocouples is first changed into alternating current of essentially the same average voltage by means of an inverter. The alternating current voltage is then increased to the desired value with a **transformer**. The high-voltage alternating current can be reconverted into direct current, if required, by a rectifier.

A thermoelectric converter is a form of **heat engine**. Heat is taken up at an upper temperature (i.e., the hot junction) and part is converted into electrical energy; the remainder is discharged (or removed) at a lower temperature (i.e., the cold ends). As with other heat engines, the **thermal efficiency** of a given thermocouple for conversion of heat into work (electrical energy) is increased by increasing the upper temperature and/or decreasing the lower temperature. The latter is usually that of the environment and is roughly constant; hence, the conversion efficiency of a given thermocouple is determined in practice mainly by the temperature of the hot junction.

Assuming given upper and lower temperatures, the thermoelectric conversion efficiency (or voltage) depends on the nature of the thermocouple materials. In the earliest devices, wires of different metals were used. As a result of a better understanding of thermoelectric phenomena, more efficient thermocouples consisting of an n (or negative) and a p (or positive) **semiconductor** have been developed.

Lead telluride, a compound of lead and tellurium, containing small amounts of either bismuth (n-type) or sodium (p-type), has been commonly used in recent times for thermoelectric converters. However, the proportion of heat supplied that is converted into electrical energy is only about 5 to 7 percent. Efforts are being made to produce more efficient thermocouple materials; semiconductors based on silicon-germanium and on compounds of selenium (i.e., selenides) appear to be promising for high-temperature applications.

Thermoelectric generators have been built with power outputs ranging from a few watts to a few kilowatts (see **Watt**). The source of heat is immaterial and any available fuel may be used. Consequently, thermoelectric generators have been employed in areas outside the regular electric power distribution system. An important application is the use of radioactive decay heat (see **Radioactivity**) to generate power in space and other remote locations (see **Radioisotope Thermoelectric Generator**). If the efficiency could be increased substantially, thermoelectric conversion might become practical for generating electricity using heat supplied by **solar energy**.

TIDAL POWER

Electric power generated by using the rise and fall of the ocean water level associated with tides. The energy contained in flowing water as it rises or falls can be converted into rotational energy in a **hydraulic turbine** and then into electrical energy in the conventional way by means of an attached generator (see **Electric Power Generation**).

Characteristics of Tides

Tides are produced mainly by the gravitational attraction of the moon and the sun on the solid earth and the oceans. About 70 percent of the tide-producing force is due to the moon and 30 percent to the sun. The moon is thus the major factor in tide formation.

Surface water is pulled away from the earth on the side facing the moon, and at the same time the solid earth is pulled away from the water on the opposite side. Thus, high tides occur in these two areas with low tides at intermediate points. As the earth rotates, the position of a given area relative to the moon changes, and so also do the tides. There are thus a periodic succession of high and low tides.

Although there are exceptions, two tidal cycles (i.e., two high tides and two low tides) occur during a lunar day of 24 hours 50 min. (The lunar day is the apparent time of revolution of the moon about the earth.) That is to say, the time between high tide and low tide at any given location is a little over 6 hours.

The difference between successive high and low water levels is called the *range* of the tide. Because of the changing positions of the moon and sun relative to the earth, the range varies continuously. There are, however, some characteristic features of this variation.

At times near full or new moon, when sun, moon, and earth are approximately in a line, the gravitational forces of sun and moon enhance each other. The tidal range is then exceptionally large; the high tides are higher, and the low tides are lower than average. These high tides are called *spring tides*. On the other hand, near the first and third quarters of the moon, when the sun and moon are at right angles with respect to the earth, *neap tides* occur. The tidal range is then exceptionally small; the high tides are lower and the low tides higher than average.

Superimposed on these short-term variations caused by the sun–moon system are smaller long-term variations arising from changes in the distance of the earth from the sun. Near the spring (March 21) and autumn (September 23) equinoxes, the earth is closer to the sun, whereas around the summer (June 22) and winter (December 22) solstices it is farthest from the sun. As a result, spring tides tend to be highest close to the equinoxes and lowest near the solstices.

The tidal range varies greatly with location. In the open ocean, it is commonly 2 to 3 ft (0.6 to 0.9 m), but interaction of the water with land at the shoreline can result in amplification of the range. In some places, such as shallow bays and estuaries, the amplification is considerable. As far as is known, the largest maximum tidal range, over 50 ft (15 m), occurs in the Bay of Fundy on the Atlantic Coast of Canada near Maine. Large maximum tidal ranges also occur in the Severn River estuary in England [45 ft (14 m)] and in the Rance River estuary in northwest France [40 ft (12 m)]. The only significant commercial tidal power installation in the world is at the mouth of the Rance River.

Site Requirements

The utilization of tidal energy requires construction of a barrier (or barriers) across a narrow inlet to an estuary or bay, thus forming an enclosure (or basin) in which ocean water can be impounded. Electricity can be generated by allowing water to flow through a turbine from the basin filled at high tide to the open ocean during falling tides and also as the basin is being filled from the ocean during rising tides (see below).

In each case, the maximum amount of electrical energy that can be generated depends on the product of the tidal range and the mass (or volume) of water flowing through the turbine. The volume is equal to the range multiplied by the area of the impounded water. Hence, the electrical energy is proportional to the square of the range and the area of the enclosed basin. A favorable site for a tidal power plant should then have a large tidal range, and the geographic features should permit enclosure of large areas with reasonably short dams or other barrages. Sluice gates in the dams permit water to pass to or from the enclosed basin (or basins).

The only place in the conterminous United States that appears to be suitable for tidal power development is the Passamaquoddy Bay site between Maine and the province of New Brunswick, Canada. The maximum tidal range is some 26 ft (8 m), and the initial electrical **capacity** of a tidal power plant is estimated to be about 800 megawatts (MW), where 1 MW = 1000 kW = 1 million watts (see **Watt**). This is close to the capacity of a single large coal-fired or nuclear power steam plant. However, because a tidal power plant can operate only a few hours daily, the annual output of electrical energy would, at best, be only about half that of a conventional steam plant of the same rated capacity. (The total electrical capacity of all potential sites in both the U. S. and Canadian portions of the Bay of Fundy, which includes Passamaquoddy Bay, is estimated to be almost 30,000 MW.)

The Passamaquoddy project, involving the construction of about 15 dams and four locks to permit passage of shipping, was first considered seriously in the 1920s, and some preliminary work was started in the 1930s. But the project was abandoned when further study showed it to be uneconomic. This conclusion was confirmed in subsequent investigations. However, because of changing economic conditions, there is a renewed interest in a joint U. S.-Canadian development of tidal power in the Bay of Fundy.

Utilization of Tidal Energy

The generation of electricity from water power requires that there be a difference in levels (or *head*) between which water flows (see **Hydroelectric Power**). A number of concepts have been proposed for generating electricity by utilizing the head that can be produced by the rise and fall of the tides to operate a hydraulic turbine. Of these, certain single-basin and two-basin schemes appear to be the most practical.

Single-Basin Concepts. The simplest way to generate tidal power is to use a single basin with a retaining dam in the following manner. Sluice gates in the dam are opened to permit seawater to enter the basin; at high tide the gates are closed. After a time, the outside water level falls with the outgoing tide and a head is established between the water in the basin and that in the sea. The basin water is then allowed to flow through turbines to generate electricity and is discharged to the sea. This continues until the head is too small to permit useful power generation. At low tide the sluice gates are opened and the basin is refilled by the rising tide, and the procedure is repeated. In this mode of operation, electricity would be generated only during periods of falling tide.

The operating time and electrical output of the turbine-generators can be extended by using special two-way hydraulic turbines that can operate with water flowing in either direction. Thus, electricity can be generated while the tide is falling and the water level is higher in the basin than in the sea and also while the tide is rising and the level is higher in the sea. There are, however, still inactive periods after high tide and low tide during which the water head is building up in the basin and the sea, respectively.

An improvement in the single-basin system is the pumping procedure used at the Rance River tidal-power project. Near high tide, as the basin is being filled, water is pumped from the open water into the basin to raise the level above where it would normally be at high tide. When the basin is being emptied, the additional water falls to the low tide level. Since the distance the water is raised by pumping is much less than the distance it falls while generating power, there is a net increase in the electrical output after allowing for the power used in pumping.

A drawback to one-basin schemes, even with two-way turbines and pumping, is that electric power is generated for no more than about 7 hours in two separate periods during a 12-hour 25-min tidal cycle. Furthermore, the periods of power generation coincide only occasionally with periods of peak demand. These problems are solved to some extent in the two-basin scheme described below. However, a fundamental drawback to all methods for generating tidal power is the variability in output caused by the variations in the tidal range.

Two-Basin Schemes. Two-basin schemes, such as that proposed for the Passamaquoddy project, require two separate but adjacent basins. In one basin, called the "high pool," the water level is maintained above that in the other, the "low pool." Because there is always a head between upper and lower pools, electricity can be generated continuously, although at a variable rate.

To start the two-basin operation, the low pool is allowed to empty at low tide and the sluice gates are closed. As the tide rises, the high pool is allowed to fill, and the gates are closed at high tide. Water from the high pool then flows through a turbine to the low pool, and electricity is generated. When the rising water in the low pool reaches the falling level of the open seawater, the gates are opened from this pool to the sea. The level in the low pool falls with the outgoing tide and continues to do so until the lowest level is reached, when the gates are again closed. Finally, when the rising tide brings the level of the sea up to that in the high pool, the gates of this pool are opened until the maximum water level is reached. This completes a single tidal cycle.

The operation of the two-basin scheme can be controlled so that there is a continuous water flow from the upper to the lower pool. However, since the water head between the pools varies during each tidal cycle, as well as from day to day, so also does the power generated. As is the case with single-basin schemes, the peak power generation does not often correspond in time with the peak demand. One way of improving the situation is to use off-peak power, from the tidal-power generators or from an alternative system, to pump water from the low pool to the high pool. An increased head would then be available for tidal power generation at times of peak demand (see **Pumped Storage**).

Turbines. Because tidal-power systems operate at a relatively low head, high rates of water flow are required to generate substantial amounts of power. Thus wide flow channels and large-diameter turbines are required. In the Rance River plant, for example, where the maximum operating head is roughly 27 ft (8.2 m), the internal diameter of the turbines is almost 18 ft (5.5 m). There are 24 such turbines, each with an electrical capacity of 10 megawatts. Tidal-power turbines are of the variable-pitch propeller type which are especially suited to low-head operation (see **Hydraulic Turbines**). The turbine shaft is horizontal (or almost horizontal) rather than vertical as in conventional propeller-type hydraulic turbines.

In the Rance River project, each turbine shaft is connected directly to a generator enclosed in a sealed, bulb-shaped housing. A combined turbine–generator unit is located in a horizontal water-flow channel. The variable-pitch turbine blades are designed to operate with water flowing in either direction. A turbine can also function in reverse as a pump, by supplying power to the generator which now acts as a motor (see **Electric Motor**).

The turbines originally proposed for the Passamaquoddy project were of the "tube" type located in a wide passage slightly inclined to the horizontal. With the sloping shaft, the generator could then be positioned above the water passage. Two-way operation was not required, but the design was such as to permit reverse operation of the turbine/generator as a pump/motor system.

TOPPING CYCLE

In general, the first (or top) **heat-engine** cycle of a **combined-cycle generation** system. In such a system, the heat discharged from the topping cycle is used to generate steam for operating a steam turbine. The temperature at which heat is taken up in a steam turbine is limited to about 1000°F (540°C) by the properties of materials, but much higher temperatures are attained by the combustion of **fossil fuels.** The purpose of a topping cycle is to utilize this high-temperature heat and thus increase the overall **thermal efficiency** (i.e., proportion of heat taken up that is converted into useful work or electrical energy). There is a corresponding decrease in the condenser cooling water requirement for a given power output (see **Electric Power Generation**).

The most highly developed topping cycle is the **gas turbine,** but the **diesel engine** could be used in smaller installations. The **potassium vapor cycle** and **thermionic conversion** are possibilities, but the major interest at present is in **magneto-hydrodynamic conversion,** largely because it has the potential for a high thermal efficiency using coal as the fuel.

TOTAL (OR INTEGRATED) ENERGY SYSTEM

A system designed to meet both the electrical demand (for lighting, refrigeration, domestic electrical appliances, elevators, etc.) and the thermal demand (for hot water and space heating and cooling) of a large building complex or community. A distinction is sometimes made between "total" energy systems which are independent of an external (utility) power source and the more flexible "integrated" systems which are connected with an electric utility so that power can be either bought from or sold to the utility according to circumstances.

However, the distinction between total and integrated systems is not always apparent. Thus, the Modular Integrated Utility System (MIUS) of the U. S. Department of Housing and Urban Development was intended to be self-contained, even to the extent of disposing of municipal wastes (see **Municipal Waste Fuels**); the

Integrated Community Energy System (ICES) concept of the U. S. Department of Energy, on the other hand, includes independent (or "stand-alone") as well as utility-connected systems. In the description here, the term Total Energy System (TES) is used in a general sense to include systems of both types.

A TES utilizes a **heat engine** as **prime mover;** it takes in heat at an upper temperature, converts part into useful mechanical work, and rejects (or discharges) the remainder at a lower temperature. The mechanical work drives a generator to produce electricity (see **Electric Generator**), and the thermal demand is met by the rejected heat (and sometimes other sources of waste heat.) The TES is thus a type of **cogeneration** system. The electric generation efficiency of a TES may be lower than that of a central station power plant, but utilization of the rejected heat that would otherwise be dissipated to the surroundings can result in an increase of some 25 percent in the overall fuel utilization efficiency.

The TES concept has been used, in one form or another, for many years. At one time, when electric power stations were located near cities, the warm condenser water was utilized for district space heating (see **Waste Heat Utilization**). The low cost of fuel oil and natural gas for heating purposes in the early years of the 20th century resulted in a decline in such systems. However, in the 1960s, largely under the stimulus of the natural-gas industry, the number of more localized Total Energy Systems increased. Some of these systems failed, but many were successful, and in the early 1980s, some 500 (or so) were operating in the United States. With the urgent need to conserve energy, interest in the TES concept is growing, especially for areas where fuel costs are high.

TES Types

The great majority of existing Total Energy Systems have electric power capacities of less than 10 MW (1 MW is 1

stream flow is increased. The potential (or pressure) energy of the fluid is thus partly converted into kinetic energy (or energy of motion). The fast-moving stream of fluid impinges on the curved rotor blades and causes the shaft to rotate.

Two principles—impulse and reaction—are involved in the interaction of the moving fluid stream with the rotor blades; turbines are thus categorized as impulse or reaction types. The difference in behavior arises primarily from the shapes of the passages between the rotor blades or, in other words, the shapes of these blades. In each case, the shape of the fixed blades is such as to form nozzles, as described above.

In *impulse turbines*, the spaces between the rotor blades are fairly uniform, as in Fig. 179. There is consequently almost no change in pressure as the fluid passes through these spaces. The rotor is then turned by the direct force (or impulse) of the fluid on the blades. The speed of the fluid decreases as its energy is transferred to the rotor.

In *reaction turbines*, the rotor blades are shaped to have convergent passages between them (Fig. 180). In other words,

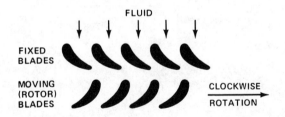

Fig. 180   Fixed and rotor blades in a reaction turbine.

the rotor blades act like moving nozzles. In addition to the partial conversion of pressure energy into energy of motion in passing between the fixed blades, additional conversion occurs in the passage of the fluid through the spaces between the rotor blades. The flow of accelerated fluid through the moving blade passages produces a reaction that causes the blades to travel in the opposite direction.

As a general rule, impulse turbines have the maximum efficiency for utilizing the fluid energy when the rotor blade rim speed is equal to roughly half the speed of the fluid impinging on the blades. With reaction turbines, the maximum occurs when the blade rim speed is the same as that of the impinging fluid. In practice, turbine blades usually exhibit both impulse and reaction characteristics with one or the other dominant. Large steam turbines with several stages (see below) may have predominantly impulse blades in the early stages when the fluid speed is highest and reaction blades in the later stages.

In order to make most efficient use of the energy of the working fluid, turbines, especially gas and steam turbines, have a series of successive stages; each stage consists of a set of fixed and rotor blades. The fluid pressure then decreases in passing from one stage to the next. The lengths of the blades are increased as the pressure of the fluid (gas or vapor) decreases and the volume increases. The overall diameter is also increased to maintain a constant torque (or rotational effect) as the force on the blades decreases (Fig. 181). A turbine is consequently commonly represented schematically as a truncated triangle, with the fluid entering at the narrow end and leaving at the wide end (see, for example, Fig. 158).

TVA AMMONIA DESULFURIZATION PROCESS

A wet, regenerative process of the Tennessee Valley Authority for **desulfurization of stack gases** by removal of **sulfur oxides.** The absorber is an

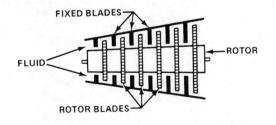

Fig. 181   Increasing turbine rotor diameter.

ammonia (NH_3) solution which is converted first into ammonium sulfite [$(NH_4)_2SO_3$] and then into the bisulfite (NH_4HSO_3) by the absorption of sulfur dioxide (SO_2) from stack (flue) gases in a scrubber. The spent solution, containing the sulfite and bisulfite, is acidified with ammonium bisulfate (NH_4HSO_4), which is effectively a mixture of ammonium sulfate [$(NH_4)_2SO_4$] and sulfuric acid (H_2SO_4). Sulfur dioxide gas is liberated, leaving ammonium sulfate. When ammonium sulfate is heated, ammonia for recycling as absorber is liberated and ammonium bisulfate for treating the spent absorbing solution remains. The sulfur dioxide gas can be recovered for use in industry, or it may be converted into elemental (solid) sulfur or sulfuric acid (see **Desulfurization: Waste Products**). A problem experienced with ammonia scrubbing is the escape of an ammonium salt particulate fume from the scrubber, especially on days of low temperature and high humidity.

U

U-GAS (UTILITY GAS) PROCESS

An Institute of Gas Technology Process for **coal gasification** by reaction with steam and oxygen gas (or air). When oxygen gas is used, the product is an **intermediate-Btu fuel gas;** with air, it is a **low-Btu fuel gas** because of the presence of inert nitrogen from the air. The intermediate-Btu gas can be converted into a **high-Btu fuel gas** (or **substitute natural gas**). If the coal has caking characteristics, it must be pretreated by heating in oxygen gas (or air) at about 800°F (430°C); the pressure is the same as in the remainder of the system.

The crushed coal, pretreated if necessary, is fed to the **fluidized-bed** gasifier. Oxygen gas (or air) and steam are introduced from below in two streams; one stream flows upward through a conical grid whereas the other enters directly from the bottom. The grid is an important feature of the gasifier. The conditions are such that the coal ash is agglomerated into larger and heavier particles that tend to fall out from the fluidized bed; the grid then facilitates their separation from the coal fines (small particles). The latter are returned to the fluidized bed by the steam-oxygen (or air) stream, whereas the ash is discharged (Fig. 182).

The temperature in the lower part of the gasifier is about 1900°F (1040°C), and the pressure may be up to 24 atm (2.4 MPa). Carbon in the coal reacts with both oxygen (alone or in air) and steam; the carbon-oxygen reaction provides the heat required for the carbon-steam reaction and for maintaining the gasifier temperature.

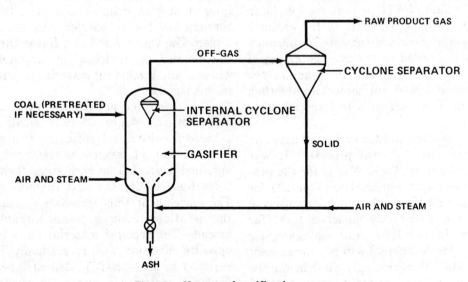

Fig. 182   U-gas coal gasification process.

Coal fines carried over with the product gas are removed in **cyclone separators,** one of which is within the gasifier. The solid matter from the external cyclone is returned to the gasifier in one of the steam and oxygen gas (or air) streams.

The gas leaving the cyclone is treated for hydrogen sulfide removal (see **Desulfurization of Fuel Gases**). The composition of the product gas depends on the nature of the feed coal and the gasification conditions. When oxygen gas is used in the process, the product contains, on the average, roughly 40 volume percent of carbon monoxide, 36 percent of hydrogen, and 6 percent of methane (dry basis) as fuel constituents; the remainder is mostly inert carbon dioxide. The **heating value** of the gas is roughly 300 Btu/cu ft (12 MJ/cu m). By subjecting the gas to the **water-gas shift reaction** followed by **methanation,** it could be converted into a high-Btu fuel gas.

If air is used instead of oxygen in the coal gasification process, about half the product is inert nitrogen gas. The heating value is then about 150 Btu/cu ft (6 MJ/cu m).

URANIUM ISOTOPE ENRICHMENT

The increase in the proportion of the lighter, fissile uranium-235 isotope from its normal value of 0.71 percent by weight in natural uranium. The fuel in the common **light-water reactors** consists of uranium (dioxide) enriched to the average extent of about 2.5 to 3 percent in uranium-235. Uranium of higher enrichment (more than 90 percent) is used in a few special reactors.

The main method for enriching uranium is the *gaseous diffusion* process. It was developed during World War II for the production of highly enriched uranium-235 for weapons, and the same principle is used to obtain the moderately enriched fuels for reactors. In the 1980s, the *gas centrifuge* method for uranium isotope enrichment will be utilized increasingly, first in Europe and later in the United States. A third

technique, that of *laser enrichment,* is promising but is still in the experimental stage.

Gaseous Diffusion

The molecules (or ultimate particles) of a gas (or vapor) are constantly in random motion with the average speed, at a given temperature, depending only on their mass. In a mixture of gases (or vapors), the average speed of the lighter molecules is greater than that of the heavier molecules. Consequently, if the mixture is in contact with a porous barrier, the lighter molecules will strike the barrier and pass through it more frequently than the heavier ones. This is the basis for the gaseous diffusion process of isotope enrichment.

The only compound of uranium suitable for gaseous diffusion is uranium hexafluoride. Although it is a solid at ordinary temperatures, it volatilizes readily above $133.5°F$ ($56.4°C$). The vapor is a mixture of uranium-235 hexafluoride ($^{235}UF_6$) and uranium-238 hexafluoride ($^{238}UF_6$) molecules; since the $^{235}UF_6$ molecules are the lighter, they will pass (or diffuse) more readily through a porous barrier.

The separation unit consists of a number of cylindrical (or tubular) barriers in a vessel called a *converter.* Uranium hexafluoride vapor feed flows through the interior of each cylinder and part diffuses through the porous barrier into the outer region. The vapor that has diffused through the barrier is enriched in uranium-235 whereas the remaining material is depleted in this isotope.

The extent of enrichment in a single converter (or stage) is quite small. To achieve a substantial enrichment, a series (or *cascade*) of stages is required; the enriched product from each stage then provides the feed for the next (higher) stage. The enrichment thus increases steadily as the hexafluoride vapor passes through the cascade. The depleted material flows in the opposite direction and is gradually impoverished in uranium-235. The stripped (or depleted) material is eventually discharged as *tails*.

A few stages of a diffusion cascade are outlined in Fig. 183. The stages above the point where the main feed enters (shown at right) are called the enriching section because the proportion of uranium-235 increasingly exceeds that in the normal uranium hexafluoride feed. In the stripping section, below the feed point, the proportion of uranium-235 becomes increasingly less than the normal value. As seen in the figure, the feed for each stage in the cascade is a mixture of the diffused (enriched) material from the stage immediately below and the undiffused (depleted) material from the stage immediately above. The cascade operates continuously with natural uranium hexafluoride feed supplied (at the right) and enriched product drawn off at one end (top) and depleted waste (or tails) at the other end (bottom).

The number of stages in a cascade depends on the isotopic composition of the feed material, generally normal uranium hexafluoride, the enriched product (e.g., 2.5 to 3 percent uranium-235), and the tails (commonly 0.2 percent). For these particular conditions more than 1500 stages are required. The number of stages would be smaller if the uranium-235 content of the tails were increased, but then a larger amount of this valuable isotope would go into the waste. In order to increase the output of a gaseous diffusion plant, each stage of the cascade consists of a number of converters in parallel, rather than the single unit shown in Fig. 183.

Gas Centrifuge

Theoretical studies of uranium isotope enrichment by the gas-centrifuge method were conducted in the United States during World War II, and some efforts were made to develop suitable centrifuges. However, the project was abandoned in favor of the gaseous diffusion process because the latter appeared to present fewer engineering problems. One of the considerations was that materials with adequate strength were not readily available for construction of

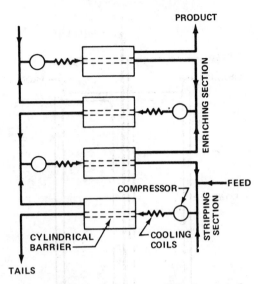

Fig. 183 **Portion of a gaseous-diffusion cascade.**

high-speed centrifuge machines. In later years, however, interest in the gas centrifuge was revived, especially in Europe, as a means for producing on a limited scale moderately enriched uranium for use as nuclear fuel. A major stimulus in the United States has been the expectation that gas centrifuges will require only about 5 percent of the operating energy of an equivalent gaseous diffusion cascade.

The basis of the gas-centrifuge technique for producing enriched uranium is that if a gas (or vapor) consisting of molecules with different masses is centrifuged (i.e., spun rapidly), the heavier molecules will move toward the outer circumference whereas the lighter ones tend to collect near the center. Consequently, when normal uranium hexafluoride vapor is centrifuged, material drawn off from the central region will be somewhat enriched in the lighter isotope, uranium-235.

The isotope separation effectiveness of the centrifuge can be enhanced by circulating the gas in the rapidly spinning centrifuge rotor, known as the bowl. In one type of gas centrifuge, the circulation is achieved by maintaining a lower temperature at the top than at the bottom of the bowl (Fig. 184). The uranium hexafluoride vapor feed is introduced at the axis of the

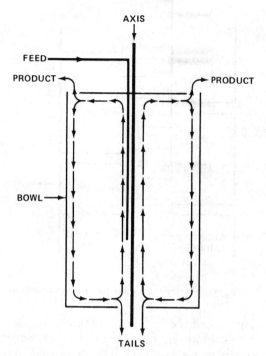

Fig. 184   Gas-centrifuge method for isotope enrichment.

bowl and it circulates, as shown by the arrows. Enriched product is drawn off at the top of the bowl and the depleted material (tails) from the bottom.

In order to obtain the desired extent of uranium-235 enrichment, the gas-centrifuge process is operated as a cascade similar to that used in the gaseous diffusion method. However, the degree of enrichment per stage is higher for the gas centrifuge; hence, a smaller number of stages (perhaps one-tenth) is required than in a gaseous diffusion cascade. On the other hand, the capacity of a centrifuge bowl is generally much less than that of a gaseous diffusion converter unit; the number of bowls per stage is consequently larger for the same output.

Laser Methods

Laser methods for isotope enrichment were apparently proposed in the U.S.S.R. in 1970 and first demonstrated in 1972 in the United States by the enrichment of deuterium, the heavier isotope of hydrogen. The use of lasers for uranium-235 enrich-

ment, which is still in the experimental stage, is of interest for several reasons. The energy utilization is expected to be less and capital and land requirements smaller than for other enrichment methods. Furthermore, the procedure should be economically applicable to the enrichment of the large quantities of gaseous diffusion plant tails now in storage.

Every substance has a characteristic spectrum consisting of a number (often a very large number) of lines representing the frequencies (or wavelengths) at which the substance absorbs light (ultraviolet, visible, or infrared). Each frequency corresponds to a specific amount of energy that is absorbed by the substance in a transition from one permitted energy state (or energy level) to a higher permitted energy state. The higher energy states are called *excited states.*

Several different kinds of energy states are possible but only two are of interest here; they are electronic states (in atoms and molecules) and vibrational states (in molecules only). The *electronic states* are the permitted excitation states of a single electron. There is a limit to the number of such states, and when the excitation energy exceeds the upper limit, the electron is removed completely. This process, referred to as *ionization,* results in the formation of a free negative electron and a positively charged atomic or molecular residue, called an *ion.* Ionization is more likely to occur in atoms; in molecules, breaking of a chemical linkage (dissociation) is often more probable (see below).

The *vibrational states* represent the permitted energies of vibration with respect to one another of two atoms of a molecule. In a molecule containing more than two atoms, there are several different possible vibrational modes, but the most important here are those in which the vibrating atoms are connected by a chemical bond. For example, in uranium hexafluoride (UF_6), each fluorine atom is bonded to the central uranium atom and the U–F vibrations are the significant ones. Although there are six such vibrational modes, the energy states are essentially the same in each mode; hence, only one need be considered.

An increase in vibrational energy from one state to a higher excited state represents a greater amplitude of U–F vibration. When a sufficiently high vibrational energy is attained, the amplitude becomes so large that the two atoms fly apart. This behavior, involving the complete breakage of a chemical bond, is called *dissociation*. With sufficient vibrational energy, for example, the breakage of a U–F bond results in dissociation of the UF_6 molecule into UF_5 and a fluorine (F) atom. The dissociation energy for a given bond is generally less in higher electronic states. In fact, in some excited electronic states dissociation occurs instantaneously.

Laser methods for isotope enrichment are based on the following facts. (1) Different isotopic species, either atomic or molecular, of a given element have small but finite differences in their spectral lines; in other words, equivalent transitions from one energy state to another (electronic or vibrational) require different amounts of excitation energy. (2) A laser is a high-intensity source of light that can be tuned so that emission occurs over a very narrow frequency range; that is to say, the laser light "particles" (or *photons*) have an almost exact single energy value. It is thus possible, in principle, to excite one isotopic species by means of a precisely tuned laser while leaving unaffected another isotopic species of the same substance.

Two general procedures are under study for enrichment in uranium-235 (and other isotopes). One involving the use of atoms is called the *photoionization method* (i.e., ionization by light). Uranium atoms present in the vapor produced by heating molten uranium metal to about 4170°F (2300°C) are irradiated with a laser beam tuned to a frequency (i.e., energy) that can cause electronic excitation in uranium-235 atoms but not in uranium-238. A second laser beam, which does not need to be selective, then provides enough energy to remove the excited electrons completely from the uranium-235 atoms (i.e., ionization occurs). Alternatively, the second laser may excite an electron to a high level and rely on an electric field to remove it from the atom.

Once ionization has occurred, the positively charged uranium-235 ions can be separated from the uranium-238 (and other atoms), which have not been ionized, by means of a magnetic field alone or in combination with an electric field. The ions are diverted by these fields toward a negatively charged collector plate where uranium-235 atoms are deposited as solid.

The second method of laser enrichment is based on *photodissociation* (i.e., dissociation by light) of molecules of uranium-235 hexafluoride. A tuned infrared laser beam causes vibrational excitation of these molecules in the vapor but not those of the uranium-238 compound. Sufficient vibrational energy could result in dissociation into uranium-235 pentafluoride ($^{235}UF_5$) and fluorine. Another suggestion is to use an ultraviolet (high-energy) laser to cause electronic excitation of the vibrationally excited uranium-235 hexafluoride; the electronically excited state may then dissociate immediately. Once it is formed, the $^{235}UF_5$ tends to separate as a solid, leaving the unaffected $^{238}UF_6$ molecules in the vapor.

A problem in the photodissociation procedure is that the initial uranium hexafluoride molecules (i.e., before laser irradiation) have a range of vibrational energies; a single laser frequency is then able to excite only a relatively few molecules, so that the efficiency is very small. It is necessary, therefore, to devise a means for getting a large proportion of the molecules in a single vibrational state prior to laser irradiation. One proposal is to accelerate the uranium hexafluoride vapor through a supersonic expansion nozzle. The temperature falls so that nearly all the molecules occupy the lowest vibrational state, but the vapor does not condense although it normally would when cooled.

See also **Separative Work Units.**

V

VACUUM CARBONATE PROCESS

A process for removing hydrogen sulfide from fuel gases by chemical absorption in a weakly alkaline aqueous solution of sodium carbonate (see **Desulfurization of Fuel Gases**). The impure feed gas entering at the bottom of the absorber column encounters a downflow of the carbonate solution at ambient temperature (see Fig. 26). The hydrogen sulfide, together with other **acid gases,** especially hydrogen cyanide and some carbon dioxide, is absorbed, and the clean gas is withdrawn from the top of the column. The spent absorber solution is transferred to the regenerator column, called the "actifier," where the acid gases are released by steam stripping in a partial vacuum. The vacuum is maintained by steam jets that also remove the hydrogen sulfide gas. The regenerated carbonate solution is recycled to the absorber column.

VAPOR (OR CONDENSING) TURBINE

A **turbine** operating on a **Rankine** (condensing) **cycle** in which part of the heat supplied by a vapor is converted into mechanical work of rotation (see **Heat Engine**). The **steam turbine** is by far the most common type of vapor or condensing turbine. However, the same principle may be employed in turbines operated at either lower or higher temperatures. Although these turbines are still of limited use, they have potential for future applications.

For operation at lower temperatures than in a steam turbine [e.g., below about 300°F (150°C)], the working fluid may be ammonia, propane, isobutane, or a Freon or similar (halogenated hydrocarbon) compound. Such low-temperature turbines have been developed for converting heat derived from solar energy, either directly (see **Solar Energy: Direct Thermal Applications**) or indirectly (see **Ocean Thermal Energy Conversion**) into mechanical work and then into electricity by means of a generator. Another possibility is to use the heat discharged in steam-turbine condenser water to operate a low-temperature turbine (see **Bottoming Cycle; Waste Heat Utilization**).

Low-temperature vapor turbines are simple closed-cycle systems, similar to the steam-turbine cycle in Fig. 158 (see also Fig. 97). Because of the low vapor intake temperature, compared to a steam turbine, the **thermal efficiency** (i.e., the fraction of the heat supplied that is converted into useful work) is inevitably low (e.g., 10 percent or less). If the working fluid is ammonia, propane, or isobutane, the turbine exhaust pressure is above atmospheric [e.g., about 6 atm (0.6 MPa)] to permit condensation of the vapor by the available water supply. Some Freons (e.g., R-12 and R-113), however, can be condensed to liquid at pressures below 1 atm (0.1 MPa) at normal temperature.

In low-temperature turbines the intake pressures are lower and the exhaust pressures are higher than in steam turbines. Furthermore, the heat available per unit mass of vapor is lower in the former case. As a result, low-temperature turbines would have considerably lower power ratings than steam turbines of the same size.

Potassium, which boils at 1400°F (760°C) at atmospheric pressure (and higher at higher pressures), has been proposed as the working substance for a high-temperature vapor turbine (see **Potassium Vapor Cycle**). The vapor inlet temperature in a potassium turbine would thus be well above the maximum practical temperature (1000°F or 540°C) of steam turbines. The potential use of a potassium vapor turbine is in a high-temperature cycle to precede a conventional steam-turbine cycle (see **Topping Cycle**).

A potassium turbine would operate as a simple, closed-cycle system in which the condenser, where the exhaust vapor is converted into a liquid, is actually a steam generator. The heat discharged in the turbine exhaust is utilized in a **heat exchanger** to produce superheated steam at about the same high temperature and pressure as a fossil-fuel boiler. The steam would then be used to drive a steam turbine in the usual manner, with steam reheating and regenerative feedwater heating.

VISCOSITY

A measure of the internal friction or flow resistance of a fluid (liquid or gas).

The *dynamic viscosity* is the force that must be applied per unit area to permit two layers, unit distance apart, to move with unit velocity relative to each other. The dynamic viscosity is expressed in terms of the poise (conventional metric units) or pascal seconds (SI system). The *kinematic viscosity* is the dynamic viscosity divided by the density (mass per unit volume); it is expressed in stokes (metric units) or square meters per second (SI system). The viscosity of a liquid usually decreases as the temperature is increased; with gases, the reverse is true.

Viscosity of liquids is commonly measured by determining the time required for a given volume to flow through a tube or orifice of known internal diameter at a specified temperature. The time is directly proportional to the kinematic viscosity. In the petroleum industry, kinematic viscosities are measured at a standard temperature with a standard instrument. In the United States, the Saybolt viscosimeter is used, and the results for lighter oils are expressed in seconds Saybolt Universal (SSU) at 122°F (50°C). For measurements with the more viscous oils, the instrument has a larger orifice; the flow time is then given in seconds Saybolt Furol (SSF) at 122°F (50°C).

W

WASTE HEAT UTILIZATION

As used here, refers to the beneficial use of the heat discharged in the **steam turbine** condenser cooling water from an **electric power generation** plant. From about 50 to 65 percent of the heat supplied either by combustion of a **fossil fuel** (coal, fuel oil, or natural gas) or by fission of a nuclear fuel (see **Nuclear Power Reactor**) in existing **steam-electric plants** is removed by the cooling water. This heat is dissipated to the environment by direct discharge of the warmed water to the ocean or a river or by passage through a **cooling lake** or **cooling tower.** Various possibilities are being considered for making use of the large amount of heat that is lost in this way.

Fossil fuels can be utilized more efficiently, with less heat discharged for a given total electric power output, if the steam turbine is preceded by another **heat engine** that operates at a higher temperature. The heat discharged from the high-temperature cycle would then be used to generate steam for a conventional turbine. Some possibilities under consideration are mentioned in the section on **topping cycles.** Another approach is to add a **bottoming cycle** to the steam turbine. The heat discharged from the latter would be used to operate a low-temperature turbine.

Instead of converting the waste heat into mechanical work in a heat engine, it may be more practical to utilize the heat directly. The more promising uses fall into three categories: (1) space heating and cooling, (2) agriculture, including raising livestock, and (3) aquaculture.

Space Heating and Cooling

Warm water can be distributed to homes, offices, stores, and industry, as it is in several European countries and to some extent in the United States, for space heating and domestic use. This is referred to as district heating. Cooling could also be achieved with hot water [near 212°F (100°C)] using absorption cooling (see **Refrigeration**). However, to be useful for space heating and cooling, the temperature of the water should be 120 to 212°F (49 to 100°C), whereas the average temperature of steam-turbine condenser discharge water is only about 105°F (40°C).

It is possible to operate a turbine at a higher discharge temperature, but this would result in a lower **thermal efficiency** (i.e., proportion of heat converted into mechanical work or electrical energy). In addition, the accompanying increase in the exhaust steam pressure would result in a smaller power output for a given turbine. An alternative approach, which might be preferred because it could be more readily adapted to a changing heat demand, would be to draw off steam at an intermediate stage of the turbine and use it to increase the temperature of the condenser discharge.

It is unlikely that the heated water from an electric power plant will provide district heating and cooling for an existing community because of the high cost of installing the required underground distribution system. But in developing a new residential and commercial complex, the hot-water pipes could be laid at the same time as the service-water lines. It is improbable that such a complex would be built close to a steam-electric plant (or vice versa), but it has been estimated that water at a temperature approaching 212°F (100°C) could be transmitted in wide pipes for distances up to 40 miles (65 km) without serious heat losses.

Warm Water in Agriculture

Water at a temperature of 75 to 115°F (24 to 46°C), which is within the range of condenser discharges, could be used to heat greenhouses in winter. Flowers, tomatoes, and cucumbers have been grown in this way throughout the year. Other agricultural uses of warm water that have been tested are for protecting fruit trees from frost by spraying, for irrigation, and for soil warming to extend the growing seasons of trees and vegetables.

Warm Water in Aquaculture

Fish farming is already being practiced to some extent in the United States, but it could be greatly extended by using warm water from power plant discharges. Catfish and tilapia, an excellent food fish from Africa, exhibit optimum growth, relative to nutrients supplied, at a temperature of about 90°F (32°C). If this temperature could be maintained throughout the year, the fish yield from a given pond could be increased. Lower temperatures are required for spawning and egg development; consequently, fingerlings would have to be raised in a hatchery and transferred to the warmer water at the appropriate stage of development.

An essential requirement for economic aquaculture is a supply of low-cost nutrients. It has been suggested that various small plant and animal organisms, upon which fish normally feed, could be produced in warm water ponds supplied with various animal and food processing wastes.

Several edible marine shellfish (e.g., shrimp, oyster, and lobster) also exhibit maximum growth in water at temperatures somewhat above normal. Experiments conducted in the United States, the United Kingdom, and Japan have shown that shellfish can be cultivated successfully under controlled conditions in seawater that has been warmed by condenser discharges.

See also **Combined-Cycle Generation; Cogeneration; Total Energy System.**

WATER GAS

An **intermediate-Btu fuel gas** produced by the interaction of **coke** (or coal) with steam at a temperature of about 1800°F (980°C). Water gas was formerly used for heating purposes, but it has been replaced by **natural gas.** The coke was first heated by combustion in a stream of air; the air "blow" stage was then terminated and superheated steam introduced to react with the hot carbon (in the coke) in the "run" stage. This reaction absorbs heat and the temperature falls. At a certain point the steam flow was stopped and replaced by air, and so on (see **Coal Gasification**).

Water gas, produced in the run stage, consisted of roughly equal proportions (by volume) of combustible carbon monoxide and hydrogen with about 5 percent of inert nitrogen (from the air) and 5 percent of carbon dioxide (from combustion) on a dry basis. The **heating value** was about 300 Btu/cu ft (11 MJ/cu m). It was sometimes called "blue water gas" or "blue gas," because it burns with a bluish flame. In "carburetted water gas," the blue gas was enriched with simple gaseous **hydrocarbons,** such as **methane** and propane, made by the high-temperature breakdown (**cracking**) of heavy oils. The heating value of the gas depended on the extent of enrichment, but it was commonly around 540 Btu/cu ft (20 MJ/cu m). Carburetted water gas was often used to supplement **coal gas** (town gas).

WATER-GAS SHIFT REACTION

A process for converting a mixture of carbon monoxide (CO) and steam (H_2O) into carbon dioxide (CO_2) and hydrogen (H_2). The conversion is accompanied by the liberation of heat; thus

$$CO + H_2O = CO_2 + H_2$$
$$+ \; 620 \; \text{Btu/lb CO (1440 kJ/kg)}$$

The water-gas shift is used to remove carbon monoxide from the **synthesis gas** formed in the manufacture of hydrogen from coal, natural gas, or petroleum products (see **Hydrogen Production**). It is also used to adjust the ratio of carbon monoxide to hydrogen (1 to 3) in the mixture for producing **methane** (see **Methanation**) or (1 to 2) for methanol production (see **Alcohol Fuels**). The carbon dioxide formed in the water-gas shift can be readily removed with an alkaline absorber.

A mixture of synthesis gas with steam at a temperature of about 750°F (400°C) and a pressure up to 27 atm (2.7 MPa) is passed over a cobalt molybdate (CoO–MoO_3) catalyst. This catalyst has the advantage of not being readily poisoned (i.e., inactivated) by sulfur compounds in the synthesis gas. Heat evolved in the water-gas shift raises the temperature of the gases leaving the reactor. The composition (i.e., ratio of carbon monoxide to hydrogen) of the product gas depends on the composition of the synthesis gas and the proportion of steam in the reactor feed mixture.

WATT

The international metric unit (SI system) of **power** (i.e., the rate at which work is done or energy transformed or transferred). The watt is defined as work done at a rate of 1 **joule** per second. It is equal to the power developed in an electric circuit in which a current of 1 ampere flows under an electromotive force (or voltage) of 1 volt. The power in watts in an electric circuit is thus equal to the product of the current in amperes and the voltage in volts; for this reason, the watt has long been used to express electric power (i.e., the rate of generating or utilizing electrical energy).

When large numbers of watts (W) are involved, the multiple units *kilowatt* (kW) and *megawatt* (MW) are used, where 1 kW = 1000 W and 1 MW = 1000 kW = 1 million W.

The following are the equivalent of 1 W in other units: 1.34×10^{-3} (i.e., $1/746$) horsepower; 9.48×10^{-4} Btu/sec or 3.413 Btu/hr (see **British thermal unit**); and 0.239 international calorie/sec.

Work done or energy transformed or transferred is determined by the product of the power and the time for which the power is operative. (In more precise terms, when the power varies with time, as it does with an alternating current, the energy is equal to the integral of the power over the operating time.) In electric power systems, the energy generated or used is commonly stated in *kilowatt-hour* (kW-hr) units; the electrical energy is then the product (or integral) of the power in kW and the operating time in hours. Since 1 W is equivalent to 3.413 Btu/hr, as given above, it follows that the energy of 1 kW-hr is equivalent to $3.413 \times 1000 = 3413$ Btu.

WELLMAN–GALUSHA PROCESS

A commercial process developed by the McDowell-Wellman Company in the United States for **coal gasification** with steam and air (usually) or oxygen gas. Several Wellman-Galusha gasifiers are operating in the United States and elsewhere, using steam and air to generate a **low-Btu fuel gas** for local use. Tests have been made with oxygen gas in place of air; the product is then an **intermediate-Btu fuel gas**. The latter can be converted into a **high-Btu fuel gas** (or **substitute natural gas**). The

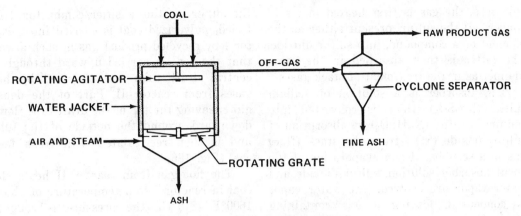

Fig. 185   Wellman–Galusha process.

Wellman-Galusha gasifier can be adapted for use with both caking and noncaking coals.

Crushed coal is fed continuously into the top of the **fixed-bed** gasifier and a mixture of steam and air is supplied through a rotating grate near the bottom (Fig. 185). The grate, which is mounted eccentrically, distributes the steam-air mixture through the coal bed and also forces the coal ash residue into an ash cone at the lower end of the gasifier. The gasifier is surrounded by a cooling-water jacket, and air passing over the heated water provides the steam required for the process.

Wellman-Galusha gasifiers are available with or without an agitator. The agitator, consisting of a pair of slowly rotating horizontal arms, increases the capacity of the gasifier by maintaining uniformity in the coal bed and reducing channeling. It also makes possible operation with caking coals.

As the feed coal descends through the coal bed, it interacts with steam in the upper (gasification) zone at about 1200°F (650°C). The remaining **char** enters the lower (combustion) zone, where the temperature may reach 1800°F (980°C) as a result of the reaction between carbon and oxygen (in the air). This provides the heat required for the carbon (coal)-steam gasification process. The pressure in the standard Wellman-Galusha gasifiers is close to

atmospheric, but tests at pressures up to 20 atm (2 MPa), to increase the capacity, have been successful.

The gas leaving the gasifier is passed through a **cyclone separator** to remove carried over coal particles. The product gas is then cooled and treated for hydrogen sulfide removal if necessary (see **Desulfurization of Fuel Gases**). With air as the combustion medium, the resulting gas has the following approximate average composition on a dry basis: carbon monoxide 22 volume percent, hydrogen 16 percent, and methane 3 percent as the fuel constituents; the remainder is mainly inert nitrogen (roughly 50 percent) from the air and carbon dioxide. The **heating value** of the gas is approximately 150 Btu/cu ft (5.6 MJ/cu m).

If oxygen is used instead of air, the product gas contains little or no nitrogen, and its heating value is roughly 300 Btu/cu ft (11 MJ/m^3). By subjecting this gas to the **water-gas shift reaction** and **methanation** the product would be a high-Btu fuel gas.

WELLMAN–LORD SO$_2$ RECOVERY PROCESS

A wet, regenerative process primarily for the **desulfurization of stack gases** by removal of **sulfur oxides**. The process is also used for removing sulfur and sulfur compounds from **Claus process** off-gas; in this case, the gas is first heated in air to convert all the sulfur, present either as the element or a compound, into sulfur dioxide. The latter is then absorbed in the same manner as in the treatment of stack gases.

The absorber is a solution of sodium sulfite (Na_2SO_3) which is converted into sodium bisulfite ($NaHSO_3$) by absorption of sulfur dioxide (SO_2) from the stack (flue) gas in a scrubber. Upon evaporation of the spent absorber solution, sulfur dioxide and water vapor are evolved. The water vapor is condensed, leaving a gas containing about 90 percent of sulfur dioxide. The solid remaining in the evaporator is regenerated sodium sulfite which is dissolved in the water condensate and recycled to the scrubber. Sodium sulfate (Na_2SO_4) and other sulfur salts gradually build up in the solution as a result of oxidation by air and absorption of sulfur trioxide (SO_3). Losses of absorber sustained in this manner must be made up, and the waste products must be disposed of safely. The recovered sulfur dioxide may be liquefied and used in industry or converted into elemental (solid) sulfur or sulfuric acid (see **Desulfurization: Waste Products**).

WESTINGHOUSE COAL GASIFICATION PROCESS

A Westinghouse Electric Corporation process for **coal gasification** with steam and air. The product is a **low-Btu fuel gas** to be used locally, especially for a **combined-cycle generation** electric power plant. Provided the coal feed is dried, caking coals, as well as noncaking coals, can be gasified without pretreatment. Although the Westinghouse process was designed for gasification with air (and steam), tests are being made with oxygen gas. The product would then be an **intermediate-Btu fuel gas** which could be converted into a **high-Btu fuel gas** (or **substitute natural gas**).

The process is conducted in two separate **fluidized-bed** reactors; one is called the devolatilizer/desulfurizer (I) and the other is the gasifier/combustor (II). Dried, pulverized coal is carried into reactor I by recycled product gas in such a way that the coal is carried upward through a central draft tube and fluidized by hot gases from reactor II. Part of the dense **char** leaving the top of the draft tube flows downward around the outside of the tube and is then recirculated with fresh feed coal (Fig. 186).

The hot gas from reactor II heats the coal in reactor I to a temperature of about 1600°F (870°C); the pressure is roughly 12 atm (1.2 MPa). The coal is devolatilized (i.e., volatile matter is driven off), and the sulfur present is converted into hydrogen sulfide. Dolomite (calcium and magnesium carbonates) or lime (calcium oxide), called the absorber, may be added to reactor I to remove the hydrogen sulfide. The spent absorber could be regenerated, if desired, and returned to the reactor (see **ATGAS/PATGAS Process**).

The accumulated char remaining after devolatilization of the coal in reactor I passes to reactor II where it is fluidized by air and steam. Combustion of part of the carbon in the char with oxygen (in the air) in the lower, combustor region produces hot gas (carbon monoxide and some dioxide) at a temperature of 1800 to 2000°F (980 to 1090°C); the pressure is around 12 atm (1.2 MPa). This gas provides the heat required for the carbon–steam reaction in the upper, gasifier region of reactor II. The resulting gas passes to reactor I, as stated above, and the residual agglomerated ash is removed from the bottom of II.

The raw product gas leaving reactor I includes the fuel gases formed from the coal by devolatilization (mostly hydrogen and methane), combustion (carbon monoxide), and gasification (carbon monoxide and hydrogen), in addition to inert nitrogen (from the air) and carbon dioxide. After removal of char particles in a **cyclone separator**, the resulting gas contains, on the average, 20 volume percent of carbon monoxide, 15 percent of hydrogen, and 3

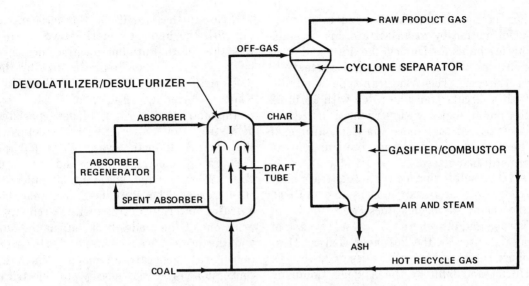

Fig. 186   Westinghouse coal gasification process.

percent of methane (dry basis); roughly 50 percent of the gas is nitrogen and 10 percent carbon dioxide. The **heating value** is about 150 Btu/cu ft (5.6 MJ/cu m).

WIND ENERGY CONVERSION

Conversion of the kinetic energy (i.e., energy of motion) of the wind into mechanical energy that can be utilized to perform useful work or to generate electricity. Winds arise primarily from temperature differences of the earth's surface resulting from unequal exposure to (or absorption of) solar radiation. Air heated by contact with a warmer surface tends to rise, and cooler air from a less heated surface flows in to take its place; the resulting airflow constitutes wind. Wind energy is thus a form of **solar energy**. Like direct solar energy, wind energy is highly variable, but in a different way.   For example, strong winds often occur at night when there is no solar radiation and in the winter when solar radiation is limited. To an extent, therefore, direct solar energy and wind energy are complementary, since one may be available when the other is not.

Most machines for converting wind energy into mechanical energy consist basically of a number of sails, vanes, or blades radiating from a hub or central axis. The axis may be horizontal, as in the more familiar windmills, or vertical, as it is in some cases.   When the wind blows against the vanes or sails they rotate about the axis and the rotational motion can be made to perform useful work.   Wind-energy conversion devices are commonly known as *wind turbines* because they convert the energy of the windstream into energy of rotation; the component which rotates is called the *rotor* (see **Turbine**). The terms turbine and rotor are, however, often regarded as being synonymous.

Because wind turbines produce rotational motion, wind energy is readily converted into electrical energy by connecting the turbine to an **electric generator** (see also **Electric Power Generation**).   The combination of wind turbine and generator is sometimes referred to as an *aerogenerator*. A step-up transmission is usually required to match the relatively slow speed of the wind rotor to the higher speed of an electric generator.

Historical Background

Wind turbines of various designs have been in use for about two thousand years, especially for milling grain and pumping water. (For pumping, the rotational motion of the turbine is changed to up-and-down motion by means of a crank or similar mechanism.) During the 18th and 19th centuries windmills were used extensively in Europe. The American-type windmill, with a circular fan-like rotor with 16 to 32 flat metal blades, was introduced in 1883. It is estimated that more than six million of these wind turbines were built and many are still being used.

The use of wind power to generate electricity was proposed in Denmark in 1890, and many aerogenerators were built in Europe and elsewhere. Of these, two are of special interest: the mill near Gedser, Denmark, and the machine designed by U. Hutter and built by the Allgeier Company in West Germany. Both systems were completed in 1957 and operated successfully for about 10 years. In each case, operation was terminated largely because of the low cost and ready availability of fossil fuels at the time. (The Gedser machine was reactivated in 1977.)

The Gedser wind turbine was of simple design. It had three paddle-shaped blades of uniform width, except for semicircular brake flaps at the ends to stop the rotation or slow it down at high wind speeds. The electric power output was 200 kilowatts (kW) (see **Watt**) at a wind speed of 33.5 miles/hr (15 m/sec). A somewhat similar machine was completed in 1977 on Cuttyhunk Island off the Massachusetts coast. A larger, three-bladed, turbine designed to generate up to 3000 kW of electric power has been built for a utility near Palm Springs, California.

The Hutter–Allgeier wind turbine was designed for a lower wind speed and lower power but with a higher overall efficiency than the Gedser mill. The rotor had two variable-pitch, aerodynamically efficient blades similar to aircraft propeller blades. The rated power output was 100 kW at a wind speed of 18 miles/hr (8 m/sec). The Hutter–Allgeier machine is of special interest because it had a strong influence on the design of large wind turbines in the United States in recent years (see later).

Prior to the late 1970s, the aerogenerator with the highest electric power output was the Smith–Putnam machine, designed by P. C. Putnam and built in 1941 by the S. Morgan Smith Company on Grandpa's Knob, Vermont. Each of the two rectangular-shaped rotor blades consisted of a steel frame covered with a steel skin; the overall diameter was 175 ft (53 m). The rated electric power output was 1250 kW at a wind speed of 30 miles/hr (13.4 m/sec). The Putnam–Smith machine operated intermittently until March 1945 when one of the blades broke off at the hub and the project was terminated. However, there was general agreement that the conversion of wind energy to electrical energy on a fairly large scale was technologically feasible although not economic at the time.

The changing economics of energy in the 1970s has led to a reconsideration of wind as a possible energy source. In 1972, a Solar Energy Panel, organized by the National Science Foundation (NSF) and the National Aeronautics and Space Administration (NASA) in the United States, suggested that wind energy merited further study.

The national Wind Energy Conversion (WEC) program, initiated by the NSF and NASA in 1973, is now under the direction of the U. S. Department of Energy in cooperation with NASA. In brief, the purposes of this program are to determine the potential for using the wind as an energy source, to identify the more promising applications of wind turbines in various power ranges, and to encourage the development of techniques which could make wind-generated electric power economically competitive with power from more conventional energy sources. Some aspects of the WEC program are described below.

Wind Energy Utilization

Site Selection. The power available in the wind increases rapidly with the wind speed; hence, WEC machines should be located preferably in areas where the winds are strong and persistent. Although daily winds at a given site may be highly variable, the monthly and especially annual average speeds are remarkably constant from year to year. The major contribution to the wind power available at a given site is actually made by winds with speeds above the average. Nevertheless, the most suitable sites for wind turbines would be found in areas where the annual average wind speeds are known to be moderately high or high.

In addition to the requirement of a reasonably high wind speed (e.g., more than 10 miles/hr or 4.5 m/sec), a suitable site for an aerogenerator should have some of the following features:

1. No tall obstructions for some distance in the upwind direction (i.e., the direction from which the wind is blowing).
2. Open plain or open shoreline.
3. Top of a smooth, well-rounded hill with gentle slopes on a flat plain or on an island in a lake or the sea.
4. A narrow, mountain gap through which wind is channeled. (In some cases the situation can be improved by relatively minor changes in the terrain.)

Potential Applications. Wind-turbine generators have been built in a wide range of power outputs from a kilowatt or so to a few thousand kilowatts. Machines of low power can generate sufficient electricity for space heating and cooling of homes and for operating domestic appliances. Low-power WEC generators have been used for many years for the corrosion protection of buried metal pipelines. Applications for somewhat more powerful turbines, up to about 50 kW, are for operating irrigation pumps, navigational signals (e.g., lighthouses and buoys), and remote communication, relay, and weather stations, and for offshore oil-drilling platforms.

Aerogenerators in the intermediate power range, roughly 100 to 250 kW, can supply electricity to isolated populations (e.g., on islands), to farm cooperatives, and to small industries. Finally, the largest WEC generators, with rated powers of a few thousand kilowatts, are usually planned for interconnection with an electric utility system. Present indications are that the optimum economic diameter of a wind turbine with a two-bladed, propeller type rotor is about 350 ft (107 m); the electric power output would range from 2000 to 5000 kW or 2 to 5 megawatts (MW), where 1 MW is 1000 kW or 1 million watts.

Energy Storage. Operation of a wind turbine is not practical at very high or very low wind speeds. Consequently, if other sources, such as electric utility power, are not available, some form of **energy storage** is required. When the power generated exceeds the demand, the excess energy would be stored for use at other times. For WEC machines of low and intermediate electric power, battery storage is convenient (see **Storage Battery**).

For wind turbines with power outputs up to about 20 kW, direct-current generators can be used to charge batteries directly. For higher powers, alternating-current generators are required and the current must be rectified for battery charging. Direct current from the batteries can be utilized to heat water for space heating and for domestic hot water, and to operate lights and small tools and appliances. Conversion into alternating current, by means of an inverter, may be necessary for larger tools and appliances and for television sets.

Other kinds of storage may be more desirable in agricultural operations. For example, if the wind energy is to be used for heating greenhouses or drying crops, it can be stored as hot water. Either direct or alternating current may then be used in resistance heaters without the need for batteries. Alternatively, the mechanical motion produced by the wind turbine can be converted directly into heat by frictional effects, such as by churning water.

For high-power aerogenerators that are integrated with an electric utility, a favorable situation would be operation of several wind turbines in connection with a **hydroelectric power** plant. If the total power, wind and hydroelectric, being generated should exceed the demand, the hydroelectric plant can be partly shut down; alternatively, the excess power could be used to pump water from an auxiliary reservoir at the bottom of the dam back into the main reservoir. In this way, the overall capacity of the hydroelectric system would be increased.

WEC Demonstration Program. In addition to encouraging the commercial production of WEC generators with low, moderate, and high power outputs, the Department of Energy (with NASA) is sponsoring a program for the development of large aerogenerators. A number of such generators at a suitable location would constitute a wind-power "farm," which would be connected into an electric utility.

The first stage in the development program was the 100-kW experimental WEC generator designed and built by NASA at the Lewis Research Center near Plum Brook, Ohio. The turbine, which started operation in September 1975, resembles the Hutter–Allgeier machine mentioned earlier. It has two aerodynamically efficient, propeller-type blades made of aluminum (frame and skin). The pitch angle (i.e., the angle between the blade and the plane of rotation) can be changed automatically to maintain an essentially constant rotation rate regardless of the wind speed. The overall diameter of the area swept by the blades, called the rotor diameter, is 125 ft (37.5 m).

Aerogenerators like the Plum Brook machine were subsequently built at Clayton, New Mexico; Block Island, Rhode Island; Culebra Island, Puerto Rico, and Oahu, Hawaii. The rated output in each case is roughly twice that at Plum Brook (i.e., about 200 kW) because of the higher average wind speeds.

WIND

Fig. 187   Outline of large aerogenerator.

The next stage in the large wind turbine demonstration program represented a substantial increase in electric power output to 2 MW. The machine of the Plum Brook type, with a rotor about 200 ft (61 m) in diameter, was completed near Boone, North Carolina, in 1979. However, it suffered some blade damage in 1981 when it shut down automatically in a high wind.

The general outline of the generators referred to above is shown in Fig. 187. The blades are attached to a hub which is connected to a *nacelle* mounted on top of a tower. (The term nacelle derives from the name for the housing containing the engines of an aircraft.) The height of the tower is such as to leave a clearance of about 50 ft (16 m) between the ground and the rotating blade tip. The nacelle contains the generator and a gearbox, which increases the slow rotation rate of the blades (e.g., 40 rpm) to the high rate (e.g., 1800 rpm) of the generator. The gearbox output also operates the blade-pitch control mechanism to maintain an essentially constant rotation of the generator regardless of changes in the wind speed. A yaw control mounted below the nacelle causes slow rotation of the nacelle in order to keep the blades facing into the wind when it changes direction.

The largest turbines in the WEC demonstration program in the early 1980s are the three machines, each with a rated electric power output of 2.5 MW, near Goldendale, Washington. The first started operating at the end of 1980 and the others in 1981. The generators are connected into the Bonneville Power Administration system and are the first in the United States to supply commercial electric power from the wind.

A special feature of these large turbines is the rotor design. Instead of two separate blades, the rotor is a single piece 300 ft (91.5 m) long with a hub at its midpoint. The pitch of the central section of the blade is fixed, but 45-ft (13.7-m) long tips at each end are adjusted automatically during operation to maintain an essentially constant rate of rotation (17.5 rpm) as the wind speed changes. The 200-ft (61-m) high tower is also unusual in being cylindrical rather than having a lattice structure.

The next stage in the WEC program for high-power aerogenerators is to be a 4-MW system for locations where the average annual wind speed is 14 miles/hr (9.7 m/sec).

Wind-Energy Turbine Principles

Wind Speed and Rotor Diameter. The **power** (i.e., energy per unit time) available per unit cross-sectional area in a freely flowing wind is equal to half the density (*d*) multiplied by the cube of the wind speed (*V*); that is,

$$\text{Wind power per unit area} = \frac{1}{2} V^3 d$$

(If d is expressed in kilograms per cubic meter and V in meters per second, the wind power is obtained in **watts** per square meter.) As a result of its proportionality to the cube of the wind speed, the available power increases rapidly with the wind speed. For example, at a free-flow wind speed of 10 miles/hr (4.5 m/sec), the available power is 5.4 watts/sq ft (59 watts/sq m), assuming a near sea-level air density of 1.2 kg/cu m. However, if the wind speed is doubled to 20 miles/hr (9.0 m/sec), the power would be eight times as great, that is, 43 watts/sq ft (470 watts/sq m). The air density decreases with increasing altitude, and the wind power, for a given wind speed, also decreases somewhat; thus, at an altitude of 5000 ft (1520 m), the power is about 83 percent of the sea-level value.

The wind power actually available to a turbine is equal to the power per unit area multiplied by the area swept out by the rotor. If the diameter of the rotor is D meters, the swept area is $\frac{1}{4} \pi D^2$ sq m, and the total available power in watts is $(\frac{1}{2} V^3 d)(\frac{1}{4} \pi D^2)$; that is,

$$\text{Available wind power} = \frac{1}{8} \pi D^2 V^3 d \text{ watts}$$

Consequently, for a given wind speed, the available power is proportional to the square of the rotor diameter. Thus, doubling the diameter of the rotor will result in a four-fold increase in the available wind power.

The combined effects of wind speed and rotor diameter variations are shown in Fig. 188. Wind machines intended for generating substantial amounts of power should have large rotors and be located in areas of high wind speeds. Where low or moderate powers are adequate, these requirements can be relaxed.

Rotor Performance: The Power Coefficient. The physical conditions in a wind turbine are such that only a fraction of the

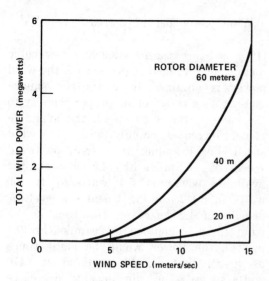

Fig. 188 Dependence of wind-rotor power on wind speed and rotor diameter.

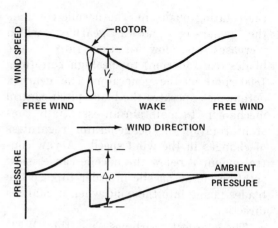

Fig. 189 Conditions in traversing a wind rotor.

available wind power can be converted into useful power. As the free windstream encounters and passes through a rotor, the wind transfers some of its energy to the rotor and its speed decreases to a minimum in the rotor wake. Subsequently, the windstream regains energy from the surrounding air and at a sufficient distance from the rotor the free-wind speed is restored (Fig. 189, upper curve).

While the wind speed is decreasing, as just described, the air pressure in the windstream changes in a different manner (Fig. 189, lower curve). It first increases as the wind approaches the rotor and then drops sharply by an amount Δp as it passes through and energy is transferred to the rotor. Finally, the pressure increases to the ambient atmospheric pressure.

The power extracted by the rotor is equal to the product of the wind speed as it passes through the rotor (i.e., V_r in Fig. 189) and the pressure drop Δp. In order to maximize the rotor power it would therefore be desirable to have both wind speed and pressure drop as large as possible. However, as V is increased, for a given value of the free-wind speed (and air density), Δp increases at first, passes through a maximum, and then decreases. Hence, for the specified free-wind speed, there is a maximum value of the rotor power.

The fraction of the free-flow wind power that can be extracted by a rotor is called the *power coefficient*; thus,

Power coefficient

$$= \frac{\text{Power of wind rotor}}{\text{Power available in the wind}}$$

where the power available is calculated from the air density, rotor diameter, and free-wind speed as shown above. The maximum theoretical power coefficient is equal to $^{16}/_{27}$ or 0.593. This value cannot be exceeded by a rotor in a free-flow windstream. (It can be exceeded under special conditions, as will be seen later.)

An ideal rotor, with propeller-type blades of proper aerodynamic design, would have a power coefficient approaching 0.59. But such a rotor would not be strong enough to withstand the stresses to which it is subjected when rotating at a high rate in a high-speed windstream. For the best practical rotors, the power coefficient is about 0.4 to 0.45, so that the rotors cannot use more than 40 to 45 percent of the available wind power. In the conversion into electric power, some of the rotor energy is lost and the overall electric power coefficient of an aerogenerator (i.e., electric power generated/available wind power) in practice is about 0.35 (35 percent).

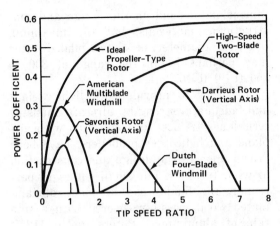

Fig. 190 Dependence of power coefficient on tip speed ratio.

Rotor Tip Speed Ratio. An important design consideration, which is related to the attainable power coefficient, is the ratio of the rotor blade tip speed to the wind speed; in brief, this quantity is called the *tip speed ratio.* (If f is the rotation frequency, i.e., rotations per second, of a rotor of diameter D meters, the tip speed is $\pi f D$ m/sec; if V m/sec is the wind speed, the tip speed ratio is $\pi f D / V$.) For an ideal propeller-type wind rotor, the power coefficient is expected to increase with the tip speed ratio toward a limiting value of 0.593, as shown by the uppermost curve in Fig. 190. For an actual rotor, however, the situation is somewhat different. At the higher wind speeds (and higher rates of rotation) the frictional drag of the air tends to reduce the power coefficient. Hence, as the tip speed ratio increases, the power coefficient passes through a maximum and then decreases. This general behavior is observed for both horizontal-axis and vertical-axis machines.

The dependence of the power coefficient on the tip speed ratio for some common rotor types is indicated in Fig. 190. (The Savonius and Darrieus rotors are vertical-axis machines described later.) It is seen that the two-bladed propeller type of rotor can attain a much higher power coefficient (i.e., it is more efficient) than the American multiblade windmill and the classical

Dutch four-bladed windmill. In practice, two-bladed propeller (horizontal-axis) rotors are found to attain a maximum power coefficient of 0.4 to 0.45 at a tip speed ratio in the range of roughly 6 to 10.

Horizontal-Axis Machines

The common wind turbine with a horizontal (or almost horizontal) axis is simple in principle, but the design of a complete system, especially a large one that will produce electric power economically, is complex. Not only must the individual components, such as the rotor, transmission, generator, and tower, be as efficient as possible, but these components must function effectively in combination. Some of the main design considerations are outlined below.

Number of Blades. Wind turbines have been built with up to six propeller-type blades, but two- and three-bladed propellers are most common. A one-bladed rotor, with a balancing counterweight, has some advantages, including lower weight and cost and simpler controls, over the multiblade type. However, starting requires high wind speeds, and vibration and other forces can be large.

Turbines with three blades have been used in several WEC machines in order to avoid the vibrations experienced with two-bladed rotors. These vibrations are related to the turning (or yawing) of the rotor in order to face it into the wind. It appears, however, that the problem can be overcome by controlling the yaw rate. Because they cost less to fabricate in large sizes and are capable of operating with a high tip speed ratio, two-bladed systems are receiving major attention.

Blade Design. Wind-turbine blades have an airfoil-type cross section (Fig. 191) and a variable pitch. They are slightly twisted from the outer tip to the root (i.e., where the blade is attached to the hub) to reduce the tendency for the rotor to stall. In a few devices, such as the Gedser and

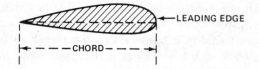

Fig. 191 **Cross section of airfoil-type wind-rotor blade.**

Smith-Putnam machines, the blades had a constant chord length (i.e., constant distance from one edge to the other). As a general rule, however, better performance is obtained with blades that are narrower at the tip than at the root (see Fig. 187).

In the larger two-bladed wind turbines, the blades are inclined at a small angle, called the *coning angle*, to the vertical. This design decreases the bending load on the roots of the blades and lessens the danger of fouling the supporting tower under severe wind conditions. In the very large rotors consisting of a single piece mounted at the midpoint, there is no coning angle.

A fundamental problem in wind-turbine design is to allow for the many forces to which the blades are subjected during normal operation. In addition to vibrational stresses resulting from rotation, there are several extraneous forces. These arise from wind turbulence, directional changes in the wind, variations of wind speed with height, wind gusts, gravitational forces, the presence of the tower (see below), etc. The blades must be constructed and attached to the hub in such a manner as to withstand these forces. Consequently, aerodynamic performance is sacrificed to some extent in the design of a rotor with adequate strength.

Although the wind power of a rotor increases with the square of the swept diameter, there is a practical limit to the size. In the first place, the mass of a blade increases rapidly with its dimensions. As a result, the wind power available per unit mass decreases as the dimensions of the rotor are increased; the tip speed ratio tends to decrease correspondingly. Furthermore, the wind drag forces generally increase with the blade size. The limiting

dimensions depend on the design and constructional materials, but the maximum practical diameter of a two-bladed rotor may perhaps be in the range of 300 to 350 ft (91 to 107 m).

In order to provide both light weight and adequate strength, aircraft industry techniques, as used in connection with airplane and helicopter propellers, have been drawn upon in the design and construction of wind turbines. For small rotors in particular, the blades can be made of laminated wood, possibly covered with a thin skin of aluminum. Rotors up to 112 ft (34 m) in diameter have been fabricated from plastic reinforced with glass fiber (e.g., the Hutter-Allgeier machine). More conventional aircraft-type blades are usually made from aluminum (or an aluminum alloy); they consist of a long central span with ribs at intervals, covered with an aluminum skin. (The original aluminum blades of the 200-kW experimental WEC generator at Block Island, Rhode Island, were replaced by wooden blades.) The very largest rotor blades have been made of steel to provide adequate strength.

Yaw Control. For localities with the prevailing wind in one direction, the design of a turbine can be greatly simplified. The rotor can be in a fixed orientation with the swept area perpendicular to the predominant wind direction. Such a machine is said to be *yaw-fixed.* Most wind turbines, however, are *yaw-active*; that is to say, as the wind direction changes, a motor rotates the turbine slowly about the vertical (or yaw) axis so as to face the blades into the wind. The area of the windstream swept by the wind is then a maximum.

In the smaller turbines, yaw action is controlled by a tail vane, similar to that in the American pumping windmill. In larger machines, a servomechanism operated by a wind-direction sensor controls the yaw motor that keeps the turbine properly oriented.

Wind Speed Variation. Wind speed variability must be considered, especially in the design of the larger WEC machines. For a given turbine, there is a minimum wind speed called the *cut-in speed* at which rotation is allowed to start. The cut-in speed for the Gedser mill was 11 miles/hr (5 m/sec); for the Hutter–Allgeier rotor it was 5.6 miles/hr (2.5 m/sec); and for the Plum Brook machine it is 8 miles/hr (3.6 m/sec).

At higher wind speeds, the speed of rotation of the blades (and of the hub to which they are attached) tends to increase. In fact, for a given wind speed, the rate of rotation must exceed a certain value if a stall, with loss of power, is to be avoided. However, a changing speed of rotation will generally not provide the optimum conditions for generating constant-frequency alternating current.

One solution to this problem is to allow the turbine speed to vary and to couple the hub with a mechanical or electrical system which can generate a current of constant frequency from a variable-speed input. For example, variable-frequency alternating current may be generated and rectified to produce direct current. The direct current can then be used to generate constant-frequency alternating current by means of an inverter. If the electricity produced by the WEC machine is to be used for battery charging, a direct-current generator (up to about 20 kW) may be connected directly to the rotor hub; the speed of rotation of the turbine is then immaterial provided it is not so low as to cause stalling.

Although it results in some loss of efficiency, an alternative solution, used in the Hutter–Allgeier and Plum Brook (and similar) aerogenerators, has been preferred. Between the cut-in wind speed and the rated wind speed, at which the generator produces its rated power, the rotation rate is maintained constant by varying the output of the generator. At wind speeds exceeding the rated value (e.g., 18 miles/hr or 8 m/sec for the Plum Brook machine), the rotor speed is held constant by automatic adjustment of the pitch of the blades. At very high wind speeds the blades

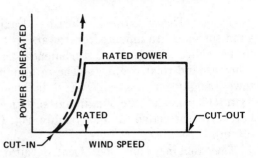

Fig. 192 **Power generated and wind speed.**

are feathered, as in an aircraft, and rotation ceases. The wind speed at which this occurs (60 miles/hr or 27 m/sec in the Plum Brook machine) is called the *cut-out speed* (or *furling speed*). Over the range from the rated wind speed to the cut-out speed, the electric power output of the generator is essentially constant.

The variation of the generated power with the wind speed is shown (approximately) by the full lines in Fig. 192; the dashed curve represents the theoretical variation without any control of the rotation rate of the turbine (i.e., power proportional to the cube of the wind speed). At wind speeds above the rated value, part of the available wind power is not used (or "spilled") as a consequence of changing the pitch angle of the rotor. Consequently, in this respect, the system is somewhat less efficient than one in which the rotor speed is allowed to increase with wind speed, within limits. However, machines of the latter type are less easily adapted to connection into an electric utility grid.

By designing the system to suit the wind conditions at a particular site, the wind power lost by spilling can be relatively small. In order to take advantage of winds with speeds above average, the rated wind speed should be greater than the average value but not so high that the generator is often operating below the rated power. The ratio of rated to average wind speeds should be about 1.8 for an average speed of 11 miles/hr (5 m/sec) decreasing to 1.5 for an average speed of 22 miles/hr (10 m/sec).

Turbine-Tower System. Horizontal-axis wind turbines are mounted on towers so as to be above the level of turbulence and other ground-related effects. The minimum tower height for a small WECS is about 33 ft (10 m), and the maximum practical height is estimated to be roughly 200 ft (61 m).

The turbine may be located either upwind or downwind of the tower. In the upwind location (i.e., the wind encounters the turbine before reaching the tower), the wake of the passing rotor blades causes repeated changes in the wind forces on the tower. As a result, the tower will tend to vibrate and may eventually be damaged. On the other hand, if the turbine is downwind from the tower as in Fig. 186, the tower vibrations are less but the blades are now subjected to severe alternating forces as they pass through the tower wake.

Both upwind and downwind locations have been used in WEC devices. In the Gedser mill (and some smaller machines in Denmark), for example, the turbine was upwind of the tower, but downwind rotors have been preferred in the United States, especially for the larger aerogenerators. Although other forces acting on the blades of these large machines are significant, tower effects are still important and tower design is an essential aspect of the overall system design.

Vertical-Axis Machines

Vertical-axis rotors have the great advantage of not having to be turned into the windstream as the wind direction changes. Because their operation is independent of wind direction, vertical-axis machines are called *panemones* (from Greek words meaning "all winds"). Until recent times, these devices received little attention for the generation of electric power, partly because vertical-axis rotors have lower power coefficients at high tip speed ratios (see Fig. 190). In addition, vertical-axis turbines are said to be more difficult to control in strong winds. It is

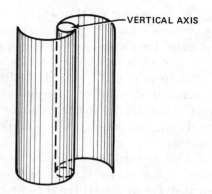

Fig. 193   Savonius rotor.

now realized, however, that the lower efficiency of panemones may be more than offset by the simpler design and consequent lower construction (and maintenance) costs. Elimination of the need for yawing into the wind, for example, results in decreased stresses on blades, bearings, and other components. Moreover, the transmission and generator are on (or near) the ground rather than at the top of a tall tower.

The Savonius Rotor. The Savonius rotor, sometimes referred to as the S-rotor because of its shape, is a simple vertical-axis, self-starting machine patented by the Finnish engineer S. J. Savonius in 1929. It consists essentially of two half-cylinders attached to a vertical axis and facing in opposite directions to form a two-vaned rotor (Fig. 193). The addition of a third vane results in smoother operation, but a somewhat higher wind speed is required to start the rotation. The ratio of the height to the overall diameter of the machine can be varied, but it is generally less than three to one. The power coefficient of the S-rotor is low, but it might possibly be improved by changes in the design, number, and arrangement of the vanes. These matters need to be investigated.

The Darrieus Rotor. The vertical-axis rotor, invented by G. J. M. Darrieus in France and patented in 1931, has two or three thin, curved ("egg-beater") blades with airfoil cross section and constant chord length (Fig. 194). The tip speed ratio

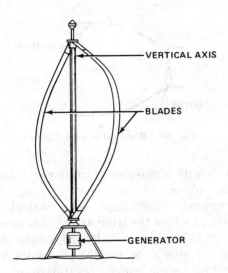

Fig. 194 Darrieus rotor.

and power coefficient are considerably better than those of the S-rotor but are still below the values for a modern horizontal-axis, two-bladed propeller rotor. However, the design of the complete turbine–generator system is simpler for the Darrieus rotor. The generator is located on (or near) the ground, and a strong supporting tower is not required. Thus, the overall costs should be less than for an equivalent horizontal-axis system.

One of the reported drawbacks of the Darrieus rotor has been that it is not self-starting. Tests indicate that, with small machines, the problems can be solved by attaching S-rotors at the top and bottom of the vertical (rotational) axis. This approach does not appear to be feasible with larger machines, but if the wind-power system is connected to a utility grid, the generator can serve as a motor to start the turbine. The (alternating-current) load can also provide a means for controlling the speed of the rotor regardless of the wind speed, so that variable-pitch blades are not required. At very high wind speeds, stalling occurs and the rotation stops automatically.

During the early 1970s, studies of the Darrieus design were initiated in Canada and later in the United States. A low-power (few kilowatt) machine was built and operated successfully at the Canadian National Research Council laboratories and a 200-kW aerogenerator was constructed on the Magdalen Islands in the Gulf of St. Lawrence. The system operated from July 1977 to July 1978 when the rotor was damaged in an accident; it was then rebuilt with some modifications. The world's largest Darrieus generator, with a diameter of 213 ft (65 m), is due for completion in 1983 for the Hydro-Quebec utility. The rated electric power output is 3.8 MW.

In the United States, a comprehensive investigation of the Darrieus turbine is being made for the Department of Energy at the Sandia National Laboratories. In 1974 a two-bladed rotor 16 ft (5 m) in diameter generated 1 kW of electricity at a rated wind speed of 15 miles/hr (6.7 m/sec). Subsequently, a machine with a diameter of almost 56 ft (17 m) has been operated.

The Department of Energy has contracted with industry for the construction of a number of Darrieus WEC generators, based on the work done at the Sandia National Laboratories. The machines have two 100-ft (30.5-m) long aluminum blades with an airfoil-type cross section and a uniform chord length of 24 in. (0.61 m); the rotor diameter is 56 ft (17 m), and the height is 82.5 ft (25 m). Each generator has a rated electric power output of 100 kW. If these machines operate successfully, larger Darrieus rotors with higher power outputs will be constructed.

Advanced WEC Concepts

In addition to the established WEC designs, as described in the preceding sections, some advanced ideas having the potential for generating electricity at a lower cost are being studied.

The Giromill. The "cyclogiro" principle, first developed in connection with the aerodynamics of aircraft propellers, is being applied in the Giromill concept by the McDonnell Douglas Corporation. The Giromill would be a vertical-axis rotor consisting of two, three, or more vertical blades, with a symmetrical airfoil cross section, attached to horizontal arms radiating (at top, middle, and bottom) from the axis of rotation. As the device is rotated by the wind, a blade-modulator control system would orient the blades continuously with respect to the wind direction. In this way, the rotation rate could be maintained constant regardless of the wind speed. At high wind speeds, the controls would be released to permit the individual blades to revolve freely while the rotor remained almost stationary.

Application of cyclogiro vortex theory to the Giromill indicates that the power coefficient would exceed the conventional 0.593 maximum. The reason is that the area of the windstream available to the rotor is substantially larger than the area actually swept by the blades. A three-bladed rotor with a diameter of 100 ft (30.5 m) and a height of 150 ft (46 m) is expected to produce 100 kW of electric power at a wind speed of 15 miles/hour (6.7 m/sec).

Diffuser-Augmented Wind Turbine. The diffuser-augmented wind turbine (DAWT) concept of the Grumman Aerospace Corporation is based on the observation made in England in the 1950s that the output of a conventional (horizontal-axis) wind turbine could be increased by means of a divergent shroud (or *diffuser*) around and downstream from the rotor. (The diffuser converts kinetic energy of the wind into pressure energy, that is, it causes an increase in pressure.) A simple form of DAWT is shown in Fig. 195. The diffuser makes it possible for both the airflow rate through the rotor and the pressure drop across it to be greater than the optimum values in an unshrouded rotor. Hence, the power that can be extracted from the wind, which is determined by the product of these two quantities, as seen earlier, can be increased correspondingly. The rotor power may be up to about four times as great as without a diffuser.

In a modified scheme, the DAWT rotor, with fixed-pitch blades, is preceded by a

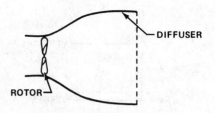

Fig. 195   Wind rotor with diffuser.

stator with a number of stationary blades. The stator blades have trailing (i.e., downwind) edge flaps that control the angle at which the windstream impinges on the turbine blades. By adjusting these flaps it should be possible to maintain a constant rotor speed regardless of the free-wind speed. The DAWT is expected to be capable of extracting wind power over a wider range of wind speeds than a conventional turbine.

WINKLER PROCESS

A commercial process developed in Germany in the 1920s and licensed to Davy Powergas, Inc., in the United States, for **coal gasification** with steam and oxygen gas (or air). With oxygen gas, the product is an **intermediate-Btu fuel gas**; when oxygen is replaced by air, a **low-Btu fuel gas**, containing inert nitrogen from the air, is produced. The intermediate-Btu gas can be converted into a **high-Btu fuel gas** (or **substitute natural gas**). More than 100 Winkler gasifier units have been operated in Europe and Asia, and some may be built in the United States. Most commercial plants use lignite and subbituminous non-caking coals (see **Coal**); caking bituminous coals may require pretreatment by heating in air at about 750°F (400°C).

The crushed coal is fed to the bottom of the **fluidized-bed** gasifier where it is fluidized by the upward flow of steam and oxygen gas or air (Fig. 196). The coal particles, steam, and oxygen gas (or air) are mixed intimately, and the coal is rapidly gasified. The temperature in the gasification region ranges from 1800 to 2100°F (980 to 1150°C), depending on the nature of the coal; reac-

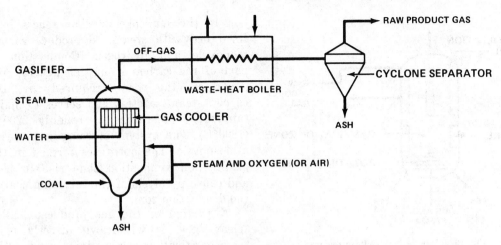

Fig. 196 Winkler process.

tive coals, like lignite, require lower gasification temperatures than less reactive varieties.

A secondary supply of steam and oxygen gas (or air) is introduced just above the fluidized bed to gasify unreacted particles of coal (and **char**) leaving the bed. Part of the carbon in the coal interacts with steam to produce carbon monoxide and hydrogen; the heat required for the reaction is provided by the combustion of carbon in oxygen (or air). Commercial Winkler units operate at close to atmospheric pressure, but tests are being made at pressures up to 15 atm (1.5 MPa) to increase the capacity.

The larger and heavier ash particles remaining after gasification of the coal fall through the fluidized bed and are discharged from the bottom of the gasifier. The lighter particles of ash (fly ash) and unreacted char, however, are carried over with the product gas. If the gasifier temperature is high, the fly-ash particles may soften (or melt) and deposit in the gas exit duct. To prevent this, the gas is partially cooled by a radiant-heat steam boiler in the upper part of the gasifier. When reactive coals are used, the gasifier temperature is reduced and the steam boiler is not required.

The raw product gas leaving the top of the gasifier is passed through a waste-heat (**heat exchanger**) boiler to lower the tem-

perature and then through a **cyclone separator** and a water scrubber to remove fly ash (and char particles). After extraction of hydrogen sulfide (see **Desulfurization of Fuel Gases**), the gas contains roughly 40 volume percent of carbon monoxide, 35 percent of hydrogen, and 3 percent of methane (dry basis) as the fuel constituents; most of the remainder is inert carbon dioxide. The **heating value** is approximately 270 Btu/cu ft (10 MJ/cu m). By subjecting this gas to the **water-gas shift reaction** and **methanation,** the product would be a high-Btu fuel gas.

If air is used instead of oxygen gas in the gasifier, more than half of the product is inert nitrogen and the heating value is around 120 Btu/cu ft (4.5 MJ/cu m).

WOODALL–DUCKHAM/GAS INTEGRALE (WD/GI) PROCESS

A commercial process developed by Il Gas Integrale in Italy and licensed to Babcock Contractors, Inc. [formerly Woodall-Duckham (USA) Ltd.] for **coal gasification** with steam and oxygen gas (or air). Over 100 WD/GI gasifier units are in operation in Europe, and they are being considered for the United States. With oxygen gas, the product is an **intermediate-Btu fuel gas**, but if air is used the presence of inert nitrogen makes

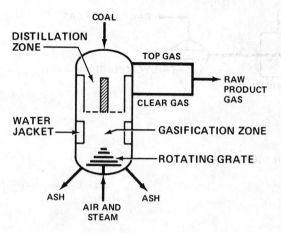

Fig. 197   Woodall-Duckham process.

the product a **low-Btu fuel gas**. The intermediate-Btu gas can be converted into a **high-Btu fuel gas** (or **substitute natural gas**). The WD/GI process has been operated with noncaking and moderately caking coals, and tests are being made with highly caking, bituminous coals.

Coal is introduced at the top of the gasifier which has a rotating grate at the bottom (Fig. 197). The grate, composed of concentric rings, serves to distribute the steam and oxygen gas (or air) introduced from below and also to remove residual coal ash. The gasifier consists of two zones; a lower **gasification/combustion zone**, surrounded by a water jacket, and an upper distillation zone. The water jacket provides the steam used in the gasification process.

As the coal descends through the distillation zone, partial **pyrolysis** (i.e., destructive distillation) occurs with the formation of a **char** (from noncaking coals) or semicoke (from moderately caking coals). In the gasification/combustion zone, the car-

bon in the semicoke or char reacts in a fixed bed with steam, to produce carbon monoxide and hydrogen. Combustion of part of the carbon in oxygen gas (or air) generates the heat required by the carbon-steam reaction and serves to maintain the temperature at roughly 2200°F (1200°C). The system pressure is close to atmospheric. The hot gases formed in the gasification/combustion zone rise to heat and cause pyrolysis of the incoming coal in the distillation zone.

A portion of the gas produced, called "clear" gas, is withdrawn directly from the gasification/combustion zone. The remainder, called "top" (or "mixed") gas, which includes gas (and oil and tar) from the pyrolysis of the coal, is withdrawn from the top of the distillation zone. The top gas is passed through oil and tar separators and is mixed with the clear gas to form the raw product gas. This is then treated for sulfur removal (see **Desulfurization of Fuel Gases**).

When the combustion gas is oxygen, the product contains roughly 38 volume percent of carbon monoxide, 38 percent of hydrogen, and 2 to 3 percent of methane (dry basis) as fuel constituents; most of the remainder is inert carbon dioxide. The **heating value** is about 270 Btu/cu ft (10 MJ/cu m). By subjecting this gas to the **water-gas shift reaction** and **methanation**, the product would be a high-Btu fuel gas.

If air is used instead of oxygen in the gasification process, about half of the product gas is inert nitrogen (from the air). The heating value is then approximately 140 Btu/cu ft (5.2 MJ/cu m).

Z

ZINC CHLORIDE CATALYTIC PROCESS

A Conoco Coal Development Company process for **coal liquefaction** by high-temperature treatment in the presence of hydrogen gas and zinc chloride ($ZnCl_2$) as a catalyst. The product is a liquid **hydrocarbon** mixture, with a large fraction distilling in the **gasoline** range. Although the process was successful on a bench scale, it proved difficult to scale it up to a larger scale; consequently, further development was terminated, at least temporarily.

The dried pulverized coal is made into a slurry with an oil produced in the process and fed to a high-temperature reactor with hydrogen gas and zinc chloride. The coal is broken down (see **Pyrolysis**) and hydrogenated, and the resulting vapors are condensed and fractionated by **distillation**. The fractions are **naphtha**, medium **fuel oil**, part of which is used to form the feed slurry, and a heavier fuel oil. The solid residue in the reactor, consisting mainly of **char**, zinc chloride, and ash, is heated with air in a **fluidized-bed** combustor to burn off the carbon and recover the catalyst.